T0332073

Control of Linear Systems

Trackability and Tracking of General Linear Systems

Control of Linear Systems

Trackability and Tracking of General Linear Systems

Lyubomir T. Gruyitch

CRC Press
Taylor & Francis Group
Boca Raton London New York

CRC Press is an imprint of the
Taylor & Francis Group, an **informa** business

Contents

Preface

On the state of the art

There are dynamical physical systems called **plants** (e.g., from submarines to space vehicles, from tool machines to mobile robots, from distillation columns to power plants,...), (Definition 18, Section 1.5), which have to obey the demand that their real dynamical behavior deviates very little from their required, i.e., **desired**, dynamical behavior. The control task is to force such a system (i.e., plant) to satisfy the demand. The small deviations of variables of such a system (plant) justify the linearization of their nonlinear mathematical models. Another reason to study the linear systems is that the results on them are the first useful information about the nonlinear dynamical system (plant) behavior. The third reason is methodological and pedagogical. Starting to teach students about control and to explain to them the basic control goal, phenomena, concepts, problems, and methods is best to begin with the lectures on the linear control systems. Control philosophy is essentially different from the philosophy of mathematics, physics, chemistry, mechanics, thermodynamics, fluids, electrotechnics, electronics,..., which are incorporated in the modeling of the physical systems as an introductory stage to the control science and engineering. Control is the core of the existence of any organized system, not only of technical, engineering ones, but also of economics and finanacial systems, of biological systems including human organism, of society and state. Control principles are general and valid for all of them.

This book concerns the fundamental topics of *the* **continuous-*time time*-invariant linear control systems**, in the sequel called for short *control systems* or just *systems*. It shows that in the framework of the linear systems there are still problems left untouchable but solvable.

Initial (state and output) conditions contain all significant information on the system history, i.e., on the external actions and on their consequences upon the system until the initial moment. The initial conditions are not predictable, are not known a priori and they are not touchable. The claim that initial conditions do not appear in the complex domain is wrong. The Laplace transform of derivatives of a variable contains the initial conditions. They are not *time* dependent. *The initial conditions do exist in the complex domain.* They appear in the Laplace transforms of the derivatives of the system input, of the system state and of the system output. In order to avoid the mathematical problem originated by the existence of the initial conditions and in order to study effectively the system described completely, in the forced regime under arbitrary initial conditions, the superposition principle has been exploited to justify the assumption on the zero initial conditions of the system in a forced regime. The assumption has been unconditionally accepted and has governed the courses and the research on the linear *time*-invariant continuous-*time* control systems and dynamical systems in general [170], [175].

The fact is that *the system transfers simultaneously the influence of the input variables, i.e., of the input vector, and of the initial conditions on its state and on its output.*

Every dynamical system has its, i.e., every dynamical plant and every control system have their (internal and/or output) dynamics that determines, respectively, its/their (internal and/or output) dynamical situation called *state* and (internal and/or output) *state variables*, i.e., its/their (internal and/or output) *state vector* all that *regardless of the existence or the nonexistence of the input derivatives.* The physically unjustified mathematical condition has been accepted that the concept of state has a sense if and only if the derivatives of the input vector do not influence the system (internal and/or output) dynamics, or at least that they do not appear in the mathematical model of the system. Although this condition is physically unjustifiable, it has become very useful to develop the effective mathematical machinery for the related studies. This is the reason for which the theory and the practice have been very well developed only for one class of the physical control systems. In order to treat other classes of the physical dynamical or control systems their mathematical models should be transformed formally mathematically with the full loss of the physical sense of the new (mathematical) variables.

The concept of the state, the state variables, the state vector, and the state space is well defined and widely effectively directly used only in the framework of the dynamical systems in general, and of the control systems in particular, described by the first order vector linear differential state equation and by the algebraic vector linear output equation, which are called **Input-State-Output** (abbreviated: *ISO*) (or: **State-space**) (**control**) **systems**.

Their mathematical models do not contain any derivative of the input vector function $\mathbf{I}(.)$.

There is a fundamental lacuna in the control theory due to the nonexistence of the clear, well defined, concept of the state also for the systems subjected to the influence of the input vector derivatives so that the physical meaning and sense of the system variables is preserved.

The *time* domain mathematical models and their studies enable the direct insight in the dynamical phenomena of the systems. The *time* domain is adequate for explaining physical phenomena of, and processes in, the systems. It is the adequate setting for definitions of dynamical properties of the systems. Unfortunately, the direct mathematical treatment of the systems in the *time* domain is not always very effective for the study of the system qualitative dynamical properties.

The Laplace transform of the *time* domain system description is the basis for getting the complex domain mathematical descriptions of the system. The Laplace transform induced the fundamental dynamical system characteristic. It is well known, in general, as the transfer function matrix $G(s)$ of a *Multiple-Input Multiple-Output* $(MIMO)$ system, or, in a simpler case, as the transfer function $G(s)$ of a *Single-Input Single-Output* $(SISO)$ system. $G(s)$ enables us to study simply various system qualitative dynamical properties [e.g., the system response *under all zero initial conditions*, system *controllability* and *observability*, *Lyapunov stability* of a completely controllable and observable system, *Bounded-Input Bounded-Output* $(BIBO)$ stability under *all zero initial conditions*]. It is the powerful mathematical system characteristic to treat dynamical properties of the system, by the definition, under *all zero initial conditions*. Its crucial property is its total independence of the system inputs, and, naturally, of initial conditions because it is defined *for all zero initial conditions*. The system itself completely determines $G(s)$ so that $G(s)$ describes in the complex domain the manner how the system transfers in the course of *time* the influence only of the external action through the system to its output variables (i.e., to its output vector). $G(s)$ does not describe in the complex domain how the system transfers the influence of the initial conditions on the system output behavior.

An additional crucial lacuna of the systems and control theory is the nonexistence of a complex domain method for the effective treatment of the systems subjected simultaneously to both input actions and initial conditions. It is a consequence of the stringent assumption on all zero initial conditions under which the system transfer function $G(s)$ is only valid. The same holds for the system matrix $P(s)$.

Another consequence is the inherent gap between the Lyapunov stability definitions and the Lyapunov stability criteria expressed in terms of the system transfer function matrix $G(s)$. The definition of the former holds exclusively for the free regime under arbitrary initial conditions, but the definition of the latter is valid exclusively for the forced regime under all zero initial conditions. This is a conceptual paradox. It confuses the students. The paradox is overcome and the complete solution is given in [159], [170].

Besides, the criterion for the pole-zero cancellation is unclear if the initial conditions are not equal to zero. The books [159], [170] present the full solution.

On the book

The author, in addition to the analysis of the scientific papers listed in the bibliography of the book, consulted in particular the books by the following authors in the course of preparing and/or writing this book: B. D. O. Anderson and J. B. Moore [4], P. J. Antsaklis and A. N. Michel [7], [8], S. Barnett [16], Y. Bar-Shalom and T. E. Fortmann [18], Y. Bar-Shalom and X. R. Li [20], Y. Bar-Shalom, X. R. Li and T. Kirubarajan [21], Y. Bar-Shalom, P. K Willet, and X. Tian [22], A. Benzaoiua, F. Mesquine and M. Benhayoun [24], L. D. Berkovitz [25], L. D. Berkovitz and N. G. Medhin [26], S. P. Bhattacharyya, A. Datta and L. H. Keel [32], D. Biswa [33], S. S. Blackman [34], S. S. Blackman and R. Popoli, [35], J. H. Blakelock [36], P. Borne, G. Dauphin-Tanguy, J.-P. Richard, F. Rotella and I. Zambettakis [37], W. L. Brogan [40], G. S. Brown and D. P. Campbell [41], F. M. Callier and C. A. Desoer [44], [43], C.-T. Chen [48], H. Chestnut and R. W. Mayer [50], M. J. Corless and A. E. Frazho [53], J. J. D'Azzo and C. H. Houpis [63], J. J. D'Azzo, C. H. Houpis and S. N. Sheldon [64], C. A. Desoer [66], C. A. Desoer and M. Vidyasagar [69], V. Dragan and T. Morozan , B. Etkin [77], F. W. Fairman [80], Y. Feng and M. Yagoubi [86], F. R. Gantmacher [94], [95], G. C. Goodwin [102], Ly. T. Gruyitch [159], [170], [175], [181], [188], M. Haidekker [196], J. P. Hespanh [199], C. H. Houpis and S. N. Sheldon [200], M. Haidekker [196], D. G. Hull [202], E. Jarzębowska , T. Kailath [205], D. E. Kirk [212], B. Kisačanin.and G. C. Agarwal [213], B. C. Kuo [219], [220], H. Kwakernaak and R. Sivan [221], P. Lancaster and M. Tismenetsky [222], A. M. Lyapunov [230], L. A. MacColl [231], J. M. Maciejowski [232], J. L. Melsa and D. G. Schultz [239], R. K. Miller and A. N. Michel [243], T. Nambu [258], K. Ogata [264], [265], D. H. Owens [266], H. M. Power and R. J. Simpson [274], Z. Qu and D. M. Dawson [275], H. H. Rosenbrock [283], A. Sinha [289], R. E. Skelton [290], J. C. West [309], D. M. Wiberg [310], R. L. Williams II and D. A. Lawrence [311], W. A. Wolovich [312], W. M.

Wonham [313] and B.-T. Yazdan [318].

This book is complementary to them and/or extends, broadens, and generalizes inherently their parts that are related to the system transfer functions and/or to the dynamical system state concept, and/or to tracking issues and/or to the control synthesis. For the analogous extensions and generalizations to the observability and controllability of the general linear (*time*-invariant continuous-*time*) systems see the accompanying book [171].

The book treats general aspects of the fundamental control issues for the following five classes of the **systems**.

- *The Input-Output (IO) systems* described by the ν-*th* order linear vector differential equation expressed in terms of the output vector $\mathbf{Y} \in \mathfrak{R}^N$. This class of the control systems has been only partially studied; it has been studied by formally mathematically transforming its mathematical model into the form of the *ISO* systems. The variables of such transformations lose the physical sense of the system variables if the system is subjected to actions of the derivative(s) of the input vector. The book resolves this lacuna of the control theory.

- *The (first order) Input-State-Output (ISO) systems* determined by the first order linear vector differential equation expressed in terms of the state vector $\mathbf{X}$–*the state equation*, $\mathbf{X} \in \mathfrak{R}^n$, and by the algebraic vector equation expressed in terms of the output $\mathbf{Y}$–*the output equation*. They are well known as *the state-space systems*. They contain only one derivative that is the first derivative of the state vector. They do not contain any derivative of the input vector.

- *The (first order) Extended-Input-State-Output (EISO) systems* determined by the first order linear vector differential equation expressed in terms of the state vector $\mathbf{X}$–*the state equation*, $\mathbf{X} \in \mathfrak{R}^n$, and by the algebraic vector equation expressed in terms of the output $\mathbf{Y}$–*the output equation*. They contain the first order derivative $\mathbf{X}^{(1)}$ of the state vector $\mathbf{X}$ and μ derivatives of the input vector $\mathbf{I}$ only in the state vector equation, $\mu > 0$ (because for $\mu = 0$ the *EISO* system becomes the *ISO* system). The *EISO* systems have not been studied so far.

- *The Higher order Input-State-Output (HISO) systems* characterized by the α-*th* order, $\alpha > 1$, linear vector differential equation expressed in terms of the vector $\mathbf{R}$–*the state equation*, $\mathbf{R} \in \mathfrak{R}^\rho$, and by the linear vector algebraic equation of the output vector $\mathbf{Y}$– *the output equation*. This class of the control systems has not been studied so far.

- *The Input-Internal and Output state (IIO) systems* characterized by the α-*th* order linear vector differential equation expressed in terms of the internal dynamics vector $\mathbf{R}$–*the internal dynamics, i.e., the internal state, equation,* and by the ν-*th* order linear vector (differential if $\nu > 0$, algebraic if $\nu = 0$) equation expressed in terms of the output vector $\mathbf{Y}$– *the output (state if $\nu > 0$) equation.* The books [159], [170] introduced and initiated the study of this class of the dynamical systems. However, the *IIO control systems* have not been studied so far.

  Notice that if $\nu = 0$ then the *IIO* system becomes the *HISO* system, which explains why $\nu > 0$ is accepted for the *IIO* systems.

The existence of the actions of the input vector derivatives on the system is the reason, justification and need **to extend and to generalize the state concept of dynamical, hence of control, systems**. It is done in the books [171], [170]. *The existing state concept is the special case of the general state concept defined in the book.* This led to the crucial new stability results [170], to the inherently new results on observability and controllability in [171] and leads in this book to the new fundamental results on the system trackability, tracking and control synthesis.

This book presents the advanced self-contained original study of the fundamental qualitative dynamical properties of all five classes of control systems:

- *Tracking,*

- *Trackability,*

and

- *Control synthesis,*

under arbitrary and unknown initial conditions and disturbances. These topics together with mathematical models of the systems and mathematical preliminaries constitute the main body of the book. Various subsidiary statements, results, and rigorous detailed proofs form the book.

By its definition, the system transfer function matrix $G(s)$ is not applicable as soon as any initial condition is not equal to zero. It fails to express in the complex domain how the initial conditions influence the system output behavior in the course of *time*. This posed the following basic question and opened the following fundamental problem:

Problem 1 *[159], [170] Does a linear time-invariant dynamical system in general and/or control system in particular have a complex domain characteristic mathematically expressed in the form of a matrix such that it satisfies the following two conditions:*

1. It describes in the complex domain how the system simultaneously transfers in the course of time the influences of both the input vector action and of all initial conditions on the system state or output response, and

2. It is completely determined only by the system itself meaning its full independence of both the input vector and the vector of all initial conditions?

The reply is affirmative [159], [170], [171]. Such system characteristic is **the system full (complete) transfer function matrix** $F(s)$ [159], [170], [171]. We will use it throughout the book. For its definition, determination, usage and applications see [170]. It enabled us to discover and prove in [159], [170] new results on the system matrix, on the system equivalence, on the Lyapunov stability properties, and on the BI stability properties of the linear *time*-invariant continuous-*time* dynamical systems in general, hence control systems in particular. All that has been recently also done in the accompanying book for *time*-invariant **discrete-*time* systems** [42].

The book broadens in the sequel the concept of the system full transfer function matrix $F(s)$ to the control systems. The familiarity with the reference [170] is very helpful to follow this book; for some parts of the book such familiarity is advantageous.

In fact the book [170] appears as the prerequisite for easy following this book. In other words, this book continues the book [170].

The goal of the book is to contribute to the advancement of the linear control systems theory and the corresponding university courses, to open new directions for research and for applications in the framework of *time*-invariant continuous-*time* linear control systems. It represents a further development of the existing linear control systems theory that is not repeated herein. *The contributions of the book largely and crucially are beyond the existing control theory.*

In gratitude

The author expresses his gratitude to Ms. Nora Konopka, Global Editorial Director-Engineering, for her formidable, exceptionally careful leading the publication process during which she proposed to divide the original manuscript of 633 pages into two books, this one and the accompanying book [171], under their titles as they are published.

Ms. Michele Dimont, Project Editor, for the devoted leading of the book editing,

Ms. Vanessa Garrett. Editorial Assistant – Engineering, for the careful and effective administrative work,

all with CRC Press/Taylor & Francis.

The author is grateful to Mr. George Pearson with MacKichan Company for his very kind and effective assistance in improving my usage of the excellent Scientific Work Place for scientific works.

Belgrade, September 28, 2017, March 13, April 23, May 20, 2018.

Lyubomir T. Gruyitch, the author

Part I

SYSTEM CLASSES

Chapter 1

Introduction

1.1 *Time*

All processes, motions and movements, all behaviors of the systems and their responses, as well as all external actions on the systems, occur and propagate in *time*. It is natural from the physical point of view to study the systems directly in the temporal domain. This requires to be clear how we understand what is *time* and what are its properties, which we explain in brief as follows (for the more complete analysis see: [170], [167], [181], [182], [183]).

Definition 2 *Time*
*Time (i.e., **the temporal variable**) denoted by t or by τ is an independent scalar physical variable such that:*
*- Its value called **instant** or **moment** determines uniquely **when** somebody or something started/interrupted to exist,*
*- Its values determine uniquely **since when and until when** somebody or something existed/exists or will exist,*
*- Its values determine uniquely **how long** somebody or something existed/exists or will exist,*
*- Its values determine uniquely whether an event E_1 occurs **then when** another event E_2 has not yet happened, or the event E_1 takes place just **then when** the event E_2 happens, or the event E_1 occurs **then when** the event E_2 has already happened,*
*- Its value **occupies (covers, encloses, imbues, impregnates, is over and in, penetrates) equally** everybody and everything (i.e., beings, objects, energy, matter, and space) **everywhere and always**, and*

- Its value has been, is, and will be **permanently changing smoothly, strictly monotonously continuously, equally** *in all spatial directions and their senses, in and around everybody and everything,* **independently** *of everybody and everything (i.e., independently of beings, objects, energy, matter, and space),* **independently** *of all other variables, independently of all happenings, movements and processes.*

Time is a basic and elementary constituent of the existence of everybody and of everything [181], [182], [183].

All human trials during millenniums have failed to explain, to express, the nature, the phenomenon, of *time* in terms of other well defined notions, in terms of other physical variables and phenomena [182, Axiom 25, p. 52], [183, Axiom 25, p. 53], [181]. The nature of *time*, the physical content of it, cannot be explained in terms of other basic constituents of the existence (in terms of energy, matter, space) or in terms of other physical phenomena or variables. *Time* has its own, original, nature that we can only call it *the nature of time*, i.e., *the temporal nature* or *the time nature* [167], [181], [182], [183].

An arbitrary value of *time* t (τ), i.e., **an arbitrary instant** or **moment**, is denoted also by t (or by τ), respectively. It is an **instantaneous** (**momentous**) and **elementary** *time* value. It can happen exactly once and then it is the same everywhere for, and in, everybody and everything (i.e., for, and in, beings, energy, matter, objects, and space), for all other variables, for all happenings, for all movements, for all processes, for all biological, economical, financial, physical and social systems. It is not repeatable. Nobody and nothing can influence the flow of instants [182, Axiom 25, p. 52], [183, Axiom 25, p. 53], [181].

The physical dimension of *time* is denoted by $[T]$, where T stands for *time*, t $[T]$. It cannot be expressed in terms of the physical dimension of another variable. Its physical dimension is one of the basic physical dimensions. It is used to express the physical dimensions of the most of the physical variables. A selected unity 1_t of *time* can be arbitrarily chosen and then fixed. If it is *second* s then $1_t = s$, which we denote by $t \langle 1_t \rangle = t \langle s \rangle$.

There can be assigned *exactly one* (which is denoted by $\exists!$) real number to every moment (instant), and vice versa. The numerical value **num** t of the moment t is a real number and dimensionless, *num* $t \in \Re$ and *num* t $[-]$, where $\Re$ is the set of all real numbers.

Theorem 3 *Universal time speed law [182, Theorem 45, p. 98], [183, Theorem 45, p. 98]*

Time is the unique physical variable such that the speed v_t (v_τ) of the evolution (of the flow) of its values and of its numerical values:

a)   Is invariant with respect to a choice of a relative zero moment t_{zero}, of an initial moment t_0, of a time scale and of a time unit 1_t, i.e., invariant relative to a choice of a time axis, invariant relative to a selection of spatial coordinates, invariant relative to everybody and everything, and

b)   Its value (its numerical value) is invariant and equals one arbitrary time unit per the same time unit (equals one), respectively,

$$v_t = 1[TT^{-1}]\left\langle 1_t 1_t^{-1}\right\rangle = 1[TT^{-1}]\left\langle 1_\tau 1_\tau^{-1}\right\rangle = v_\tau, \; numv_t = numv_\tau = 1,$$
$$(1.1)$$

relative to arbitrary time axes T and T_τ, i.e., its numerical value equals 1 (one) with respect to all time axes (with respect to any accepted relative zero instant t_{zero}, any chosen initial instant t_0, any time scale and any selected time unit 1_t), with respect to all spatial coordinate systems, with respect to all beings and all objects.

The uniqueness of *time*, the constancy and the invariance of the *time* speed determine that *time* itself is not relative and cannot be relative (for more details, proofs and explanations see the books [165], [167], [181], [182], [183], [263]).

Time set $\mathcal{T}$ is the set of all moments. It is open, unbounded and connected set. It is in the biunivoque (one-to-one) correspondence with the set $\mathfrak{R}$ of all real numbers,

$$\mathcal{T} = \{t : num\, t \in \mathfrak{R}, \; dt > 0, \; t^{(1)} \equiv 1\},$$
$$\forall t \in \mathcal{T}, \; \exists! x \in \mathfrak{R} \Longrightarrow x = num\, t$$
$$and \; \forall x \in \mathfrak{R}, \; \exists! t \in \mathcal{T} \Longrightarrow num\, t = x,$$
$$num\, inf\mathcal{T} = num\, t_{inf} = -\infty \notin \mathcal{T} \; and \; num\, sup\mathcal{T} = num\, t_{sup} = \infty \notin \mathcal{T}.$$
$$(1.2)$$

The rule of the correspondence determines an **accepted relative zero numerical *time* value** t_{zero}, a ***time* scale** and a ***time* unit** denoted by 1_t (or by 1_τ). The *time* unit can be ... , millisecond, second, minute, hour, day, ... , *which Newton explained by clarifying the sense of relative time* [263, I of Scholium, p. 8]. Unfortunately, this fact has been ignored in the modern physics and science.

Note 4 *Choice of the relative zero moment t_{zero} and the initial moment t_0*

We accept herein the relative zero moment t_{zero} to have the zero numerical value, num $t_{zero} = 0$ because we deal with the time-invariant systems. Besides, we adopt it for t_{zero} to be also the initial moment t_0, $t_0 = t_{zero}$, num $t_0 = 0$, in view of the time-invariance of the systems to be studied. This determines the subset $\mathfrak{T}_0$ of the time set $\mathfrak{T}$,

$$\mathfrak{T}_0 = \{t : t \in \mathfrak{T}, \ numt \in [0, \infty[\}.$$

Sometimes, we will denote the initial moment explicitly by t_0 but it will mean that num $t_0 = 0$.

Note 5 *We usually use the letters t and τ to designate time itself and an arbitrary moment, as well as the numerical value of the arbitrary moment with respect to the chosen zero instant, e.g., $t = 0$ is used in the sense numt $= 0$. From the physical point of view this is incorrect. The numerical value num t of the instant t is a real number without a physical dimension, while the instant t is a temporal value that has the physical dimension - the temporal dimension T of time. We overcome that by using the normalized, dimensionless, mathematical temporal variable, denoted by $\bar{t}$ and defined by*

$$\bar{t} = \frac{t}{1_t}[-],$$

so that the time set $\mathcal{T}$ is to be replaced by

$$\overline{\mathcal{T}} = \{\bar{t}[-] : \bar{t} = num\bar{t} = num \ t \in \mathfrak{R}, \ d\bar{t} > 0, \ \bar{t}^{(1)} \equiv 1\}.$$

With this in mind we will use in the sequel the letter t also for $\bar{t}$, and $\mathcal{T}$ also for $\overline{\mathcal{T}}$. Hence,

$$t[-] = numt[-].$$

Between any two different instants $t_1 \in \mathcal{T}$ and $t_2 \in \mathcal{T}$ there is a third instant $t_3 \in \mathcal{T}$, either $t_1 < t_3 < t_2$ or $t_2 < t_3 < t_1$. The *time* set $\mathcal{T}$ is **continuum**. It is called also **the continuous-*time* set**. This book is on *continuous-time* systems and their control.

1.2   *Time*, physical principles, and systems

The following general principles hold for every physical variable. *Time*, being also a physical variable, satisfies also the same principles that are expressed in the following forms [182, pp. 136-146], [183, pp. 126-136]:

Principle 6 *Physical Continuity and Uniqueness Principle-***PCUP** *Scalar form*

A physical variable can change its value from one value to another one only by passing through every intermediate value, and it possesses a unique local instantaneous real value in any place (in any being or in any object) at any moment.

Principle 7 *Physical Continuity and Uniqueness Principle-***PCUP** *Matrix and vector form*

A vector physical variable or a matrix (vector) of physical variables can change, respectively, its vector or matrix (vector) value from one vector or matrix (vector) value to another one only by passing elementwise through every intermediate vector or matrix (vector) value, and it possesses a unique local instantaneous real vector or matrix (vector) value in any place [i.e., in any being or in any object] at any moment, respectively.

Principle 8 *Physical Continuity and Uniqueness Principle-***PCUP** *System form*

The system physical variables (including those their derivatives or integrals, which are also physical variables), can change, respectively, their (scalar or vector or matrix) values from one (scalar or vector or matrix) value to another one only by passing elementwise through every intermediate (scalar or vector or matrix) value, and they possess unique local instantaneous real (scalar or vector or matrix) values in any place at any moment.

The *PCUP* appears important for an accurate modeling physical systems.

Corollary 9 *Mathematical model of a physical variable, mathematical model of a physical system and PCUP*

a)   For a mathematical (scalar or vector) variable to be, respectively, an adequate description of a physical (scalar or vector) variable it is necessary that it obeys the Physical Continuity and Uniqueness Principle.

b)   For a mathematical model of a physical system to be an adequate description of the physical system it is necessary that all its system variables obey the Physical Continuity and Uniqueness Principle; i.e., that the mathematical model obeys the Physical Continuity and Uniqueness Principle.

The properties of *time* and the common properties of the physical variables expressed by *PCUP* (Principle 6 through Principle 8) lead to

Principle 10 *Time Continuity and Uniqueness Principle-TCUP*

Any (scalar or vector) physical variable (any vector/matrix of physical variables) can change, respectively, its (scalar/vector/matrix) value from one (scalar/vector/matrix) value to another one only continuously in time by passing (elementwise) through every intermediate (scala /vector/matrix) value, and it possesses a unique local instantaneous real (scalar/vector/matrix) value in any place (in any being or in any object) at any moment.

Definition 11 *The system form of the TCUP means that all system variables satisfy the TCUP.*

The *TCUP* is very useful for the stability study of dynamic systems and for their control synthesis. This was shown in [161], [163], [168], [177], [184], [186], [187].

Corollary 12 *Mathematical representation of a physical variable, mathematical model of a physical system and TCUP*

a)   For a mathematical (scalar or vector) variable to be, respectively, an adequate description of a physical (scalar or vector) variable it is necessary that it obeys the Time Continuity and Uniqueness Principle.

b)   For a mathematical model of a physical system to be an adequate description of the physical system it is necessary that its system variables obey the Time Continuity and Uniqueness Principle; or equivalently, that the mathematical model obeys the Time Continuity and Uniqueness Principle.

c)   For a mathematical model of a physical system to be an adequate description of the physical system it is necessary that its solutions are unique and continuous in time.

For more on *time*, on its consistent physical and mathematical relativity theory, and on its relationship to systems, see [165], [167], [181], [182], [183].

1.3   Notational preliminaries

Lower case ordinary letters denote scalars, bold (lower case and capital, Greek and Roman) letters signify vectors, capital italic letters stand for matrices, and we use capital 𝔉𝔯𝔞𝔨𝔱𝔲𝔯 letters for sets and spaces. For example, the identity matrix of the dimension i is denoted by I_i,

$$I_i = diag\{1 \quad 1 \quad ... \quad 1\} \in \mathfrak{R}^{i \times i}, \ I_n = I \in \mathfrak{R}^{n \times n}. \tag{1.3}$$

The variables in the mathematical models are dimensionless because their values are normalized relative to their characteristic values. Throughout the book we accept the following condition to hold:

Condition 13 **Normalized variables**
The value of every variable Z appearing in a system mathematical model is dimensionless normalized physical variable Z_{Ph} relative to some its characteristic value Z_{PhCh} (e.g., nominal value Z_{PhN} or the unit value 1_Z):

$$Z \ [-] = \frac{Z_{Ph} \ [Z_{Ph}]}{Z_{PhCh} \ [Z_{Ph}]}. \tag{1.4}$$

Note 14 **Useful simple vector notation** *[115], [159], [170]*
Instead of using, for example,

$$\mathbf{Y}^{\mp}(s) = F(s) \bullet$$

$$\bullet \left[\begin{array}{ccccccc} I^{\mp^T}(s) & I^T(0^{\mp}) & .. & I^{(\mu-1)^T}(0^{\mp}) & Y^T(0^{\mp}) & .. & Y^{(\nu-1)^T}(0^{\mp}) \end{array} \right]^T,$$

the following simple vector notation enabled us to define and use effectively the system full transfer function matrix $F(s)$:

$$\mathbf{Y}^{\mp}(s) = F(s)\mathbf{V}(s),$$

$$\mathbf{V}(s) = \left[\begin{array}{c} \mathbf{I}^{\mp}(s) \\ \mathbf{C}_0^{\mp} \end{array} \right], \ \mathbf{C}_0^{\mp} = \left[\begin{array}{c} \mathbf{I}^{\mu-1}(0^{\mp}) \\ \mathbf{Y}^{\nu-1}(0^{\mp}) \end{array} \right],$$

$$\mathbf{I}^{\mu-1}(0^{\mp}) = \left[\begin{array}{c} I(0^{\mp}) \\ I^{(1)}(0^{\mp}) \\ ... \\ I^{(\mu-1)}(0^{\mp}) \end{array} \right], \ \mathbf{Y}^{\nu-1}(0^{\mp}) = \left[\begin{array}{c} Y(0^{\mp}) \\ Y^{(1)}(0^{\mp}) \\ ... \\ Y^{(\nu-1)}(0^{\mp}) \end{array} \right],$$

by introducing the general compact vector notation

$$\mathbf{Y}^k = \left[\begin{array}{c} \mathbf{Y} \\ \mathbf{Y}^{(1)} \\ ... \\ \mathbf{Y}^{(k)} \end{array} \right] = \left[\begin{array}{c} \mathbf{Y}^{(0)} \\ \mathbf{Y}^{(1)} \\ ... \\ \mathbf{Y}^{(k)} \end{array} \right] \in \mathfrak{R}^{(k+1)N}, \ k \in \{0, 1, ...\}, \ \mathbf{Y}^0 = \mathbf{Y}. \tag{1.5}$$

It is different from the k-th derivative $\mathbf{Y}^{(k)}$ of $\mathbf{Y}$:

$$\mathbf{Y}^{(k)} = \frac{d^k \mathbf{Y}}{dt^k} \in \mathfrak{R}^N, \ k \in \{1, ...\}, \ \mathbf{Y}^k \neq \mathbf{Y}^{(k)}.$$

This permits to express $\sum_{i=0}^{i=\nu} A_i \mathbf{Y}^{(i)}(t)$ as follows,

$$A_i \in \mathfrak{R}^{N \times N}, \quad \sum_{i=0}^{i=\nu} A_i \mathbf{Y}^{(i)}(t) = \left[A_0 \vdots A_1 \vdots ... \vdots A_\nu \right] \begin{bmatrix} \mathbf{Y}^{(0)}(t) \\ \mathbf{Y}^{(1)}(t) \\ ... \\ \mathbf{Y}^{(k)}(t) \end{bmatrix},$$

i.e., in the compact form by introducing the extended system matrix $A^{(\nu)}$ composed of the system matrices $A_i \in \mathfrak{R}^{N \times N}$, $i \in \{0, 1, ..., \nu\}$,

$$A^{(\nu)} = \left[A_0 \vdots A_1 \vdots ... \vdots A_\nu \right] \in \mathfrak{R}^{N \times (\nu+1)N}, \tag{1.6}$$

$$A^{(\nu)} \neq A^\nu = \underbrace{AA...A}_{\nu-times} \in \mathfrak{R}^{N \times N} \tag{1.7}$$

so that

$$\sum_{i=0}^{i=\nu} A_i \mathbf{Y}^{(i)}(t) = A^{(\nu)} \mathbf{Y}^\nu(t). \tag{1.8}$$

Let $\mathbb{C}$ be the set of all complex numbers s We use also the complex matrix function $S_i^{(k)}(.) : \mathbb{C} \longrightarrow \mathbb{C}^{i(k+1) \times i}$ of s,

$$S_i^{(k)}(s) = \left[s^0 I_i \vdots s^1 I_i \vdots s^2 I_i \vdots ... \vdots s^k I_i \right]^T \in \mathbb{C}^{i(k+1) \times i},$$

$$(k, i) \in \{(\mu, M), \ (\nu, N)\}, \ rank \ S_i^{(k)}(s) \equiv i \ on \ \mathbb{C}, \tag{1.9}$$

in order to set

$$\sum_{i=0}^{i=\nu} A_i s^i = \sum_{i=0}^{i=\nu} A_i s^i I_N$$

into the compact form $A^{(\nu)} S_N^{(\nu)}(s)$,

$$\sum_{i=0}^{i=\nu} A_i s^i = A_0 s^0 I_N + A_1 s^1 I_N + ... + A_\nu s^\nu I_N = A^{(\nu)} S_N^{(\nu)}(s) \in \mathbb{C}^{N \times N}. \tag{1.10}$$

We also introduce another subsidiary matrix function denoted as

$Z_k^{(\varsigma-1)}(.) : \mathbb{C} \to \mathbb{C}^{(\varsigma+1)k \times \varsigma k}$ and defined by:

$$Z_k^{(\varsigma-1)}(s) = \begin{bmatrix} O_k & O_k & O_k & \dots & O_k \\ s^0 I_k & O_k & O_k & \dots & O_k \\ \dots & \dots & \dots & \dots & \dots \\ s^{\varsigma-1} I_k & s^{\varsigma-2} I_k & s^{\varsigma-3} I_k & \dots & s^0 I_k \end{bmatrix}, \varsigma \geq 1,$$

$$\varsigma = 1 \Longrightarrow Z_k^{(1-1)}(s) = Z_k^{(0)}(s) = s^0 I_k = I_k,$$

$$Z_k^{(\varsigma-1)}(s) \in \mathbb{C}^{(\varsigma+1)k \times \varsigma k}, \ (\varsigma, k) \in \{(\mu, M), \ (\nu, N)\}, \tag{1.11}$$

where the final entry of $Z_k^{(\varsigma-1)}(s)$ is always $s^0 I_k$. This function permits to put initial conditions induced by the Laplace transform $\mathcal{L}\{\mathbf{Y}^\varsigma(t)\}$ of $\mathbf{Y}^\varsigma(t)$ in the compact form [159], [170]:

$$\mathcal{L}\{\mathbf{Y}^\varsigma(t)\} = s^\varsigma \mathbf{Y}(s) - Z_N^{(\varsigma-1)}(s)\mathbf{Y}^{\varsigma-1}(0). \tag{1.12}$$

Note 15 *Higher order and/or higher dimension of the system, more advantageous the new notation.*

Furthermore, we use the symbolic vector notation and operations in the elementwise sense as follows:

-   The zero and unit vectors,

$$\mathbf{0}_N = [0 \ 0 \ \dots 0]^T \in \mathfrak{R}^N, \ \mathbf{1}_N = [1 \ 1 \ \dots 1]^T \in \mathfrak{R}^N, \tag{1.13}$$

-   The matrix E associated elementwise with the vector $\mathbf{e}$,

$$\mathbf{e} = [e_1 \ e_2 \ \dots \ e_N]^T \in \mathfrak{R}^N \Longrightarrow E = diag\{e_1 \ e_2 \ \dots \ e_N\} \in \mathfrak{R}^{N \times N},$$

-   The vector and matrix elementwise absolute values,

$$|\mathbf{e}| = [|e_1| \ |e_2| \ \dots \ |e_N|]^T, \ |E| = diag\{|e_1| \ |e_2| \ \dots \ |e_N|\}, \tag{1.14}$$

-   The elementwise vector equality and inequality,

$$\mathbf{w} = [w_1 \ w_2 \ \dots \ w_N]^T, \ \mathbf{w} = \mathbf{e} \Longleftrightarrow w_i = e_i, \ \forall i = 1, 2, ..., N,$$
$$\mathbf{w} \neq \mathbf{e} \Longleftrightarrow w_i \neq e_i, \ \forall i = 1, 2, ..., N,$$

The definition of the scalar sign function used herein reads:

-   $sign(.) : \mathfrak{R} \to \{-1, 0, 1\}$ *the signum scalar function,*

$$signe = e |e|^{-1} \ if \ e \neq 0, \ signe = 0 \ if \ e = 0. \tag{1.15}$$

Other notation is defined at its first appearance in the text and in Appendix A.

1.4   Compact, simple, and elegant calculus

The introduction and definition of:

- The extended vector $\mathbf{Y}^k \in \mathfrak{R}^{(k+1)N}$ (1.5), which is composed of the vector $\mathbf{Y}$ and its derivatives up to the order k,

- The extended matrix $A^{(\nu)} \in \mathfrak{R}^{N\times(\nu+1)N}$ (1.6), the entries of which are submatrices A_i, $i = 0, 1, 2, .., \nu$,

- The complex matrix functions $S_i^{(k)}(s) \in \mathbb{C}^{\,i(k+1)\times i}$ (1.9) and $Z_k^{(s-1)}(.)$: $\mathbb{C} \rightarrow \mathbb{C}^{(s+1)k\times sk}$ (1.11),

and the introduction and definition of the vectors and matrices:

- $\mathbf{0}_N$, $\mathbf{1}_N$ (1.13), $|\mathbf{e}|$ and $|\mathbf{E}|$ (1.14),

enable us to develop a compact, simple and elegant calculus.
The matrix differential equation:

$$\sum_{i=0}^{i=\nu} A_i \mathbf{Y}^{(i)}(t) = \sum_{i=0}^{i=\mu\leq\nu} B_i \mathbf{I}^{(i)}(t),$$

$$A_i \in \mathfrak{R}^{N\times N}, \ \mathbf{Y} \in \mathfrak{R}^N, \ B_i \in \mathfrak{R}^{N\times M}, \ \mathbf{I} \in \mathfrak{R}^M, \tag{1.16}$$

has the equivalent compact form in the *time* domain [159], [170]:

$$A^{(\nu)}\mathbf{Y}^\nu(t) = B^{(\mu)}\mathbf{I}^\mu(t),$$

$$A^{(\nu)} \in \mathfrak{R}^{N\times(\nu+1)N}, \ \mathbf{Y}^\nu \in \mathfrak{R}^{(\nu+1)N}, \ B^{(\mu)} \in \mathfrak{R}^{N\times(\mu+1)M}, \ \mathbf{I}^\mu \in \mathfrak{R}^{(\mu+1)M}. \tag{1.17}$$

Comment 16 *Compact form of the linear differential equation*
   Equation (1.17) is differential, not algebraic, equation that is the compact form of the original differential Equation (1.16).
   If $N \leq M$ and rank $A^{(\nu)} = N$ then the matrix $\left(A^{(\nu)}\left(A^{(\nu)}\right)^T\right)$ is nonsingular and the right inverse $\left(A^{(\nu)}\right)^T\left(A^{(\nu)}\left(A^{(\nu)}\right)^T\right)^{-1}$ of $A^{(\nu)}$ is well defined. If Equation (1.17) had been algebraic and treated as algebraic then we would have been formally able to solve it for $\mathbf{Y}^\nu(t)$:

$$\mathbf{Y}^\nu(t) = \left(A^{(\nu)}\right)^T\left(A^{(\nu)}\left(A^{(\nu)}\right)^T\right)^{-1}B^{(\mu)}\mathbf{I}^\mu(t),$$

but this would not have been a solution to Equation (1.17) because it is differential, not algebraic, equation.

However, the application of the Laplace transform to Equation (1.17) together with $S_i^{(k)}(s) \in \mathbb{C}^{i(k+1) \times i}$ (1.9) and $Z_k^{(\varsigma-1)}(.): \mathbb{C} \to \mathbb{C}^{(\varsigma+1)k \times \varsigma k}$ (1.11), transforms the problem of solving the differential equation to the task to solve the following algebraic equation [159], [170]:

$$A^{(\nu)} S_N^{(\nu)}(s) \mathbf{Y}(s) = B^{(\mu)} S_M^{(\mu)} \mathbf{I}(s) + A^{(\nu)} Z_N^{(\nu-1)}(s) \mathbf{Y}_0^{\nu-1} - B^{(\mu)} Z_M^{(\mu-1)}(s) \mathbf{I}_0^{\mu-1}.$$

The solution reads:

$$\mathbf{Y}(s) = \left(A^{(\nu)} S_N^{(\nu)}(s) \right)^{-1} \left\{ \begin{array}{c} B^{(\mu)} S_M^{(\mu)} \mathbf{I}(s) + A^{(\nu)} Z_N^{(\nu-1)}(s) \mathbf{Y}_0^{\nu-1} - \\ -B^{(\mu)} Z_M^{(\mu-1)}(s) \mathbf{I}_0^{\mu-1} \end{array} \right\},$$

and

$$\mathbf{Y}(t) = \mathcal{L}^{-1} \left\{ \left(A^{(\nu)} S_N^{(\nu)}(s) \right)^{-1} \left\{ \begin{array}{c} B^{(\mu)} S_M^{(\mu)} \mathbf{I}(s) + A^{(\nu)} Z_N^{(\nu-1)}(s) \mathbf{Y}_0^{\nu-1} - \\ -B^{(\mu)} Z_M^{(\mu-1)}(s) \mathbf{I}_0^{\mu-1} \end{array} \right\} \right\},$$

where $\mathcal{L}^{-1}\{.\}$ denotes the inverse Laplace transform.

The compact, simple and elegant calculus is the basis for all calculations in the book. It is effectively applicable not only to linear continuous-time systems [159], [170], but also to linear discrete-time systems [42] and to nonlinear dynamical systems [175].

1.5 *Time* and system behavior

Time is a basic constituent of the environment of every dynamical physical system. *The time field is the temporal environment, i.e., the time environment, of the system* [181], [182], [183].

A *time*-dependent variable will be denoted for short by the corresponding letter, e.g., scalar variables by $D, I, R, S, U, Y, ...$ and vector variables by $\mathbf{D}, \mathbf{I}, \mathbf{R}, \mathbf{S}, \mathbf{U}, \mathbf{Y}, ...$ From the mathematical point of view they are functions, e.g., $D = D(.): \mathfrak{T} \longrightarrow \mathfrak{R}^1$, $\mathbf{D} = \mathbf{D}(.): \mathfrak{T} \longrightarrow \mathfrak{R}^d$.

A variation of the value of every *time*-dependent variable is in *time*.

As usual, $\mathfrak{R}_+$ is the set of all nonnegative real numbers, $\mathfrak{R}^+$ is the set of all positive real numbers, $\mathfrak{R}^k$ is the k dimensional real vector space, the elements of which are $k-$dimensional real valued vectors, where k is any natural number. Notice that $\mathfrak{R}^1 \neq \mathfrak{R}$.

There are three substantial characteristic groups of the variables that are associated with the dynamical system in general. Their definitions follow

by referring to [48, Definition 3-6, p. 83], [170], [175], [219, p. 105], [244, 2.
Definition, p. 380], [245, 2. Definition, p. 380], [264, p. 4], [265, p. 664].

Note 17 *The capital letters D, I, R, S, U, Y (and **D**, **I**, **R**, **S**, **U**, **Y**) denote the total scalar (vector) values of the variables $D(.)$, $I(.)$, $R(.)$, $S(.)$,
$U(.)$, $Y(.)$ (of the vector variables **D**(.), **I**(.), **R**(.), **S**(.), **U**(.), **Y**(.)) relative to their total zero scalar (vector) value, if it exists, or relative to their
accepted zero scalar (vector) value, respectively.*

A characteristic of the dynamical systems is their *dynamical behavior*.
The dynamical system can possess the explicit internal dynamics and the
implicit output dynamics or explicit both the internal and output dynamics.
A special family of the dynamical systems are *plants*, i.e., *objects*.

Definition 18 *Plant (object)*
   *A **plant** $\mathcal{P}$ (i.e., an **object** $\mathcal{O}$) is a system that should under specific
conditions called **nominal** (**nonperturbed**) realize its demanded dynamical
behavior and under other (nonnominal, perturbed, real) conditions should
realize its dynamical behavior sufficiently close to its demanded dynamical
behavior over some (bounded or unbounded) time interval.*

The physical nature of a plant can be anyone.

Definition 19 *Input variables, input vector and input space*
   *A variable that acts on the system and its influence is essential for the
system behavior is the **system input variable** denoted by $I \in \mathfrak{R}$. The system
can be under the action of several mutually independent input variables I_1,
I_2, ..., I_M. They compose **the system input vector** (for short, **input**)*

$$\mathbf{I} = [I_1 \ \ I_2 \ \ ... \ \ I_M]^T \in \mathfrak{R}^M, \tag{1.18}$$

*which is an element of **the input space** $\mathfrak{R}^M$.*
   ***The instantaneous values** of the variables I_i and $\mathbf{I}$ at an instant $t \in \mathfrak{T}$
are $I_i(t)$ and $\mathbf{I}(t)$, respectively.*
   *The capital letters I and $\mathbf{I}$ denote the total (scalar, vector) values of the
variable I and the vector $\mathbf{I}$ relative to their total zero (scalar, vector) value, if
it exists, or relative to their accepted zero (scalar, vector) value, respectively.*

The (left, right) Laplace transforms $\mathcal{L}^\mp\{I_i(t)\} = I_i^\mp(s)$, $i = 1, 2, ..., M$,
of the entries $I_i(t)$ form the (left, right) Laplace transform $\mathcal{L}^\mp\{\mathbf{I}(t)\} = \mathbf{I}^\mp(s)$
of the input vector function $\mathbf{I}(.): \mathfrak{T}_0 \to \mathfrak{R}^M$,

$$\mathcal{L}^\mp\{\mathbf{I}(t)\} = \mathbf{I}^\mp(s) = \left[I_1^\mp(s) \ I_2^\mp(s) \ \ ... \ \ I_M^\mp(s) \right]^T \in \mathbb{C}^M. \tag{1.19}$$

We introduce the complex matrix function $I^{\mp}(.) : \mathbb{C} \longrightarrow \mathbb{C}^{M \times M}$,

$$I^{\mp}(s) = diag\left\{I_1^{\mp}(s) \ \ I_2^{\mp}(s) \ \ \ldots \ I_M^{\mp}(s)\right\} \in \mathbb{C}^{M \times M}. \tag{1.20}$$

It and the unit vector $\mathbf{1}_M$ (1.13) permit us to represent $\mathbf{I}^{\mp}(s)$ into the following form

$$\mathbf{I}^{\mp}(s) = \underbrace{diag\left\{I_1^{\mp}(s) \ \ I_2^{\mp}(s) \ \ \ldots \ I_M^{\mp}(s)\right\}}_{I^{\mp}(s)}\mathbf{1}_M = I^{\mp}(s)\mathbf{1}_M. \tag{1.21}$$

Let the family $\mathfrak{L}$ of *time* dependent bounded input vector functions $\mathbf{I}(.) : \mathfrak{T}_0 \to \mathfrak{R}^M$ be such that their Laplace transforms are strictly proper real rational vector functions of the complex variable s,

$$\mathfrak{L} = \left\{\mathbf{I}(.) : \begin{array}{c} \exists \gamma(\mathbf{I}) \in \mathfrak{R}^+ \Longrightarrow \|\mathbf{I}(t)\| < \gamma(\mathbf{I}), \ \forall t \in \mathfrak{T}_0, \\ \mathcal{L}^{\mp}\{\mathbf{I}(t)\} = \mathbf{I}^{\mp}(s), \ (1.19)\text{-}(1.21) \Longrightarrow \\ \displaystyle I_k^{\mp}(s) = \frac{\sum_{j=0}^{j=\zeta_k} a_{kj}s^j}{\sum_{j=0}^{j=\psi_k} b_{kj}s^j}, 0 \leq \zeta_k < \psi_k, \ \forall k = 1, 2, ..., M, \end{array}\right\}. \tag{1.22}$$

We demand that the left Laplace transform $\mathbf{I}^-(s)$, or the right Laplace transform $\mathbf{I}^+(s)$, or just the Laplace transform $\mathbf{I}(s)$ of the input vector function $\mathbf{I}(.) \in \mathfrak{L}$, is strictly proper. It guarantees that the original $\mathbf{I}(t)$ does not contain an impulse component.

Notice that the zero input vector function $\mathbf{I}(.), \mathbf{I}(t) \equiv \mathbf{0}_M$, belongs to $\mathfrak{L}$.

$\mathfrak{C}^{ki} = \mathfrak{C}^k(\mathfrak{R}^i)$ is *the family of all functions defined and k-times continuously differentiable on* $\mathfrak{R}^i$, and

$\mathfrak{C}^k = \mathfrak{C}^k(\mathfrak{T}_0)$ is *the family of all functions defined, continuous and k-times continuously differentiable on* $\mathfrak{T}_0$, $\mathfrak{C} = \mathfrak{C}^0(\mathfrak{T}_0)$,

$\mathfrak{J}^k$ is a given, or to be determined, family of all bounded and *k-times* continuously differentiable permitted input vector functions $\mathbf{I}(.) \in \mathfrak{C}^k \cap \mathfrak{L}$,

$$\mathfrak{J}^k \subset \mathfrak{C}^k \cap \mathfrak{L}. \tag{1.23}$$

$\mathfrak{J}^0 = \mathfrak{J}$ is the family of all bounded continuous permitted input vector functions $\mathbf{I}(.) \in \mathfrak{C} \cap \mathfrak{L}$,

$$\mathfrak{J} \subset \mathfrak{C} \cap \mathfrak{L}. \tag{1.24}$$

Definition 20 $\mathfrak{J}_-^k$ *is a subfamily of* $\mathfrak{J}^k$, $\mathfrak{J}_-^k \subset \mathfrak{J}^k$, *such that the real part of every pole of the Laplace transform* $\mathbf{I}(s)$ *of every* $\mathbf{I}(.) \in \mathfrak{J}_-^k$ *is negative,* $\mathfrak{J}_- = \mathfrak{J}_-^0$.

Definition 21 *Disturbance variable and disturbance vector*

*An input variable D of a system that acts on the system without using any information about the system demanded dynamical behavior or by using it in order to perturb the system behavior is **the disturbance variable** (for short: **disturbance**) for the system.*

*If there are several, e.g., d, disturbance variables D_1, D_2, ... , D_d, then they are entries of **the disturbance vector** (for short: **disturbance**) **D**,*

$$\mathbf{D} = \left[D_1 \vdots D_2 \vdots ... \vdots D_d \right]^T \in \mathfrak{R}^d. \tag{1.25}$$

*The instantaneous values of the variables D_i and **D** at an instant $t \in \mathfrak{T}$ are $D_i(t)$ and $\mathbf{D}(t)$, respectively.*

A disturbance action on a physical system most often is not physically rejectable. The disturbance acts on the system at best independently of the system behavior, because if the disturbance exploits the information about the system demanded behavior in order to perturb the system behavior then it is **an enemy disturbance**. In order to stop the disturbance action on the system its source should be often destroyed, which is rarely possible. The physical nature of disturbances can be anyone.

$\mathfrak{D}^k$ is a given, or to be determined, family of all bounded and $(k + 1)$-*times* continuously differentiable on $\mathfrak{T}$ total disturbance vector functions $\mathbf{D}(.) \in \mathfrak{J}^{k+1}$, (1.23), such that they and their first $k + 1$ derivatives obey $TCUP$ (Principle 10),

$$\mathfrak{D}^k \subseteq \mathfrak{J}^{(k+1)}. \tag{1.26}$$

$\mathfrak{D}^0 = \mathfrak{D}$ is the family of all bounded continuous and continuously differentiable total disturbance vector functions $\mathbf{D}(.) \in \mathfrak{D}$ such that they and their first derivatives obey $TCUP$,

$$\mathfrak{D} \subseteq \mathfrak{J}. \tag{1.27}$$

$\mathfrak{D}^k_-$ is a subfamily of $\mathfrak{D}^k$, $\mathfrak{D}^k_- \subset \mathfrak{D}^k$, such that the real part of every pole of the Laplace transform $\mathbf{D}(s)$ of every $\mathbf{D}(.) \in \mathfrak{D}^k_-$ is negative, $\mathfrak{D}_- = \mathfrak{D}^0_-$,

The system output behavior is determined by the temporal evolution of its *output variables and their derivatives*, in the sense of the following definitions:

Definition 22 *Output variables, output vector, output space, and response*

A variable $Y \in \mathfrak{R}$ is an **output variable** *of the system if and only if its values result from the system behavior, they are (directly or indirectly) measurable, and we are interested in them.*

The number N is the maximal number of linearly independent output variables Y_1, Y_2, ..., Y_N on $\mathfrak{T}$ of the system. They form the **output vector** **Y** *of the system, which is element of* **the output space $\mathfrak{R}^N$:**

$$\mathbf{Y} = [Y_1 \ Y_2 \ ...Y_N]^T \in \mathfrak{R}^N. \tag{1.28}$$

The time evolution $\mathbf{Y}(t)$ of the output vector $\mathbf{Y}$ takes place, i.e. the output vector $\mathbf{Y}$ propagates, in the integral output space $\mathcal{I}$,

$$\mathcal{I} = \mathfrak{T} \times \mathfrak{R}^N. \tag{1.29}$$

The instantaneous values of the variables Y_i and $\mathbf{Y}$ at an instant $t \in \mathfrak{T}$ are $Y_i(t)$ and $\mathbf{Y}(t)$, respectively.

The time variation $\mathbf{Y}(t)$ of the system output vector $\mathbf{Y}$ is the **system (output) response.**

The plant desired output behavior *is denoted by $\mathbf{Y}_d(t)$.*

Note 23 *There are systems, the output variable of which is fed back to the system input. Such output variable is also the system input variable, and such system has its own (local) feedback.*

$\mathfrak{Y}_d^k$ is a given, or to be determined, family of all bounded and $(k + 1)$-times continuously differentiable realizable total desired output vector functions $\mathbf{Y}_d(.) \in \mathfrak{J}^{k+1}$, (1.23), such that they and their first $k + 1$ derivatives obey $TCUP$, i.e., $\mathfrak{Y}_d^k$ is a given, or to be determined, family of all bounded continuously differentiable realizable total desired extended output vector functions $\mathbf{Y}_d^k(.) \in \mathfrak{C}^1$ such that they obey $TCUP$,

$$\mathfrak{Y}_d^k \subseteq \mathfrak{J}^{k+1}, \tag{1.30}$$

$\mathfrak{Y}_d^0 = \mathfrak{Y}_d$,

$$\mathfrak{Y}_d \subseteq \mathfrak{J}. \tag{1.31}$$

$\mathfrak{Y}_{d0}^k$ is the set of the desired output initial conditions $\mathbf{Y}_{d0}^k = \mathbf{Y}_d^k(t_0)$ of $\mathbf{Y}_d^k(t)$ of every $\mathbf{Y}_d(.) \in \mathfrak{Y}_d^k$,

$$\mathfrak{Y}_{d0}^k = \left\{ \mathbf{Y}_{d0}^k : \mathbf{Y}_{d0}^k = \mathbf{Y}_d^k(t_0), \ \mathbf{Y}_d(.) \in \mathfrak{Y}_d^k \right\}. \tag{1.32}$$

If and only if $\mathbf{Y}_d(.) \in \mathfrak{Y}_d^k$ then $\mathbf{Y}_{d0}^k = \mathcal{Y}_d^k \left(t_0; t_0; \mathbf{Y}_{d0}^k \right) \in \mathfrak{Y}_{d0}^k$.

$\mathfrak{Y}_{d-}^k$ is a subfamily of $\mathfrak{Y}_d^k$, $\mathfrak{Y}_{d-}^k \subset \mathfrak{Y}_d^k$, such that the real part of every pole of the Laplace transform $\mathbf{Y}_d(s)$ of every $\mathbf{Y}_d(.) \in \mathfrak{Y}_{d-}^k$ is negative, $\mathfrak{Y}_{d-} = \mathfrak{Y}_{d-}^0$,

Definition 24 *Realizability of* $\mathbf{Y}_d(.)$ *on* $\mathfrak{T}_0 \times \mathfrak{D}^i \times \mathfrak{U}^\mu \times \mathfrak{Y}_d^k$ *[175, Definition 111, p.46]*

i)   The desired response $\mathbf{Y}_d(.) \in \mathfrak{Y}_d^k$ *of the plant is* **realizable** *on* $\mathfrak{T}_0 \times \mathfrak{D}^i \times \mathfrak{U}^\mu \times \mathfrak{Y}_d^k$ *if and only if for every* $[\mathbf{D}(.), \mathbf{Y}_d(.)] \in \mathfrak{D}^i \times \mathfrak{Y}_d^k$ *there exists a control vector function* $\mathbf{U}(.) \in \mathfrak{U}^\mu$ *defined on* $\mathfrak{T}_0$ *such that* $\mathbf{Y}_d(.)$ *is the unique plant output response through* $\mathbf{Y}_{d0}$ *on* $\mathfrak{T}_0$ *under the action of any* $\mathbf{D}(.) \in \mathfrak{D}^i$.

ii) If and only if additionally to i) the control vector $\mathbf{U}$ *can act on every entry* $Y_{dk}(.)$, $\forall k = 1, 2, .., n$, *of* $\mathbf{Y}_d$ *mutually independently when the plant is under the influence of any disturbance* $\mathbf{D}(.) \in \mathfrak{D}^i$ *then* $\mathbf{Y}_d(.)$ *is* **elementwise realizable** *on* $\mathfrak{T}_0 \times \mathfrak{D}^i \times \mathfrak{U}^\mu \times \mathfrak{Y}_d^k$.

iii) If and only if additionally to i) the dimension r *of the control vector* $\mathbf{U}$ *is the least number of the entries* $U_j(.)$, $j = 1, 2, ..., r$, *of* $\mathbf{U}$ *that satisfies i) then the control is* **minimal** *for the realizability of* $\mathbf{Y}_d(.)$ *on* $\mathfrak{T}_0 \times \mathfrak{D}^i \times \mathfrak{U}^\mu \times \mathfrak{Y}_d^k$.

A (physical and a mathematical) dynamical system can be subjected to the action of the input vector derivatives $\mathbf{I}^{(l)}(t)$, $l \in \{1, 2, ...\}$. The system internal and output dynamical behavior depend then not only on the input vector $\mathbf{I}(t)$ but also on all its derivatives acting on the system. This is reality that inspires us, justifies and demands us to generalize the concept of the dynamical system *state* as follows.

Definition 25 *State of a dynamical system*

The (internal, output) state of a physical dynamical system at a moment $\tau \in \mathfrak{T}$ *is, respectively, the system (internal, output) dynamical physical situation at the moment* τ, *which, together with the input vector and its derivatives acting on the system at any moment* $(t \geq \tau) \in \mathfrak{T}$, *determines uniquely the system behavior [i.e., the system (internal, output) state and the system output response] for all* $(t > \tau) \in \mathfrak{T}$, *respectively.*

The (internal, output) state of a mathematical dynamical system at a moment $\tau \in \mathfrak{T}$ *is, respectively, the minimal amount of information about the system at the moment* τ, *which, together with information about the action on the system (about the system input vector and its derivatives acting on the system) at any moment* $(t \geq \tau) \in \mathfrak{T}$, *determines uniquely* *the system behavior (i.e., the system (internal, output) state and its output response) for all $(t > \tau) \in \mathfrak{T}$, respectively.*

*The minimal number $n_{(.)}$ of linearly independent variables $S_{(.)i}$ on $\mathfrak{T}$, $i = 1, 2, \ldots, n_{(.)}$, the values $S_{(.)i}(\tau)$ of which are at every moment $\tau \in \mathfrak{T}$ in the biunivoque correspondence with the system (internal: $(\cdot) = I$, output: $(\cdot) = O$) state at the same moment τ, is **the state dimension** and the variables $S_{(.)i}$, $i = 1, 2, \ldots, n_{(.)}$, are, respectively, **the (internal: $(\cdot) = I$, output: $(\cdot) = O$) state variables of the system.** They compose, respectively, **the (internal: $(\cdot) = I$, output: $(\cdot) = O$) state vector $S_{(.)}$ of the system**,*

$$\mathbf{S}_{(\cdot)} = \begin{bmatrix} S_{(\cdot)1} & S_{(\cdot)2} & \ldots S_{(\cdot)n_{(.)}} \end{bmatrix}^T \in \mathfrak{R}^n, \quad (.) = \, , I, \, O. \qquad (1.33)$$

*The space $\mathfrak{R}^{n(\cdot)}$ is, respectively, **the (internal: $(.) = I$, output: $(\cdot) = O$) state space of the system.***

*The state vector function $\mathbf{S}(.): \mathfrak{T} \longrightarrow \mathfrak{R}^n$ is **the motion of the system.***

The instantaneous value of the (internal, output) state vector function $\mathbf{S}_{(\cdot)}(.)$ at an instant $t \in \mathfrak{T}$ is the instantaneous (internal, output) state vector $\mathbf{S}_{(\cdot)}(t)$ at the instant t, respectively

***The plant desired state behavior** is denoted by $\mathbf{S}_d(t)$.*

This definition broadens and generalizes the well known and commonly accepted definition of the state of the dynamical in general, control in particular, systems.

In what follows the term *mathematical system* denotes the accepted mathematical model (description) of the corresponding physical system.

The system explicit internal dynamics variable is its *internal (dynamics) state variable S_I*. This is typical for the *ISO, EISO*, and *HISO* systems.

The *IO* and *IIO* systems possess the explicit output dynamics, too. The *IO* systems internal dynamics has not been well directly studied. The system output dynamics variable is its *output (dynamics) state variable S_O*.

The *IO* system internal dynamics is simultaneously its output dynamics so that $S_I = S_O = S$, where S is the system *full state variable S_F, $S_F = S$*.

The internal dynamics of the *ISO, EISO*, and *HISO* systems determines completely their output dynamics in the free regime so that for them $S_I = S_O = S_F = S$, too.

The *IIO* system internal dynamics and output dynamics are explicit and different so that $S_I \neq S_O$ and the full state variable is the vector variable

$$\mathbf{S}_F = \mathbf{S} = \begin{bmatrix} \mathbf{S}_I^T : \mathbf{S}_O^T \end{bmatrix}^T.$$

The properties of the system determine the form and the character of the system state vector **S**:

- **The Input-Output** (IO) **systems** are described by the ν-th order linear vector differential *input-output, i.e., the output state,* equation of the output vector $\mathbf{Y} \in \mathfrak{R}^N$,

$$\mathbf{Y} = \begin{bmatrix} Y_1 \vdots Y_2 \vdots \ldots \vdots Y_N \end{bmatrix}^T \in \mathfrak{R}^N, \; Y_i \in \mathfrak{R}, \; i = 1, 2, \ldots, N. \qquad (1.34)$$

  Their extended output vector $\mathbf{Y}^{\nu-1}$,

$$\mathbf{Y}^{\nu-1} = \begin{bmatrix} \mathbf{Y}^T \vdots \mathbf{Y}^{(1)^T} \vdots \ldots \vdots \mathbf{Y}^{(\nu-1)^T} \end{bmatrix}^T \in \mathfrak{R}^{\nu N}, \; n = \nu N, \qquad (1.35)$$

  is their state vector $\mathbf{S}_{IO}$, which is also their internal state vector $\mathbf{S}_{IOI}$, their output state vector $\mathbf{S}_{IOO}$ and their full state vector $\mathbf{S}_F$,

$$\mathbf{S}_{IOI} = \mathbf{S}_{IOO} = \mathbf{S}_{IOF} = \mathbf{S}_{IO} = \mathbf{Y}^{\nu-1} \in \mathfrak{R}^n, \; n = \nu N. \qquad (1.36)$$

- **The Input-State-Output** (ISO) **systems** are determined by the first order linear vector differential equation in the vector $\mathbf{X}$ (1.38)–*the (internal) state equation,* by *the algebraic output vector equation* of the output vector $\mathbf{Y}$, and the only derivative in them is the first derivative of the state vector. Their state vector $\mathbf{S}_{ISO}$ is the vector $\mathbf{X}$, which is also their internal state vector $\mathbf{S}_{ISOI}$ and their full state vector $\mathbf{S}_{ISOF}$:

$$\mathbf{S}_{ISOI} = \mathbf{S}_{ISOF} = \mathbf{S}_{ISO} = \mathbf{X} \in \mathfrak{R}^n. \qquad (1.37)$$

  They do not possess the output state vector $\mathbf{S}_O$ because they do not have an independent output dynamics. Their output equation does not contain any derivative of the output vector.

- **The Extended Input-State-Output** ($EISO$) **systems** are determined by the first order linear vector differential equation in the vector $\mathbf{X}$ (1.38),

$$\mathbf{X} = \begin{bmatrix} X_1 \vdots X_2 \vdots \ldots \vdots X_n \end{bmatrix}^T \in \mathfrak{R}^n, \; X_i \in \mathfrak{R}, \; \forall i = 1, 2, \ldots, n, \qquad (1.38)$$

which is *the (internal) state equation,* by *the algebraic output vector equation* of the output vector $\mathbf{Y}$, and, in addition to the first derivative of the state vector, there are derivatives of the input vector only in the state equation. Their state vector $\mathbf{S}_{EISO}$ is the vector $\mathbf{X}$ (1.38) that is also the internal state vector $\mathbf{S}_{EISOI}$, and the full state vector $\mathbf{S}_{EISOF}$:

$$\mathbf{S}_{EISOI} = \mathbf{S}_{EISOF} = \mathbf{S}_{EISO} = \mathbf{X} \in \mathfrak{R}^n, \tag{1.39}$$

They do not possess the output state vector $\mathbf{S}_O$ for the same reason for which the *ISO* systems do not have the output state vector.

Note 26 *On the highest derivative of the input vector*
In order to avoid the problem of the appearance of impulse discontinuities in the system behavior the systems theory and the control theory restrict the order of the highest derivative of the input vector to be at most equal to the system order. However, the problem of the appearance of impulse discontinuities in the system behavior does not exist if the input vector function is defined and continuously differentiable μ-times, where μ is the order of the highest input vector derivative acting on the system. For its physical origin see in the sequel Note 69 (Subsection 2.1.1).

- **The Higher Order-Input-State-Output (*HISO*) systems** are characterized by the α-*th* order linear vector differential equation, i.e., *the α-th order (internal) state equation,* in *the substate vector* $\mathbf{R}$,

$$\mathbf{R} = \left[R_1 \vdots R_2 \vdots ... \vdots R_\rho \right]^T \in \mathfrak{R}^\rho, \ R_i \in \mathfrak{R}, \ i = 1, 2, ..., \rho, \tag{1.40}$$

and are additionally determined by *the algebraic output vector equation* of the output vector $\mathbf{Y}$. Their internal state vector $\mathbf{S}_{HISOI}$ is the extended vector $\mathbf{R}^{\alpha-1}$,

$$\mathbf{R}^{\alpha-1} = \left[\mathbf{R}^T \vdots \mathbf{R}^{(1)^T} \vdots ... \vdots \mathbf{R}^{(\alpha-1)^T} \right]^T \in \mathfrak{R}^{\alpha\rho}, \ n = \alpha\rho, \tag{1.41}$$

which is also their full state vector $\mathbf{S}_f$,

$$\mathbf{S}_{HISOI} = \mathbf{S}_{HISOf} = \mathbf{S}_{HISO} = \mathbf{R}^{\alpha-1} \in \mathfrak{R}^n, \ n = \alpha\rho. \tag{1.42}$$

They do not possess the output state vector $\mathbf{S}_O$. The derivatives of the input vector can exist only in the state equation.

- **The Input-Internal and Output state** (IIO) **systems** are characterized by the α-*th* order linear vector differential equation, i.e., by *the α-th order internal state equation*, in *the substate vector* $\mathbf{R}$, and by the linear output vector ν-*th* order differential equation, i.e., by *the output state equation* of the output vector $\mathbf{Y}$. Their extended vector $\mathbf{R}^{\alpha-1}$ (1.42) is their internal state vector $\mathbf{S}_{IIOI}$ (1.43),

$$\mathbf{S}_{IIOI} = \mathbf{R}^{\alpha-1} \in \mathfrak{R}^{n_I}, \; n_I = \alpha\rho, \qquad (1.43)$$

and their output state vector $\mathbf{S}_{IIOO}$ is the extended output vector $\mathbf{Y}^{\nu-1}$,

$$\mathbf{S}_{IIOO} = \mathbf{Y}^{\nu-1} = \left[\mathbf{Y}^T \; \vdots \; \mathbf{Y}^{(1)^T} \; \vdots \; ... \; \vdots \; \mathbf{Y}^{(\nu-1)^T} \right]^T \in \mathfrak{R}^{n_O}, \; n_O = \nu N. \qquad (1.44)$$

Their full state vector $\mathbf{S}_{IIOf}$, which is their state vector $\mathbf{S}_{IIO}$, is composed of their internal state vector $\mathbf{S}_{IIOI} = \mathbf{R}^{\alpha-1}$ and of their output state vector $\mathbf{S}_{IIO} = \mathbf{Y}^{\nu-1}$,

$$\mathbf{S}_{IIOf} = \left[\begin{array}{c} \mathbf{S}_{IIOI} \\ \mathbf{S}_{IIOO} \end{array} \right] = \left[\begin{array}{c} \mathbf{R}^{\alpha-1} \\ \mathbf{Y}^{\nu-1} \end{array} \right] = \mathbf{S}_{IIO} \in \mathfrak{R}^n, \; n = \alpha\rho + \nu N. \quad (1.45)$$

Comment 27 *The state variables and the state vectors defined by (1.34)–(1.45) have the full physical sense (for more details see Note 42 in Section 2.1) and Note 59 in Section 3.1.*

Definition 28 *System state, motion and response*
   The system state vector $S(t)$ at a moment $t \in \mathfrak{T}$ is the vector value of the system motion $S(.; t_0; S_0; \mathbf{I})$ at the same moment t:

$$\mathbf{S}(t) \equiv \mathcal{S}(t; t_0; \mathbf{S}_0; \mathbf{I}) \implies \mathbf{S}(t_0) \equiv \mathcal{S}(t_0; t_0; \mathbf{S}_0; \mathbf{I}) \equiv \mathbf{S}_0.$$

1.6   *Time* and control

Definition 29 *Control variable and control vector*
   *An input variable U of a system (e.g., of a plant) that acts, together with its μ derivatives, on the system by using information about the system demanded behavior in order to force the system to realize its demanded behavior under the system nominal conditions and to force the system real behavior to be sufficiently close to the system demanded behavior under perturbed conditions is **the control variable** for the system.*

*If and only if there are several, e.g., r, control variables $U_1, U_2, \dots, U_r$, then they form **the control vector** (for short: **control**) $\mathbf{U}$,*

$$\mathbf{U} = \left[U_1 \vdots U_2 \vdots \dots \vdots U_r \right]^T \in \mathfrak{R}^r, \tag{1.46}$$

and together with their μ derivatives that act on the system form the extended control vector $\mathbf{U}^\mu$,

$$\mathbf{U}_i^\mu = \left[U_i^i \vdots U_i^{(1)} \vdots \dots \vdots U_i^{(\mu)} \right]^T \in \mathfrak{R}^{\mu+1}, \quad i = 1, 2, \dots, r, \tag{1.47}$$

$$\mathbf{U}^\mu = \left[\mathbf{U}_1^{(\mu)} \vdots \mathbf{U}_2^{(\mu)} \vdots \dots \vdots \mathbf{U}_r^{(\mu)} \right]^T \in \mathfrak{R}^{(\mu+1)r}. \tag{1.48}$$

The instantaneous values of the control variables U_i and of the control vector $\mathbf{U}$ at an instant $t \in \mathfrak{T}$ are $U_i(t)$ and $\mathbf{U}(t)$, respectively.

*A system that creates, generates, the control for the given system is **the controller** $\mathcal{C}$ for the given system. Its output vector $\mathbf{Y}_C$ is the control vector $\mathbf{U}$, $\mathbf{Y}_C = \mathbf{U}$.*

The physical nature of a control variable can be anyone.

Note 30 *Rejection or compensation?*

In this book we accept to use the term "compensation (for disturbance action)" rather than the term "rejection (the disturbance action)" for the reasons explained in [175, Remark 134, p. 62] and [188, Remark 234, pp. 169, 170].

Definition 31 *Control system*

*The system composed of a plant and of its controller is **the control system** $\mathcal{CS}$ of the plant.*

*If and only if the controller uses only information about the plant demanded behavior (and possibly about the disturbance) in order to act on the plant then the control system of the plant is under a) **open loop without disturbance compensation,**, under b) **with the direct disturbance compensation**, respectively, Figure 1.1.*

*If and only if the controller uses only information about the plant demanded behavior and about the plant real behavior in order to get information about the deviation of the latter from the former then the control system of the plant is **closed loop (feedback) with the indirect disturbance compensation**, Figure 1.2.*

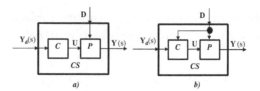

Figure 1.1: The structural scheme of the open loop control system (*CS*): (a) without the disturbance (**D**) compenstaion and (b) with the direct disturbance (**D**) compenstaion.

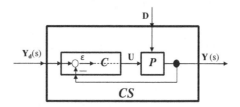

Figure 1.2: Structural scheme of the closed loop control system (CS) with the indirect disturbance compensation.

If and only if the controller uses information about the plant demanded behavior, about the disturbance, about the difference between the plant demanded behavior and the plant real behavior to get information about the deviation of the latter from the former in order to act on the plant then the control system of the plant is **combined with the direct and indirect disturbance compensation**, *respectively, Figure 1.3.*

Note 32 *Plant input and output vectors*

The plant $\mathcal{P}$ input vectors are in general the disturbance vector **D** and the control vector **U**,

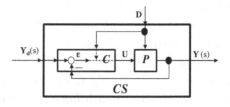

Figure 1.3: Structural scheme of the control system (CS) with the combined (direct and indirect) disturbance compensation.

so that the plant input vector $\mathbf{I}_P$ has two subvectors:

$$\mathbf{I}_P = \begin{bmatrix} \mathbf{D}^T & \mathbf{U}^T \end{bmatrix}^T \in \mathfrak{R}^{d+r}. \tag{1.49}$$

The plant $\mathcal{P}$ output vector $\mathbf{Y}_p$ is in general denoted by $\mathbf{Y}$,

$$\mathbf{Y}_p = \mathbf{Y} \in \mathfrak{R}^N. \tag{1.50}$$

Note 33 *Controller input and output vectors*

The controller $\mathcal{C}$ input vectors are in general the disturbance vector $\mathbf{D}$, the plant output vector $\mathbf{Y}$, and the plant desired output vector $\mathbf{Y}_d$, so that the controller input vector in general

$$\mathbf{I}_C = \begin{bmatrix} \mathbf{D}^T & \mathbf{Y}^T & \mathbf{Y}_d^T \end{bmatrix}^T \in \mathfrak{R}^{d+2N}, \tag{1.51}$$

The feedback controller $\mathcal{C}_f$ input vectors are in general the plant output vector $\mathbf{Y}$ and the plant desired output vector $\mathbf{Y}_d$ so that the feedback controller input vector

$$\mathbf{I}_{Cf} = \begin{bmatrix} \mathbf{Y}^T & \mathbf{Y}_d^T \end{bmatrix}^T \in \mathfrak{R}^{2N}, \tag{1.52}$$

Usually we treat mathematically *the output error vector* $\mathbf{e}$,

$$\mathbf{e} = \mathbf{Y}_d - \mathbf{Y}, \tag{1.53}$$

as the feedback controller input vector,

$$\mathbf{I}_{Cf} = \mathbf{e}, \tag{1.54}$$

although the controller receives the signals on $\mathbf{Y}_d$ and $\mathbf{Y}$, determines their difference, i.e., the output error vector $\mathbf{e} = \mathbf{Y}_d - \mathbf{Y}$, and creates the error signal $\xi_{\mathbf{e}}$ usually proportional to $\mathbf{e}$, $\xi_{\mathbf{e}} = k_{\mathbf{e}}\mathbf{e}$.

The controller $\mathcal{C}$ output vector vector $\mathbf{Y}_C$ in general and the feeback controller $\mathcal{C}_f$ output vector $\mathbf{Y}_{Cf}$ in particular are the control vector $\mathbf{U}$,

$$\mathbf{Y}_C = \mathbf{Y}_{Cf} = \mathbf{U} \in \mathfrak{R}^r. \tag{1.55}$$

Note 34 *Control system input and output vectors*

The control system $\mathcal{CS}$ input vectors and the closed loop, i.e., feedback, control system $\mathcal{CS}_f$ input vectors are in general the disturbance vector $\mathbf{D}$ and the plant desired output vector $\mathbf{Y}_d$ so that the control system input vector

$$\mathbf{I}_{CS} = \mathbf{I}_{CSf} = \begin{bmatrix} \mathbf{D}^T & \mathbf{Y}_d^T \end{bmatrix}^T \in \mathfrak{R}^{d+2N}. \tag{1.56}$$

The control system $\mathcal{CS}$ output vector $\mathbf{Y}_C$ and the closed loop, i.e., feedback, control system $\mathcal{CS}_f$ output vector $\mathbf{Y}_{Cf}$ are the same and are the plant output vector $\mathbf{Y}_P$,

$$\mathbf{Y}_{CS} = \mathbf{Y}_{Cf} = \mathbf{Y}_P \in \mathfrak{R}^N. \tag{1.57}$$

1.7   System transfer function matrices

Let σ and ω be real numbers, or real valued scalar variables, and $j = \sqrt{-1}$ be the imaginary unit, $j \in \mathbb{C}$:

$$s = (\sigma + j\omega) \in \mathbb{C}, \ \sigma \in \mathfrak{R}, \ \omega \in \mathfrak{R}. \tag{1.58}$$

The set of all values of the complex variable s determines the complex domain $\mathbb{C}$, $s \in \mathbb{C}$. It enables us very effective and simple studies of the systems.

The basis for the effective and complete analysis and synthesis of the systems in the complex domain is *the system full transfer function matrix $F(s)$*. It was discovered, defined and determined for *time*-invariant continuous-*time* linear systems in [115], and for *time*-invariant discrete-*time* linear systems in [164]. Its usage started in these references and continued in [42], [111], [112], [113], [114], [159], [161], [162], [164], [170], [171], [180], [186] for the analysis of the system whole output response. *The reader will benefit if she/he is familiar with their definitions and their derivations given in [170].* We present only the following general definition and explanation of $F(s)$.

Definition 35 *The full (complete) input-output (IO) transfer function matrix of the system in general [159], [170]*

The full (complete) input-output (IO) transfer function matrix of the system, which is denoted by $F(s)$, $F(s) \in \mathbb{C}^{N \times (M+\varsigma)}$, is the complex matrix value of the system full (complete) input-output (IO) matrix transfer function $F(.)$, $F(.) : \mathbb{C} \to \mathbb{C}^{N \times (M+\varsigma)}$, which is a matrix function of the complex variable s such that it determines uniquely the (left, right) Laplace transform $\mathbf{Y}^{(\mp)}(s)$ of the system output $\mathbf{Y}(t)$ as a homogenous linear function of the (left, right) Laplace transform $\mathbf{I}^{(\mp)}(s)$ of the system input vector $\mathbf{I}(t)$ for an arbitrary variation of $\mathbf{I}(t)$, for arbitrary initial vector values

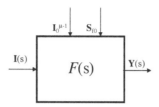

Figure 1.4: The full block of the system.

$\mathbf{I}_{0\mp}^{\mu-1}$ and $\mathbf{S}_{f0\mp}$ *of the extended input vector* $\mathbf{I}^{\mu-1}(t)$ *and of the full state vector* $\mathbf{S}_f(t)$ *at* $t = 0^{\mp}$, *respectively:*

$$\mathbf{Y}^{(\mp)}(s) = F(s) \left[\left(\mathbf{I}^{(\mp)}(s) \right)^T \ \vdots \ \left(\mathbf{I}_{0\mp}^{\mu-1} \right)^T \ \vdots \ \mathbf{S}_{f0\mp}^T \right]^T, \qquad (1.59)$$

Figure 1.4.

In order to show clearly the complete analogy of the system *full transfer function matrix* $F(s)$ with the system transfer function matrix $G(s)$, and of the system *full block diagram* induced by $F(s)$ and the classical system block induced by $G(s)$, we introduce the action vector function $\mathbf{V}(.) : \mathfrak{T}_0 \longrightarrow \mathfrak{R}^{M+\varsigma}$,

$$\mathbf{V}(t) = \left[\begin{array}{c} \mathbf{I}(t) \\ \delta^{\mp}(t)\mathbf{C}_0 \end{array} \right] \in \mathfrak{R}^{M+\varsigma} \qquad (1.60)$$

which comprises the input vector function $\mathbf{I}(t)$ and the vector $\mathbf{C}_0 \in \mathfrak{R}^{\varsigma}$ of all initial conditions: of the input $(\mathbf{I}_{0\mp}, \mathbf{I}_{0\mp}^{(1)}, ..., \mathbf{I}_{0\mp}^{(\mu-1)}, \text{ i.e., } \mathbf{I}_{0\mp}^{\mu-1})$, of the (full) state, i.e., of the internal state, $(\mathbf{R}_{0\mp}, \mathbf{R}_{0\mp}^{(1)}, ..., \mathbf{R}_{0\mp}^{(\alpha-1)}, \text{ i.e., } \mathbf{R}_{0\mp}^{\alpha-1}, \text{ or } \mathbf{X}_{0\mp})$ and of the output state $(\mathbf{Y}_{0\mp}, \mathbf{Y}_{0\mp}^{(1)}, ..., \mathbf{Y}_{0\mp}^{(\nu-1)}, \text{ i.e., } \mathbf{Y}_{0\mp}^{\nu-1})$ of the system, in general,

$$\mathbf{C}_0^{\mp} = \mathbf{C}_0^{\mp} \left(\mathbf{I}_{0\mp}^{\mu-1}, \mathbf{R}_{0\mp}^{\alpha-1}, \mathbf{X}_{0\mp}, \mathbf{Y}_{0\mp}^{\nu-1} \right). \qquad (1.61)$$

Left/right Laplace transform of $\mathbf{V}(.) : \mathfrak{T} \longrightarrow \mathfrak{R}^{M+\varsigma}$ is $\mathbf{V}(.) : \mathbb{C} \longrightarrow \mathbb{C}^{M+\varsigma}$,

$$\mathbf{V}^{\mp}(s) = \left[\begin{array}{c} \mathbf{I}^{\mp}(s) \\ \mathbf{C}_0^{\mp} \end{array} \right] \in \mathbb{C}^{M+\varsigma}, \qquad (1.62)$$

respectively.

The equivalent definition to Definition 35 reads:

Definition 36 *The full (complete) input-output (IO) transfer function matrix of the system in general [159], [170]*

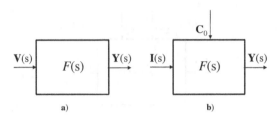

Figure 1.5: (a) The full block of the system in the compact form. (b) The full block of the system in the slightly extended form.

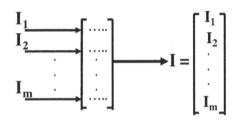

Figure 1.6: The vector generator symbolic block.

The full (complete) input-output (IO) transfer function matrix of the system, which is denoted by $F(s)$, $F(s) \in \mathbb{C}^{N \times (M+\varsigma)}$, is the complex matrix value of *the system full (complete) input-output (IO) matrix transfer function* $F(.)$, $F(.): \mathbb{C} \to \mathbb{C}^{N \times (M+\varsigma)}$, which is a matrix function of the complex variable s such that it determines uniquely the (left, right) Laplace transform $\mathbf{Y}^{(\mp)}(s)$ of the system output $\mathbf{Y}(t)$ as a homogenous linear function of the (left, right) Laplace transform $\mathbf{V}^{(\mp)}(s)$ of the overall system action vector $\mathbf{V}(t)$ for its arbitrary value and its variation,

$$\mathbf{Y}^{(\mp)}(s) = F(s)\mathbf{V}^{(\mp)}(s), \tag{1.63}$$

Figure 1.5.

The preceding definitions establish the solution to Problem 1 (Preface).

1.8   Full block diagrams: control systems

1.8.1   Full block diagrams

Let *the vector generator* creates its output vector $\mathbf{I}$ if its entries are the input vectors $\mathbf{I}_1$, $\mathbf{I}_2$, ... $\mathbf{I}_m$. Figure 1.6 shows the vector generator symbolic block.

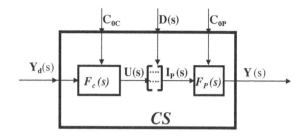

Figure 1.7: Full block diagram of the open loop control system (with the direct disturbance compensation).

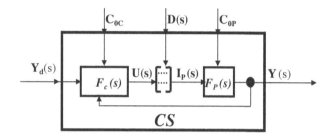

Figure 1.8: Full block diagram of the closed loop control system (with the indirect disturbance compensation).

Let us illustrate in principle how we determine the control system full transfer function matrix $F_{CS}(s)$ in terms of:

- **The plant full transfer function matrix** $F_P(s)$ assumed given (it is determined for every type of the plant separately in Part I),

- **The controller full transfer function matrix** $F_C(s)$ accepted known (it is determined for every kind of the controller separately in Part I),

- **The Laplace transforms of their input vectors**, and

- **The vectors C_{0P} and C_{0C} of all the initial conditions acting on the plant $\mathcal{P}$ and on the controller $\mathcal{C}$**, respectively.

We do this in what follows for:

- **The open loop control system** (with the direct disturbance compensation), the full block diagram of which is in Figure 1.7,

- **The closed loop control system** (with the indirect disturbance compensation), the full block diagram of which is in Figure 1.8,

- **The combined loops control system** (with the direct and indirect

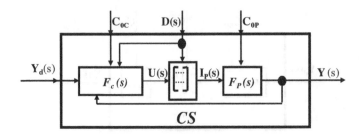

Figure 1.9: Full block diagram of the combined loops control system (with the direct and indirect disturbance compensation).

disturbance compensation), the full block diagram of which is in Figure 1.8.

1.8.2 $F_{CS}(s)$ of the open loop control system

We refer to Definition 36, (Section 1.7), and Figure 1.7:

$$\mathbf{Y}(s) = F_P(s)\mathbf{V}_P(s), \quad \mathbf{V}_P(s) = \begin{bmatrix} \mathbf{I}_P^T(s) & \mathbf{C}_{0P}^T \end{bmatrix}^T,$$
$$\mathbf{I}_P(s) = \begin{bmatrix} \mathbf{D}^T(s) & \mathbf{U}^T(s) \end{bmatrix}^T, \tag{1.64}$$

and introduce the plant transfer function $G_{PD}(s)$ relative to $\mathbf{D}(s)$, the plant transfer function $G_{PU}(s)$ relative to $\mathbf{U}(s)$ and the plant transfer function $G_{P0}(s)$ relative to $\mathbf{C}_{0P}$ so that

$$F_P(s) = \begin{bmatrix} G_{PU}(s) & \vdots & G_{PU}(s) & \vdots & G_{P0}(s) \end{bmatrix},$$
$$\mathbf{Y}(s) = F_P(s)\mathbf{V}_P(s), \quad \mathbf{V}_P(s) = \begin{bmatrix} \mathbf{D}^T(s) & \mathbf{U}^T(s) & \mathbf{C}_{0P}^T \end{bmatrix}^T,$$
$$\mathbf{Y}(s) = \begin{bmatrix} G_{PD}(s) & \vdots & G_{PU}(s) & \vdots & G_{P0}(s) \end{bmatrix} \begin{bmatrix} \mathbf{D}(s) \\ \mathbf{U}(s) \\ \mathbf{C}_{0P} \end{bmatrix}, \tag{1.65}$$

and

$$F_C(s) = \begin{bmatrix} G_{CY_d}(s) & \vdots & G_{C0}(s) \end{bmatrix},$$
$$\mathbf{U}(s) = F_C(s)\mathbf{V}_c(s), \quad \mathbf{V}_c(s) = \begin{bmatrix} \mathbf{Y}_d^T(s) & \mathbf{C}_{0C}^T \end{bmatrix}^T \implies$$
$$\mathbf{U}(s) = \begin{bmatrix} G_{CY_d}(s) & \vdots & G_{C0}(s) \end{bmatrix} \begin{bmatrix} \mathbf{Y}_d^T(s) & \mathbf{C}_{0C}^T \end{bmatrix}^T. \tag{1.66}$$

We replace $\mathbf{U}(s)$ from (1.66) into (1.65):

$$
\mathbf{Y}(s) = \left\{ \begin{array}{c} \left[G_{PD}(s) \vdots G_{PU}(s) \vdots G_{P0}(s) \right] \bullet \\ \mathbf{D}(s) \\ \bullet \left[\left[G_{CY_d}(s) \vdots G_{C0}(s) \right] \left[\begin{array}{c} \mathbf{Y}_d(s) \\ \mathbf{C}_{0C} \end{array} \right] \right] \\ \mathbf{C}_{0P} \end{array} \right\} =
$$
$$
= F_{CS}(s)\,\mathbf{V}_{CS}(s),
\tag{1.67}
$$

where, [159], [170],

$$
F_{CS}(s) = \left[G_{PD}(s) \vdots G_{PU}(s)\,G_{CY_d}(s) \vdots G_{CS0}(s) \right],
\tag{1.68}
$$

$$
\mathbf{V}_{CS}(s) = \left[\begin{array}{c} \mathbf{D}(s) \\ \mathbf{Y}_d(s) \\ \mathbf{C}_{0CS} \end{array} \right],\quad \mathbf{C}_{0CS} = \left[\begin{array}{c} \mathbf{C}_{0C} \\ \mathbf{C}_{0P} \end{array} \right],
\tag{1.69}
$$

$$
G_{CS0}(s) = \left[G_{PU}(s)\,G_{C0}(s) \vdots G_{P0}(s) \right],
\tag{1.70}
$$

$$
\mathbf{Y}(s) = \left[G_{PD}(s) \vdots G_{PU}(s)\,G_{CY_d}(s) \vdots G_{CS0}(s) \right] \left[\begin{array}{c} \mathbf{D}(s) \\ \mathbf{Y}_d(s) \\ \mathbf{C}_{0CS} \end{array} \right].
\tag{1.71}
$$

This shows in the complex domain how the system transfers in the course of *time* the influence of the action vector $\mathbf{V}_{CS}(t) = \mathcal{L}^{-1}\{\mathbf{V}_{CS}(s)\}$ on the system output vector $\mathbf{Y}(t) = \mathcal{L}^{-1}\{\mathbf{Y}(s)\}$.

Equations (1.68)–(1.71) determine $F_{CS}(s)$ and $\mathbf{Y}(s)$ in their extended forms:

$$
F_{CS}(s) = \left[G_{PD}(s) \vdots G_{PU}(s)\,G_{CY_d}(s) \vdots G_{PU}(s)\,G_{C0}(s) \vdots G_{P0}(s) \right],
\tag{1.72}
$$

$$
\mathbf{V}_{CS}(s) = \left[\begin{array}{cccc} \mathbf{D}^T(s) & \mathbf{Y}_d^T(s) & \mathbf{C}_{0C}^T & \mathbf{C}_{0P}^T \end{array}\right]^T,
\tag{1.73}
$$

$$
\mathbf{Y}(s) = \left\{ \begin{array}{c} \left[G_{PD}(s) \vdots G_{PU}(s)\,G_{CY_d}(s) \vdots G_{PU}(s)\,G_{C0}(s) \vdots G_{P0}(s) \right] \bullet \\ \bullet \left[\begin{array}{cccc} \mathbf{D}^T(s) & \mathbf{Y}_d^T(s) & \mathbf{C}_{0C}^T & \mathbf{C}_{0P}^T \end{array}\right]^T \end{array} \right\}.
\tag{1.74}
$$

Equations (1.72)–(1.74) imply the following open loop control system transfer functions:

-       Its transfer function $G_{CSD}(s) = G_{PD}(s)$ relative to the disturbance **D**

-       Its transfer function $G_{CSY_d}(s) = G_{PU}(s) G_{CY_d}(s)$ relative to the the desired output $\mathbf{Y}_d$:

-       Its transfer function $G_{CSOC}(s) = G_{PU}(s) G_{C0}(s)$ relative to the vector $\mathbf{C}_{OC}$ of the controller initial conditions and

-       Its transfer function $G_{CSOO}(s) = G_{P0}(s)$ relative to the vector $\mathbf{C}_{OP}$ of the plant initial conditions.

1.8.3 $F_{CS}(\mathbf{s})$ of the closed loop control system

We apply both Definition 36 and Figure 1.8. Equations (1.65) related to the plant rest unchanged. However, the description of $\mathbf{U}(s)$ changes because the input vectors are both $\mathbf{Y}$ and $\mathbf{Y}_d$. Hence,

$$\mathbf{U}(s) = F_C(s)\,\mathbf{V}_c(s), \quad \mathbf{V}_C(s) = \begin{bmatrix} \mathbf{Y}_d^T(s) & \mathbf{Y}^T(s) & \mathbf{C}_{0C}^T \end{bmatrix}^T \implies$$

$$\mathbf{U}(s) = F_C(s) \begin{bmatrix} \mathbf{Y}_d(s) \\ \mathbf{Y}(s) \\ \mathbf{C}_{0C} \end{bmatrix}, \quad F_C(s) = \begin{bmatrix} G_{CY_d}(s) & \vdots & -G_{CY}(s) & \vdots & G_{C0}(s) \end{bmatrix}$$

$$usually\ G_{CY_d}(s) = G_{CY}(s), \tag{1.75}$$

We replace $\mathbf{U}(s)$ from (1.75) into (1.65):

$$\mathbf{Y}(s) = \left\{ \begin{bmatrix} G_{PD}(s) & \vdots & G_{PU}(s) & \vdots & G_{P0}(s) \end{bmatrix} \bullet \right.$$
$$\mathbf{D}(s)$$
$$\left. \bullet \begin{bmatrix} \begin{bmatrix} G_{CY_d}(s) & \vdots & -G_{CY_d}(s) & \vdots & G_{C0}(s) \end{bmatrix} \begin{bmatrix} \mathbf{Y}_d(s) \\ \mathbf{Y}(s) \\ \mathbf{C}_{0C} \end{bmatrix} \end{bmatrix} \right\} =$$
$$\mathbf{C}_{0P}$$
$$= F_{CS}(s)\,\mathbf{V}_{CS}(s), \tag{1.76}$$

so that

$$\mathbf{Y}(s) =$$

$$= \begin{bmatrix} G_{PD}(s) & \vdots & G_{PU}(s) G_{CY_d}(s) & \vdots & -G_{PU}(s) G_{CY}(s) & \vdots \\ & & \vdots\ G_{PU}(s) G_{C0}(s) & \vdots & G_{P0}(s) & \end{bmatrix} \bullet$$

$$\bullet \begin{bmatrix} \mathbf{D}^T(s) & \mathbf{Y}_d^T(s) & \mathbf{Y}^T(s) & \mathbf{C}_{0C}^T & \mathbf{C}_{0P}^T \end{bmatrix}^T. \tag{1.77}$$

We solve this equation in $\mathbf{Y}(s)$:

$$\mathbf{Y}(s) = [I_N + G_{PU}(s)\,G_{CY}(s)]^{-1} \bullet$$

$$\bullet \left[G_{PD}(s) \;\vdots\; G_{PU}(s)\,G_{CY_d}(s) \;\vdots\; G_{PU}(s)\,G_{C0}(s) \;\vdots\; G_{P0}(s) \right] \bullet$$

$$\bullet \left[\; \mathbf{D}^T(s) \quad \mathbf{Y}_d^T(s) \quad \mathbf{C}_{0C}^T \quad \mathbf{C}_{0P}^T \; \right]^T. \tag{1.78}$$

This shows in the complex domain how the system transfers in the course of *time* the influence of the action vector $\mathbf{V}_{CS}(t) = \mathcal{L}^{-1}\{\mathbf{V}_{CS}(s)\}$ on the system output vector $\mathbf{Y}(t) = \mathcal{L}^{-1}\{\mathbf{Y}(s)\}$.

Equations (1.77) and (1.78) determine the closed loop control system full transfer function matrix $F_{CS}(s)$,

$$F_{CS}(s) = [I_N + G_{PU}(s)\,G_{CY}(s)]^{-1} \bullet$$

$$\bullet \left[G_{PD}(s) \;\vdots\; G_{PU}(s)\,G_{CY_d}(s) \;\vdots\; G_{PU}(s)\,G_{C0}(s) \;\vdots\; G_{P0}(s) \right],$$

its submatrices:

- Its transfer function $G_{CSD}(s)$ relative to the disturbance $\mathbf{D}$:

$$G_{CSD}(s) = [I_N + G_{PU}(s)\,G_{CY}(s)]^{-1} G_{PD}(s),$$

- Its transfer function $G_{CSY_d}(s)$ relative to the desired output $\mathbf{Y}_d$:

$$G_{CSY_d}(s) = [I_N + G_{PU}(s)\,G_{CY}(s)]^{-1} G_{PU}(s)\,G_{CY_d}(s),$$

- Its transfer function $G_{CS0C}(s)$ relative to the vector $\mathbf{C}_{0C}$ of the controller initial conditions:

$$G_{CS0C}(s) = [I_N + G_{PU}(s)\,G_{CY}(s)]^{-1} G_{PU}(s)\,G_{C0}(s),$$

- Its transfer function $G_{CS0O}(s)$ relative to the vector $\mathbf{C}_{0P}$ of the plant initial conditions:

$$G_{CS0O}(s) = [I_N + G_{PU}(s)\,G_{CY}(s)]^{-1} G_{P0}(s),$$

and the Laplace transform $\mathbf{V}_{CS}(s)$ of the action vector $\mathbf{V}_{CS}(t)$:

$$\mathbf{V}_{CS}(s) = \left[\; \mathbf{D}^T(s) \quad \mathbf{Y}_d^T(s) \quad \mathbf{C}_{0C}^T \quad \mathbf{C}_{0P}^T \; \right]^T.$$

Notice that the transfer function $G_{CSY}(s)$ of the closed loop control system relative to the real output $\mathbf{Y}$ obeys

$$G_{CSY_d}(s) = -[I_N + G_{PU}(s)\,G_{CY}(s)]^{-1} G_{PU}(s)\,G_{CY_d}(s).$$

1.8.4 $F_{CS}(s)$ of the combined loops control system

We use both Definition 36 and Figure 1.8. Equations (1.65) related to the plant rest unchanged. However, the description of $\mathbf{U}(s)$ changes because the input vectors are $\mathbf{D}$, $\mathbf{Y}$ and $\mathbf{Y}_d$. Hence,

$$\mathbf{U}(s) = F_C(s)\mathbf{V}_c(s), \quad \mathbf{V}_C(s) = \begin{bmatrix} \mathbf{D}(s) \\ \mathbf{Y}_d(s) \\ \mathbf{Y}(s) \\ \mathbf{C}_{0C} \end{bmatrix} \Longrightarrow$$

$$\mathbf{U}(s) = F_C(s)\begin{bmatrix} \mathbf{D}^T(s) & \mathbf{Y}_d^T(s) & \mathbf{C}_{0C}^T & \mathbf{C}_{0P}^T \end{bmatrix}^T,$$

$$F_C(s) = \begin{bmatrix} G_{CD}(s) \;\vdots\; G_{CY_d}(s) \;\vdots\; -G_{CY}(s) \;\vdots\; G_{C0}(s) \end{bmatrix} \qquad (1.79)$$

We replace $\mathbf{U}(s)$ from (1.79) into (1.65):

$$\mathbf{Y}(s) =$$

$$\left\{ \begin{bmatrix} G_{PD}(s) \;\vdots\; G_{PU}(s) \;\vdots\; G_{P0}(s) \end{bmatrix} \bullet \\ \mathbf{D}(s) \\ \bullet \begin{bmatrix} \begin{bmatrix} G_{CD}(s) \;\vdots\; G_{CY_d}(s) \;\vdots\; -G_{CY}(s) \;\vdots\; G_{C0}(s) \end{bmatrix} \begin{bmatrix} \mathbf{D}(s) \\ \mathbf{Y}_d(s) \\ \mathbf{Y}(s) \\ \mathbf{C}_{0C} \end{bmatrix} \end{bmatrix} \\ \mathbf{C}_{0P} \end{bmatrix} \right\},$$

$$\qquad (1.80)$$

so that

$$\mathbf{Y}(s) =$$

$$= \begin{bmatrix} G_{PD}(s) + G_{PU}(s)G_{CD}(s) \;\vdots\; G_{PU}(s)G_{CY_d}(s) \;\vdots\; \\ \;\vdots\; -G_{PU}(s)G_{CY}(s) \;\vdots\; G_{PU}(s)G_{C0}(s) \;\vdots\; G_{P0}(s) \end{bmatrix} \bullet$$

$$\bullet\begin{bmatrix} \mathbf{D}^T(s) & \mathbf{Y}_d^T(s) & \mathbf{Y}^T(s) & \mathbf{C}_{0C}^T & \mathbf{C}_{0P}^T \end{bmatrix}^T. \qquad (1.81)$$

We solve this equation in $\mathbf{Y}(s)$:

$$\mathbf{Y}(s) = [I_N + G_{PU}(s)G_{CY_d}(s)]^{-1} \bullet$$

$$\bullet\begin{bmatrix} G_{PD}(s) + G_{PU}(s)G_{CD}(s) \;\vdots\; G_{PU}(s)G_{CY_d}(s) \;\vdots\; \\ \;\vdots\; G_{PU}(s)G_{C0}(s) \;\vdots\; G_{P0}(s) \end{bmatrix} \bullet$$

$$\bullet\begin{bmatrix} \mathbf{D}^T(s) & \mathbf{Y}_d^T(s) & \mathbf{C}_{0C}^T & \mathbf{C}_{0P}^T \end{bmatrix}^T. \qquad (1.82)$$

This shows in the complex domain how the system transfers in the course of *time* the influence of the action vector $\mathbf{V}_{CS}(t) = \mathcal{L}^{-1}\{\mathbf{V}_{CS}(s)\}$ on the system output vector $\mathbf{Y}(t) = \mathcal{L}^{-1}\{\mathbf{Y}(s)\}$.

Equations (1.81) and (1.82) determine the closed loop control system full transfer function matrix $F_{CS}(s)$,

$$F_{CS}(s) = [I_N + G_{PU}(s)G_{CY}(s)]^{-1} \bullet \tag{1.83}$$

$$\bullet \left[\begin{array}{c} G_{PD}(s) + G_{PU}(s)G_{CD}(s) \vdots G_{PU}(s)G_{CY_d}(s) \vdots \\ \vdots G_{PU}(s)G_{C0}(s) \vdots G_{P0}(s) \end{array} \right],$$

its submatrices:

-    Its transfer function $G_{CSD}(s)$ relative to the disturbance $\mathbf{D}$:

$$G_{CSD}(s) = [I_N + G_{PU}(s)G_{CY}(s)]^{-1}[G_{PD}(s) + G_{PU}(s)G_{CD}(s)],$$

-    Its transfer function $G_{CSY_d}(s)$ relative to the desired output $\mathbf{Y}_d$:

$$G_{CSY_d}(s) = [I_N + G_{PU}(s)G_{CY}(s)]^{-1}G_{PU}(s)G_{CY_d}(s),$$

-    Its transfer function $G_{CS0C}(s)$ relative to the vector $\mathbf{C}_{0C}$ of the controller initial conditions:

$$G_{CS0C}(s) = [I_N + G_{PU}(s)G_{CY}(s)]^{-1}G_{PU}(s)G_{C0}(s),$$

and

-    Its transfer function $G_{CS0O}(s)$ relative to the vector $\mathbf{C}_{0P}$ of the plant initial conditions:

$$G_{CS0O}(s) = G_{P0}(s),$$

and the Laplace transform $\mathbf{V}_{CS}(s)$ of its action vector $\mathbf{V}_{CS}(t)$:

$$\mathbf{V}_{CS}(s) = \begin{bmatrix} \mathbf{D}^T(s) & \mathbf{Y}_d^T(s) & \mathbf{C}_{0C}^T & \mathbf{C}_{0P}^T \end{bmatrix}^T.$$

Notice that the transfer function $G_{CSY}(s)$ of the combined loops control system relative to the real output $\mathbf{Y}$ obeys

$$G_{CSY_d}(s) = -G_{PU}(s)G_{CY_d}(s).$$

1.9 Matrix functions and polynomials

Note 37 *On matrix functions, matrix polynomials, polynomial matrices and rational matrix functions see the book [171].*

Chapter 2

IO systems

2.1 *IO* system mathematical model

2.1.1 Time domain

This section deals with physical dynamical systems in general and control systems in particular, which are mathematically described directly in the form of a *time*-invariant linear vector **Input-Output** (*IO*) differential equation of the classical form (2.1),

$$\sum_{k=0}^{k=\nu} A_k \mathbf{Y}^{(k)}(t) = \sum_{k=0}^{k=\eta} D_k \mathbf{D}^{(k)}(t) + \sum_{k=0}^{k=\mu} B_k \mathbf{U}^{(k)}(t) = \sum_{k=0}^{k=\xi} H_k \mathbf{I}^{(k)}(t), \ \forall t \in \mathfrak{T}_0,$$

$$\nu \geq 1, \ \xi = \max(\eta, \mu), \ \mathbf{Y}^{(k)}(t) = \frac{d^k \mathbf{Y}(t)}{dt^k}, \ 0 \leq \eta \leq \nu, \ 0 \leq \mu \leq \nu,$$

$$A_k \in \mathfrak{R}^{N \times N}, \ D_k \in \mathfrak{R}^{N \times d}, \ B_k \in \mathfrak{R}^{N \times r}, \ k = 0, 1, .., \nu, \ det A_\nu \neq 0,$$

$$\eta < \nu \implies D_i = O_{N,d}, \ i = \eta + 1, \eta + 2, ..., \nu.$$

$$\mu < \nu \implies B_i = O_{N,r}, \ i = \mu + 1, \mu + 2, ..., \nu. \tag{2.1}$$

Note 38 *System, plant, and control system*
   If and only if there is $k \in \{0, 1, ..., \mu\}$ *such that* $B_k \neq O_{N,r}$ *then the IO system (2.1) becomes the IO plant (2.1) (Definition 18, Section 1.5). Otherwise, the IO system (2.1) represents the IO control system,* $\xi = \eta$ *and* $H_k \equiv \left[D_k \ \vdots \ O_{N,r} \right].$

The disturbance vector $\mathbf{D}$ (1.25) (Section 1.5) and the control vector $\mathbf{U}$ (1.46) (Section 1.6) compose the system input vector $\mathbf{I}$ (2.4) (Section 1.5):

$$\mathbf{I} = \mathbf{I}_{IO} = \begin{bmatrix} \mathbf{D} \\ \mathbf{U} \end{bmatrix} \in \Re^{d+r}, \ M = d + r. \tag{2.2}$$

We accept the following:

Condition 39 *The matrix A_ν of the IO system (2.1) is nonsingular, i.e., it obeys*

$$\det A_\nu \neq 0. \tag{2.3}$$

Note 40 *Throughout this book we accept the validity of Condition 39.*

Note 41 *The condition on the nonsingularity of the matrix A_ν imposed in Condition 39 guarantees*

$$\exists s \in \mathbb{C} \Longrightarrow \det \left(\sum_{k=0}^{k=\nu} A_k s^k \right) \neq 0,$$

and permits the solvability of the Laplace transform of (2.1) for $\mathbf{Y}(s)$ [170].
   Besides, the condition $\det A_\nu \neq 0$ is a sufficient condition, but not necessary condition, for all the output variables of the system (2.1) to have the same order ν of their highest derivatives.

$\Re^k$ is the k-dimensional real vector space, $k \in \{1, 2, ...\}$, (Section 1.5). $\mathbb{C}^k$ denotes the k-dimensional complex vector space (Section 1.5). $O_{M \times N}$ is the zero matrix in $\Re^{M \times N}$, and O_N is the zero matrix in $\Re^{N \times N}$, $O_N = O_{N \times N}$. The vector $\mathbf{0}_k \in \Re^k$ is the zero vector in $\Re^k$ and $\mathbf{1}_k \in \Re^k$ is the unit vector in $\Re^k$, Equations (1.13), Section 1.3.
   The total *input vector*

$$\mathbf{I} = [I_1 \ I_2 \ ... \ I_M]^T \in \Re^M, \tag{2.4}$$

its subvectors

$$\mathbf{D} = [D_1 \ D_2 \ ... \ D_d]^T \in \Re^d, \tag{2.5}$$
$$\mathbf{U} = [U_1 \ U_2 \ ... \ U_r]^T \in \Re^r, \tag{2.6}$$

(Definition 19), and the total *output vector*

$$\mathbf{Y} = [Y_1 \ Y_2 \ ... \ Y_N]^T \in \Re^N, \tag{2.7}$$

(Definition 22) (Section 1.5). The values I_i, D_j, U_k, and Y_l are the total values of the input and the output variables, respectively. *The total value* of a variable signifies that its value is measured with respect to its total zero, if it has the total zero value, and if it does not have the total zero value then an appropriate value is accepted to play the role of the total zero value.

The form of the system mathematical model (2.1) is too complex and makes the system study unreasonably cumbersome. We simplify it by applying the elegant and simple compact notation for the extended matrices proposed in [115] and in brief explained in Note 14 (Section 1.3). At first we introduce the extended matrices $A^{(\nu)}$, $B^{(\mu)}$, and $D^{(\eta)}$,

$$A^{(\nu)} = \left[A_0 \vdots A_1 \vdots ... \vdots A_\nu \right] \in \Re^{N \times (\nu+1)N},$$

$$B^{(\mu)} = \left[B_0 \vdots B_1 \vdots ... \vdots B_\mu \right] \in \Re^{N \times (\mu+1)r},$$

$$D^{(\eta)} = \left[D_0 \vdots D_1 \vdots ... \vdots D_\eta \right] \in \Re^{N \times (\eta+1)d}, \qquad (2.8)$$

and then the very simple extended vectors $\mathbf{D}^\eta(t)$, $\mathbf{I}^\xi(t)$, $\mathbf{U}^\mu(t)$ and $\mathbf{Y}^\nu(t)$:

$$\mathbf{D}^\eta(t) = \left[\mathbf{D}^T(t) \vdots \mathbf{D}^{(1)^T}(t) \vdots ... \vdots \mathbf{D}^{(\eta)^T}(t) \right]^T \in \Re^{(\eta+1)d}, \qquad (2.9)$$

$$\mathbf{I}^\xi(t) = \left[\mathbf{I}^T(t) \vdots \mathbf{I}^{(1)^T}(t) \vdots ... \vdots \mathbf{I}^{(\xi)^T}(t) \right]^T \in \Re^{(\xi+1)M} \qquad (2.10)$$

$$\mathbf{U}^\mu(t) = \left[\mathbf{U}^T(t) \vdots \mathbf{U}^{(1)^T}(t) \vdots ... \vdots \mathbf{U}^{(\mu)^T}(t) \right]^T \in \Re^{(\mu+1)r} \qquad (2.11)$$

$$\mathbf{Y}^\nu(t) = \left[\mathbf{Y}^T(t) \vdots \mathbf{Y}^{(1)^T}(t) \vdots ... \vdots \mathbf{Y}^{(\nu)^T}(t) \right]^T \in \Re^{(\nu+1)N}, \qquad (2.12)$$

They induce the corresponding initial vectors $\mathbf{D}_0^{\eta-1} = \mathbf{D}^{\eta-1}(0)$, $\mathbf{I}_0^{\xi-1} = \mathbf{I}^{\xi-1}(0)$, $\mathbf{U}_0^{\mu-1} = \mathbf{U}^{\mu-1}(0)$, and $\mathbf{Y}_0^{\nu-1} = \mathbf{Y}^{\nu-1}(0)$.

We repeat that the upper index ν in the parentheses in $A^{(\nu)}$ makes $A^{(\nu)}$ essentially different from the ν-*th* power A^ν of A,

$$A^{(\nu)} = \left[A_0 \vdots A_1 \vdots ... \vdots A_\nu \right] \neq A^\nu = \underbrace{AA....A}_{\nu \ times} . \qquad (2.13)$$

Notice also that for the extended vector $\mathbf{Y}^v$ the superscript ν is not in the parentheses in order to distinguish it from the ν-th derivative $d^\nu \mathbf{Y}(t)/dt^\nu$ of $\mathbf{Y}(t)$,

$$\mathbf{Y}^v(t) = \left[\mathbf{Y}^T(t) \vdots \mathbf{Y}^{(1)^T}(t) \vdots ... \vdots \mathbf{Y}^{(\nu)^T}(t)\right]^T \neq \mathbf{Y}^{(\nu)}(t) = \frac{d^\nu \mathbf{Y}(t)}{dt^\nu}. \quad (2.14)$$

The application of the above compact notation (2.8)–(2.12) to the *IO* vector differential equation (2.1) transforms it into the following simple, elegant, and compact form:

$$A^{(\nu)}\mathbf{Y}^\nu(t) = D^{(\eta)}\mathbf{D}^\eta(t) + B^{(\mu)}\mathbf{U}^\mu(t) = H^{(\mu)}\mathbf{I}^\mu(t), \ \forall t \in \mathfrak{T}_0,$$

$$H^{(\mu)} = \left[D^{(\mu)} \vdots B^{(\mu)}\right], \ \mathbf{I}^\mu(t) = \left[\left(\mathbf{D}^{(\mu)}\right)^T \vdots \left(\mathbf{U}^{(\mu)}\right)^T\right]^T. \quad (2.15)$$

Note 42 *The state vector $\mathbf{S}_{IO}$ of the IO system (2.15) is defined in (1.36) (Section 1.5) by:*

$$\mathbf{S}_{IO} = \mathbf{Y}^{\nu-1} = \left[\mathbf{Y}^T \vdots \mathbf{Y}^{(1)^T} \vdots ... \vdots \mathbf{Y}^{(\nu-1)^T}\right]^T \in \mathfrak{R}^n, \ n = \nu N, \quad (2.16)$$

This new vector notation $\mathbf{Y}^{\nu-1}$ has permitted us to define the state of the IO system (2.15) by preserving the physical sense. It enabled us to establish in [170] the direct link between the definitions of the Lyapunov and of BI stability properties with the corresponding conditions for them in the complex domain. It enables us to discover in what follows the complex domain criteria for observability, controllability, and trackability directly from their definitions. Such criteria possess the complete physical meaning.

2.1.2 Complex domain

The following complex matrix functions [115], [159], [170], the first one of which is $S_i^{(k)}(.) : \mathbb{C} \longrightarrow \mathbb{C}^{i(k+1)\times i}$, essentially simplify the system study via the complex domain,

$$S_i^{(k)}(s) = \left[s^0 I_i \vdots s^1 I_i \vdots s^2 I_i \vdots ... \vdots s^k I_i\right]^T \in \mathbb{C}^{i(k+1)\times i},$$

$$(k, i) \in \{(\mu, M), \ (\nu, N)\}. \quad (2.17)$$

The matrix I_i is the i-th order identity matrix, $I_i \in \mathfrak{R}^{i \times i}$, $I_n = I$. Another complex function is $Z_k^{(\varsigma-1)}(.) : \mathbb{C} \rightarrow \mathbb{C}^{(\varsigma+1)k \times \varsigma k}$,

$$
Z_k^{(\varsigma-1)}(s) = \begin{bmatrix} O_k & O_k & O_k & \cdots & O_k \\ s^0 I_k & O_k & O_k & \cdots & O_k \\ \cdots & \cdots & \cdots & \cdots & \cdots \\ s^{\varsigma-1} I_k & s^{\varsigma-2} I_k & s^{\varsigma-3} I_k & \cdots & s^0 I_k \end{bmatrix}, \quad \varsigma \geq 1,
$$

$$
\varsigma = 1 \Longrightarrow Z_k^{(1-1)}(s) = Z_k^{(0)}(s) = s^0 I_k = I_k,
$$

$$
Z_k^{(\varsigma-1)}(s) \in \mathbb{C}^{(\varsigma+1)k \times \varsigma k}, \quad (\varsigma, k) \in \{(\mu, M), (\nu, N)\}, \tag{2.18}
$$

where the final entry of $Z_k^{(\varsigma-1)}(s)$ is always $s^0 I_k$. They enable us to resolve effectively Fundamental problem 1 (Section).

Note 43 *[115], [159], [170] If $\varsigma \leq 0$ then the matrix $Z_k^{(\varsigma-1)}(s) = Z_k^{(-1)}(s)$ is not defined, does not exist, and should be completely omitted rather than to be replaced by the zero matrix. Derivatives exist only for natural numbers, i.e., $\mathbf{Y}^{(\varsigma)}(t)$ can exist only for $\varsigma \geq 1$. Matrix function $Z_k^{(\varsigma-1)}(.)$ is related to the Laplace transform of derivatives only.*

It is well known that the Laplace transform of

$$
\sum_{k=0}^{k=\nu} A_k \mathbf{Y}^{(k)}(t) \tag{2.19}
$$

contains the Laplace transform $\mathbf{Y}(s)$ of $\mathbf{Y}(t)$ multiplied by a matrix polynomial in s and a double sum containing the products of powers of the complex variable s and initial values of $\mathbf{Y}(t)$ and of its derivatives up to the order of $\nu - 1$, all multiplied by the corresponding system matrices. The references [115], [159], and [170] contain the proof that the simple, compact, and elegant form of the Laplace transform of (2.19) reads:

$$
\mathcal{L} \left\{ \sum_{k=0}^{k=\nu} A_k \mathbf{Y}^{(k)}(t) \right\} = A^{(\nu)} S_N^{(\nu)}(s) \mathbf{Y}(s) - A^{(\nu)} Z_N^{(\nu-1)}(s) \mathbf{Y}_0^{\nu-1}, \tag{2.20}
$$

where the matrices $S_N^{(\nu)}(s)$ and $Z_N^{(\nu-1)}(s)$ are defined in (2.17) and in (2.18), respectively. Analogously,

$$
\mathcal{L} \left\{ \sum_{k=0}^{k=\eta} D_k \mathbf{D}^{(k)}(t) \right\} = D^{(\eta)} S_d^{(\eta)}(s) \mathbf{D}(s) - D^{(\eta)} Z_d^{(\eta-1)}(s) \mathbf{D}^{\eta-1}(0), \tag{2.21}
$$

$$\mathcal{L}\left\{\sum_{k=0}^{k=\mu} B_k \mathbf{U}^{(k)}(t)\right\} = B^{(\mu)}S_r^{(\mu)}(s)\mathbf{U}(s) - B^{(\mu)}Z_r^{(\mu-1)}(s)\mathbf{U}^{\mu-1}(0). \quad (2.22)$$

Equations (2.20), (2.21), and (2.22) determine the simple compact form of the Laplace transform of (2.1), hence of (2.15):

$$A^{(\nu)}S_N^{(\nu)}(s)\mathbf{Y}(s) - A^{(\nu)}Z_N^{(\nu-1)}(s)\mathbf{Y}^{\nu-1}(0) =$$
$$= D^{(\eta)}S_d^{(\eta)}(s)\mathbf{D}(s) - D^{(\eta)}Z_d^{(\eta-1)}(s)\mathbf{D}^{\eta-1}(0)+$$
$$+ B^{(\mu)}S_r^{(\mu)}(s)\mathbf{U}(s) - B^{(\mu)}Z_r^{(\mu-1)}(s)\mathbf{U}^{\mu-1}(0). \quad (2.23)$$

This equation determines $\mathbf{Y}(s)$:

$$\mathbf{Y}(s) = \left(A^{(\nu)}S_N^{(\nu)}(s)\right)^{-1} \bullet$$

$$\bullet \left[D^{(\eta)}S_d^{(\eta)}(s)\vdots B^{(\mu)}S_r^{(\mu)}(s)\vdots - D^{(\eta)}Z_d^{(\eta-1)}(s)\vdots - B^{(\mu)}Z_r^{(\mu-1)}(s)\vdots A^{(\nu)}Z_N^{(\nu-1)}(s)\right]$$

$$\bullet \begin{bmatrix} \mathbf{D}(s) \\ \mathbf{U}(s) \\ \mathbf{D}^{\eta-1}(0) \\ \mathbf{U}^{\mu-1}(0) \\ \mathbf{Y}^{\nu-1}(0) \end{bmatrix} = F_{IO}(s)\,\mathbf{V}_{IO}(s), \quad (2.24)$$

since the inverse of $A^{(\nu)}S_N^{(\nu)}(s)$ exists due to Condition 39. The plant *full transfer function matrix* $F_{IO}(s)$ results from (2.24)

$$F_{IO}(s) = \left(A^{(\nu)}S_N^{(\nu)}(s)\right)^{-1} \bullet$$

$$\bullet \left[D^{(\eta)}S_d^{(\eta)}(s)\vdots B^{(\mu)}S_r^{(\mu)}(s)\vdots - D^{(\eta)}Z_d^{(\eta-1)}(s)\vdots - B^{(\mu)}Z_r^{(\mu-1)}(s)\vdots A^{(\nu)}Z_N^{(\nu-1)}(s)\right]$$
$$(2.25)$$

The inverse Laplace transform of $F_{IO}(s)$ is *the IO* system *full fundamental matrix* $\Psi_{IO}(t)$, $\Psi_{IO}(t) = \mathcal{L}^{-1}\{F_{IO}(s)\}$, and the inverse Laplace transform of

$$\left(A^{(\nu)}S_N^{(\nu)}(s)\right)^{-1} = \frac{adj\left(A^{(\nu)}S_N^{(\nu)}(s)\right)}{p_{IO}(s)} = p_{IO}^{-1}(s)\,adj\left(A^{(\nu)}S_N^{(\nu)}(s)\right) \quad (2.26)$$

is *the IO* system *fundamental matrix* $\Phi_{IO}(t)$ [170]:

$$\Phi_{IO}(t) = \mathcal{L}^{-1}\left\{\left(A^{(\nu)}S_N^{(\nu)}(s)\right)^{-1}\right\}. \quad (2.27)$$

Equation (2.25) discovers that the polynomial $p_{IO}(s)$,

$$p_{IO}(s) = \det\left(A^{(\nu)}S_N^{(\nu)}(s)\right), \tag{2.28}$$

is the characteristic polynomial of the IO system (2.15) and the denominator polynomial of all its transfer function matrices due to Equation (2.25) as shown also in what follows. Equation (2.25) induces also the matrix polynomial $L_{IO}(s)$ defined by

$$L_{IO}(s) = adj\left(A^{(\nu)}S_N^{(\nu)}(s)\right)B^{(\mu)}S_r^{(\mu)}(s), \ \ L_{IO}(s) \in \mathbb{C}^{N \times r}. \tag{2.29}$$

It is the numerator matrix polynomial of the plant transfer function matrix $G_{IOU}(s)$ relative to the control vector $\mathbf{U}$:

$$G_{IOU}(s) = \left(A^{(\nu)}S_N^{(\nu)}(s)\right)^{-1}B^{(\mu)}S_r^{(\mu)}(s) =$$

$$= \frac{adj\left(A^{(\nu)}S_N^{(\nu)}(s)\right)B^{(\mu)}S_r^{(\mu)}(s)}{p_{IO}(s)} = \frac{L_{IO}(s)}{p_{IO}(s)}, \tag{2.30}$$

Equation (2.24) determines also all other specific transfer function matrices of the IO system (2.15):

- With respect to the disturbance $\mathbf{D}$:

$$G_{IOD}(s) = \frac{adj\left(A^{(\nu)}S_N^{(\nu)}(s)\right)D^{(\eta)}S_d^{(\eta)}(s)}{p_{IO}(s)}, \tag{2.31}$$

- With respect to the initial conditions $\mathbf{D}^{\eta-1}(0)$ of the disturbance $\mathbf{D}$:

$$G_{IOD_0}(s) = -\frac{adj\left(A^{(\nu)}S_N^{(\nu)}(s)\right)D^{(\eta)}Z_d^{(\eta-1)}(s)}{p_{IO}(s)}, \tag{2.32}$$

- With respect to the initial conditions $\mathbf{U}^{\mu-1}(0)$ of the control $\mathbf{U}$:

$$G_{IOU_0}(s) = -\frac{adj\left(A^{(\nu)}S_N^{(\nu)}(s)\right)B^{(\mu)}Z_r^{(\mu-1)}(s)}{p_{IO}(s)}, \tag{2.33}$$

- With respect to the initial conditions $\mathbf{Y}^{\nu-1}(0)$ of the output $\mathbf{Y}$:

$$G_{IOY_0}(s) = \frac{adj\left(A^{(\nu)}S_N^{(\nu)}(s)\right)A^{(\nu)}Z_N^{(\nu-1)}(s)}{p_{IO}(s)}. \tag{2.34}$$

The Laplace transform $\mathbf{V}_{IO}(s)$ of the IO system action vector $\mathbf{V}_{IO}(t)$ reads

$$\mathbf{V}_{IO}(s) = \begin{bmatrix} \mathbf{D}(s) \\ \mathbf{U}(s) \\ \mathbf{D}^{\eta-1}(0) \\ \mathbf{U}^{\mu-1}(0) \\ \mathbf{Y}^{\nu-1}(0) \end{bmatrix} = \begin{bmatrix} \mathbf{I}_{IO}(s) \\ \mathbf{C}_{IO0} \end{bmatrix}. \tag{2.35}$$

The Laplace transform $\mathbf{I}_{IO}(s)$ of the IO system input vector $\mathbf{I}_{IO}(t)$ reads:

$$\mathbf{I}_{IO}(s) = \begin{bmatrix} \mathbf{D}(s) \\ \mathbf{U}(s) \end{bmatrix}. \tag{2.36}$$

The vector $\mathbf{C}_{IO0}$ of all IO system initial conditions has the following form:

$$\mathbf{C}_{IO0} = \begin{bmatrix} \mathbf{D}^{\eta-1}(0) \\ \mathbf{U}^{\mu-1}(0) \\ \mathbf{Y}^{\nu-1}(0) \end{bmatrix}. \tag{2.37}$$

The IO system output response $\mathbf{Y}\left(t; \mathbf{Y}_0^{\nu-1}; \mathbf{D}^{\eta}; \mathbf{U}^{\mu}\right)$ is the inverse Laplace transform of (2.24):

$$\mathbf{Y}\left(t; \mathbf{Y}_0^{\nu-1}; \mathbf{D}^{\eta}; \mathbf{U}^{\mu}\right) = \mathcal{L}^{-1}\left\{F_{IO}(s)\,\mathbf{V}_{IO}(s)\right\}. \tag{2.38}$$

Example 44 *Appendix B1. in the book [171, Example 227, pp. 243-248] contains an IO system example.*

What follows shows an important physical meaning of the system full transfer function matrix $F(s)$. For the definition, types and properties of the Dirac unit impulse $\delta(.)$, see [170].

Definition 45 *[170, Definition 181, p. 171, Note 182, p. 172] A matrix function $\mathbf{\Psi}_{IO}(.) : \mathfrak{T} \longrightarrow \mathfrak{R}^{N\times[(\mu+1)M+\nu N]}$ is **the full fundamental matrix function** of the IO system (2.15) if and only if it obeys both (i) and (ii) for an arbitrary input vector function $\mathbf{I}(.)$, Equation (2.2), and for arbitrary initial conditions $\mathbf{I}_{0-}^{\mu-1}$ and $\mathbf{Y}_{0-}^{\nu-1}$,*
    (i)

$$\mathbf{Y}(t; \mathbf{Y}_{0-}^{\nu-1}; \mathbf{I}) = \int_{0-}^{t}\left\{\mathbf{\Psi}_{IO}(\tau)\begin{bmatrix} \mathbf{I}(t-\tau) \\ \delta(t-\tau)\mathbf{I}_{0-}^{\mu-1} \\ \delta(t-\tau)\mathbf{Y}_{0-}^{\nu-1} \end{bmatrix}\right\}d\tau =$$

$$= \int_{0-}^{t}\left\{\mathbf{\Psi}_{IO}(t-\tau)\begin{bmatrix} \mathbf{I}(\tau) \\ \delta(\tau)\mathbf{I}_{0-}^{\mu-1} \\ \delta(\tau)\mathbf{Y}_{0-}^{\nu-1} \end{bmatrix}\right\}d\tau, \tag{2.39}$$

equivalently

$$\mathbf{Y}(t; \mathbf{Y}_{0^-}^{\nu-1}; \mathbf{I}) = \int_{0^-}^{t} \Gamma_{IO}(\tau)\mathbf{I}(t-\tau)d\tau + \Gamma_{IOi_0}(t)\mathbf{I}_{0^-}^{\mu-1} + \Gamma_{IOy_0}(t)\mathbf{Y}_{0^-}^{\nu-1},$$

$$\int_{0^-}^{t} [\Gamma_{IO}(\tau)\mathbf{I}(t-\tau)d\tau] = \int_{0^-}^{t} [\Gamma_{IO}(t-\tau)\mathbf{I}(\tau)d\tau],$$

$$\mathbf{I}(t-\tau) = \mathbf{I}(t,\tau), \ \Gamma_{IO}(t-\tau) = \Gamma_{IO}(t,\tau), \tag{2.40}$$

and

$$\Psi_{IO}(t) = \left[\Gamma_{IO}(t) \vdots \Gamma_{IOi_0}(t) \vdots \Gamma_{IOy_0}(t)\right],$$

$$\Gamma_{IO}(t) \in \mathfrak{R}^{N \times M}, \ \Gamma_{IOi_0}(t) \in \mathfrak{R}^{N \times \mu M}, \Gamma_{IOy_0}(t) \in \mathfrak{R}^{N \times \nu N}, \tag{2.41}$$

(ii)

$$\Gamma_{IOi_0}(0^-) = \left[\Gamma_{IOi_01} \vdots O_{N,(\mu-1)M}\right] \ where$$

$$\Gamma_{IOi_01}(0^-)\mathbf{i}_{0^-} = -\int_{0^-}^{0^-} [\Gamma_{IO}(\tau)\mathbf{i}(t-\tau)d\tau],$$

$$\Gamma_{IOy_0}(0^-) \equiv \left[I_N \vdots O_{N,(\nu-1)N}\right]. \tag{2.42}$$

Note 46 *[170, Equation (10.4), p. 172] The second Equation (2.39) under (i) of Definition 45 results from its first equation and from the properties of* $\delta(.)$:

$$\mathbf{Y}(t; \mathbf{Y}_{0^-}^{\nu-1}; \mathbf{I}) =$$

$$= \int_{0^-}^{t} \Gamma_{IO}(\tau)\mathbf{I}(t,\tau)d\tau + \Gamma_{IOi_0}(t)\mathbf{I}_{0^-}^{\mu-1} + \Gamma_{IOy_0}(t)\mathbf{Y}_{0^-}^{\nu-1}, \ t \in \mathfrak{T}_0. \tag{2.43}$$

The matrix $\Gamma_{IO}(t)$ is **the output fundamental matrix** of the IO system (2.15),

Theorem 47 *[170, Theorem 183, pp. 172, 173] (i) The full fundamental matrix function* $\Psi_{IO}(.)$ *of the IO system (2.15) is the inverse of the left Laplace transform of the system full transfer function matrix* $F_{IO}(s)$,

$$\Psi_{IO}(t) = \mathcal{L}^{-1}\{F_{IO}(s)\}. \tag{2.44}$$

(ii) The full transfer function matrix $F_{IO}(s)$ of the IO system (2.15) is the left Laplace transform of the system full fundamental matrix $\Psi_{IO}(t)$,

$$F_{IO}(s) = \mathcal{L}^{-}\{\Psi_{IO}(t)\}. \tag{2.45}$$

(iii) The submatrices $\Gamma_{IO}(t)$, $\Gamma_{IOi_0}(t)$ and $\Gamma_{IOy_0}(t)$ are the inverse Laplace transforms of $G_{IO}(s)$, $G_{IOi_0}(s)$ and $G_{IOy_0}(s)$, respectively,

$$\Gamma_{IO}(t) = \mathcal{L}^{-1}\{G_{IO}(s)\} = \mathcal{L}^{-1}\left\{\Phi_{IO}(s)\left[B^{(\mu)}S_M^{(\mu)}(s)\right]\right\},$$

$$\Gamma_{IOi_0}(t) = \mathcal{L}^{-1}\{G_{IOi_0}(s)\} = -\mathcal{L}^{-1}\left\{\Phi_{IO}(s)\left[B^{(\mu)}Z_M^{(\mu-1)}(s)\right]\right\},$$

$$\Gamma_{IOy_0}(t) = \mathcal{L}^{-1}\{G_{IOy_0}(s)\} = \mathcal{L}^{-1}\left\{\Phi_{IO}(s)\left[A^{(\nu)}Z_N^{(\nu-1)}(s)\right]\right\}, \tag{2.46}$$

*where $\Phi_{IO}(s)$ is the left Laplace transform of the IO **system fundamental matrix function** $\Phi_{IO}(.): \mathfrak{T} \longrightarrow \mathfrak{R}^{N \times N}$,*

$$\Phi_{IO}(s) = \mathcal{L}^{-}\{\Phi_{IO}(t)\}, \quad \Phi_{IO}(t) = \mathcal{L}^{-1}\{\Phi_{IO}(s)\}, \tag{2.47}$$

and

$$\Phi_{IO}(s) = \left(A^{(\nu)}S_N^{(\nu)}(s)\right)^{-1}, \quad \Phi_{IO}(t) = \mathcal{L}^{-1}\left\{\left(A^{(\nu)}S_N^{(\nu)}(s)\right)^{-1}\right\}. \tag{2.48}$$

(iv) The IO system full fundamental matrix $\Psi_{IO}(t)$ and its fundamental matrix $\Phi_{IO}(t)$ are linked as follows:

$$\Psi_{IO}(t) = \mathcal{L}^{-1}\left\{\Phi_{IO}(s)\left[B^{(\mu)}S_M^{(\mu)}(s) \vdots - B^{(\mu)}Z_M^{(\mu-1)}(s) \vdots A^{(\nu)}Z_N^{(\nu-1)}(s)\right]\right\},$$

$$\Psi_{IO}(s) = \Phi_{IO}(s)\left[B^{(\mu)}S_M^{(\mu)}(s) \vdots - B^{(\mu)}Z_M^{(\mu-1)}(s) \vdots A^{(\nu)}Z_N^{(\nu-1)}(s)\right]. \tag{2.49}$$

2.2   *IO* plant desired regime

We accept the following definition of a desired regime by following [159], [170]:

Definition 48 *Desired regime*

*A system (plant) is in **a desired** (called also: **nominal** or **nonperturbed**) regime on $\mathfrak{T}_0$ (for short: in **a desired regime**) if and only if it realizes its desired (output) response $\mathbf{Y}_d(t)$ all the time on $\mathfrak{T}_0$,*

$$\mathbf{Y}(t) = \mathbf{Y}_d(t), \ \forall t \in \mathfrak{T}_0. \tag{2.50}$$

*The terms **nominal** and **nonperturbed** are meaningful in general, i.e., for any system, e.g., for plants, controllers and control systems; while the term **desired** has the full sense only for plants.*

Proposition 49 *[170] In order for the plant to be in a desired (nominal, nonperturbed) regime, i.e.,*

$$\mathbf{Y}(t) = \mathbf{Y}_d(t), \ \forall t \in \mathfrak{T}_0,$$

it is necessary that the initial real output vector is equal to the initial desired output vector,

$$\mathbf{Y}_0 = \mathbf{Y}_{d0}.$$

The system cannot be in a nominal regime (on $\mathfrak{T}_0$) if its initial real output vector is different from the initial desired output vector:

$$\mathbf{Y}_0 \neq \mathbf{Y}_{d0} \Longrightarrow \exists \sigma \in \mathfrak{T}_0 \Longrightarrow \mathbf{Y}(\sigma) \neq \mathbf{Y}_d(\sigma).$$

The real initial output vector $\mathbf{Y}(0) = \mathbf{Y}_0$ is most often different from the desired initial output vector $\mathbf{Y}_d(0) = \mathbf{Y}_{d0}$. The system is most often in a *nondesired* (*non-nominal, perturbed, disturbed*) regime.

Definition 50 *Nominal control* $\mathbf{U}_N(.)]$ *relative to* $[\mathbf{D}(.), \mathbf{Y}_d(.)]$ *of the IO plant (2.15)*

A control vector function $\mathbf{U}^*(.)$ *of the IO plant (2.15) is* *nominal relative to* $[\mathbf{D}(.), \mathbf{Y}_d(.)]$, *which is denoted by* $\mathbf{U}_N(.)$, *if and only if*

$$\mathbf{U}(.)] = \mathbf{U}^*(.)$$

ensures that the corresponding real response $\mathbf{Y}(.) = \mathbf{Y}^*(.)$ *to the input action of* $\mathbf{D}(.)$ *on the plant obeys* $\mathbf{Y}^*(t) = \mathbf{Y}_d(t)$ *all the time as soon as all the internal and the output system initial conditions are desired (nominal, nonperturbed).*

This definition and (2.15) imply the following theorem:

Theorem 51 *[159, Theorem 50, pp. 46-48], [170, Theorem 56, pp. 49-52] In order for a control vector function* $\mathbf{U}^*(.)$ *to be nominal for the IO plant (2.15) relative to* $[\mathbf{D}(.), \mathbf{Y}_d(.)]$: $\mathbf{U}^*(.) = \mathbf{U}_N(.)$, *it is necessary and sufficient that 1) and 2) hold:*

1) rank $rank B^{(\mu)} S_r^{(\mu)}(s) = N \leq r$, *i.e.,* $rank B^{(\mu)} = N \leq r$, *and*
2) anyone of the following equations is valid:

$$B^{(\mu)} \mathbf{U}^{*}{}^{\mu}(t) = -D^{(\eta)} \mathbf{D}^{\eta}(t) + A^{(\nu)} \mathbf{Y}_d^{\nu}(t), \ \forall t \in \mathfrak{T}_0, \qquad (2.51)$$

or, equivalently in the complex domain:

$$
\mathbf{U}^*(s) = \left(B^{(\mu)} S_r^{(\mu)}(s) \right)^T \left[\left(B^{(\mu)} S_r^{(\mu)}(s) \right) \left(B^{(\mu)} S_r^{(\mu)}(s) \right)^T \right]^{-1} \bullet
$$

$$
\bullet \left\langle \begin{array}{c} B^{(\mu)} Z_r^{(\mu-1)}(s) \mathbf{U}^{*\mu-1}(0) + A^{(\nu)} \left[S_N^{(\nu)}(s) \mathbf{Y}_d(s) - Z_N^{(\nu-1)}(s) \mathbf{Y}_d^{\nu-1}(0) \right] - \\ -D^{(\eta)} \left[S_d^{(\eta)}(s) \mathbf{D}(s) - Z_d^{(\eta-1)}(s) \mathbf{D}^{\eta-1}(0) \right] \end{array} \right\rangle .
$$

$$(2.52)$$

This theorem holds for all *IO* plants (2.15).

The condition $N \leq r$ emphasizes the importance of Fundamental control principle 104, Section 7.1.

Condition 52 *The desired output response $\mathbf{Y}_d(t)$ of the IO system (2.15) is realizable, i.e., $N \leq r$. Both it and the nominal input $\mathbf{I}_N(.)$ are known.*

The compact form of the *IO* plant (2.15) in terms of the deviations follows from (2.53), (2.55), (2.56),

$$\mathbf{d} = \mathbf{D} - \mathbf{D_N}, \qquad\qquad\qquad (2.53)$$

$$\mathbf{i} = \mathbf{I} - \mathbf{I}_N \qquad\qquad\qquad (2.54)$$

$$\mathbf{u} = \mathbf{U} - \mathbf{U}_N, \qquad\qquad\qquad (2.55)$$

$$\mathbf{y} = \mathbf{Y} - \mathbf{Y}_d, \qquad\qquad\qquad (2.56)$$

and (2.15):

$$A^{(\nu)} \mathbf{y}^\nu(t) = D^{(\eta)} \mathbf{d}^\eta(t) + B^{(\mu)} \mathbf{u}^\mu(t) = H^{(\mu)} \mathbf{i}^\mu(t), \ \forall t \in \mathfrak{T}_0. \qquad (2.57)$$

Note 53 *Equation (2.57) is the IO system model determined in terms of the deviations of all variables. It has exactly the same form, the same order, and the same matrices as the system model expressed in total values of the variables (2.15). They possess the same characteristics and properties by noting once more that $\mathbf{y} = \mathbf{0}_N$ represents $\mathbf{Y} = \mathbf{Y}_d$. For example, they have the same transfer function matrices, and the stability properties of $\mathbf{y}^{\gamma-1} = \mathbf{0}_{\nu N}$ of (2.57) are simultaneously the same stability properties of $\mathbf{Y}_d^{\gamma-1}(t)$ of (2.15).*

2.3 *IO* feedback controller

2.3.1 Time domain

The plant desired output vector $\mathbf{Y}_d$ (Definition 22) and the plant real output vector $\mathbf{Y}$ (2.7) compose the *IO* feedback controller input vector $\mathbf{I}_{IOC_f}$ (1.52) (Section 1.6) so that in (2.1), i.e., in (2.15):

$$\mathbf{I} = \mathbf{I}_{IOC_f} = \begin{bmatrix} \mathbf{Y}_d \\ \mathbf{Y} \end{bmatrix} \in \mathfrak{R}^{2N}, \ M = 2N.$$

The mathematical description (2.58) of *the Input-Output (IO) feedback controller* reads:

$$\sum_{k=0}^{k=v} A_{Ck}\mathbf{U}^{(k)}(t) = \sum_{k=0}^{k=\mu} J_k \mathbf{Y}_d^{(k)}(t) - \sum_{k=0}^{k=\mu} J_k \mathbf{Y}^{(k)}(t), \ \forall t \in \mathfrak{T}_0,$$

$$v \geq 0, \ 0 \leq \mu \leq v, \ A_{Ck} \in \mathfrak{R}^{rxr}, \ det A_{Cv} \neq 0, \ J_k \in \mathfrak{R}^{rxN}, \ k = 0, 1, .., v,$$

$$v > 0, \ \mu < v \Longrightarrow J_i = O_{r,N}, \ i = \mu + 1, \ \mu + 2, ..., v. \tag{2.58}$$

Equation (1.53) transforms (2.58) into

$$\sum_{k=0}^{k=v} A_{Ck}\mathbf{U}^{(k)}(t) = \sum_{k=0}^{k=\mu} J_k \left[\mathbf{Y}_d^{(k)}(t) - \mathbf{Y}^{(k)}(t) \right] = \sum_{k=0}^{k=\mu} J_k \mathbf{e}^{(k)}(t), \forall t \in \mathfrak{T}_0.$$
$$\tag{2.59}$$

The compact form of the *IO* closed loop controller (2.59) reads:

$$A_C^{(v)}\mathbf{U}^v(t) = J^{(\mu)}\mathbf{e}^\mu(t), \ \forall t \in \mathfrak{T}_0, \tag{2.60}$$

or in terms of deviations:

$$A_C^{(v)}\mathbf{u}^v(t) = J^{(\mu)}\mathbf{e}^\mu(t), \ \forall t \in \mathfrak{T}_0. \tag{2.61}$$

2.3.2 Complex domain

The Laplace transform of (2.60) and of (2.61) has the following form in which $\mathbf{E}(s)$ is the Laplace transform of $\mathbf{e}(t)$:

$$\mathbf{U}(s) = \left(A_C^{(v)} S_r^{(v)}(s) \right)^{-1} \bullet$$

$$\bullet \left[J^{(\mu)} S_N^{(\mu)}(s) \ \vdots \ - J^{(\mu)} Z_N^{(\mu-1)}(s) \ \vdots \ A_C^{(v)} Z_r^{(v-1)}(s) \right] \bullet$$

$$\bullet \begin{bmatrix} \mathbf{E}(s) \\ \mathbf{e}^{\mu-1}(0) \\ \mathbf{U}^{v-1}(0) \end{bmatrix} = F_{IOC}(s) \, \mathbf{V}_{IOC}(s), \tag{2.62}$$

since the inverse of $A_C^{(v)} S_r^{(v)}(s)$ exists due to the condition $det A_{Cv} \neq 0$. The *IO* feedback controller full transfer function matrix $F_{IOC_f}(s)$ results from (2.62)

$$F_{IOC_f}(s) = \left(A_C^{(v)} S_r^{(v)}(s) \right)^{-1} \bullet$$

$$\bullet \left[J^{(\mu)} S_N^{(\mu)}(s) \vdots - J^{(\mu)} Z_N^{(\mu-1)}(s) \vdots A_C^{(v)} Z_r^{(v-1)}(s) \right] \tag{2.63}$$

Its inverse Laplace transform is *the IO* controller *full fundamental matrix* $\Psi_{IOC}(t)$, $\Psi_{IOC}(t) = \mathcal{L}^{-1}\{F_{IOC}(s)\}$, and the inverse Laplace transform of $\left(A_C^{(v)} S_r^{(v)}(s) \right)^{-1}$ is *the IO* controller *fundamental matrix* $\Phi_{IOC}(t)$, $\Phi_{IOC}(t) = \mathcal{L}^{-1}\left\{ \left(A_C^{(v)} S_r^{(v)}(s) \right)^{-1} \right\}$. Equation (2.63) determines the characteristic polynomial $p_{cIO}(s)$ of the *IO* controller (2.59),

$$p_{cIO}(s) = \det \left(A_C^{(v)} S_r^{(v)}(s) \right), \tag{2.64}$$

the matrix numerator polynomial $L_{cIO}(s)$ of $G_{IOCE}(s)$

$$L_{cIO}(s) = adj \left(A_C^{(v)} S_r^{(v)}(s) \right) J^{(\mu)} S_N^{(\mu)}(s) \tag{2.65}$$

and its specific transfer function matrices:
- With respect to the output error vector $\mathbf{e}$:

$$G_{IOCE}(s) = \frac{adj \left(A_C^{(v)} S_r^{(v)}(s) \right) J^{(\mu)} S_N^{(\mu)}(s)}{p_{cIO}(s)} = \frac{L_{cIO}(s)}{p_{cIO}(s)}, \tag{2.66}$$

- With respect to the extended initial error vector $\mathbf{e}_0^{\mu-1}$:

$$G_{IOCE_0}(s) = -\frac{adj \left(A_C^{(v)} S_r^{(v)}(s) \right) J^{(\mu)} Z_N^{(\mu-1)}(s)}{p_{cIO}(s)}, \tag{2.67}$$

- With respect to the extended initial control vector $\mathbf{U}_0^{\mu-1}$:

$$G_{IOU_0}(s) = \frac{adj \left(A_C^{(v)} S_r^{(v)}(s) \right) A_C^{(v)} Z_r^{(v-1)}(s)}{p_{cIO}(s)}. \tag{2.68}$$

The Laplace transform of the *IO* feedback controller action vector $\mathbf{V}_{C_f}(t)$ reads

$$\mathbf{V}_{IOC_f}(s) = \begin{bmatrix} \mathbf{E}(s) \\ \mathbf{e}^{\mu-1}(0) \\ \mathbf{U}^{v-1}(0) \end{bmatrix} = \begin{bmatrix} \mathbf{E}(s) \\ \mathbf{C}_{IOC_f0} \end{bmatrix}. \tag{2.69}$$

The Laplace transform $\mathbf{I}_{IOC_f}(s)$ of the *IO* feedback controller input vector $\mathbf{I}_{IOC_f}(t)$ is $\mathbf{e}(s)$:

$$\mathbf{I}_{IOC_f}(s) = \mathbf{E}(s). \tag{2.70}$$

The vector $\mathbf{C}_{IOC_f0}$ of all *IO* feedback controller initial conditions has the following form:

$$\mathbf{C}_{IOC_f0} = \left[\begin{array}{c} \mathbf{e}^{\mu-1}(0) \\ \mathbf{U}^{\upsilon-1}(0) \end{array} \right]. \tag{2.71}$$

The *IO* feedback controller output response $\mathbf{U}\left(t; \mathbf{U}_0^{\mu-1}; \mathbf{e}\right)$ is the inverse Laplace transform of (2.62):

$$\mathbf{U}\left(t; \mathbf{U}_0^{\mu-1}; \mathbf{e}\right) = \mathcal{L}^{-1}\left\{F_{IOC_f}(s)\,\mathbf{V}_{IOC_f}(s)\right\}. \tag{2.72}$$

2.4 Exercises

Exercise 54 *1. Select an IO physical plant.*

2. Determine its time domain IO mathematical model.

3. Determine its complex domain IO mathematical model: its full transfer function matrix and all its transfer function matrices, as well as the vectors $\mathbf{V}_{IO}(s)$ *and* $\mathbf{C}_{IOo}$.

Exercise 55 *1. Select an IO controller.*

2. Determine its time domain IO mathematical model.

3. Determine its complex domain IO mathematical model: its full transfer function matrix and all its transfer function matrices, as well as the vectors $\mathbf{V}_{IOC}(s)$ *and* $\mathbf{C}_{IOCo}$.

Exercise 56 *1. Determine the time domain IO mathematical model of the control system composed of the chosen IO system and IO controller.*

2. Determine the complex domain IO mathematical model of the control system composed of the chosen IO system and IO controller: its full transfer function matrix and all its transfer function matrices, as well as the vectors $\mathbf{V}_{IOCS}(s)$ *and* $\mathbf{C}_{IOCSo}$. *Hint: Section 1.7 and Section 1.8.1.*

Exercise 57 *Test all Lyapunov and BI stability properties of the chosen IO system, IO controller and of the control system composed of them. Hint: [170, Part III.]*

Chapter 3

ISO systems

3.1 *ISO* system mathematical model

3.1.1 Time domain

The dynamical systems theory and the control theory have been mainly developed for the linear **Input-State-Output** (*ISO*) **(dynamical, control) systems**. Their mathematical models contain *the state vector differential equation* (3.1) and *the output algebraic vector equation* (3.2),

$$\frac{d\mathbf{X}(t)}{dt} = A\mathbf{X}(t) + D\mathbf{D}(t) + B\mathbf{U}(t) = A\mathbf{X}(t) + P\mathbf{I}(t), \ \forall t \in \mathfrak{T}_0,$$

$$A \in \mathfrak{R}^{n \times n}, D \in \mathfrak{R}^{n \times d}, \mathbf{B} \in \mathfrak{R}^{n \times r}, P = \left[D \vdots \mathbf{B} \right] \in \mathfrak{R}^{n \times (d+r)}, \qquad (3.1)$$

$$\mathbf{Y}(t) = C\mathbf{X}(t) + V\mathbf{D}(t) + U\mathbf{U}(t) = C\mathbf{X}(t) + Q\mathbf{I}(t), \ \forall t \in \mathfrak{T}_0,$$

$$C \in \mathfrak{R}^{N \times n}, V \in \mathfrak{R}^{N \times d}, U \in \mathfrak{R}^{N \times r}, Q = \left[V \vdots U \right] \in \mathfrak{R}^{N \times (d+r)}. \qquad (3.2)$$

The *ISO* mathematical model (3.1), (3.2) is well known also as *the state-space system (description)*.

Note 58 *System, plant, and control system*
If and only if B$\neq O_{n,r}$ then the ISO system (3.1), (3.2) becomes the ISO plant (3.1), (3.2) (Definition 18, Section 1.5). Otherwise, the ISO system (3.1), (3.2) represents the ISO control system.

The state vector $\mathbf{S}_{ISO}$ of the *ISO* system (3.1), (3.2) is the vector **X** (1.37).

The fundamental matrix function

$$\Phi_{ISO}(.,t_0) \equiv \Phi(.,t_0) : \mathfrak{T} \longrightarrow \mathfrak{R}^{n \times n} \tag{3.3}$$

of the system (3.1), (3.2),

$$\Phi(t,t_0) = e^{At}\left(e^{At_0}\right)^{-1} = e^{A(t-t_0)} \in \mathfrak{R}^{n \times n}, \tag{3.4}$$

has the following well-known properties:

$$det\Phi(t,t_0) \neq 0, \ \forall t \in \mathfrak{T}_0, \ \forall t_0 \in \mathfrak{T}, \tag{3.5}$$

$$\Phi(t,t_0)\,\Phi(t_0,t) = e^{A(t-t_0)}e^{A(t_0-t)} \equiv e^{A0} = I_n \implies$$
$$\Phi(t_0,t) = \Phi^{-1}(t,t_0), \tag{3.6}$$

$$\Phi^{(1)}(t,t_0) = A\Phi(t,t_0) = \Phi(t,t_0)\,A. \tag{3.7}$$

By applying the classical method to solve the state equation (3.1) by its integration we determine its solution:

$$\mathbf{X}(t;t_0;\mathbf{X}_0;\mathbf{I}) = \Phi(t,t_0)\,\mathbf{X}_0 + \int_{t_0}^{t} \Phi(t,\tau)\,P\mathbf{I}(\tau)\,d\tau =$$

$$= \Phi(t,t_0)\left[\mathbf{X}_0 + \int_{t_0}^{t} \Phi(t_0,\tau)\,P\mathbf{I}(\tau)\,d\tau\right], \ \forall t \in \mathfrak{T}_0. \tag{3.8}$$

This and the system output Equation (3.2) determine the system response:

$$\mathbf{Y}(t;t_0;\mathbf{X}_0;\mathbf{I}) = C\Phi(t,t_0)\,\mathbf{X}_0 + \int_{t_0}^{t} C\Phi(t,\tau)\,P\mathbf{I}(\tau)\,d\tau + Q\mathbf{I}(t) =$$

$$= C\Phi(t,t_0)\left[\mathbf{X}_0 + \int_{t_0}^{t} \Phi(t_0,\tau)\,P\mathbf{I}(\tau)\,d\tau\right] + Q\mathbf{I}(t), \ \forall t \in \mathfrak{T}_0, \tag{3.9}$$

Note 59 *The IO system (2.1), Section 2, can be formally mathematically transformed into the equivalent ISO system (3.1), (3.2) (for such transformation in the general case of the IO system (2.15) see Appendix E.1 and for more details: [170, Appendix C.1, pp. 417-420]). The obtained state variables are without any physical meaning if $\mu > 0$ in the IO system (2.15). Also, the ISO system (3.1), (3.2) can be transformed into the IO system (2.15) (for the transformation in the general case see [170, Appendix C.2, p. 421]).*

3.1.2 Complex domain

We recall (1.3) (Section 1.3) that I is the identity matrix of the dimension n: $I_n = I$.

The application of the Laplace transform to the *ISO* system (3.1), (3.2) gives its complex domain description:

$$s\mathbf{X}(s) - \mathbf{X}_0 = A\mathbf{X}(s) + D\mathbf{D}(s) + B\mathbf{U}(s), \ \mathbf{Y}(s) = C\mathbf{X}(s) + V\mathbf{D}(s) + U\mathbf{U}(s).$$

We determine first $\mathbf{X}(s)$ from the first equation, and then replace the solution into the second equation to get the well known result for $\mathbf{Y}(s)$:

$$\mathbf{X}(s) = (sI - A)^{-1} [D\mathbf{D}(s) + B\mathbf{U}(s) + \mathbf{X}_0], \tag{3.10}$$

$$\mathbf{Y}(s) = C(sI - A)^{-1} [D\mathbf{D}(s) + B\mathbf{U}(s) + \mathbf{X}_0] + V\mathbf{D}(s) + U\mathbf{U}(s), \tag{3.11}$$

which we can set into the following forms:

$$\mathbf{X}(s) = (sI - A)^{-1} \left[D \vdots \mathrm{B} \vdots I \right] \begin{bmatrix} \mathbf{D}(s) \\ \mathbf{U}(s) \\ \mathbf{X}_0 \end{bmatrix} = F_{ISOIS}(s) \mathbf{V}_{ISO}(s), \tag{3.12}$$

$$\mathbf{Y}(s) = \left[C(sI - A)^{-1} D + V \vdots C(sI - A)^{-1} \mathrm{B} + U \vdots C(sI - A)^{-1} \right] \cdot$$

$$\cdot \begin{bmatrix} \mathbf{D}(s) \\ \mathbf{U}(s) \\ \mathbf{X}_0 \end{bmatrix} = F_{ISO}(s) \mathbf{V}_{ISO}(s), \tag{3.13}$$

where :
- $F_{ISOIS}(s)$,

$$F_{ISOIS}(s) = (sI - A)^{-1} \left[D \vdots \mathrm{B} \vdots I \right], \tag{3.14}$$

is the *ISO* plant (3.1), (3.2) *input to state (IS)* full transfer function matrix, the inverse Laplace transform of which is the plant *IS full fundamental matrix* $\Psi_{ISOIS}(t)$ [170],

$$\Psi_{ISOIS}(t) = \mathcal{L}^{-1} \{F_{ISOIS}(s)\}, \tag{3.15}$$

the resolvent matrix $(sI - A)^{-1}$ of the matrix A is the Laplace transform $\Phi(s)$ of $\Phi(t, 0)$,

$$\Phi(s) = \mathcal{L}\{\Phi(t, 0)\} = \mathcal{L}\{e^{At}\} = (sI - A)^{-1}. \tag{3.16}$$

so that its inverse Laplace transform is the *ISO* system *fundamental matrix* $\Phi(t,0)$ (3.3),

$$\Phi(t;0) = \mathcal{L}^{-1}\{\Phi(s)\} = \mathcal{L}^{-1}\left\{(sI - A)^{-1}\right\} = e^{At}, \qquad (3.17)$$

which is nonsingular on $\mathfrak{T}$, Equation (3.5),
   - $G_{ISOISD}(s)$,

$$G_{ISOISD}(s) = (sI - A)^{-1}D,$$

is the *ISO* plant (3.1), (3.2) *disturbance to state* (*IS*) transfer function matrix,
   - $G_{ISOISU}(s)$,

$$G_{ISOISU}(s) = (sI - A)^{-1}\mathrm{B}, \qquad (3.18)$$

is the *ISO* plant (3.1), (3.2) *control to state* (*IS*) transfer function matrix,
   - $G_{ISOISX_0}(s)$,

$$G_{ISOISX_0}(s) = (sI - A)^{-1}, \qquad (3.19)$$

is the *ISO* plant (3.1), (3.2) *initial state to state* (*IS*) transfer function matrix,
   - $F_{ISO}(s)$,

$$F_{ISO}(s) =$$
$$= \left[C(sI - A)^{-1}D + V \vdots C(sI - A)^{-1}\mathrm{B} + U \vdots C(sI - A)^{-1}\right], \qquad (3.20)$$

is the *ISO* plant (3.1), (3.2) *input to output* (*IO*) *full transfer function matrix*, the inverse Laplace transform of which is the plant *IO full fundamental matrix* $\Psi_{ISO}(t)$ [170],

$$\Psi_{ISO}(t) = \mathcal{L}^{-1}\{F_{ISO}(s)\}, \qquad (3.21)$$

   - $p_{ISO}(s)$,

$$p_{ISO}(s) = \det(sI - A), \qquad (3.22)$$

is the characteristic polynomial of the *ISO* plant (3.1), (3.2) and the denominator polynomial of all its transfer function matrices,
   - $G_{ISOD}(s)$,

$$G_{ISOD}(s) = C(sI - A)^{-1}D + V =$$
$$= p_{ISO}^{-1}(s)\left[Cadj(sI - A)D + p_{ISO}(s)V\right], \qquad (3.23)$$

is the *ISO* plant (3.1), (3.2) *transfer function matrix relative to the distur-bance* **D**,

- $G_{ISOU}(s)$,

$$G_{ISOU}(s) = C(sI - A)^{-1}B + U =$$

$$= p_{ISO}^{-1}(s)\left[Cadj(sI - A)B + p_{ISO}(s)U\right] = \frac{L_{ISO}(s)}{p_{ISO}(s)}, \quad (3.24)$$

is the *ISO* plant (3.1), (3.2) *transfer function matrix relative to the control* **U**, and $L_{ISO}(s)$ is the numerator matrix polynomial of $G_{ISOU}(s)$,

$$L_{ISO}(s) = Cadj(sI_n - A)B + p_{ISO}(s)U, \ L_{ISO}(s) \in \mathbb{C}^{N \times r}, \quad (3.25)$$

which obeys

$$L_{ISO}(s) = \begin{cases} \sum_{i=0}^{i=n} L_i s^i = L_{ISO}^{(n)} S_r^{(n)}(s), \ L_i \in \mathfrak{R}^{N \times r}, \\ \forall i = 0, 1, ..., n, \Longleftrightarrow \ U \neq O_{N,r} \\ \sum_{i=0}^{i=n-1} L_i s^i = L_{ISO}^{(n-1)} S_r^{(n-1)}(s), \ L_i \in \mathfrak{R}^{N \times r}, \\ \forall i = 0, 1, ..., n-1, \Longleftrightarrow \ U = O_{N,r}, \end{cases} \quad (3.26)$$

where

$$L_{ISO}^{(n)} = \left[L_0 \ \vdots \ L_1 \ \vdots \ ... \ \vdots \ L_n\right] \in \mathfrak{R}^{N \times (n+1)r} \Longleftrightarrow U \neq O_{N,r}, \quad (3.27)$$

$$L_{ISO}^{(n-1)} = \left[L_0 \ \vdots \ L_1 \ \vdots \ ... \ \vdots \ L_{n-1}\right] \in \mathfrak{R}^{N \times nr} \Longleftrightarrow U = O_{N,r}. \quad (3.28)$$

$$L_{ISO} = \left\{ \begin{matrix} L_{ISO}^{(n)} \Longleftrightarrow U \neq O_{N,r}, \\ L_{ISO}^{(n-1)} \Longleftrightarrow U = O_{N,r}. \end{matrix} \right\}, \quad (3.29)$$

- $G_{ISOX_0}(s)$,

$$G_{ISOX_0}(s) = C(sI - A)^{-1} = p_{ISO}^{-1}(s) Cadj(sI - A), \quad (3.30)$$

is the *ISO* plant (3.1), (3.2) *transfer function matrix relative to the initial state* **X**$_0$,

- $\mathbf{V}_{ISO}(s)$ and $\mathbf{C}_{ISO0}$,

$$\mathbf{V}_{ISO}(s) = \begin{bmatrix} \mathbf{I}_{ISO}(s) \\ \mathbf{C}_{ISO0} \end{bmatrix}, \ \mathbf{I}_{ISO}(s) = \begin{bmatrix} \mathbf{D}(s) \\ \mathbf{U}(s) \end{bmatrix}, \ \mathbf{C}_{ISO0} = \mathbf{X}_0, \quad (3.31)$$

are the Laplace transform of the action vector $\mathbf{V}_{ISOP}(t)$ and the vector $\mathbf{C}_{ISO0}$ of all plant initial conditions, respectively.

For an example of the *ISO* system (3.1), (3.2) see the book [171].

3.2   *ISO* plant desired regime

The following definition clarifies the meaning of the nominal input control vector $\mathbf{U}_N$ and of the nominal state vector $\mathbf{X}_N$ with respect to a chosen or given disturbance vector function $\mathbf{D}(.)$ and the desired output function $\mathbf{Y}_d(.)$ of the *ISO* plant (3.1), (3.2).

Definition 60 *A functional vector control-state pair* $[\mathbf{U}^*(.), \mathbf{X}^*(.)]$ *is **nominal** for the ISO plant (3.1), (3.2) **relative to the functional pair***

$$[\mathbf{D}(.), \mathbf{Y}_d(.)],$$

which is denoted by $[\mathbf{U}_N(.), \mathbf{X}_N(.)]$, *if and only if* $[\mathbf{U}(.), \mathbf{X}(.)] = [\mathbf{U}^*(.), \mathbf{X}^*(.)]$ *ensures that the corresponding real response* $\mathbf{Y}(.) = \mathbf{Y}^*(.)$ *of the plant obeys* $\mathbf{Y}^*(t) = \mathbf{Y}_d(t)$ *all the time,*

$$[\mathbf{U}^*(.), \mathbf{X}^*(.)] = [\mathbf{U}_N(.), \mathbf{X}_N(.)] \Longleftrightarrow \langle \mathbf{Y}^*(t) = \mathbf{Y}_d(t), \ \forall t \in \mathfrak{T}_0 \rangle.$$

The nominal motion $\mathbf{X}_N(.; \mathbf{X}_{N0}; \mathbf{D}; \mathbf{U}_N)$, $\mathbf{X}_N(0; \mathbf{X}_{N0}; \mathbf{D}; \mathbf{U}_N) \equiv \mathbf{X}_{N0}$, *is the desired motion* $\mathbf{X}_d(.; \mathbf{X}_{d0}; \mathbf{D}; \mathbf{U}_N)$ *of the ISO plant (3.1), (3.2) **relative to the functional vector pair*** $[\mathbf{D}(.), \mathbf{Y}_d(.)]$, *for short: **the desired motion of the system**,*

$$\mathbf{X}_d(.; \mathbf{X}_{d0}; \mathbf{D}; \mathbf{U}_N) \equiv \mathbf{X}_N(.; \mathbf{X}_{N0}; \mathbf{D}; \mathbf{U}_N),$$
$$\mathbf{X}_d(0; \mathbf{X}_{d0}; \mathbf{D}; \mathbf{U}_N) \equiv \mathbf{X}_{d0} \equiv \mathbf{X}_{N0}. \tag{3.32}$$

Notice that the full system matrix [170, Section 11.2, pp. 192-199]

$$\begin{bmatrix} -\mathbf{B} & sI - A \\ U & C \end{bmatrix} \in \mathbb{C}^{(n+N)\times(r+n)} \tag{3.33}$$

is a rectangular matrix in general.

Definition 60 and (3.1), (3.2) imply the following theorem:

Theorem 61 *In order for the vector control-state pair* $[\mathbf{U}^*(.), \mathbf{X}^*(.)]$ *to be nominal for the ISO plant (3.1), (3.2) relative to the functional vector pair* $[\mathbf{D}(.), \mathbf{Y}_d(.)]$, $[\mathbf{U}^*(.), \mathbf{X}^*(.)] = [\mathbf{U}_N(.), \mathbf{X}_N(.)]$, *it is necessary and sufficient that it obeys the following equations:*

$$-\mathbf{B}\mathbf{U}^*(t) + \frac{d\mathbf{X}^*(t)}{dt} - A\mathbf{X}^*(t) = D\mathbf{D}(t), \ \forall t \in \mathfrak{T}_0, \tag{3.34}$$

$$U\mathbf{U}^*(t) + C\mathbf{X}^*(t) = \mathbf{Y}_d(t) - V\mathbf{D}(t), \ \forall t \in \mathfrak{T}_0, \tag{3.35}$$

or equivalently in the complex domain,

$$\left[\begin{array}{cc} -B & sI - A \\ U & C \end{array}\right]\left[\begin{array}{c} \mathbf{U}^*(s) \\ \mathbf{X}^*(s) \end{array}\right] = \left[\begin{array}{c} \mathbf{X}_0^* + DD(s) \\ \mathbf{Y}_d(s) - VD(s) \end{array}\right]. \tag{3.36}$$

Let us consider the existence of the solutions of the equations (3.34), (3.35), or equivalently of (3.36). There are $(n + r)$ unknown variables and $(N+n)$ equations. The unknown variables are the entries of $\mathbf{U}^*(s) \in \mathbb{C}^r$ and of $\mathbf{X}^*(s) \in \mathbb{C}^n$. There are $(n+r)$ unknown variables and $(N+n)$ equations. The unknown variables are the entries of $\mathbf{U}^*(s) \in \mathbb{C}^r$ and of $\mathbf{X}^*(s) \in \mathbb{C}^n$.

Claim 62 *In order to exist a nominal functional vector control-state pair*

$$[\mathbf{U}_N(.), \mathbf{X}_N(.)]$$

for the ISO plant (3.1), (3.2) relative to its desired response $\mathbf{Y}_d(.)$ it is necessary and sufficient that $N \leq r$. Then, the functional vector control-state pair $[\mathbf{U}_N(.), \mathbf{X}_N(.)]$ is nominal relative to the desired response $\mathbf{Y}_d(.)$ of the plant (3.1), (3.2) in view of Theorem 61.

Proof. The dimension of the matrix (3.33) is $(n + N) \times (r + n)$. It is well known (e.g., [7, p. 115]) that for Equation (3.36) to have a solution it is necessary and sufficient that the rank of the matrix (3.33) is equal to $n + N$, which is possible if and only if $n + N \leq r + n$, i.e., if and only if $N \leq r$. ∎

The condition $N \leq r$ agrees with Fundamental control principle 104, Section 7.1.

Claim 62 resolves completely the problem of the existence of a nominal functional vector control-state pair $[\mathbf{U}_N(.), \mathbf{X}_N(.)]$ for the *ISO* plant (3.1), (3.2) relative to the functional vector pair $[\mathbf{D}(.), \mathbf{Y}_d(.)]$.

Condition 63 *The desired output response of the ISO plant (3.1), (3.2) is realizable, i.e., $N \leq r$. The nominal control-state pair $[\mathbf{U}_N(.), \mathbf{X}_N(.)]$ is known.*

The *ISO* plant description in terms of the deviations (3.37),

$$\mathbf{x} = \mathbf{X} - \mathbf{X}_N = \mathbf{X} - \mathbf{X}_d, \tag{3.37}$$

(2.56), and (2.53), (2.55) (Section 2.2) reads:

$$\frac{d\mathbf{x}(t)}{dt} = A\mathbf{x}(t) + D\mathbf{d}(t) + B\mathbf{u}(t), \quad \forall t \in \mathfrak{T}_0, \tag{3.38}$$

$$\mathbf{y}(t) = C\mathbf{x}(t) + V\mathbf{d}(t) + U\mathbf{u}(t), \quad \forall t \in \mathfrak{T}_0, \tag{3.39}$$

due to (3.1) and (3.2).

3.3　*ISO* feedback controller

3.3.1　Time domain

When we set

$$\mathbf{I} = \mathbf{I}_{Cf} = \begin{bmatrix} \mathbf{Y}_d \\ \mathbf{Y} \end{bmatrix} \in \mathfrak{R}^{2N}, \ M = 2N,$$

$$P = \begin{bmatrix} J \vdots & -J \end{bmatrix} \in \mathfrak{R}^{n \times 2N}, \ Q = \begin{bmatrix} V \vdots & -V \end{bmatrix} \in \mathfrak{R}^{r \times 2N},$$

into (3.1), (3.2) the result is the *ISO* controller description in terms of the total coordinates:

$$\frac{d\mathbf{X}_C(t)}{dt} = A_C \mathbf{X}_C(t) + J \left[\mathbf{Y}_d(t) - \mathbf{Y}(t) \right], \ \forall t \in \mathfrak{T}_0, \tag{3.40}$$

$$\mathbf{U}(t) = C \mathbf{X}_C(t) + V \left[\mathbf{Y}_d(t) - \mathbf{Y}(t) \right], \ \forall t \in \mathfrak{T}_0. \tag{3.41}$$

The *ISO* controller description in terms of the deviations (3.37), (2.53), (1.53) (Section 1.6), (2.56) and (2.55) reads:

$$\frac{d\mathbf{x}_C(t)}{dt} = A_C \mathbf{x}_C(t) + J\mathbf{e}(t), \ \forall t \in \mathfrak{T}_0, \tag{3.42}$$

$$\mathbf{u}(t) = C\mathbf{x}_C(t) + V\mathbf{e}(t), \ \forall t \in \mathfrak{T}_0, \tag{3.43}$$

due to (3.40) and (3.41).

3.3.2　Complex domain

The application of the Laplace transform to the *ISO* controller (3.40), (3.41), i.e., (3.42), (3.42) gives its complex domain description:

$$s\mathbf{X}_C(s) - \mathbf{X}_{C0} = A_C \mathbf{X}_C(s) + J \left[\mathbf{Y}_d(s) - \mathbf{Y}(s) \right],$$

$$\mathbf{U}(s) = C\mathbf{X}_C(s) + V \left[\mathbf{Y}_d(s) - \mathbf{Y}(s) \right].$$

We determine first $\mathbf{X}_C(s)$ from the first equation, and then replace the solution into the second equation to get the result for $\mathbf{U}(s)$:

$$\mathbf{X}_C(s) = (sI - A_C)^{-1} \left(J \left[\mathbf{Y}_d(s) - \mathbf{Y}(s) \right] + \mathbf{X}_{C0} \right), \tag{3.44}$$

$$\mathbf{U}(s) = C (sI - A_C)^{-1} \left(J \left[\mathbf{Y}_d(s) - \mathbf{Y}(s) \right] + \mathbf{X}_{C0} \right) + V \left[\mathbf{Y}_d(s) - \mathbf{Y}(s) \right], \tag{3.45}$$

which we can set into the following forms by using (1.53):

$$\mathbf{X}_C(s) = (sI - A_C)^{-1} \begin{bmatrix} J \vdots I \end{bmatrix} \begin{bmatrix} \mathbf{E}(s) \\ \mathbf{X}_{C0} \end{bmatrix} = F_{ISOC_fIS}(s) \, \mathbf{V}_{ISOC_f}(s), \quad (3.46)$$

$$\mathbf{U}(s) = \begin{bmatrix} C(sI - A_C)^{-1} J + V \vdots C(sI - A_C)^{-1} \end{bmatrix} \begin{bmatrix} \mathbf{E}(s) \\ \mathbf{X}_{C0} \end{bmatrix} =$$
$$= F_{ISOC_f}(s) \, \mathbf{V}_{ISOC_f}(s), \quad (3.47)$$

where:
   - $F_{ISOC_fIS}(s)$,

$$F_{ISOC_fIS}(s) = \begin{bmatrix} (sI - A_C)^{-1} J \vdots (sI - A_C)^{-1} \end{bmatrix}, \quad (3.48)$$

is the *ISO* controller (3.40) and (3.41) *input to state (IS) full transfer function matrix*, the inverse Laplace transform of which is the controller *IS full fundamental matrix* $\Psi_{ISOC_fIS}(t)$,

$$\Psi_{ISOC_fIS}(t) = \mathcal{L}^{-1}\{F_{ISOC_fIS}(s)\}, \quad (3.49)$$

   - $G_{ISOCIS}(s)$,

$$G_{ISOCIS}(s) = C(sI - A_C)^{-1} J \quad (3.50)$$

is the *ISO* controller (3.40) and (3.41) *transfer function matrix relating the controller state* $\mathbf{X}_C$ *to the input* $\mathbf{e}$,
   - $G_{ISOCISX_0}(s)$,

$$G_{ISOCISX_{C0}}(s) = C(sI - A_C)^{-1} \quad (3.51)$$

is the *ISO* controller (3.40) and (3.41) *transfer function matrix relating the controller state* $\mathbf{X}_C$ *to the initial state* $\mathbf{X}_{C0}$,
   - $\mathbf{V}_{ISOCIS}(s)$ and $\mathbf{C}_{ISOCIS}$,

$$\mathbf{V}_{ISOCIS}(s) = \begin{bmatrix} \mathbf{I}_{C_f}(s) \\ \mathbf{C}_{ISOCIS0} \end{bmatrix}, \quad \mathbf{I}_{C_f}(s) = \mathbf{E}(s), \quad \mathbf{C}_{ISOCIS0} = \mathbf{X}_{C0}, \quad (3.52)$$

are the Laplace transform of the action vector $\mathbf{V}_{ISOCIS}(t)$ and the vector $\mathbf{C}_{ISOCIS0}$ of all initial conditions, respectively,
   - $F_{ISOC}(s)$,

$$F_{ISOC}(s) = \begin{bmatrix} C(sI - A_C)^{-1} J + V \vdots C(sI - A_C)^{-1} \end{bmatrix}, \quad (3.53)$$

is the ISO controller (3.40) and (3.41) *input to output (IO) full transfer function matrix*, the inverse Laplace transform of which is the system IO *full fundamental matrix* $\Psi_{ISO}(t)$ [170],

$$\Psi_{ISOC}(t) = \mathcal{L}^{-1}\{F_{ISOC}(s)\},\qquad(3.54)$$

- $p_{cISO}(s)$,

$$p_{cISO}(s) = \det(sI - A_C),\qquad(3.55)$$

is the characteristic polynomial of the ISO controller (3.40), (3.41) and the denominator polynomial of all its transfer function matrices,
   - $L_{cISO}(s)$,

$$L_{cISO}(s) = C\,adj\,(sI - A_C)\,J + p_{cISO}(s)\,V,\qquad(3.56)$$

is the numerator matrix polynomial of $G_{ISOCE}(s)$,
   - $G_{ISOCE}(s)$,

$$G_{ISOCE}(s) = C\,(sI - A_C)^{-1}\,J + V =$$

$$= p_{cISO}^{-1}(s)\,[Cadj\,(sI - A_C)\,J + p_{cISO}(s)\,V] = \frac{L_{cISO}(s)}{p_{cISO}(s)},\qquad(3.57)$$

is the ISO controller (3.40), (3.41) *transfer function matrix relating the controller output* $\mathbf{U}$ *to the input* $\mathbf{e}$,
   - $G_{ISOCX_0}(s)$,

$$G_{ISOCX_{C0}}(s) = C\,(sI - A_C)^{-1} = p_{cISO}^{-1}(s)\,Cadj\,(sI - A_C),\qquad(3.58)$$

is the ISO controller (3.40) and (3.41) *transfer function matrix relating the controller output* $\mathbf{U}$ *to the initial state* $\mathbf{X}_{C0}$,
   - $\mathbf{V}_{ISOC}(s)$ and $\mathbf{C}_{ISOC}$,

$$\mathbf{V}_{ISOC}(s) = \begin{bmatrix} \mathbf{I}_{C_f}(s) \\ \mathbf{C}_{ISOC_f0} \end{bmatrix},\ \mathbf{I}_{C_f}(s) = \begin{bmatrix} \mathbf{Y}_d(s) \\ \mathbf{Y}(s) \end{bmatrix},\ \mathbf{C}_{ISOC_f0} = \mathbf{X}_{C0},\quad(3.59)$$

are the Laplace transform of the action vector $\mathbf{V}_{ISOC}(t)$ and the vector $\mathbf{C}_{ISOC_f0}$ of all initial conditions, respectively.

3.4   Exercises

Exercise 64 *1. Select a physical ISO plant.*
   2. Determine its time domain ISO mathematical model.
   3. Determine its complex domain ISO mathematical model: its full transfer function matrix and all its transfer function matrices, as well as the vectors $\mathbf{V}_{ISO}(s)$ *and* $\mathbf{C}_{ISOP_0}$.

Exercise 65 *1. Select an ISO controller.*

2. Determine its time domain ISO mathematical model.

3. Determine its complex domain ISO mathematical model: its full transfer function matrix and all its transfer function matrices, as well as the vectors $\mathbf{V}_{ISOC}(s)$ and $\mathbf{C}_{ISOCo}$.

Exercise 66 *1. Determine the time domain ISO mathematical model of the control system composed of the chosen ISO plant and ISO controller.*

2. Determine its complex domain ISO mathematical model of the control system composed of the chosen ISO plant and ISO controller: its full transfer function matrix and all its transfer function matrices, as well as the vectors $\mathbf{V}_{ISOCS}(s)$ and $\mathbf{C}_{ISOCSo}$. Hint: Section 1.7 and Subsection 1.8.1.

Exercise 67 *Test all Lyapunov and BI stability properties of the chosen ISO plant, ISO controller and of the control system composed of them. Hint: [170, Part III.]*

Chapter 4

EISO systems

4.1 *EISO* system mathematical model

4.1.1 Time domain

A slightly more general class than the *ISO* systems (3.1), (3.2) is the family of **the *Extended Input-State-Output* systems (*EISO* systems)** described in terms of the total coordinates by

$$\frac{d\mathbf{X}(t)}{dt} = A\mathbf{X}(t) + D^{(\mu)}\mathbf{D}^{\mu}(t) + \mathrm{B}^{(\mu)}\mathbf{U}^{\mu}(t) = A\mathbf{X}(t) + P^{(\mu)}\mathbf{I}^{\mu}(t), \; \forall t \in \mathfrak{T}_0,$$

$$A \in \mathfrak{R}^{n \times n}, \; D^{(\mu)} \in \mathfrak{R}^{n \times (\mu+1)d}, \; \mathrm{B}^{(\mu)} \in \mathfrak{R}^{n \times (\mu+1)r}, \; P^{(\mu)} \in \mathfrak{R}^{n \times (\mu+1)M}, \quad (4.1)$$

$$\mathbf{Y}(t) = C\mathbf{X}(t) + V\mathbf{D}(t) + U\mathbf{U}(t) = C\mathbf{X}(t) + Q\mathbf{I}(t), \; \forall t \in \mathfrak{T}_0.$$

$$C \in \mathfrak{R}^{N \times n}, \; V \in \mathfrak{R}^{N \times d}, \; U \in \mathfrak{R}^{N \times r}, \; Q \in \mathfrak{R}^{N \times M}. \quad (4.2)$$

The overall input mathematical data of the *EISO* system are the input vector $\mathbf{I}$ and the matrix $P^{(\mu)}$ related to the extended input vector $\mathbf{I}^{\mu}$:

$$\mathbf{I} = \mathbf{I}_{EISO} = \begin{bmatrix} \mathbf{D}^T & \mathbf{U}^T \end{bmatrix}^T \in \mathfrak{R}^{d+r}, \; M = d + r,$$

$$P^{(\mu)} = \begin{bmatrix} P_0 & \vdots & P_1 & \vdots & \dots & \vdots & P_{\mu} \end{bmatrix} \in \mathfrak{R}^{n \times (\mu+1)M},$$

$$P^{(\mu)} = \begin{bmatrix} D^{(\mu)} & \vdots & \mathrm{B}^{(\mu)} \end{bmatrix} \in \mathfrak{R}^{n \times (\mu+1)(d+r)}, \; Q = \begin{bmatrix} V & \vdots & U \end{bmatrix} \in \mathfrak{R}^{N \times (d+r)},$$

$$\mathrm{B}^{(\mu)} = \begin{bmatrix} \mathrm{B}_0 \vdots \mathrm{B}_1 \vdots \dots \vdots \mathrm{B}_{\mu} \end{bmatrix} \in \mathfrak{R}^{n \times (\mu+1)r}, \; D^{(\mu)} = \begin{bmatrix} D_0 \vdots D_1 \vdots \dots \vdots D_{\mu} \end{bmatrix} \in \mathfrak{R}^{n \times (\mu+1)d}.$$

$$(4.3)$$

If $\mu = 0$ then the *EISO* system (4.1), (4.2) becomes the *ISO* system (3.1), (3.2) (Section 3.1). In order to present one physical origin of the *EISO* system (4.1), (4.2) we discover it in the physical *IO* systems (2.15), (Section 2.1), as shown in the following:

Theorem 68 *The EISO form (4.1), (4.2) of the IO system (2.15)*

The IO system (2.15) can be transformed into the EISO form (4.1), (4.2) by preserving the physical meaning of all variables, where we should distinguish the case $\nu > 1$ from the case $\nu = 1$ in the IO system (2.15), for which

$$\nu N = n, \tag{4.4}$$

with the following choice of the system physical substate vectors $\mathbf{X}_i$ of the system physical state vector $\mathbf{X}$:

$$\left\{ \begin{array}{c} \nu > 1 \Longrightarrow \mathbf{X}_i = \mathbf{Y}^{(i-1)} \in \mathfrak{R}^N, \ \forall i = 1, 2, ..., \nu, \ i.e., \\ \mathbf{X} = \left[\mathbf{X}_1^T \vdots \mathbf{X}_2^T \vdots ... \vdots \mathbf{X}_\nu^T \right]^T = \left[\mathbf{Y}^T \vdots \mathbf{Y}^{(1)T} \vdots ... \vdots \mathbf{Y}^{(\nu-1)^T} \right]^T = \\ = \mathbf{Y}^{\nu-1} \in \mathfrak{R}^n \end{array} \right\}$$

$$\nu = 1 \Longrightarrow \mathbf{X}_1 = \mathbf{X} = \mathbf{Y} \in \mathfrak{R}^N, \tag{4.5}$$

and with the following matrices of the EISO form (4.1), (4.2) in terms of the matrices A_k, $k = 0, 1, ..., \nu$, of the IO system (2.15):

$$\nu > 1 \Longrightarrow A =$$

$$\left[\begin{array}{ccccc} O_N & I_N & ... & O_N & O_N \\ O_N & O_N & ... & O_N & O_N \\ O_N & O_N & ... & I_N & O_N \\ ... & ... & ... & ... & ... \\ O_N & O_N & ... & O_N & I_N \\ -A_\nu^{-1}A_0 & -A_\nu^{-1}A_1 & ... & -A_\nu^{-1}A_{\nu-2} & -A_\nu^{-1}A_{\nu-1} \end{array} \right] \in \mathfrak{R}^{\nu N \times \nu N},$$

$$\nu = 1 \Longrightarrow A = -A_1^{-1}A_0 \in \mathfrak{R}^{N \times N}, \tag{4.6}$$

$$P^{(\mu)} = \left\{ \begin{array}{c} \left[\begin{array}{c} O_{(\nu-1)N,(\mu+1)M} \\ A_\nu^{-1}H^{(\mu)} \end{array} \right] \in \mathfrak{R}^{n \times (\mu+1)M}, \ \nu > 1, \\ A_1^{-1}H^{(\mu)} \in \mathfrak{R}^{N \times (\mu+1)M}, \ \nu = 1, \end{array} \right\}, \tag{4.7}$$

or, equivalently,

$$P^{(\mu)} = \left\{ \begin{array}{c} \left[\begin{array}{c} O_{(\nu-1)N,N} \\ I_N \end{array} \right] A_\nu^{-1}H^{(\mu)}, \ \nu > 1, \\ I_N A_1^{-1}H^{(\mu)}, \ \nu = 1, \end{array} \right\}, \tag{4.8}$$

$$P_{inv} = \left\{ \begin{array}{c} \left[\begin{array}{c} O_{(\nu-1)N,N} \\ I_N \end{array} \right] \in \mathfrak{R}^{n \times N}, \ \nu > 1, \\ I_N \in \mathfrak{R}^{N \times N}, \ \nu = 1, \end{array} \right\}, \tag{4.9}$$

i.e.,

$$P^{(\mu)} = P_{inv} A_\nu^{-1} H^{(\mu)} \in \Re^{n \times (\mu+1)M} \tag{4.10}$$

$$C = [I_N \ \ O_N \ \ O_N \ \ O_N \ \ ... \ \ O_N] \in \Re^{N \times n}, \ Q = O_{N,M} \in \Re^{N \times M}. \tag{4.11}$$

The proof of this theorem is in Appendix D.1.

Note 69 *The substate vectors* $\mathbf{X}_i$ *and the state vector* $\mathbf{X}$ *composed of them and all defined by (4.5) have the full physical meaning. They are the system output vector* $\mathbf{Y}$ *and its derivatives. The EISO system (4.1), (4.2) determined by (4.4) - (4.11) retains the full physical sense as the original IO system (2.15). They have the same properties.*

The EISO form (4.1), (4.2) of the original IO system (2.15) differs from the well known ISO form (3.1), (3.2), i.e., (E.2), (E.3) (Appendix E.1), of the IO system (2.15) for the preservation of the derivatives of the input vector in the state equation (4.1), which has not been accepted so far: Equation (3.1). The physical nature of the IO system (2.15) introduces the derivatives of the input vector in the state equation. The formal mathematical transformation given by Equations (E.4)–(E.10) (Section E.1) ignores the explicit action of the input vector derivatives on the physical state of the IO system (2.15).

The existing formal mathematical transformation of the IO system (2.15) into the ISO form (3.1), (3.2) loses the physical sense if $\mu > 0$ *so that the chosen state variables and the state vector are physically meaningless.*

This book develops the state theory for the IO systems (2.15) by exploiting their EISO form (4.1), (4.2), (4.4)–(4.11) in order to preserve the full physical sense of the original IO system (2.15). A useful tool to achieve this is the new simple compact calculus based on the compact notation

$$\left[\mathbf{Y}^T \ \vdots \ \mathbf{Y}^{(1)T} \ \vdots \ ... \ \vdots \ \mathbf{Y}^{(\nu-1)^T} \right]^T = \mathbf{Y}^{\nu-1},$$

which enabled us to define the physical (and mathematical) state vector of the IO systems (2.15) in the form $\mathbf{X} = \mathbf{Y}^{\nu-1}$.

Note 70 *The matrix* $P_{inv}^{(\mu)}$ *(4.7), (4.8), (4.10) is the invariant submatrix of the matrix* $P^{(\mu)}$. *It is invariant relative to both all matrices* A_i, $i = 0, 1, 2, ..., \nu$, *and all submatrices* H_k *of* $H^{(\mu)}$, $k = 0, 1, 2, ..., \mu$, *of the original IO system (2.15). In other words, the matrix* P_{inv} *is independent of both all matrices* A_i, $i = 0, 1, 2, ..., \nu$, *and all matrices* H_k, $k = 0, 1, 2, ..., \mu$.

Note 71 *Let $\nu > 1.$ Then $O_{(\nu-1)N,M}$ is $(\nu - 1)\,N \times M$ zero matrix.*

If and only if $\nu = 1$, then the matrix $O_{(\nu-1)N,M}$ becomes formally $O_{0,M}$ that does not exist. Then it should be simply omitted.

Conclusion 72 *For the existence of the $(\nu - 1)\,N \times M$ zero matrix $O_{(\nu-1)N,M}$ to exist it is necessary and sufficient that the natural number ν obeys $\nu > 1$:*

$$\exists O_{(\nu-1)N,M} \in \mathfrak{R}^{(\nu-1)N\times M} \iff \nu \in \{2, 3, ..., n, ...\}\,. \qquad (4.12)$$

By referring to the well known form of the solution of the *ISO* systems (3.1), (3.2) we easily show that the solution of the *EISO* system (4.1), (4.2) is determined by

$$\mathbf{X}(t; t_0; \mathbf{X}_0; \mathbf{I}^{\mu}) = e^{A(t-t_0)}\mathbf{X}_0 + \int_{t_0}^{t} e^{A(t-\tau)} P^{(\mu)}\mathbf{I}^{\mu}(\tau)\, d\tau, \qquad (4.13)$$

$$= e^{A(t-t_0)}\left[\mathbf{X}_0 + \int_{t_0}^{t} e^{A(t_0-\tau)} P^{(\mu)}\mathbf{I}^{\mu}(\tau)\, d\tau\right], \ \forall t \in \mathfrak{T}_0, \qquad (4.14)$$

or equivalently by

$$\mathbf{X}(t; t_0; \mathbf{X}_0; \mathbf{I}^{\mu}) = \Phi\,(t, t_0)\,\mathbf{X}_0 + \int_{t_0}^{t} \Phi\,(t, \tau)\,P^{(\mu)}\mathbf{I}^{\mu}(\tau)\, d\tau = \qquad (4.15)$$

$$= \Phi\,(t, t_0)\left[\mathbf{X}_0 + \int_{t_0}^{t} \Phi\,(t_0, \tau)\,P^{(\mu)}\mathbf{I}^{\mu}(\tau)\, d\tau\right], \ \forall t \in \mathfrak{T}_0, \qquad (4.16)$$

for $\Phi\,(t, t_0)$ (3.3) (Section 3.1), i.e.,

$$\Phi\,(t, t_0) = \Phi\,(t, 0)\,\Phi^{-1}\,(t_0, 0) = e^{At}\left(e^{At_0}\right)^{-1} = e^{A(t-t_0)} \in \mathfrak{R}^{n\times n}. \qquad (4.17)$$

These equations and Equation (4.2) determine the *EISO* system response to the initial state vector $\mathbf{X}_0$ and to the extended input vector function $\mathbf{I}^{\mu}(.)$:

$$\mathbf{Y}(t; t_0; \mathbf{X}_0; \mathbf{I}^{\mu}) = C\Phi\,(t, t_0)\,\mathbf{X}_0 + \int_{t_0}^{t} C\Phi\,(t, \tau)\,P^{(\mu)}\mathbf{I}^{\mu}(\tau)\, d\tau + Q\mathbf{I}(t) =$$

$$\qquad (4.18)$$

$$= C\Phi\,(t, t_0)\left[\mathbf{X}_0 + \int_{t_0}^{t} \Phi\,(t_0, \tau)\,P^{(\mu)}\mathbf{I}^{\mu}(\tau)\, d\tau\right] + Q\mathbf{I}(t), \ \forall t \in \mathfrak{T}_0. \qquad (4.19)$$

4.1.2   Complex domain

The Laplace transform of the *EISO* system (4.1), (4.2) relative to $D(t)$ and $\mathbf{U}(t)$ reads:

$$s\mathbf{X}(s) - \mathbf{X}_0 = A\mathbf{X}(s) + \left\{ \begin{array}{l} D^{(\mu)}S_d^{(\mu)}(s)\mathbf{D}(s) - D^{(\mu)}Z_d^{(\mu-1)}(s)\mathbf{D}_0^{\mu-1} + \\ +\mathrm{B}^{(\mu)}S_r^{(\mu)}(s)\mathbf{U}(s) - \mathrm{B}^{(\mu)}Z_r^{(\mu-1)}(s)\mathbf{U}_0^{\mu-1} \end{array} \right\},$$

$$\text{(4.20)}$$

$$\mathbf{Y}(s) = C\mathbf{X}(s) + V\mathbf{D}^{(\mu)}(s) + V\mathbf{U}(s). \tag{4.21}$$

These equations lead to:

$$\mathbf{X}(s) = (sI - A)^{-1} \left[\begin{array}{l} D^{(\mu)}S_d^{(\mu)}(s) \vdots \mathrm{B}^{(\mu)}S_r^{(\mu)}(s) \vdots - \\ D^{(\mu)}Z_d^{(\mu-1)}(s) \vdots - \mathrm{B}^{(\mu)}Z_r^{(\mu-1)}(s) \vdots I \end{array} \right]$$

$$\bullet \left[\mathbf{D}^T(s) \vdots \mathbf{U}^T(s) \vdots \mathbf{D}^{(\mu)\mu-1^T}_0 \vdots \mathbf{U}_0^{\mu-1^T} \vdots \mathbf{X}_0^T \right]^T =$$

$$= F_{EISOIS}(s)\, \mathbf{V}_{EISO}(s), \tag{4.22}$$

$$\mathbf{Y}(s) = \left(\left[\begin{array}{c} C(sI - A)^{-1}D^{(\mu)}S_d^{(\mu)}(s) + V \vdots \\ \vdots C(sI - A)^{-1}\mathrm{B}^{(\mu)}S_r^{(\mu)}(s) + U \vdots \\ \vdots - C(sI - A)^{-1}D^{(\mu)}Z_d^{(\mu-1)}(s) \vdots \\ \vdots - C(sI - A)^{-1}\mathrm{B}^{(\mu)}Z_r^{(\mu-1)}(s) \vdots \\ \vdots C(sI - A)^{-1} \end{array} \right] \bullet \right.$$

$$\left. \bullet \left[\mathbf{D}^T(s) \vdots \mathbf{U}^T(s) \vdots \mathbf{D}_0^{\mu-1^T} \vdots \mathbf{U}_0^{\mu-1^T} \vdots \mathbf{X}_0^T \right]^T \right) =$$

$$= F_{EISO}(s)\, \mathbf{V}_{EISO}(s), \tag{4.23}$$

where
- $F_{EISOIS}(s)$,

$$F_{EISOIS}(s) = (sI - A)^{-1} \bullet$$

$$\bullet \left[D^{(\mu)}S_d^{(\mu)}(s) \vdots \mathrm{B}^{(\mu)}S_r^{(\mu)}(s) \vdots - D^{(\mu)}Z_d^{(\mu-1)}(s) \vdots - \mathrm{B}^{(\mu)}Z_r^{(\mu-1)}(s) \vdots I \right],$$

$$\text{(4.24)}$$

is the $EISO$ system (4.1), (4.2) *input to state (IS) full transfer function matrix*, the inverse Laplace transform of which is the plant IS *full fundamental matrix* $\Psi_{EISOIS}(t)$,

$$\Psi_{EISOIS}(t) = \mathcal{L}^{-1}\{F_{EISOIS}(s)\}, \tag{4.25}$$

- $G_{EISOISD}(s)$,

$$G_{EISOISD}(s) = C(sI - A)^{-1} D^{(\mu)} S_d^{(\mu)}(s) \tag{4.26}$$

is the $EISO$ plant (4.1), (4.2) *transfer function matrix relating the state to the disturbance* $D^{(\mu)}$,
   - $G_{EISOISU}(s)$,

$$G_{EISOISU}(s) = C(sI - A)^{-1} B^{(\mu)} S_r^{(\mu)}(s) \tag{4.27}$$

is the $EISO$ plant (4.1), (4.2) *transfer function matrix relating the state to the control* $\mathbf{U}$,
   - $G_{EISOISD_0}(s)$,

$$G_{EISOISD_0}(s) = -C(sI - A)^{-1} D^{(\mu)} Z_d^{(\mu-1)}(s) \tag{4.28}$$

is the $EISO$ plant (4.1), (4.2) *transfer function matrix relating the state to the initial disturbance* $D^{(\mu)}$,
   - $G_{EISOISU_0}(s)$,

$$G_{EISOISU_0}(s) = -C(sI - A)^{-1} B^{(\mu)} Z_r^{(\mu-1)}(s) \tag{4.29}$$

is the $EISO$ plant (4.1), (4.2) *transfer function matrix relating the state to the initial control* $\mathbf{U}_0^{(\mu)}$,
   - $G_{EISOISX_0}(s)$,

$$G_{EISOISX_0}(s) = C(sI - A)^{-1} \tag{4.30}$$

is the $EISO$ plant (4.1), (4.2) *transfer function matrix relating the state to the initial state* $\mathbf{X}_0$,
   - $\mathbf{V}_{EISO}(s)$ and $\mathbf{C}_{EISO0}$,

$$\mathbf{V}_{EISO}(s) = \begin{bmatrix} \mathbf{I}_{EISO}(s) \\ \mathbf{C}_{EISO0} \end{bmatrix}, \quad \mathbf{I}(s) = \mathbf{I}_{EISO}(s) = \begin{bmatrix} \mathbf{D}(s) \\ \mathbf{U}(s) \end{bmatrix},$$

$$\mathbf{C}_{EISO0} = \begin{bmatrix} \left(\mathbf{D}_0^{\mu-1}\right)^T & \left(\mathbf{U}_0^{\mu-1}\right)^T & \mathbf{X}_0^T \end{bmatrix}^T, \tag{4.31}$$

are the Laplace transform of the action vector $\mathbf{V}_{EISO}(t)$ and the vector $\mathbf{C}_{EISO0}$ of all initial conditions, respectively,
 - $F_{EISO}(s)$

$$
F_{EISO}(s) = \begin{bmatrix}
C(sI - A)^{-1} D^{(\mu)} S_d^{(\mu)}(s) + V \;\vdots \\
\vdots \; C(sI - A)^{-1} B^{(\mu)} S_r^{(\mu)}(s) + U \;\vdots \\
\vdots \; -C(sI - A)^{-1} D^{(\mu)} Z_d^{(\mu-1)}(s) \;\vdots \\
\vdots \; -C(sI - A)^{-1} B^{(\mu)} Z_r^{(\mu-1)}(s) \;\vdots \\
\vdots \; C(sI - A)^{-1}
\end{bmatrix}, \tag{4.32}
$$

is the *EISO* plant (4.1), (4.2) *input to output (IO) full transfer function matrix* $F_{EISO}(s)$ *relative to the input pair* $[D^{(\mu)}(t),\ \mathbf{U}(t)]$ *and the initial vectors* $D^{(\mu)}{}_0^{\mu-1}$, $\mathbf{U}_0^{\mu-1}$ *and* $\mathbf{X}_0$.

The inverse Laplace transform of $F_{EISO}(s)$ is the plant *IO full fundamental matrix* $\Psi_{EISO}(t)$ [170],

$$
\Psi_{EISO}(t) = \mathcal{L}^{-1}\{F_{EISO}(s)\}, \tag{4.33}
$$

 - $p_{EISO}(s)$, for short $p(s)$,

$$
p(s) = p_{EISO}(s) = \det(sI - A), \tag{4.34}
$$

is the characteristic polynomial of the *EISO* plant (4.1), (4.2) and the denominator polynomial of all its transfer function matrices,
 - $G_{EISOD}(s)$, for short $G_{D^{(\mu)}}(s)$,

$$
G_{D^{(\mu)}}(s) = G_{EISOD}(s) = C(sI - A)^{-1} D^{(\mu)} S_d^{(\mu)}(s) + V =
$$
$$
= p^{-1}(s) \underbrace{\left[Cadj(sI - A)^{-1} D^{(\mu)} S_d^{(\mu)}(s) + p(s) V\right]}_{L_{D^{(\mu)}}(s)}, \tag{4.35}
$$

is the *EISO* plant (4.1), (4.2) *transfer function matrix relating the output to the disturbance* $D^{(\mu)}$,
 - $G_{EISOU}(s)$, for short $G_U(s)$,

$$
G_U(s) = G_{EISOU}(s) = C(sI - A)^{-1} B^{(\mu)} S_r^{(\mu)}(s) + U =
$$
$$
= p^{-1}(s) \underbrace{\left[Cadj(sI - A) B^{(\mu)} S_r^{(\mu)}(s) + p(s) U\right]}_{L_U(s)} = \frac{L_U(s)}{p(s)} \in \mathbb{C}^{N \times r}, \tag{4.36}
$$

is the *EISO* plant (4.1), (4.2) *transfer function matrix relating the output to the control* **U**, *and* $L_{UEISO}(s)$, *for short* $L_U(s)$, *is the numerator matrix polynomial of* $G_U(s)$,

$$L_U(s) = L_{UEISO}(s) = C adj(sI - A) B^{(\mu)} S_r^{(\mu)}(s) + p(s) U =$$

$$= \sum_{i=0}^{i=n+\mu} L_{Ui} s^i = L_U S_N^{(n+\mu)}(s) \in \mathbb{C}^{N \times r}, \; L_U = \left[L_{U0} \vdots \cdots \vdots L_{U(n+\mu)} \right],$$

$$(4.37)$$

which obeys

$$L_U(s) = C adj(sI_n - A) B^{(\mu)} S_r^{(\mu)}(s) + p(s) U =$$

$$= \left\{ \begin{array}{c} \sum_{i=0}^{i=q} L_i s^i = L^{(q)} S_r^{(q)}(s), \; L_i \in \mathfrak{R}^{N \times r}, \\[2mm] q = n - 1 + \mu, \\ \forall i = 0, 1, ..., q, \Longleftrightarrow U \neq O_{N,r} \\[2mm] \sum_{i=0}^{i=q-1} L_i s^i = L^{(q-1)} S_r^{(q-1)}(s), \; L_i \in \mathfrak{R}^{N \times r}, \\[2mm] \forall i = 0, 1, ..., q - 1, \Longleftrightarrow U = O_{N,r}, \end{array} \right\} \quad (4.38)$$

where

$$L^{(q)} = \left[L_0 \vdots L_1 \vdots ... \vdots L_q \right] \in \mathfrak{R}^{N \times (q+1)r} \iff U \neq O_{N,r}, \quad (4.39)$$

$$L^{(q-1)} = \left[L_0 \vdots L_1 \vdots ... \vdots L_{q-1} \right] \in \mathfrak{R}^{N \times \kappa r} \iff U = O_{N,r} \quad (4.40)$$

and

$$L_U = \left\{ \begin{array}{c} L^{(q)} \Longleftrightarrow U \neq O_{N,r}, \\ L^{(q-1)} \Longleftrightarrow U = O_{N,r} \end{array} \right\} \quad (4.41)$$

- $G_{EISOD_0}(s)$, for short $G_{D_0}(s)$,

$$G_{D_0}(s) = G_{EISOD_0}(s) = -C(sI - A)^{-1} D^{(\mu)} Z_d^{(\mu-1)}(s) =$$

$$= -p^{-1}(s) \left[C adj(sI - A) D^{(\mu)} Z_d^{(\mu-1)}(s) \right], \quad (4.42)$$

is the *EISO* plant (4.1), (4.2) *transfer function matrix relating the output to the initial extended disturbance vector* $D^{(\mu)\mu-1}_0$,

- $G_{EISOU_0}(s)$, for short $G_{U_0}(s)$,

$$G_{U_0}(s) = G_{EISOU_0}(s) = -C(sI - A)^{-1} B^{(\mu)} Z_r^{(\mu-1)}(s) =$$
$$= -p^{-1}(s) \left[C\,adj\,(sI - A)\,B^{(\mu)} Z_r^{(\mu-1)}(s) \right], \tag{4.43}$$

is the *EISO* plant (4.1), (4.2) *transfer function matrix relating the output to the extended initial control* $\mathbf{U}_0^{\mu-1}$,

- $G_{EISOX_0}(s)$, for short $G_{X_0}(s)$,

$$G_{EISOX_0}(s) = G_{X_0}(s) = C(sI - A)^{-1} =$$
$$= p^{-1}(s)\,C\,adj\,(sI - A), \tag{4.44}$$

is the *EISO* plant (4.1), (4.2) *transfer function matrix relating the output to the initial state* $\mathbf{X}_0$.

If we consider the whole extended vectors $D^\mu(t)$ and $\mathbf{U}^\mu(t)$ as the system input vectors then other forms of the system transfer function matrices result. To show that let

$$V^{(\mu)} = \left[V \vdots O_{N,d} \vdots \cdots \vdots O_{N,d} \right] \in \Re^{N\times(\mu+1)d}, \tag{4.45}$$

$$U^{(\mu)} = \left[U \vdots O_{N,r} \vdots \cdots \vdots O_{N,r} \right] \in \Re^{N\times(\mu+1)r}. \tag{4.46}$$

The Laplace transform of the *EISO* system (4.1), (4.2) relative to the extended vectors $D^{(\mu)\mu}(t)$ and $\mathbf{U}^\mu(t)$ reads:

$$s\mathbf{X}(s) - \mathbf{X}_0 = A\mathbf{X}(s) + D^{(\mu)}\mathcal{L}\{\mathbf{D}^\mu(t)\} + B^{(\mu)}\mathcal{L}\{\mathbf{U}^\mu(t)\},$$
$$\mathbf{Y}(s) = C\mathbf{X}(s) + V^{(\mu)}\mathcal{L}\{\mathbf{D}^\mu(t)\} + U^{(\mu)}\mathcal{L}\{\mathbf{U}^\mu(t)\},$$

so that the Laplace transform $\mathbf{Y}(s)$ of the system output vector $\mathbf{Y}(t)$ can be set also in the following form:

$$\mathbf{Y}(s) = \left(\begin{bmatrix} \left(C(sI-A)^{-1}D^{(\mu)} + V^{(\mu)}\right) \vdots \\ \vdots \left(C(sI-A)^{-1}B^{(\mu)} + U^{(\mu)}\right) \vdots \\ \vdots C(sI-A)^{-1} \end{bmatrix} \bullet \right.$$
$$\left. \bullet \left[\mathcal{L}\{\mathbf{D}^\mu(t)\}^T \vdots \mathcal{L}\{\mathbf{U}^\mu(t)\}^T \vdots \mathbf{X}_0^T \right]^T \right) =$$
$$= F_{EISOU^\mu}(s)\,\mathbf{V}_{EISOU^\mu}(s), \tag{4.47}$$

where:

- $F_{EISOU^\mu}(s)$,

$$F_{EISO^\mu}(s) = \begin{bmatrix} C(sI-A)^{-1}D^{(\mu)}+V^{(\mu)} & \vdots \\ \vdots & C(sI-A)^{-1}B^{(\mu)}+U^{(\mu)} & \vdots \\ & \vdots & C(sI-A)^{-1} \end{bmatrix}, \qquad (4.48)$$

is the *EISO* plant (4.1), (4.2) *input to output (IO) full transfer function matrix $F_{EISO^\mu}(s)$ relative to the extended input pair $[D^{(\mu)\mu}(t), \mathbf{U}^\mu(t)]$ and the initial state vector $\mathbf{X}_0$,*

- $G_{EISOD^\mu}(s)$, for short $G_{D^{(\mu)}\mu}(s)$,

$$G_{D^{(\mu)}\mu}(s) = G_{EISOD^\mu}(s) = C(sI-A)^{-1}D^{(\mu)}+V^{(\mu)} =$$
$$= p^{-1}(s)\underbrace{\left[Cadj(sI-A)^{-1}D^{(\mu)}+p(s)V^{(\mu)}\right]}_{N_{D^{(\mu)}}(s)} \in \mathbb{C}^{N\times d}, \qquad (4.49)$$

is the *EISO* plant (4.1), (4.2) *transfer function matrix relating the output to the extended disturbance $D^{(\mu)\mu}$,*

- and $G_{EISOU^\mu}(s)$, for short $G_{U^\mu}(s)$,

$$G_{U^\mu}(s) = G_{EISOU^\mu}(s) = C(sI-A)^{-1}B^{(\mu)}+U^{(\mu)} = \qquad (4.50)$$
$$= p^{-1}(s)\underbrace{\left[Cadj(sI-A)B^{(\mu)}+p(s)U^{(\mu)}\right]}_{N_U(s)} \Longrightarrow N_U(s) \in \mathbb{C}^{N\times(\mu+1)r} \Longrightarrow$$

$$(4.51)$$

$$G_{U^\mu}(s) = \frac{N_U(s)}{p(s)} \in \mathbb{C}^{N\times r}, \quad S_r^{(\mu)}(s) \in \mathbb{C}^{(\mu+1)r\times r}, \qquad (4.52)$$

$$G_U(s) = G_{U^\mu}(s)S_r^{(\mu)}(s), \qquad (4.53)$$

is the *EISO* plant (4.1), (4.2) *transfer function matrix $G_{EISOU^\mu}(s)$, for short $G_{U^\mu}(s)$, relating the output to the extended control $\mathbf{U}^\mu$,* and the polynomial matrix $L_{EISOU^\mu}(s)$, for short $L_{U^\mu}(s)$, is the numerator matrix polynomial of $G_{U^\mu}(s)$, which obeys Equation 4.51 in which

$$N_U(s) = C \, adj\,(sI - A)\, B^{(\mu)} + p(s)\, U^{(\mu)} =$$

$$= \left\{ \begin{array}{c} \displaystyle\sum_{i=0}^{i=n} N_{Ui} s^i = N_U^{(n)} S_{(\mu+1)r}^{(n)}(s), \ N_{Ui} \in \mathfrak{R}^{N \times (\mu+1)r}, \\[4mm] \forall i = 0, 1, ..., n, \Longleftrightarrow U \ne O_{N,r} \\[4mm] \displaystyle\sum_{i=0}^{i=n-1} N_{Ui} s^i = N_U^{(n-1)} S_{(\mu+1)r}^{(n-1)}(s), \ N_{Ui} \in \mathfrak{R}^{N \times (\mu+1)r}, \\[4mm] \forall i = 0, 1, ..., n-1, \Longleftrightarrow U = O_{N,r}, \end{array} \right\},$$

$$S_{(\mu+1)r}^{(n)}(s) \in \mathbb{C}^{(n+1)(\mu+1)r \times (\mu+1)r}, \ \ S_{(\mu+1)r}^{(n-1)}(s) \in \mathbb{C}^{n(\mu+1)r \times (\mu+1)r}, \quad (4.54)$$

where

$$N_U^{(n)} = \left[N_{U0} \vdots N_{U1} \vdots ... \vdots N_{Un} \right] \in \mathfrak{R}^{N \times (n+1)(\mu+1)r} \iff U \ne O_{N,r}, \quad (4.55)$$

$$N_U^{(n-1)} = \left[N_{U0} \vdots N_{U1} \vdots ... \vdots N_{U,n-1} \right] \in \mathfrak{R}^{N \times n(\mu+1)r} \iff U = O_{N,r}, \quad (4.56)$$

$$N_U = \left\{ \begin{array}{c} N_U^{(n)} \iff U \ne O_{N,r}, \\[2mm] N_U^{(n-1)} \iff U = O_{N,r}. \end{array} \right\}, \quad (4.57)$$

Theorem 73 *Properties of the EISO system (4.1), (4.2), (4.4)–(4.11)*

The EISO system (4.1), (4.2), (4.4)–(4.11) possesses the following properties:

I)   a) If $\nu = 1$ and $0 \le \mu \le 1$, the matrix $\left[(sI - A) \vdots P_{inv} \right]$ has the full rank n for every complex number s and for every matrix A (4.6) including every eigenvalue $s_i(A)$ of the matrix A (4.6), and for every matrix A (4.6):

$$rank \left[(sI - A) \vdots P_{inv} \right] = n, \ \forall (s, A) \in \mathbb{C} \times \mathfrak{R}^{n \times n}. \quad (4.58)$$

The rank n of the matrix $\left[(sI - A) \vdots P^{(\mu)} \right]$ is invariant and full relative to every $(s, A) \in \mathbb{C} \times \mathfrak{R}^{n \times n}$, A given by (4.6).

b) If $\nu = 1$ and $0 \le \mu \le 1$ then for the matrix $\left[(sI - A) \vdots P^{(\mu)} \right]$ to have the full rank n for every complex number s including every eigenvalue

$s_i(A)$ of the matrix A, and for every matrix A, it is necessary and sufficient that the matrix $H^{(1)}$ has the full rank $n = N$:

$$rank H^{(1)} = n = N. \tag{4.59}$$

II)   *a) If $\nu > 1$ and $0 \le \mu < \infty$, then the matrix $\left[(sI - A) \vdots P_{inv}\right]$ has the full rank n for every complex number s including every eigenvalue $s_i(A)$ of the matrix A and for every matrix $A \in \mathfrak{R}^{n \times n}$:*

$$rank \left[(sI - A) \vdots P_{inv}\right] = n, \ \forall (s, A) \in \mathbb{C} \times \mathfrak{R}^{n \times n}. \tag{4.60}$$

The rank of the matrix $\left[(sI - A) \vdots P_{inv}\right]$ is invariant and full relative to every $(s, A) \in \mathbb{C} \times \mathfrak{R}^{n \times n}$.

     b) If $\nu > 1$ and $0 \le \mu < \infty$, then for the matrix $\left[(sI - A) \vdots P^{(\mu)}\right]$ to have the full rank n for every complex number s including every eigenvalue $s_i(A)$ of the matrix A and for every matrix $A \in \mathfrak{R}^{n \times n}$:

$$rank \left[(sI - A) \vdots P^{(\mu)}\right] = n, \ \forall (s, A) \in \mathbb{C} \times \mathfrak{R}^{n \times n}. \tag{4.61}$$

it is necessary and sufficient that the extended matrix $H^{(\mu)}$ has the full rank N,

$$rank H^{(\mu)} = N. \tag{4.62}$$

This theorem is proved in Appendix D.2.

Comment 74 *If $\det A_\nu \ne 0$ then the statements of this theorem largely simplify the verification of the rank of the complex valued matrices*

$$\left[(sI - A) \vdots P_{inv}\right] \ and \ \left[(sI - A) \vdots P^{(\mu)}\right].$$

While the rank of the matrix $\left[(sI - A) \vdots P_{inv}\right]$ is invariantly equal to n, the rank of the matrix $\left[(sI - A) \vdots P^{(\mu)}\right]$ is not invariant.

However, if the rank of the matrix $H^{(\mu)}$ is full, $rank H^{(\mu)} = N$, then the rank of the matrix $\left[(sI - A) \vdots P^{(\mu)}\right]$ is independent of both A and $s \in \mathbb{C}$, i.e., it is also invariant.

For an example of the *EISO* system (4.1), (4.2) see the book [171].

4.2 EISO plant desired regime

Definition 60, (Section 3.2), slightly changes its formulation as follows.

Definition 75 *A functional vector control-state pair* $[\mathbf{U}^*(.), \mathbf{X}^*(.)]$ *is* ***nominal*** *for the EISO plant (4.1), (4.2)* ***relative to the functional pair***

$$[\mathbf{D}(.), \mathbf{Y}_d(.)],$$

which is denoted by $[\mathbf{U}_N(.), \mathbf{X}_N(.)]$, *if and only if* $[\mathbf{U}(.), \mathbf{X}(.)] = [\mathbf{U}^*(.), \mathbf{X}^*(.)]$ *ensures that the corresponding real response* $\mathbf{Y}(.) = \mathbf{Y}^*(.)$ *of the plant obeys* $\mathbf{Y}^*(t) = \mathbf{Y}_d(t)$ *all the time,*

$$[\mathbf{U}^*(.), \mathbf{X}^*(.)] = [\mathbf{U}_N(.), \mathbf{X}_N(.)] \Longleftrightarrow \langle \mathbf{Y}^*(t) = \mathbf{Y}_d(t), \ \forall t \in \mathfrak{T}_0 \rangle.$$

The nominal motion $\mathbf{X}_N(.; \mathbf{X}_{N0}; D^\mu; \mathbf{U}_N^\mu)$, $\mathbf{X}_N(0; \mathbf{X}_{N0}; D^\mu; \mathbf{U}_N^\mu) \equiv \mathbf{X}_{N0}$, *is the desired motion* $\mathbf{X}_d(.; \mathbf{X}_{d0}; D^\mu; \mathbf{U}_N^\mu)$ *of the EISO plant (4.1), (4.2) relative to the functional vector pair* $[\mathbf{D}(.), \mathbf{Y}_d(.)]$, *for short:* ***the desired motion of the system,***

$$\mathbf{X}_d(.; \mathbf{X}_{d0}; \mathbf{D}^\mu; \mathbf{U}_N^\mu) \equiv \mathbf{X}_N(.; \mathbf{X}_{N0}; \mathbf{D}^\mu; \mathbf{U}_N^\mu),$$
$$\mathbf{X}_d(0; \mathbf{X}_{d0}; \mathbf{D}^\mu; \mathbf{U}_N^\mu) \equiv \mathbf{X}_{d0} \equiv \mathbf{X}_{N0}. \tag{4.63}$$

Definition 75 and the system description (4.1), (4.2) imply the following theorem:

Theorem 76 *In order for the functional vector pair* $[\mathbf{U}^*(.), \mathbf{X}^*(.)]$ *to be nominal for the EISO plant (4.1), (4.2) relative to the functional vector pair* $[\mathbf{D}^{(\mu)}(.), \mathbf{Y}_d(.)]$, $[\mathbf{U}^*(.), \mathbf{X}^*(.)] = [\mathbf{U}_N(.), \mathbf{X}_N(.)]$, *it is necessary and sufficient that it obeys the following equations:*

$$-\mathbf{B}^{(\mu)}\mathbf{U}^{*\mu}(t) + \frac{d\mathbf{X}^*(t)}{dt} - A\mathbf{X}^*(t) = D^{(\mu)}\mathbf{D}^\mu(t), \ \forall t \in \mathfrak{T}_0, \tag{4.64}$$

$$U\mathbf{U}^*(t) + C\mathbf{X}^*(t) = \mathbf{Y}_d(t) - V\mathbf{D}(t), \ \forall t \in \mathfrak{T}_0, \tag{4.65}$$

or equivalently,

$$\begin{bmatrix} -\mathbf{B}^{(\mu)}S_r^{(\mu)}(s) & sI - A \\ U & C \end{bmatrix} \begin{bmatrix} \mathbf{U}^*(s) \\ \mathbf{X}^*(s) \end{bmatrix} =$$

$$= \begin{bmatrix} \mathbf{X}_0^* + \mathbf{B}^{(\mu)}Z_r^{(\mu-1)}(s)\mathbf{U}_0^{\mu-1} + D^{(\mu)} \begin{Bmatrix} S_d^{(\mu)}(s)\mathbf{D}(s) - \\ -Z_d^{(\mu-1)}(s)\mathbf{D}_0^{\mu-1} \end{Bmatrix} \\ \mathbf{Y}_d(s) - V\mathbf{D}(s) \end{bmatrix}. \tag{4.66}$$

This theorem opens the problem of the conditions for the existence of the solutions of the equations (4.64), (4.65), or equivalently of (4.66). There are $(r + n)$ unknown variables $\mathbf{U}^*(s) \in \mathbb{C}^r$ and $\mathbf{X}^*(s) \in \mathbb{C}^n$ and $(N + n)$ equations so that the following holds:

Claim 77 *In order to exist a nominal functional vector pair*

$$[\mathbf{U}_N(.), \mathbf{X}_N(.)]$$

for the EISO system (4.1), (4.2) relative to the functional vector pair

$$[\mathbf{D}(.), \mathbf{Y}_d(.)]$$

it is necessary and sufficient that $N \leq r$.

The proof of this claim is analogous to the proof of Claim 62 (Section 3.2). The condition emphasizes the significance of Fundamental control principle 104, Section 7.1.

Claim 77 provides the full solution to the problem of the existence of a nominal functional vector pair $[\mathbf{U}_N(.), \mathbf{X}_N(.)]$ for the *EISO* plant (4.1), (4.2) relative to the functional vector pair $[\mathbf{D}(.), \mathbf{Y}_d(.)]$.

Condition 78 *The desired output response of the EISO system (4.1), (4.2) is realizable, i.e., $N \leq r$. The nominal control-state pair $[\mathbf{U}_N(.), \mathbf{X}_N(.)]$ is known.*

The *EISO* plant description in terms of the deviations (3.37), (2.53), (2.56) and (2.55) reads:

$$\frac{d\mathbf{x}(t)}{dt} = A\mathbf{x}(t) + D^{(\mu)}\mathbf{d}^{\mu}(t) + B^{(\mu)}\mathbf{u}^{\mu}(t), \ \forall t \in \mathfrak{T}_0, \qquad (4.67)$$

$$\mathbf{y}(t) = C\mathbf{x}(t) + V\mathbf{d}(t) + U\mathbf{u}(t), \ \forall t \in \mathfrak{T}_0. \qquad (4.68)$$

4.3   *EISO* feedback controller

4.3.1   Time domain

When we set

$$\mathbf{I} = \mathbf{I}_{Cf} = \begin{bmatrix} \mathbf{Y}_d \\ \mathbf{Y} \end{bmatrix} \in \mathfrak{R}^{2N}, \ M = 2N,$$

$$P^{(\mu)} = \begin{bmatrix} J \ \vdots \ -J \end{bmatrix} \in \mathfrak{R}^{n \times 2N}, \ Q = \begin{bmatrix} W \ \vdots \ -W \end{bmatrix} \in \mathfrak{R}^{r \times 2N},$$

into (4.1), (4.2) the result is the *EISO* controller description in terms of the total coordinates:

$$\frac{d\mathbf{X}_C(t)}{dt} = A_C \mathbf{X}_C(t) + J \left[\mathbf{Y}_d^{(\mu)}(t) - \mathbf{Y}^{(\mu)}(t) \right], \ \forall t \in \mathfrak{T}_0, \tag{4.69}$$

$$\mathbf{U}(t) = C\mathbf{X}_C(t) + W \left[\mathbf{Y}_d(t) - \mathbf{Y}(t) \right], \ \forall t \in \mathfrak{T}_0. \tag{4.70}$$

The *EISO* controller description in terms of the deviations (3.37), (2.53), (1.53) (Section 1.6), (2.56), and (2.55) reads:

$$\frac{d\mathbf{x}_C(t)}{dt} = A_C \mathbf{x}_C(t) + J\mathbf{e}^{(\mu)}(t), \ \forall t \in \mathfrak{T}_0, \tag{4.71}$$

$$\mathbf{u}(t) = C\mathbf{x}_C(t) + W\mathbf{e}(t), \ \forall t \in \mathfrak{T}_0. \tag{4.72}$$

4.3.2 Complex domain

The application of the Laplace transform to the *EISO* controller (4.69), (4.70), i.e., (4.71), (4.72), gives its complex domain description:

$$s\mathbf{X}_C(s) - \mathbf{X}_{C0} = A_C \mathbf{X}_C(s) + J^{(\mu)} S_N^{(\mu)}(s) \left[\mathbf{Y}_d(s) - \mathbf{Y}(s) \right] -$$

$$- J^{(\mu)} Z_N^{(\mu-1)}(s) \left[\mathbf{Y}_{d0}^{(\mu-1)} - \mathbf{Y}_0^{(\mu-1)} \right], \tag{4.73}$$

$$\mathbf{U}(s) = C\mathbf{X}_C(s) + W \left[\mathbf{Y}_d(s) - \mathbf{Y}(s) \right]. \tag{4.74}$$

We determine first $\mathbf{X}_C(s)$ from the first equation, and then replace the solution into the second equation to get the result for $\mathbf{U}(s)$, which we set into the compact form by using (1.53):

$$\mathbf{X}_C(s) = (sI - A_C)^{-1} \left[J^{(\mu)} S_N^{(\mu)}(s) \ \vdots \ - J^{(\mu)} Z_N^{(\mu-1)}(s) \ \vdots \ I \right] \begin{bmatrix} \mathbf{E}(s) \\ \mathbf{e}_0^{\mu-1} \\ \mathbf{X}_{C0} \end{bmatrix} =$$

$$= F_{EISOC_f IS}(s) \, \mathbf{W}_{EISOC_f}(s), \tag{4.75}$$

where:

- $F_{EISOC_f IS}(s)$,

$$F_{EISOC_f IS}(s) = (sI - A_C)^{-1} \left[J^{(\mu)} S_N^{(\mu)}(s) \ \vdots \ - J^{(\mu)} Z_N^{(\mu-1)}(s) \ \vdots \ I \right], \tag{4.76}$$

is the *EISO* controller (4.69), (4.69) *input to state* (*IS*) *full transfer function matrix*, the inverse Laplace transform of which is the controller *IS full fundamental matrix* $\Psi_{EISOC_f IS}(t)$ [170],

$$\Psi_{EISOC_f IS}(t) = \mathcal{L}^{-1} \left\{ F_{EISOC_f IS}(s) \right\}, \tag{4.77}$$

- $G_{EISOCIS}(s)$,

$$G_{EISOCIS}(s) = C(sI - A_C)^{-1} J^{(\mu)} S_N^{(\mu)}(s) \qquad (4.78)$$

is the *EISO* controller (3.40) and (3.41) *transfer function matrix relating the state to the input* **e**,
- $G_{EISOCE_0}(s)$,

$$G_{EISOCE_0}(s) = -C(sI - A_C)^{-1} J^{(\mu)} Z_N^{(\mu-1)}(s), \qquad (4.79)$$

is the *EISO* controller (3.40) and (3.41) *transfer function matrix relating the state to the initial error*,
- $G_{EISOCX_0}(s)$,

$$G_{EISOCX_{C0}}(s) = C(sI - A_C)^{-1}, \qquad (4.80)$$

is the *EISO* controller (3.40) and (3.41) *transfer function matrix relating the state to the initial state* $\mathbf{X}_{C0}$,
- $V_{EISOCIS}(s)$ and $\mathbf{C}_{EISOCIS}$,

$$\mathbf{V}_{EISOCIS}(s) = \begin{bmatrix} \mathbf{I}_{C_f}(s) \\ \mathbf{C}_{EISOCIS0} \end{bmatrix}, \quad \mathbf{I}_{C_f}(s) = \mathbf{E}(s), \quad \mathbf{C}_{EISOCIS0} = \mathbf{X}_{C0},$$
$$(4.81)$$

are the Laplace transform of the action vector $V_{EISOCIS}(t)$ and the vector $\mathbf{C}_{EISOCIS0}$ of all initial conditions acting on the state, respectively,

Furthermore, the equations (4.75), i.e., (4.73), and (4.74) furnish

$$\mathbf{U}(s) = \left\{ C(sI - A_C)^{-1} J^{(\mu)} S_N^{(\mu)}(s) + W \right\} \underbrace{\left[\mathbf{Y}_d(s) - \mathbf{Y}(s) \right]}_{\mathbf{e}(s)} -$$

$$-C(sI - A_C)^{-1} \left(J^{(\mu)} Z_N^{(\mu-1)}(s) \underbrace{\left[\mathbf{Y}_{d0}^{(\mu-1)} - \mathbf{Y}_0^{(\mu-1)} \right]}_{\mathbf{e}_0^{\mu-1}} - \mathbf{X}_{C0} \right), \qquad (4.82)$$

i.e.,

$$\mathbf{U}(s) = \begin{bmatrix} C(sI - A_C)^{-1} J^{(\mu)} S_N^{(\mu)}(s) + W & \vdots \\ \vdots \; -C(sI - A_C)^{-1} J^{(\mu)} Z_N^{(\mu-1)}(s) & \vdots \; C(sI - A_C)^{-1} \end{bmatrix} \bullet$$

$$\bullet \begin{bmatrix} \mathbf{E}^T(s) & \left(\mathbf{e}_0^{\mu-1} \right)^T & \mathbf{X}_{C0}^T \end{bmatrix}^T = F_{EISOC}(s) \mathbf{V}_{EISOC}(s), \qquad (4.83)$$

which implies:
- $F_{EISOC}(s)$,

$$F_{EISOC}(s) = \begin{bmatrix} C(sI - A_C)^{-1} J^{(\mu)} S_N^{(\mu)}(s) + W & \vdots \\ \vdots - C(sI - A_C)^{-1} J^{(\mu)} Z_N^{(\mu-1)}(s) & \vdots & C(sI - A_C)^{-1} \end{bmatrix},$$
(4.84)

is the *EISO* controller (4.69) and (4.70) *input to output (IO) full transfer function matrix*, the inverse Laplace transform of which is the system *IO full fundamental matrix* $\Psi_{EISOC}(t)$ [170],

$$\Psi_{EISOC}(t) = \mathcal{L}^{-1}\{F_{EISOC}(s)\},$$
(4.85)

- $p_{cEISO}(s)$,

$$p_{cEISO}(s) = \det(sI - A_C),$$
(4.86)

is the characteristic polynomial of the *EISO* controller (4.69), (4.70) and the denominator polynomial of all its transfer function matrices,
- $L_{cEISO}(s)$,

$$L_{cEISO}(s) = C adj(sI - A_C) J + p_{cEISO}(s) V,$$
(4.87)

is the numerator matrix polynomial of $G_{EISOCE}(s)$,
- $G_{EISOCE}(s)$,

$$G_{EISOCE}(s) = C(sI - A_C)^{-1} J^{(\mu)} S_N^{(\mu)}(s) + W =$$
$$= p_{cEISO}^{-1}(s)\left[C adj(sI - A_C) J^{(\mu)} S_N^{(\mu)}(s) + p_{cEISO}(s) W\right] = \frac{L_{cEISO}(s)}{p_{cEISO}(s)},$$
(4.88)

is the *EISO* controller (4.69) and (4.70) *transfer function matrix relating the output* **U** *to the input* **e**,
- $G_{EISOCE_0}(s)$,

$$G_{EISOCE_0}(s) = -C(sI - A_C)^{-1} J^{(\mu)} Z_N^{(\mu-1)}(s) =$$
$$= -p_{cEISO}^{-1}(s)\left[C adj(sI - A_C) J^{(\mu)} Z_N^{(\mu-1)}(s)\right],$$
(4.89)

is the *EISO* controller (4.69) and (4.70) *transfer function matrix relating the output* **U** *to the initial error* $\mathbf{e}_0$,
- $G_{EISOCX_0}(s)$,

$$G_{EISOCX_0}(s) = C(sI - A_C)^{-1} = p_{cEISO}^{-1}(s) C adj(sI - A_C),$$
(4.90)

is the *EISO* controller (4.69) and (4.70) *transfer function matrix relating the output* **U** *to the initial state* $\mathbf{X}_{C0}$,

  - $V_{EISOC}(s)$ and $\mathbf{C}_{EISOC}$,

$$\mathbf{V}_{EISOC}(s) = \left[\begin{array}{c} \mathbf{I}_{EISOC_f}(s) \\ \mathbf{C}_{EISOC_f0} \end{array} \right], \ \mathbf{I}_{EISOC_f}(s) = \left[\begin{array}{c} \mathbf{Y}_d(s) \\ \mathbf{Y}(s) \end{array} \right],$$

$$\mathbf{C}_{EISOC_f0} = \mathbf{X}_{C0}, \tag{4.91}$$

are the Laplace transform of the action vector $V_{EISOC}(t)$ and the vector $\mathbf{C}_{EISOC_f0}$ of all initial conditions acting on the output, respectively.

4.4   Exercises

Exercise 79 *1. Select a physical EISO plant.*

2. Determine its time domain EISO mathematical model.

3. Determine its complex domain EISO mathematical model: its full transfer function matrix and all its transfer function matrices, as well as the vectors $V_{EISO}(s)$ *and* $\mathbf{C}_{EISO0}$.

Exercise 80 *1. Select a EISO controller.*

2. Determine its time domain EISO mathematical model.

3. Determine its complex domain EISO mathematical model: its full transfer function matrix and all its transfer function matrices, as well as the vectors $V_{EISOC}(s)$ *and* $\mathbf{C}_{EISOC0}$.

Exercise 81 *1. Determine the time domain EISO mathematical model of the control system composed of the chosen EISO plant and EISO controller.*

2. Determine its complex domain EISO mathematical model of the control system composed of the chosen EISO plant and EISO controller: its full transfer function matrix and all its transfer function matrices, as well as the vectors $V_{EISOCS}(s)$ *and* $\mathbf{C}_{EISOCS0}$. *Hint: Section 1.7 and Section 1.8.1.*

Exercise 82 *Test all Lyapunov and BI stability properties of the chosen EISO plant, of the chosen EISO controller and of the control system composed of them. Hint: [170, Part III.]*

Chapter 5

HISO systems

5.1 *HISO* system mathematical model

5.1.1 Time domain

The linear **Higher order Input-State-Output** (*HISO*) **(dynamical,
control) systems** have not been studied so far. Their mathematical models
contain *the α-th* order *linear differential state vector equation* (5.1) and *the
linear algebraic output vector equation* (5.2),

$$A^{(\alpha)}\mathbf{R}^{\alpha}(t) = D^{(\mu)}\mathbf{D}^{\mu}(t) + B^{(\mu)}\mathbf{U}^{\mu}(t) = H^{(\mu)}\mathbf{I}^{\mu}(t), \ \forall t \in \mathfrak{T}_0,$$

$$A^{(\alpha)} \in \mathfrak{R}^{\rho\times(\alpha+1)\rho}, \ \mathbf{R}^{\alpha} \in \mathfrak{R}^{(\alpha+1)\rho}, \ D^{(\mu)}\in\mathfrak{R}^{\rho\times(\mu+1)d}, \ B^{(\mu)}\in\mathfrak{R}^{\rho\times(\mu+1)r},$$

$$H^{(\mu)} = \left[D^{(\mu)}\vdots B^{(\mu)}\right] \in\mathfrak{R}^{\rho\times(\mu+1)(d+r)}, \ \mathbf{I}^{\mu} = \left[(\mathbf{D}^{\mu})^{T}\vdots(\mathbf{U}^{\mu})^{T}\right]^{T}\in\mathfrak{R}^{(\mu+1)(d+r)},$$

$$\tag{5.1}$$

$$\mathbf{Y}(t) = R^{(\alpha)}\mathbf{R}^{\alpha}(t) + V\mathbf{D}(t) + U\mathbf{U}(t) = R^{(\alpha)}\mathbf{R}^{\alpha}(t) + Q\mathbf{I}(t), \ \forall t \in \mathfrak{T}_0,$$

$$V \in \mathfrak{R}^{N\times d}, \ U\in\mathfrak{R}^{N\times r}, \ Q = \left[V\vdots U\right]\in\mathfrak{R}^{N\times(d+r)},$$

$$R^{(\alpha)} = \left[R_0\vdots R_1\vdots...\vdots R_{\alpha-1}\vdots O_{N,\rho}\right], \ R_\alpha = O_{N,\rho}. \tag{5.2}$$

The zero matrix value of $R_{y\alpha}$, $R_{y\alpha} = O_{N,\rho}$, ensures that the highest derivative $R^{(\alpha)}$ of the vector $\mathbf{R}$ does not act on the system output vector $\mathbf{Y}$.

The output vector $\mathbf{Y}$ depends only linearly and algebraically on the state
vector $\mathbf{S} = \mathbf{R}^{\alpha-1} \in \mathfrak{R}^{\alpha\rho}$ and on the output vector $\mathbf{I}$. It does not depend on
the vector $R^{(\alpha)}$ because it does not depend on the whole extended vector $\mathbf{R}^{\alpha}$
due to $R_{y\alpha} = O_{N,\rho}$.

Note 83 *The state vector* $\mathbf{S}_{HISO}$ *of the HISO system (5.1), (5.2), is defined in (1.40), (1.42) (Section 1.5) by:*

$$\mathbf{S}_{HISO} = \mathbf{R}^{\alpha-1} = \left[\mathbf{R}^T \,\vdots\, \mathbf{R}^{(1)^T} \,\vdots\, ... \,\vdots\, \mathbf{R}^{(\alpha-1)^T}\right]^T \in \mathfrak{R}^n,\ n = \alpha\rho, \qquad (5.3)$$

This new vector notation $\mathbf{R}^{\alpha-1}$ *has permitted us to define the state of the HISO system (5.1), (5.2), by preserving the physical sense. It enables us to discover in what follows the complex domain criteria for observability, controllability and trackability directly from their definitions. Such criteria possess the complete physical meaning.*

5.1.2 Complex domain

After the application of the Laplace transform to Equations (5.1) and (5.2) they are transformed into:

$$\mathbf{R}(s) = \left(A^{(\alpha)} S_\rho^{(\alpha)}(s)\right)^{-1} \bullet$$

$$\bullet \left[\begin{array}{c} D^{(\mu)} S_d^{(\mu)}(s) \,\vdots\, B^{(\mu)} S_r^{(\mu)}(s) \,\vdots\, -D^{(\mu)} Z_d^{(\mu-1)}(s) \,\vdots\, A^{(\alpha)} Z_\rho^{(\alpha-1)}(s) \,\vdots\, \\ \vdots\, -B^{(\mu)} Z_r^{(\mu-1)}(s) \end{array}\right] \bullet$$

$$\bullet \left[\mathbf{D}^T(s) \,\vdots\, \mathbf{U}^T(s) \,\vdots\, \left(\mathbf{D}_0^{\mu-1}\right)^T \,\vdots\, \left(\mathbf{R}_0^{\alpha-1}\right)^T \,\vdots\, \left(\mathbf{U}_0^{\mu-1}\right)^T\right]^T =$$

$$= F_{HISOIS}(s)\, \mathbf{V}_{HISOIS}(s), \qquad (5.4)$$

$$\mathbf{Y}(s) = F_{HISO}(s)\, \mathbf{V}_{HISO}(s) =$$

$$= \left[\begin{array}{c} \left(R^{(\alpha)} S_\rho^{(\alpha)}(s)\left(A^{(\alpha)} S_\rho^{(\alpha)}(s)\right)^{-1} D^{(\mu)} S_d^{(\mu)}(s) + V\right)^T \,\vdots\, \\ \vdots\, \left(R^{(\alpha)} S_\rho^{(\alpha)}(s)\left(A^{(\alpha)} S_\rho^{(\alpha)}(s)\right)^{-1} B^{(\mu)} S_r^{(\mu)}(s) + U\right)^T \,\vdots\, \\ \vdots\, \left(-R^{(\alpha)} S_\rho^{(\alpha)}(s)\left(A^{(\alpha)} S_\rho^{(\alpha)}(s)\right)^{-1} D^{(\mu)} Z_M^{(\mu-1)}(s)\right)^T \,\vdots\, \\ \vdots\, \left(\begin{array}{c} R^{(\alpha)} S_\rho^{(\alpha)}(s)\left(A^{(\alpha)} S_\rho^{(\alpha)}(s)\right)^{-1} A^{(\alpha)} Z_\rho^{(\alpha-1)}(s) - \\ -R^{(\alpha)} Z_\rho^{(\alpha-1)}(s) \end{array}\right)^T \,\vdots\, \\ \vdots\, \left(-R^{(\alpha)} S_\rho^{(\alpha)}(s)\left(A^{(\alpha)} S_\rho^{(\alpha)}(s)\right)^{-1} B^{(\mu)} Z_r^{(\mu-1)}(s)\right)^T \end{array}\right]^T \bullet$$

$$\bullet \left[\mathbf{D}^T(s) \,\vdots\, \mathbf{U}^T(s) \,\vdots\, \left(\mathbf{D}_0^{\mu-1}\right)^T \,\vdots\, \left(\mathbf{R}_0^{\alpha-1}\right)^T \,\vdots\, \left(\mathbf{U}_0^{\mu-1}\right)^T\right]^T, \qquad (5.5)$$

where:
- $F_{HISOIS}(s)$,

$$F_{HISOIS}(s) = \left(A^{(\alpha)}S_\rho^{(\alpha)}(s)\right)^{-1} \begin{bmatrix} \left[D^{(\mu)}S_d^{(\mu)}(s)\right]^T \ \vdots \\ \vdots \ \left[B^{(\mu)}S_r^{(\mu)}(s)\right]^T \ \vdots \\ \vdots \ \left[-D^{(\mu)}Z_d^{(\mu-1)}(s)\right]^T \ \vdots \\ \vdots \ \left[A^{(\alpha)}Z_\rho^{(\alpha-1)}(s)\right]^T \ \vdots \\ \vdots \ \left[-B^{(\mu)}Z_r^{(\mu-1)}(s)\right]^T \end{bmatrix}^T , \qquad (5.6)$$

is the *HISO* plant (5.1), (5.2) *input to state* (*IS*) *full transfer function matrix*, the inverse Laplace transform of which is the plant *IS full fundamental matrix* $\Psi_{HISOIS}(t)$ [170],

$$\Psi_{HISOIS}(t) = \mathcal{L}^{-1}\{F_{HISOIS}(s)\}, \qquad (5.7)$$

and the inverse Laplace transform of $\left(A^{(\alpha)}S_\rho^{(\alpha)}(s)\right)^{-1}$ is *the HISO system fundamental matrix* $\Phi(t)$,,

$$\Phi(t) = \mathcal{L}^{-1}\left\{\left(A^{(\alpha)}S_\rho^{(\alpha)}(s)\right)^{-1}\right\}, \qquad (5.8)$$

- $G_{HISOISD}(s)$,

$$G_{HISOISD}(s) = \left(A^{(\alpha)}S_\rho^{(\alpha)}(s)\right)^{-1} D^{(\mu)}S_d^{(\mu)}(s), \qquad (5.9)$$

is the *HISO* system (5.1), (5.2) *disturbance to state* (*IS*) *transfer function matrix*,
- $G_{HISOISU}(s)$,

$$G_{HISOISU}(s) = \left(A^{(\alpha)}S_\rho^{(\alpha)}(s)\right)^{-1} B^{(\mu)}S_r^{(\mu)}(s), \qquad (5.10)$$

is the *HISO* system (5.1), (5.2) *control to state* (*IS*) *transfer function matrix*,
- $G_{HISOISD_0}(s)$,

$$G_{HISOISD_0}(s) = -\left(A^{(\alpha)}S_\rho^{(\alpha)}(s)\right)^{-1} D^{(\mu)}Z_d^{(\mu-1)}(s), \qquad (5.11)$$

is the *HISO* system (5.1), (5.2) *initial extended disturbance* $\mathbf{D}_0^{\mu-1}$ *to the state IS transfer function matrix,*

- $G_{HISOISR_0}(s)$,

$$G_{HISOISR_0}(s) = \left(A^{(\alpha)}S_\rho^{(\alpha)}(s)\right)^{-1}A^{(\alpha)}Z_\rho^{(\alpha-1)}(s), \qquad (5.12)$$

is the *HISO* system (5.1), (5.2) *initial state vector* $\mathbf{R}_0^{\alpha-1}$ *to the state IS transfer function matrix,*

- $G_{HISOISU_0}(s)$,

$$G_{HISOISU_0}(s) = -\left(A^{(\alpha)}S_\rho^{(\alpha)}(s)\right)^{-1}B^{(\mu)}Z_r^{(\mu-1)}(s), \qquad (5.13)$$

is the *HISO* system (5.1), (5.2) *initial extended control* $\mathbf{U}_0^{\mu-1}$ *to the state IS transfer function matrix,*

- $F_{HISO}(s)$,

$$F_{HISO}(s) =$$

$$\begin{bmatrix}
\left(R^{(\alpha)}S_\rho^{(\alpha)}(s)\left(A^{(\alpha)}S_\rho^{(\alpha)}(s)\right)^{-1}D^{(\mu)}S_d^{(\mu)}(s) + V\right)^T & \vdots \\
\vdots \left(R^{(\alpha)}S_\rho^{(\alpha)}(s)\left(A^{(\alpha)}S_\rho^{(\alpha)}(s)\right)^{-1}B^{(\mu)}S_r^{(\mu)}(s) + U\right)^T & \vdots \\
\vdots \left(-R^{(\alpha)}S_\rho^{(\alpha)}(s)\left(A^{(\alpha)}S_\rho^{(\alpha)}(s)\right)^{-1}D^{(\mu)}Z_d^{(\mu-1)}(s)\right)^T & \vdots \\
\vdots \begin{pmatrix} R^{(\alpha)}S_\rho^{(\alpha)}(s)\left(A^{(\alpha)}S_\rho^{(\alpha)}(s)\right)^{-1}A^{(\alpha)}Z_\rho^{(\alpha-1)}(s) - \\ -R^{(\alpha)}Z_\rho^{(\alpha-1)}(s) \end{pmatrix}^T & \vdots \\
\vdots \left(-R^{(\alpha)}S_\rho^{(\alpha)}(s)\left(A^{(\alpha)}S_\rho^{(\alpha)}(s)\right)^{-1}B^{(\mu)}Z_r^{(\mu-1)}(s)\right)^T &
\end{bmatrix}^T , \qquad (5.14)$$

is the *HISO* plant (5.1), (5.2) *input to output (IO) full transfer function matrix,* the inverse Laplace transform of which is the system *IO full fundamental matrix* $\Psi_{HISO}(t)$,

$$\Psi_{HISO}(t) = \mathcal{L}^{-1}\{F_{HISO}(s)\}, \qquad (5.15)$$

- $p_{HISO}(s)$,

$$p_{HISO}(s) = \det\left(A^{(\alpha)}S_\rho^{(\alpha)}(s)\right), \qquad (5.16)$$

is the characteristic polynomial of the *HISO* plant (5.1), (5.2) and the denominator polynomial of all its transfer function matrices,

- $G_{HISOD}(s)$,

$$G_{HISOD}(s) = R^{(\alpha)} S_\rho^{(\alpha)}(s) \left(A^{(\alpha)} S_\rho^{(\alpha)}(s) \right)^{-1} D^{(\mu)} S_d^{(\mu)}(s) + V =$$

$$= p_{HISO}^{-1}(s) \begin{bmatrix} R^{(\alpha)} S_\rho^{(\alpha)}(s) adj \left(A^{(\alpha)} S_\rho^{(\alpha)}(s) \right) D^{(\mu)} S_d^{(\mu)}(s) + \\ + p_{HISO}(s) V \end{bmatrix}, \qquad (5.17)$$

is the *HISO* plant (5.1), (5.2) *transfer function matrix relative to the disturbance* **D**,
   - $G_{HISOU}(s)$,

$$G_{HISOU}(s) = \left[R^{(\alpha)} S_\rho^{(\alpha)}(s) \left(A^{(\alpha)} S_\rho^{(\alpha)}(s) \right)^{-1} B^{(\mu)} S_r^{(\mu)}(s) + U \right] =$$

$$= p_{HISO}^{-1}(s) \bullet \begin{bmatrix} R^{(\alpha)} S_\rho^{(\alpha)}(s) adj \left(A^{(\alpha)} S_\rho^{(\alpha)}(s) \right) B^{(\mu)} S_r^{(\mu)}(s) + \\ + p_{HISO}(s) U \end{bmatrix} =$$

$$= p_{HISO}^{-1}(s) L_{HISO}(s), \qquad (5.18)$$

is the *HISO* plant (5.1), (5.2) *transfer function matrix relative to the control* **U**, and $L_{HISO}(s)$ is the numerator matrix polynomial of $G_{HISOU}(s)$,

$$L_{HISO}(s) = \begin{bmatrix} R^{(\alpha)} S_\rho^{(\alpha)}(s) adj \left(A^{(\alpha)} S_\rho^{(\alpha)}(s) \right) B^{(\mu)} S_r^{(\mu)}(s) + \\ + p_{HISO}(s) U \end{bmatrix} \in \mathbb{C}^{N \times r},$$

$$(5.19)$$

which obeys

$$L_{HISO}(s) = \begin{cases} \sum_{i=0}^{i=n} L_i s^i = L_{HISO}^{(n)} S_r^{(n)}(s), \ L_i \in \mathfrak{R}^{N \times r}, \\ n = \alpha + \alpha\rho - 1 + \mu, \\ \forall i = 0, 1, ..., \xi, \Longleftrightarrow U \neq O_{N,r} \\ \sum_{i=0}^{i=n-1} L_i s^i = L_{HISO}^{(n-1)} S_r^{(n-1)}(s), \ L_i \in \mathfrak{R}^{N \times r}, \\ \forall i = 0, 1, ..., n-1, \Longleftrightarrow U = O_{N,r}, \end{cases} \qquad (5.20)$$

where

$$L_{HISO}^{(n)} = \begin{bmatrix} L_0 \vdots L_1 \vdots ... \vdots L_n \end{bmatrix} \in \mathfrak{R}^{N \times (n+1)r} \Longleftrightarrow U \neq O_{N,r}, \qquad (5.21)$$

$$L_{HISO}^{(n-1)} = \begin{bmatrix} L_0 \vdots L_1 \vdots ... \vdots L_{n-1} \end{bmatrix} \in \mathfrak{R}^{N \times nr} \Longleftrightarrow U = O_{N,r}. \qquad (5.22)$$

and

$$L_{HISO} = \left\{ \begin{array}{l} L_{HISO}^{(n)} \Longleftrightarrow U \neq O_{N,r}, \\ L_{HISO}^{(n-1)} \Longleftrightarrow U = O_{N,r} \end{array} \right\}, \qquad (5.23)$$

- $G_{HISOD_0}(s)$,

$$G_{HISOD_0}(s) = -p_{HISO}^{-1}(s) \bullet$$
$$\bullet \left[R^{(\alpha)} S_\rho^{(\alpha)}(s) adj \left(A^{(\alpha)} S_\rho^{(\alpha)}(s) \right) D^{(\mu)} Z_d^{(\mu-1)}(s) \right] \qquad (5.24)$$

is the *HISO* plant (5.1), (5.2) *transfer function matrix relative to the initial extended disturbance* $\mathbf{D}_0^{\mu-1}$,

- $G_{HISOR_0}(s)$,

$$G_{HISOR_0}(s) = p_{HISO}^{-1}(s) \bullet$$
$$\bullet \left(\begin{array}{c} R^{(\alpha)} S_\rho^{(\alpha)}(s) adj \left(A^{(\alpha)} S_\rho^{(\alpha)}(s) \right) A^{(\alpha)} Z_\rho^{(\alpha-1)}(s) - \\ -p_{HISO}(s) R^{(\alpha)} Z_\rho^{(\alpha-1)}(s) \end{array} \right) \qquad (5.25)$$

is the *HISO* plant (5.1), (5.2) *transfer function matrix relative to the initial state vector* $\mathbf{R}_0^{\alpha-1}$,

- $G_{HISOU_0}(s)$,

$$G_{HISOU_0}(s) = -p_{HISO}^{-1}(s) \bullet$$
$$\bullet \left[R^{(\alpha)} S_\rho^{(\alpha)}(s) adj \left(A^{(\alpha)} S_\rho^{(\alpha)}(s) \right) B^{(\mu)} Z_r^{(\mu-1)}(s) \right] \qquad (5.26)$$

is the *HISO* plant (5.1), (5.2) *transfer function matrix relative to the extended initial control vector* $\mathbf{U}_0^{\mu-1}$,

- $\mathbf{V}_{HISO}(s)$ and $\mathbf{C}_{HISO0}$,

$$\mathbf{V}_{HISO}(s) = \mathbf{V}_{HISOIS}(s) = \left[\begin{array}{ccc} \mathbf{D}^T(s) & \mathbf{U}^T(s) & \mathbf{C}_{HISO0} \end{array} \right]^T, \qquad (5.27)$$

$$\mathbf{C}_{HISO0} = \mathbf{C}_{HISOIS0} = \left[\begin{array}{ccc} \left(\mathbf{D}_0^{\mu-1} \right)^T & \left(\mathbf{R}_0^{\alpha-1} \right)^T & \left(\mathbf{U}_0^{\mu-1} \right)^T \end{array} \right]^T, \qquad (5.28)$$

are the Laplace transform of the action vector $\mathbf{V}_{HISO}(t)$ and the vector $\mathbf{C}_{HISO0}$ of all initial conditions, respectively.

For an example of the *HISO* system (5.1), (5.2) see the book [171].

5.2 The *HISO* plant desired regime

Definition 60, (Section 3.2), takes the special form for the *HISO* plant (5.1), (5.2).

Definition 84 *The functional vector control-state pair* $\left[\mathbf{U}*(.), \mathbf{R}^{*^{\alpha-1}}(.)\right]$ *is nominal for the HISO plant (5.1), (5.2) relative to the functional vector pair* $[\mathbf{D}(.), \mathbf{Y}_d(.)]$*, which is denoted by* $\left[\mathbf{U}_N(.), \mathbf{R}_N^{\alpha-1}(.)\right]$*, if and only if*

$$\left[\mathbf{U}(.), \mathbf{R}^{\alpha-1}(.)\right] = \left[\mathbf{I}*(.), \mathbf{R}^{*^{\alpha-1}}(.)\right]$$

ensures that the corresponding real response $\mathbf{Y}(.) = \mathbf{Y}*(.)$ *of the system obeys* $\mathbf{Y}*(t) = \mathbf{Y}_d(t)$ *all the time,*

$$\left[\mathbf{U}*(.), \mathbf{R}^{*^{\alpha-1}}(.)\right] = \left[\mathbf{U}_N(.), \mathbf{R}_N^{\alpha-1}(.)\right] \Longleftrightarrow \langle \mathbf{Y}*(t) = \mathbf{Y}_d(t), \ \forall t \in \mathfrak{T}_0 \rangle.$$

The system motion $\mathbf{R}_N^{\alpha-1}(.; \mathbf{R}_{N0}^{\alpha-1}; \mathbf{D}; \mathbf{U}_N)$*,* $\mathbf{R}_N^{\alpha-1}(0; \mathbf{R}_{N0}^{\alpha-1}; \mathbf{D}; \mathbf{U}_N) \equiv \mathbf{R}_{N0}^{\alpha-1}$*, is the desired motion* $\mathbf{R}_d^{\alpha-1}(.; \mathbf{R}_{d0}^{\alpha-1}; \mathbf{D}; \mathbf{U}_N)$ *of the HISO plant (5.1), (5.2) relative to the functional vector pair* $[\mathbf{D}(.), \mathbf{Y}_d(.)]$*, for short: the system desired motion,*

$$\mathbf{R}_d^{\alpha-1}(t; \mathbf{R}_{d0}^{\alpha-1}; \mathbf{D}; \mathbf{U}_N) \equiv \mathbf{R}_N^{\alpha-1}(t; \mathbf{R}_{N0}^{\alpha-1}; \mathbf{D}; \mathbf{U}_N),$$
$$\mathbf{R}_d^{\alpha-1}(0; \mathbf{R}_{d0}^{\alpha-1}; \mathbf{D}; \mathbf{U}_N) \equiv \mathbf{R}_{d0}^{\alpha-1} \equiv \mathbf{R}_{N0}^{\alpha-1}. \tag{5.29}$$

Let

$$\mathbf{v}_1(s) = \left\{ \begin{array}{c} B^{(\mu)} Z_r^{(\mu-1)}(s) \mathbf{U}_0^{\mu-1} - A^{(\alpha)} Z_\rho^{(\alpha-1)}(s) \mathbf{R}_0^{\alpha-1} - \\ -D^{(\nu)} S_d^{(\mu)}(s) \mathbf{D}(s) + D^{(\nu)} Z_d^{(\mu-1)}(s) \mathbf{D}_0^{\mu-1} \end{array} \right\}, \tag{5.30}$$

$$\mathbf{v}_2(s) = \mathbf{Y}_d(s) + R_y^{(\alpha)} Z_\rho^{(\alpha-1)}(s) \mathbf{R}_0^{\alpha-1} - V\mathbf{D}(s). \tag{5.31}$$

Definition 84, the system description (5.1), (5.2), Equations (5.30) and (5.31) imply:

Theorem 85 *In order for a functional vector pair* $[\mathbf{I}*(.), \mathbf{R}*(.)]$ *to be nominal for the HISO plant (5.1), (5.2) relative to the functional vector pair* $[\mathbf{D}(.), \mathbf{Y}_d(.)]$*,*

$$\left[\mathbf{U}*(.), \mathbf{R}^{*^{\alpha-1}}(.)\right] = \left[\mathbf{U}_N(.), \mathbf{R}_N^{\alpha-1}(.)\right],$$

it is necessary and sufficient that it obeys the following equations:

$$B^{(\mu)} \mathbf{U}^{*'}(t) - A^{(\alpha)} \mathbf{R}^{*^\alpha}(t) = -D^{(\mu)} \mathbf{D}^\mu(t), \ \forall t \in \mathfrak{T}_0, \tag{5.32}$$

$$UU^*(t) + R_y^{(\alpha)}\mathbf{R}^{*\alpha}(t) = \mathbf{Y}_d(t) - V\mathbf{D}(t), \ \forall t \in \mathfrak{T}_0, \qquad (5.33)$$

or equivalently,

$$\begin{bmatrix} B^{(\mu)}S_r^{(\mu)}(s) & -A^{(\alpha)}S_\rho^{(\alpha)}(s) \\ U & R_y^{(\alpha)}S_\rho^{(\alpha)}(s) \end{bmatrix} \begin{bmatrix} \mathbf{U}^*(s) \\ \mathbf{R}^*(s) \end{bmatrix} = \begin{bmatrix} \mathbf{v}_1(s) \\ \mathbf{v}_2(s) \end{bmatrix}. \qquad (5.34)$$

Let us consider the existence of the solutions of Equations (5.32), (5.33), i.e., of (5.34). The *HISO* plant (5.1), (5.2) contains $(r+\rho)$ unknown variables and $(N+\rho)$ equations. The unknown variables are the entries of $\mathbf{U}^*(s) \in \mathbb{C}^r$ and of $\mathbf{X}^*(s) \in \mathbb{C}^n$.

Claim 86 *In order to exist a nominal functional vector control-state pair*

$$\left[\mathbf{U}_N(.), \mathbf{R}_N^{\alpha-1}(.)\right]$$

for the HISO plant (5.1), (5.2), relative to the functional vector pair distur-bance - desired output $\left[\mathbf{D}(.), \mathbf{Y}_d(.)\right]$, *it is necessary and sufficient that* $N \leq r$.

The proof of this claim is analogous to the proof of Claim 62 (Section 3.2).

Claim 86 presents the complete solution to the problem of the existence of a nominal functional vector pair $\left[\mathbf{U}_N(.), \mathbf{R}_N^{\alpha-1}(.)\right]$ for the *HISO* plant (5.1), (5.2) relative to the functional vector pair $\left[\mathbf{D}(.), \mathbf{Y}_d(.)\right]$.

Condition 87 *The desired output response of the HISO plant (5.1), (5.2) is realizable, i.e.,* $N \leq r$. *The nominal functional vector pair* $\left[\mathbf{U}_N(.), \mathbf{R}_N^{\alpha-1}(.)\right]$ *is known.*

The *HISO* plant description in terms of the deviations (5.35),

$$\mathbf{r} = \mathbf{R} - \mathbf{R}_N, \qquad (5.35)$$

(2.53), (2.56), and (2.55) reads:

$$A^{(\alpha)}\mathbf{r}^\alpha(t) = D^{(\mu)}\mathbf{d}^\mu(t) + B^{(\mu)}\mathbf{u}^\mu(t), \ \forall t \in \mathfrak{T}_0, \qquad (5.36)$$

$$\mathbf{y}(t) = R^{(\alpha)}\mathbf{R}^\alpha(t) + V\mathbf{d}(t) + U\mathbf{u}(t), \ \forall t \in \mathfrak{T}_0. \qquad (5.37)$$

5.3 *HISO* feedback controller

5.3.1 Time domain

The *HISO* feedback controller description in terms of the total coordinates reads:

$$A_C^{(\alpha)} \mathbf{R}_C^{\alpha}(t) = J^{(\mu)} \mathbf{Y}_d^{\mu}(t) - J^{(\mu)} \mathbf{Y}^{\mu}(t) = J^{(\mu)} \mathbf{e}^{\mu}(t), \ \forall t \in \mathfrak{J}_0, \tag{5.38}$$

$$\mathbf{U}(t) = R_C^{(\alpha-1)} \mathbf{R}_C^{\alpha-1}(t) + W^{(\mu)} \mathbf{e}^{\mu}(t), \ \forall t \in \mathfrak{J}_0. \tag{5.39}$$

The *HISO* feedback controller description in terms of the deviations (2.56), (2.55), (5.35), and (1.53) reads:

$$A_C^{(\alpha)} \mathbf{r}_C^{\alpha}(t) = J^{(\mu)} \mathbf{e}^{\mu}(t), \ \forall t \in \mathfrak{T}_0, \tag{5.40}$$

$$\mathbf{u}(t) = R_C^{(\alpha-1)} \mathbf{r}_C^{\alpha-1}(t) + W^{(\mu)} \mathbf{e}^{\mu}(t), \ \forall t \in \mathfrak{T}_0, \tag{5.41}$$

due to Equations (5.38) and (5.39).

5.3.2 Complex domain

The complex domain equivalent equations to Equations (5.38), (5.39) have the following form:

$$\mathbf{R}_C(s) = \left(A_C^{(\alpha)} S_\rho^{(\alpha)}(s) \right)^{-1} \bullet$$

$$\bullet \left[J^{(\mu)} S_N^{(\mu)}(s) \ \vdots - J^{(\mu)} Z_N^{(\mu-1)}(s) \ \vdots \ A_C^{(\alpha)} Z_\rho^{(\alpha-1)}(s) \right] \bullet$$

$$\bullet \left[\ \mathbf{E}^T(s) \quad \left(\mathbf{e}_0^{\mu-1} \right)^T \quad \left(\mathbf{R}_{C0}^{\alpha-1} \right)^T \ \right]^T = F_{HISOCIS}(s) \, \mathbf{W}_{HISOCIS}(s).$$

$$\tag{5.42}$$

$$\mathbf{U}(s) =$$

$$= \begin{bmatrix} \left(R_C^{(\alpha)} S_\rho^{(\alpha)}(s) \left(A_C^{(\alpha)} S_\rho^{(\alpha)}(s) \right)^{-1} J^{(\mu)} S_N^{(\mu)}(s) + W^{(\mu)} S_N^{(\mu)}(s) \right)^T \\ \left(-R_C^{(\alpha)} S_\rho^{(\alpha)}(s) \left(A_C^{(\alpha)} S_\rho^{(\alpha)}(s) \right)^{-1} J^{(\mu)} Z_N^{(\mu-1)}(s) - W^{(\mu)} Z_N^{(\mu-1)}(s) \right)^T \\ \left(R_C^{(\alpha)} S_\rho^{(\alpha)}(s) \left(A_C^{(\alpha)} S_\rho^{(\alpha)}(s) \right)^{-1} A_C^{(\alpha)} Z_\rho^{(\alpha-1)}(s) - R_C^{(\alpha)} Z_\rho^{(\alpha-1)}(s) \right)^T \end{bmatrix}^T \bullet$$

$$\bullet \left[\ \mathbf{E}^T(s) \quad \left(\mathbf{e}_0^{\mu-1} \right)^T \quad \left(\mathbf{R}_{C0}^{\alpha-1} \right)^T \ \right]^T = F_{HISOC}(s) \, \mathbf{W}_{HISOC}(s), \tag{5.43}$$

where:

- $F_{HISOCIS}(s)$,

$$F_{HISOCIS}(s) = \left(A^{(\alpha)} S_\rho^{(\alpha)}(s)\right)^{-1} \bullet$$

$$\bullet \left[J^{(\mu)} S_N^{(\mu)}(s) \vdots - J^{(\mu)} Z_N^{(\mu-1)}(s) \vdots A_C^{(\alpha)} Z_\rho^{(\alpha-1)}(s) \right], \qquad (5.44)$$

is the *HISO* feedback controller (5.38), (5.39) *input to state (IS) full transfer function matrix*, the inverse Laplace transform of which is the controller *IS full fundamental matrix* $\Psi_{HISOCIS}(t)$ [170],

$$\Psi_{HISOCIS}(t) = \mathcal{L}^{-1}\left\{F_{HISOCIS}(s)\right\}, \qquad (5.45)$$

- $F_{HISOCISE}(s)$,

$$F_{HISOCISE}(s) = \left(A^{(\alpha)} S_\rho^{(\alpha)}(s)\right)^{-1} J^{(\mu)} S_N^{(\mu)}(s), \qquad (5.46)$$

is the *HISO* feedback controller (5.38), (5.39) *IS transfer function matrix relative to the output error* **e**,

- $F_{HISOCISE_0}(s)$,

$$F_{HISOCISE_0}(s) = -\left(A^{(\alpha)} S_\rho^{(\alpha)}(s)\right)^{-1} J^{(\mu)} Z_N^{(\mu-1)}(s), \qquad (5.47)$$

is the *HISO* feedback controller (5.38), (5.39) *IS transfer function matrix relative to the extended initial output error* $\mathbf{e}_0^{\mu-1}$,

- $F_{HISOCISR_0}(s)$,

$$F_{HISOCISR_{C0}}(s) = \left(A^{(\alpha)} S_\rho^{(\alpha)}(s)\right)^{-1} A_C^{(\alpha)} Z_\rho^{(\alpha-1)}(s), \qquad (5.48)$$

is the *HISO* feedback controller (5.38), (5.39) *IS transfer function matrix relative to the initial state vector* $\mathbf{R}_{C0}^{\alpha-1}$,

- $F_{HISOC}(s)$,

$$F_{HISOC}(s) =$$

$$= \begin{bmatrix} \left(R_C^{(\alpha)} S_\rho^{(\alpha)}(s) \left(A_C^{(\alpha)} S_\rho^{(\alpha)}(s)\right)^{-1} J^{(\mu)} S_N^{(\mu)}(s) + W^{(\mu)} S_N^{(\mu)}(s)\right)^T \\ \left(-R_C^{(\alpha)} S_\rho^{(\alpha)}(s) \left(A_C^{(\alpha)} S_\rho^{(\alpha)}(s)\right)^{-1} J^{(\mu)} Z_N^{(\mu-1)}(s) - W^{(\mu)} Z_N^{(\mu-1)}(s)\right)^T \\ \left(R_C^{(\alpha)} S_\rho^{(\alpha)}(s) \left(A_C^{(\alpha)} S_\rho^{(\alpha)}(s)\right)^{-1} A_C^{(\alpha)} Z_\rho^{(\alpha-1)}(s) - R_C^{(\alpha)} Z_\rho^{(\alpha-1)}(s)\right)^T \end{bmatrix}^T,$$

$$(5.49)$$

is the *HISO* feedback controller (5.38), (5.39) *input to output (IO) full transfer function matrix*, the inverse Laplace transform of which is the system *IO full fundamental matrix* $\Psi_{HISOC}(t)$ [170],

$$\Psi_{HISOC}(t) = \mathcal{L}^{-1}\{F_{HISOC}(s)\}, \qquad (5.50)$$

- $p_{cHISO}(s)$,

$$p_{cHISO}(s) = \det\left(A_C^{(\alpha)} S_\rho^{(\alpha)}(s)\right), \qquad (5.51)$$

is the characteristic polynomial of the *HISO* controller (5.38), (5.39) and the denominator polynomial of all its transfer function matrices,
- $L_{cHISO}(s)$,

$$L_{cHISO}(s) = \left(\begin{array}{c} R_C^{(\alpha)} S_\rho^{(\alpha)}(s) adj\left(A_C^{(\alpha)} S_\rho^{(\alpha)}(s)\right) J^{(\mu)} S_N^{(\mu)}(s) + \\ + p_{cHISO}(s) W^{(\mu)} S_N^{(\mu)}(s) \end{array}\right), \qquad (5.52)$$

is the numerator matrix polynomial of $G_{HISOCE}(s)$,
- $G_{HISOCE}(s)$,

$$G_{HISOCE}(s) = \left(\begin{array}{c} R_C^{(\alpha)} S_\rho^{(\alpha)}(s)\left(A_C^{(\alpha)} S_\rho^{(\alpha)}(s)\right)^{-1} J^{(\mu)} S_N^{(\mu)}(s) + \\ + W^{(\mu)} S_N^{(\mu)}(s) \end{array}\right) =$$

$$= p_{cHISO}^{-1}(s) \bullet$$

$$\bullet \left(\begin{array}{c} R_C^{(\alpha)} S_\rho^{(\alpha)}(s) adj\left(A_C^{(\alpha)} S_\rho^{(\alpha)}(s)\right) J^{(\mu)} S_N^{(\mu)}(s) + \\ + p_{cHISO}(s) W^{(\mu)} S_N^{(\mu)}(s) \end{array}\right) = p_{cHISO}^{-1}(s) L_{cHISO}(s),$$

$$(5.53)$$

is the *HISO* feedback controller (5.38), (5.39) *transfer function matrix relative to the output error* **e**,
- $G_{HISOCE_0}(s)$,

$$G_{HISOCE_0}(s) = -\left(\begin{array}{c} R_C^{(\alpha)} S_\rho^{(\alpha)}(s)\left(A_C^{(\alpha)} S_\rho^{(\alpha)}(s)\right)^{-1} J^{(\mu)} Z_N^{(\mu-1)}(s) + \\ + W^{(\mu)} Z_N^{(\mu-1)}(s) \end{array}\right) =$$

$$= p_{cHISO}^{-1}(s)\left(\begin{array}{c} -R_C^{(\alpha)} S_\rho^{(\alpha)}(s) adj\left(A_C^{(\alpha)} S_\rho^{(\alpha)}(s)\right) J^{(\mu)} Z_N^{(\mu-1)}(s) - \\ - p_{cHISO}(s) W^{(\mu)} Z_N^{(\mu-1)}(s) \end{array}\right) \qquad (5.54)$$

is the *HISO* feedback controller (5.38), (5.39) *transfer function matrix relative to the initial extended error vector* $\mathbf{e}_0^{\mu-1}$,

- $G_{HISOCR_{C0}}(s)$,

$$G_{HISOCR_{C0}}(s) = \left(\begin{array}{c} R_C^{(\alpha)} S_\rho^{(\alpha)}(s) \left(A_C^{(\alpha)} S_\rho^{(\alpha)}(s) \right)^{-1} A_C^{(\alpha)} Z_\rho^{(\alpha-1)}(s) - \\ -R_C^{(\alpha)} Z_\rho^{(\alpha-1)}(s) \end{array} \right) =$$

$$= p_{cHISO}^{-1}(s) \left(\begin{array}{c} R_C^{(\alpha)} S_\rho^{(\alpha)}(s) adj \left(A_C^{(\alpha)} S_\rho^{(\alpha)}(s) \right) A_C^{(\alpha)} Z_\rho^{(\alpha-1)}(s) - \\ -p_{cHISO}(s) R_C^{(\alpha)} Z_\rho^{(\alpha-1)}(s) \end{array} \right) \qquad (5.55)$$

is the *HISO* feedback controller (5.38), (5.39) *transfer function matrix relative to the initial state vector* $\mathbf{R}_{C0}^{\alpha-1}$,
    - $V_{HISOC}(s)$ and $\mathbf{C}_{HISOC0}$,

$$\mathbf{V}_{HISOC}(s) = \mathbf{V}_{HISOCIS}(s) = \left[\begin{array}{cc} \mathbf{E}^T(s) & \mathbf{C}_{HISOC0}^T \end{array} \right]^T, \qquad (5.56)$$

$$\mathbf{C}_{HISOC0} = \mathbf{C}_{HISOCIS0} = \left[\begin{array}{cc} \left(\mathbf{e}_0^{\mu-1} \right)^T & \left(\mathbf{R}_{C0}^{\alpha-1} \right)^T \end{array} \right]^T, \qquad (5.57)$$

are the Laplace transform of the action vector $V_{HISOC}(t)$ and the vector $\mathbf{C}_{HISOC0}$ of all initial conditions, respectively.

5.4   Exercises

Exercise 88 *1. Select a physical HISO plant.*
    2. Determine its time domain HISO mathematical model.
    3. Determine its complex domain HISO mathematical model.

Exercise 89 *1. Select a HISO controller.*
    2. Determine its time domain HISO mathematical model.
    3. Determine its complex domain HISO mathematical model.

Exercise 90 *1. Determine the time domain HISO mathematical model of the control system composed of the chosen HISO plant and HISO controller.*
    2. Determine the complex domain HISO mathematical model of the control system composed of the chosen HISO plant and HISO controller: its full transfer function matrix and all its transfer function matrices, as well as the vectors $V_{HISOCS}(s)$ and $\mathbf{C}_{HISOCS0}$. Hint: Section 1.7 and Section 1.8.1.

Exercise 91 *Test Lyapunov and BI stability properties of the chosen HISO plant, HISO controller and of theeir control system. Hint: [170, Part III.]*

Chapter 6

IIO systems

6.1 *IIO* system mathematical model

6.1.1 Time domain

The general description, in terms of the total vector coordinates, of *time-invariant* continuous-*time* linear **Input-Internal and Output state systems**, for short *IIO* **systems**, without a delay, has the following general form:

$$A^{(\alpha)}\mathbf{R}^{\alpha}(t) = D^{(\mu)}\mathbf{D}^{\mu}(t) + B^{(\mu)}\mathbf{U}^{\mu}(t) = H^{(\mu)}\mathbf{I}^{\mu}(t), \ \forall t \in \mathfrak{T}_0, \qquad (6.1)$$

$$E^{(\nu)}\mathbf{Y}^{\nu}(t) = \left\{ \begin{array}{c} R_y^{(\alpha-1)}\mathbf{R}^{\alpha-1}(t) + V^{(\mu)}\mathbf{D}^{\mu}(t) + U^{(\mu)}\mathbf{U}^{\mu}(t) = \\ = R_y^{(\alpha-1)}\mathbf{R}^{\alpha-1}(t) + Q^{(\mu)}\mathbf{I}^{\mu}(t) \end{array} \right\}, \ \forall t \in \mathfrak{T}_0, \qquad (6.2)$$

where

$$\mathbf{I} = \mathbf{I}_{IIO} = \begin{bmatrix} \mathbf{D}^T & \mathbf{U}^T \end{bmatrix}^T \in \mathfrak{R}^{d+r}, \ M = d + r,$$

$$H = \begin{bmatrix} D \vdots B \end{bmatrix} \in \mathfrak{R}^{\rho \times (d+r)}, \ Q = \begin{bmatrix} V \vdots U \end{bmatrix} \in \mathfrak{R}^{N \times (d+r)},$$

$$R_y^{(\alpha-1)} = \begin{bmatrix} R_{y0} \vdots R_{y1} \vdots ... \vdots R_{y,\alpha-1} \end{bmatrix} \in \mathfrak{R}^{N \times \alpha \rho}. \qquad (6.3)$$

Note 92 *If $\nu = 0$ then the IIO system (6.1), (6.2) reduces to the HISO system (5.1), (5.2), Chapter 5.*

We continue to treat the IIO system (6.1), (6.2) with $\nu > 0$.

Condition 93 *The matrices A_α and E_ν obey:*

$$det A_\alpha \neq 0, \quad which \ implies \ \exists s \in \mathbb{C} \Longrightarrow det \left[\sum_{k=0}^{k=\alpha} s^k A_k \right] \neq 0,$$

$$det E_\nu \neq 0, \quad which \ implies \ \exists s \in \mathbb{C} \Longrightarrow det \left[\sum_{k=0}^{k=\nu} s^k E_k \right] \neq 0, \qquad (6.4)$$

and

$$\nu \in \{1, 2, ...\} . \qquad (6.5)$$

Note 94 *We accept the validity of Condition 93 in the sequel.*

The left-hand side of Equation (6.1) describes *the internal dynamics of the system,* i.e. *the internal state of the system (Definition 25, Section 1.5),* and the left-hand side of Equation (6.2) describes *the output dynamics, i.e., the output state* of the system if and only if $\nu > 0$.

Note 95 *The state vector $\mathbf{S}_{IIO}$ of the IIO system (6.1), (6.2) is defined in Equation (1.45) (Section 1.5) by:*

$$\mathbf{S}_{IIO} = \left[\begin{array}{c} \mathbf{R}^{\alpha-1} \\ \mathbf{Y}^{\nu-1} \end{array} \right] = \left[\begin{array}{c} \mathbf{R}^T \ \vdots \ \mathbf{R}^{(1)^T} \ \vdots \ ... \ \vdots \ \mathbf{R}^{(\alpha-1)^T} \\ \mathbf{Y}^T \ \vdots \ \mathbf{Y}^{(1)^T} \ \vdots \ ... \ \vdots \ \mathbf{Y}^{(\nu-1)^T} \end{array} \right]^T \in \mathfrak{R}^n,$$

$$n = \alpha\rho + \nu N, \qquad (6.6)$$

The new vector notation $\mathbf{R}^{\alpha-1}$ and $\mathbf{Y}^{\nu-1}$ has permitted us to define the state of the IIO system (6.1), (6.2) by preserving the physical sense. It enabled us to establish in [170] the direct link between the definitions of the Lyapunov and of BI stability properties with the corresponding conditions for them in the complex domain. It enables us to discover in what follows the complex domain criteria for observability, controllability and trackability directly from their definitions. Such criteria possess the complete physical meaning.

The extended vector $\mathbf{R}^{\alpha-1}$ is the IIO system internal state vector $\mathbf{S}_{IIOI}$. The extended vector $\mathbf{Y}^{\nu-1}$ is the IIO system output state vector $\mathbf{S}_{IIOO}$. They compose the IIO system (full) state vector $\mathbf{S}_{IIO}$,

$$\mathbf{S}_{IIO} = \left[\begin{array}{c} \mathbf{R}^{\alpha-1} \\ \mathbf{Y}^{\nu-1} \end{array} \right] = \left[\begin{array}{c} \mathbf{S}_{IIOI} \\ \mathbf{S}_{IIOO} \end{array} \right] . \qquad (6.7)$$

6.1.2 Complex domain

We transform the equations (6.1), (6.2) by applying the Laplace transform, into

$$\mathbf{R}(s) = F_{IIOIS}(s)\,\mathbf{V}_{IIOIS}(s),\tag{6.8}$$

$$\mathbf{Y}(s) = F_{IIO}(s)\,\mathbf{V}_{IIO}(s),\tag{6.9}$$

where:

- $F_{IIOIS}(s)$,

$$F_{IIOIS}(s) = \begin{bmatrix} G_{IIOISD}^{T}(s) \\ G_{IIOISU}^{T}(s) \\ G_{IIOISD_0}^{T}(s) \\ G_{IIOISR_0}^{T}(s) \\ G_{IIOISU_0}^{T}(s) \end{bmatrix}^{T}\tag{6.10}$$

is the *IIO* system (6.1), (6.2) *input to state (IS) full transfer function matrix*, the inverse Laplace transform of which is the plant *IS full fundamental matrix* $\Psi_{IIOIS}(t)$ [170],

$$\Psi_{IIOIS}(t) = \mathcal{L}^{-1}\{F_{IIOIS}(s)\},\tag{6.11}$$

and the inverse Laplace transform of $\left(A^{(\alpha)}S_{\rho}^{(\alpha)}(s)\right)^{-1}$ is *the IIO plant IS fundamental matrix* $\Phi_{IIOIS}(t)$, $\Phi_{IIOIS}(t) = \mathcal{L}^{-1}\left\{\left(A^{(\alpha)}S_{\rho}^{(\alpha)}(s)\right)^{-1}\right\}$ [170],

- $G_{IIOISD}(s)$,

$$G_{IIOISD}(s) = \left(A^{(\alpha)}S_{\rho}^{(\alpha)}(s)\right)^{-1}D^{(\mu)}S_{d}^{(\mu)}(s),\tag{6.12}$$

is the *IIO* plant (6.1), (6.2) *disturbance to internal state (IS) transfer function matrix*,

- $G_{IIOISU}(s)$,

$$G_{IIOISU}(s) = \left(A^{(\alpha)}S_{\rho}^{(\alpha)}(s)\right)^{-1}B^{(\mu)}S_{r}^{(\mu)}(s),\tag{6.13}$$

is the *IIO* plant (6.1), (6.2) *control to internal state (IS) transfer function matrix*,

- $G_{IIOISD_0}(s)$,

$$G_{IIOISD_0}(s) = -\left(A^{(\alpha)}S_{\rho}^{(\alpha)}(s)\right)^{-1}D^{(\mu)}Z_{d}^{(\mu-1)}(s),\tag{6.14}$$

is the *IIO* plant (6.1), (6.2) *initial disturbance to internal state (IS) transfer function matrix,*

- $G_{IIOISR_0}(s)$,

$$G_{IIOISR_0}(s) = \left(A^{(\alpha)}S_\rho^{(\alpha)}(s)\right)^{-1} A^{(\alpha)}Z_\rho^{(\alpha-1)}(s), \tag{6.15}$$

is the *IIO* plant (6.1), (6.2) *initial internal state to internal state (IS) transfer function matrix,*

- $G_{IIOISU_0}(s)$,

$$G_{IIOISU_0}(s) = \left(A^{(\alpha)}S_\rho^{(\alpha)}(s)\right)^{-1} B^{(\mu)}Z_r^{(\mu-1)}(s), \tag{6.16}$$

is the *IIO* plant (6.1), (6.2) *initial control to internal state (IS) transfer function matrix,*

$\mathbf{V}_{IIOIS}(s)$,

$$\mathbf{V}_{IIOIS}(s) = \left[\mathbf{D}^T(s) \vdots \mathbf{U}^T(s)\vdots \mathbf{C}_{IIOIS0}^T\right]^T, \tag{6.17}$$

is the Laplace transform of the action vector $\mathbf{V}_{IIOIS}(t)$, and $\mathbf{C}_{IIOIS0}$,

$$\mathbf{C}_{IIOIS0} = \left[\left(\mathbf{D}_0^{\mu-1}\right)^T \vdots \left(\mathbf{R}_0^{\alpha-1}\right)^T \vdots \left(\mathbf{U}_0^{\mu-1}\right)^T\right]^T, \tag{6.18}$$

is the vector of all initial conditions acting on the system internal state,

- $F_{IIO}(s)$,

$$F_{IIO}(s) = \begin{bmatrix} G_{IIOD}^T(s) \\ G_{IIOU}^T(s) \\ G_{IIOD_0}^T(s) \\ G_{IIOR_0}^T(s) \\ G_{IIOU_0}^T(s) \\ G_{IIOY_0}^T(s) \end{bmatrix}^T, \tag{6.19}$$

is the *IIO* plant (6.1), (6.2) *input to output (IO) full transfer function matrix,* the inverse Laplace transform of which is the plant *IO full fundamental matrix* $\Psi_{IIO}(t)$,

$$\Psi_{IIO}(t) = \mathcal{L}^{-1}\{F_{IIO}(s)\}, \tag{6.20}$$

- $p_{IIO}(s)$,

$$p_{IIO}(s) = \det\left(E^{(\nu)}S_N^{(\nu)}(s)\right)\det\left(A^{(\alpha)}S_\rho^{(\alpha)}(s)\right), \tag{6.21}$$

is the characteristic polynomial of the *IIO* plant (6.1), (6.2) and the denominator polynomial of all its transfer function matrices

- $G_{IIOD}(s)$,

$$G_{IIOD}(s) =$$

$$= \left[\begin{array}{c} \left(E^{(\nu)}S_N^{(\nu)}(s)\right)^{-1} R_y^{(\alpha-1)} S_\rho^{(\alpha)}(s) \left(A^{(\alpha)}S_\rho^{(\alpha)}(s)\right)^{-1} H^{(\mu)}S_d^{(\mu)}(s) \\ + \left(E^{(\nu)}S_N^{(\nu)}(s)\right)^{-1} V^{(\mu)}S_d^{(\mu)}(s) \end{array} \right] =$$

$$= p_{IIO}^{-1}(s) \bullet$$

$$\bullet \left[\begin{array}{c} adj\left(E^{(\nu)}S_N^{(\nu)}(s)\right) R_y^{(\alpha-1)} S_\rho^{(\alpha)}(s) adj\left(A^{(\alpha)}S_\rho^{(\alpha)}(s)\right) H^{(\mu)}S_d^{(\mu)}(s) \\ + \left[\det\left(A^{(\alpha)}S_\rho^{(\alpha)}(s)\right)\right] adj\left(E^{(\nu)}S_N^{(\nu)}(s)\right) V^{(\mu)}S_d^{(\mu)}(s) \end{array} \right],$$

$$\tag{6.22}$$

is the *IIO* plant (6.1), (6.2) *IO transfer function matrix relative to the disturbance* **D**,

- $G_{IIOU}(s)$,

$$G_{IIOU}(s) =$$

$$= \left[\begin{array}{c} \left(E^{(\nu)}S_N^{(\nu)}(s)\right)^{-1} R_y^{(\alpha-1)} S_\rho^{(\alpha)}(s) \left(A^{(\alpha)}S_\rho^{(\alpha)}(s)\right)^{-1} B^{(\mu)}S_r^{(\mu)}(s)+ \\ + \left(E^{(\nu)}S_N^{(\nu)}(s)\right)^{-1} U^{(\mu)}S_r^{(\mu)}(s), \end{array} \right] =$$

$$p_{IIO}^{-1}(s) \bullet$$

$$\bullet \left[\begin{array}{c} adj\left(E^{(\nu)}S_N^{(\nu)}(s)\right) R_y^{(\alpha-1)} S_\rho^{(\alpha)}(s) adj\left(A^{(\alpha)}S_\rho^{(\alpha)}(s)\right) B^{(\mu)}S_r^{(\mu)}(s) \\ + \left[\det\left(A^{(\alpha)}S_\rho^{(\alpha)}(s)\right)\right] adj\left(E^{(\nu)}S_N^{(\nu)}(s)\right) U^{(\mu)}S_r^{(\mu)}(s), \end{array} \right]$$

$$\tag{6.23}$$

is the *IIO* plant (6.1), (6.2) *IO transfer function matrix relative to the control* **U**, and $L_{IIO}(s)$ is the numerator matrix polynomial of $G_{IIOU}(s)$,

$$L_{IIO}(s) =$$

$$= \left[\begin{array}{c} adj\left(E^{(\nu)}S_N^{(\nu)}(s)\right) R_y^{(\alpha-1)} S_\rho^{(\alpha)}(s) adj\left(A^{(\alpha)}S_\rho^{(\alpha)}(s)\right) B^{(\mu)}S_r^{(\mu)}(s) \\ + \left[\det\left(A^{(\alpha)}S_\rho^{(\alpha)}(s)\right)\right] adj\left(E^{(\nu)}S_N^{(\nu)}(s)\right) U^{(\mu)}S_r^{(\mu)}(s), \end{array} \right],$$

$$L_{IIO}(s) \in \mathbb{C}^{N \times r} \tag{6.24}$$

which obeys

$$L_{IIO}(s) = \begin{cases} \sum_{i=0}^{i=\xi} L_i s^i = L_{IIO}^{(\xi)} S_r^{(\xi)}(s), \ L_i \in \mathfrak{R}^{N \times r}, \\ \forall i = 0,1,...,\xi, \Longleftrightarrow \ U \neq O_{N,r}, \\ \sum_{i=0}^{i=\xi-1} L_i s^i = L_{IIO}^{(\xi-1)} S_r^{(\xi-1)}(s), \ L_i \in \mathfrak{R}^{N \times r}, \\ \forall i = 0,1,...,\xi-1, \Longleftrightarrow \ U = O_{N,r}, \end{cases} \tag{6.25}$$

where

$$\xi = N\nu - 1 + \alpha + \alpha\rho - 1 + \mu, \tag{6.26}$$

$$L_{IIO}^{(\xi)} = \left[L_0 \vdots L_1 \vdots ... \vdots L_\xi \right] \in \mathfrak{R}^{N \times (\xi+1)r} \Longleftrightarrow U \neq O_{N,r}, \tag{6.27}$$

$$L_{IIO}^{(\xi-1)} = \left[L_0 \vdots L_1 \vdots ... \vdots L_{\xi-1} \right] \in \mathfrak{R}^{N \times \xi r} \Longleftrightarrow U = O_{N,r}, \tag{6.28}$$

and

$$L_{IIO} = \begin{cases} L_{IIO}^{(\xi)} \Longleftrightarrow U \neq O_{N,r}, \\ L_{IIO}^{(\xi-1)} \Longleftrightarrow U = O_{N,r}, \end{cases} \tag{6.29}$$

- $G_{IIOD_0}(s)$,

$$G_{IIOD_0}(s) = -p_{IIO}^{-1}(s) \bullet$$
$$\bullet \left[\begin{array}{l} adj\left(E^{(\nu)} S_N^{(\nu)}(s) \right) R_y^{(\alpha-1)} S_\rho^{(\alpha)}(s) adj\left(A^{(\alpha)} S_\rho^{(\alpha)}(s) \right) D^{(\mu)} Z_d^{(\mu-1)}(s) \\ + \left(\det A^{(\alpha)} S_\rho^{(\alpha)}(s) \right) adj\left(E^{(\nu)} S_N^{(\nu)}(s) \right) V^{(\mu)} Z_d^{(\mu-1)}(s), \end{array} \right]$$
$$\tag{6.30}$$

is the *IIO* plant (6.1), (6.2) *IO transfer function matrix relative to the initial extended disturbance* $\mathbf{D}_0^{\mu-1}$,

- $G_{IIOR_0}(s)$,

$$G_{IIOR_0}(s) = p_{IIO}^{-1}(s) \bullet$$
$$\bullet \left[\begin{array}{l} adj\left(E^{(\nu)} S_N^{(\nu)}(s) \right) R_y^{(\alpha-1)} S_\rho^{(\alpha)}(s) adj\left(A^{(\alpha)} S_\rho^{(\alpha)}(s) \right) A^{(\alpha)} Z_\rho^{(\alpha-1)}(s) \\ - \left[\det\left(A^{(\alpha)} S_\rho^{(\alpha)}(s) \right) \right] adj\left(E^{(\nu)} S_N^{(\nu)}(s) \right) R_y^{(\alpha-1)} Z_\rho^{(\alpha-1)}(s), \end{array} \right]$$
$$\tag{6.31}$$

is the *IIO* plant (6.1), (6.2) *IO transfer function matrix relative to the initial internal state vector* $\mathbf{R}_0^{\alpha-1}$,

- $G_{IIOU_0}(s)$,

$$G_{IIOU_0}(s) = p_{IIO}^{-1}(s) \bullet$$

$$\bullet \begin{bmatrix} -adj\left(E^{(\nu)}S_N^{(\nu)}(s)\right) R_y^{(\alpha-1)} S_\rho^{(\alpha)}(s) adj\left(A^{(\alpha)}S_\rho^{(\alpha)}(s)\right) B^{(\mu)} Z_r^{(\mu-1)}(s) \\ -\left[\det\left(A^{(\alpha)}S_\rho^{(\alpha)}(s)\right)\right] adj\left(E^{(\nu)}S_N^{(\nu)}(s)\right) U^{(\mu)} Z_r^{(\mu-1)}(s) \end{bmatrix},$$

(6.32)

is the *IO transfer function matrix relative to the initial extended control vector* $\mathbf{U}_0^{\mu-1}$ *of the IIO plant* (6.1), (6.2),
  - $G_{IIOY_0}(s)$,

$$G_{IIOY_0}(s) = \left(E^{(\nu)}S_N^{(\nu)}(s)\right)^{-1} E^{(\nu)} Z_N^{(\nu-1)}(s) \tag{6.33}$$

is the *IIO plant* (6.1), (6.2) *IO transfer function matrix relative to the extended initial output state vector* $\mathbf{Y}_0^{\nu-1}$,
  - $\mathbf{V}_{IIO}(s)$ and $\mathbf{C}_{IIO0}$,

$$\mathbf{V}_{IIO}(s) = \begin{bmatrix} \mathbf{I}_{IIO}(s) \\ \mathbf{C}_{IIO0} \end{bmatrix}, \quad \mathbf{I}_{IIO}(s) = \begin{bmatrix} \mathbf{D}(s) \\ \mathbf{U}(s) \end{bmatrix}, \quad \mathbf{C}_{IIO0} = \begin{bmatrix} \mathbf{D}_0^{\mu-1} \\ \mathbf{R}_0^{\alpha-1} \\ \mathbf{U}_0^{\mu-1} \\ \mathbf{Y}_0^{\nu-1} \end{bmatrix},$$

(6.34)

are the Laplace transform of the action vector $\mathbf{V}_{IIO}(t)$ and the vector $\mathbf{C}_{IIO0}$ of all initial conditions, respectively.

Equations (6.9), (6.19), (6.21)–(6.23), (6.30)–(6.33) determine the Laplace transform $\mathbf{Y}(s)$ of the output vector $\mathbf{Y}(t)$,

$$\mathbf{Y}(s) = G_{IIOD}(s)\,\mathbf{D}(s) + G_{IIOU}(s)\,\mathbf{U}(s) + G_{IIOD_0}(s)\,\mathbf{D}_0^{\mu-1} +$$
$$+ G_{IIOR_0}(s)\,\mathbf{R}_0^{\alpha-1} + G_{IIOU_0}(s)\,\mathbf{U}_0^{\mu-1} + G_{IIOY_0}(s)\,\mathbf{Y}_0^{\nu-1} =$$
$$= F_{IIO}(s)\,\mathbf{V}_{IIO}(s). \tag{6.35}$$

This can be set in a more compact form. Equations (6.34) and (6.35), together with

$$G_{IIO}(s) = \left[G_{IIOD}(s) \vdots G_{IIOU}(s)\right], \quad G_{IIOI_0}(s) = \left[G_{IIOD_0}(s) \vdots G_{IIOU_0}(s)\right],$$

give the compact form to the Laplace transform $\mathbf{Y}^{\mp}(s)$ of the system response $\mathbf{Y}(t; \mathbf{R}_{0-}^{\alpha-1}; \mathbf{Y}_0^{\nu-1}; \mathbf{I}^{\mu})$,

$$\mathbf{Y}^{\mp}(s) = G_{IIO}(s)\mathbf{I}(s) + G_{IIOI_0}(s)\mathbf{I}_0^{\mu-1} + G_{IIOR_0}(s)\,\mathbf{R}_{0-}^{\alpha-1} + G_{IIOY_0}(s)\,\mathbf{Y}_{0\mp}^{\nu-1}.$$

(6.36)

The inverse Laplace transform of this equation determines the IIO system (6.1), (6.2) response $\mathbf{Y}(t; \mathbf{R}_{0-}^{\alpha-1}; \mathbf{Y}_0^{\nu-1}; \mathbf{I}^\mu)$,

$$\mathbf{Y}(t; \mathbf{R}_{0-}^{\alpha-1}; \mathbf{Y}_0^{\nu-1}; \mathbf{I}^\mu) = \mathcal{L}^{-1}\left\{\mathbf{Y}^{\mp}(s)\right\} = \int_{0-}^{t} \Gamma_{IIO}(\tau)\mathbf{I}(t-\tau)d\tau +$$

$$+\Gamma_{IIOI_0}(t)\mathbf{I}_{0-}^{\mu-1} + \Gamma_{IIOR_0}(t)\mathbf{R}_{0-}^{\alpha-1} + \Gamma_{IIOY_0}(t)\mathbf{Y}_0^{\nu-1}, \qquad (6.37)$$

$$\forall t \in \mathfrak{T}_0,$$

where

$$\Gamma_{IIO}(t) = \mathcal{L}^{-1}\left\{G_{IIO}(s)\right\} =$$

$$= \mathcal{L}^{-1}\left\{ \Theta_{IIO}(s) \left[\begin{array}{c} E_\nu^{-1} R_y^{(\alpha)} S_\rho^{(\alpha)}(s) \left(A^{(\alpha)} S_\rho^{(\alpha)}(s)\right)^{-1} H^{(\mu)} S_M^{(\mu)}(s) + \\ + E_\nu^{-1} Q^{(\mu)} S_M^{(\mu)}(s) \end{array} \right] \right\},$$

$$(6.38)$$

$$\Gamma_{IIOI_0}(t) = \mathcal{L}^{-1}\left\{G_{IIOI_0}(s)\right\} =$$

$$= \mathcal{L}^{-1}\left\{ \Theta_{IIO}(s) \left[\begin{array}{c} -E_\nu^{-1} R_y^{(\alpha)} S_\rho^{(\alpha)}(s) \left(A^{(\alpha)} S_\rho^{(\alpha)}(s)\right)^{-1} H^{(\mu)} Z_M^{(\mu-1)}(s) - \\ - E_\nu^{-1} Q^{(\mu)} Z_M^{(\mu-1)}(s) \end{array} \right] \right\},$$

$$(6.39)$$

$$\Gamma_{IIOR_0}(t) = \mathcal{L}^{-1}\left\{G_{IIOR_0}(s)\right\} =$$

$$= \mathcal{L}^{-1}\left\{ \Theta_{IIO}(s) \left[\begin{array}{c} E_\nu^{-1} R_y^{(\alpha)} S_\rho^{(\alpha)}(s) \left(A^{(\alpha)} S_\rho^{(\alpha)}(s)\right)^{-1} A^{(\alpha)} Z_\rho^{(\alpha-1)}(s) - \\ - E_\nu^{-1} R_y^{(\alpha)} Z_\rho^{(\alpha-1)}(s) \end{array} \right] \right\},$$

$$(6.40)$$

$$\Gamma_{IIOY_0}(t) = \mathcal{L}^{-1}\left\{G_{IIOY_0}(s)\right\} = \mathcal{L}^{-1}\left\{ \Theta_{IIO}(s) E_\nu^{-1} E^{(\nu)} Z_N^{(\nu-1)}(s)\right\}. \quad (6.41)$$

Equations (6.38)–(6.41) define well the matrices $\Gamma_{IIO}(t)$, $\Gamma_{IIOI_0}(t)$, $\Gamma_{IIOR_0}(t)$ and $\Gamma_{IIOY_0}(t)$ in terms of the system transfer function matrices $G_{IIO}(s)$, $G_{IIOI_0}(s)$, $G_{IIOR_0}(s)$ and $G_{IIOY_0}(s)$, respectively.

For an example of the IIO system (6.1), (6.2) see the book [171].

6.2 IIO plant desired regime

We adjust Definition 60, (Section 3.2), to the IIO plant (6.1), (6.2):

Definition 96 *A functional control-state pair* $\left[\mathbf{U}*(.), \mathbf{R}^{*\alpha-1}(.)\right]$ *is **nominal** for the IIO plant (6.1), (6.2) relative to the functional vector pair* $[\mathbf{D}(.), \mathbf{Y}_d(.)]$, *which is denoted by* $[\mathbf{U}_N(.), \mathbf{R}_N^{\alpha-1}(.)]$, *if and only if* $\left[\mathbf{I}(.), \mathbf{R}^{\alpha-1}(.)\right] = \left[\mathbf{I}*(.), \mathbf{R}^{*\alpha-1}(.)\right]$ *ensures that the corresponding real response* $\mathbf{Y}(.) = \mathbf{Y}*(.)$ *of the system obeys* $\mathbf{Y}*(t) = \mathbf{Y}_d(t)$ *all the time as soon as* $\mathbf{Y}_0^{\nu-1} = \mathbf{Y}_{d0}^{\nu-1}$,

$$\left[\mathbf{I}*(.), \mathbf{R}^{*\alpha-1}(.)\right] = [\mathbf{I}_N(.), \mathbf{R}_N^{\alpha-1}(.)] \Longleftrightarrow$$
$$\Longleftrightarrow \left\langle \mathbf{Y}_0^{\nu-1} = \mathbf{Y}_{d0}^{\nu-1} \Longrightarrow \mathbf{Y}*(t) = \mathbf{Y}_d(t), \ \forall t \in \mathfrak{T}_0 \right\rangle.$$

Let

$$\mathbf{w}_1(s) = \left\{ \begin{array}{l} -D^{(\mu)} S_d^{(\mu)}(s) \mathbf{D}(s) + D^{(\mu)} Z_d^{(\mu-1)}(s) \mathbf{D}_0^{\mu-1} + \\ +B^{(\mu)} Z_r^{(\mu-1)}(s) \mathbf{U}_0^{*\mu-1} - A_P^{(\alpha)} Z_\rho^{(\alpha-1)}(s) \mathbf{R}_0^{*\alpha-1} \end{array} \right\}, \tag{6.42}$$

$$\mathbf{w}_2(s) = \left\{ \begin{array}{l} E^{(\nu)} S_N^{(\nu)}(s) \mathbf{Y}_d(s) - E^{(\nu)} Z_N^{(\nu-1)}(s) \mathbf{Y}_{d0}^{\nu-1} + \\ +R_y^{(\alpha-1)} Z_\rho^{(\alpha-1)}(s) \mathbf{R}_0^{*\alpha-1} - V^{(\mu)} S_d^{(\mu)}(s) \mathbf{D}(s) + \\ +V^{(\mu)} Z_d^{(\mu-1)}(s) \mathbf{D}_0^{\mu-1} + U^{(\mu)} Z_r^{(\nu-1)}(s) \mathbf{U}_0^{*\mu-1} \end{array} \right\}. \tag{6.43}$$

Definition 96 and the plant description (6.1), (6.2) imply the following:

Theorem 97 *In order for a functional vector pair* $\left[\mathbf{U}*(.), \mathbf{R}^{*\alpha-1}(.)\right]$ *to be nominal for the IIO plant (6.1), (6.2) relative to the functional vector pair* $[\mathbf{D}(.), \mathbf{Y}_d(.)]$,

$$\left[\mathbf{U}*(.), \mathbf{R}^{*\alpha-1}(.)\right] = [\mathbf{U}_N(.), \mathbf{R}_N^{\alpha-1}(.)],$$

it is necessary and sufficient that it obeys the following equations:

$$B^{(\mu)} \mathbf{U}^{*\mu}(t) - A^{(\alpha)} \mathbf{R}^{*\alpha}(t) = -D^{(\mu)} \mathbf{D}^\mu(t), \ \forall t \in \mathfrak{T}_0, \tag{6.44}$$

$$U^{(\mu)} \mathbf{U}^{*\mu}(t) + R^{(\alpha)} \mathbf{R}^{*\alpha}(t) = E^{(\nu)} \mathbf{Y}_d^\nu(t) - V^{(\mu)} \mathbf{D}^\mu(t), \ \forall t \in \mathfrak{T}_0, \tag{6.45}$$

or equivalently,

$$\begin{bmatrix} B^{(\mu)} S_r^{(\mu)}(s) & -A^{(\alpha)} S_\rho^{(\alpha)}(s) \\ U^{(\mu)} S_r^{(\mu)}(s) & R^{(\alpha)} S_\rho^{(\alpha)}(s) \end{bmatrix} \begin{bmatrix} \mathbf{U}*(s) \\ \mathbf{R}*(s) \end{bmatrix} = \begin{bmatrix} \mathbf{w}_1(s) \\ \mathbf{w}_2(s) \end{bmatrix}. \tag{6.46}$$

What are the conditions for the existence of the solutions of the equations (6.44), (6.45), i.e., of (6.46)? There are $(r+\rho)$ unknown variables and $(N+\rho)$ equations. The unknown variables are the entries of $\mathbf{U}*(s) \in \mathbb{C}^r$ and of $\mathbf{R}*(s) \in \mathbb{C}^\rho$.

Claim 98 *In order to exist a nominal functional vector pair*

$$\left[\mathbf{U}_N(.), \mathbf{R}_N^{\alpha-1}(.)\right]$$

for the IIO plant (6.1), (6.2) relative to the functional pair $[\mathbf{D}(.), \mathbf{Y}_d(.)]$ *it is necessary and sufficient that* $N \leq r$.

The proof of this claim follows the proof of Claim 62 (Section 3.2).

The condition $N \leq r$ is compatible with Fundamental control principle 104, Section 7.1.

Condition 99 *The desired output response of the IIO plant (6.1), (6.2) is realizable, i.e.,* $N \leq r$. *The nominal control-state pair* $\left[\mathbf{U}_N(.), \mathbf{R}_N^{\alpha-1}(.)\right]$ *is known.*

The *time* domain description of the *IIO* plant in terms of the deviations reads:

$$A^{(\alpha)}\mathbf{r}^{\alpha}(t) = D^{(\mu)}\mathbf{d}^{\mu}(t) + B^{(\mu)}\mathbf{u}^{\mu}(t), \ \forall t \in \mathfrak{T}_0, \tag{6.47}$$

$$E^{(\nu)}\mathbf{y}^{\nu}(t) = R_y^{(\alpha-1)}\mathbf{r}^{\alpha-1}(t) + V^{(\mu)}\mathbf{d}^{\mu}(t) + U^{(\mu)}\mathbf{u}^{\mu}(t), \ \forall t \in \mathfrak{T}_0. \tag{6.48}$$

6.3   *IIO* feedback controller

6.3.1   Time domain

The *time* domain description of the *IIO* controller in terms of the total coordinates reads:

$$A_C^{(\alpha)}\mathbf{R}_C^{\alpha}(t) = J^{(\mu)}\mathbf{Y}_d^{\mu}(t) - J^{(\mu)}\mathbf{Y}^{\mu}(t), \ \forall t \in \mathfrak{T}_0,$$

$$E^{(\nu)}\mathbf{U}^{\nu}(t) = R_C^{(\alpha-1)}\mathbf{R}_C^{\alpha-1}(t) + W^{(\mu)}\mathbf{Y}_d^{\mu}(t) - W^{(\mu)}\mathbf{Y}^{\mu}(t), \ \forall t \in \mathfrak{T}_0,$$

or equivalently, due to (2.56), (1.53) (Section 1.6),

$$A_C^{(\alpha)}\mathbf{R}_C^{\alpha}(t) = J^{(\mu)}\mathbf{e}^{\mu}(t), \ \forall t \in \mathfrak{T}_0, \tag{6.49}$$

$$E^{(\nu)}\mathbf{U}^{\nu}(t) = R_C^{(\alpha-1)}\mathbf{R}_C^{\alpha-1}(t) + W^{(\mu)}\mathbf{e}^{\mu}(t), \ \forall t \in \mathfrak{T}_0, \tag{6.50}$$

and in terms of the deviations in view of (5.35) and (2.56):

$$A_C^{(\alpha)}\mathbf{r}^{\alpha}(t) = J^{(\mu)}\mathbf{e}^{\mu}(t), \ \forall t \in \mathfrak{T}_0, \tag{6.51}$$

$$E^{(\nu)}\mathbf{u}^{\nu}(t) = R_C^{(\alpha-1)}\mathbf{r}^{\alpha-1}(t) + W^{(\mu)}\mathbf{e}^{\mu}(t), \ \forall t \in \mathfrak{T}_0. \tag{6.52}$$

6.3.2 Complex domain

The Laplace transform of (6.49), (6.50), or of (6.51), (6.52), leads to

$$\mathbf{R}(s) = F_{IIOCIS}(s)\, \mathbf{V}_{IIOCIS}(s), \tag{6.53}$$

$$\mathbf{U}(s) = F_{IIOC}(s)\, \mathbf{V}_{IIOC}(s), \tag{6.54}$$

where:
- $F_{IIOCIS}(s)$,

$$F_{IIOCIS}(s) =$$
$$= \left[G_{IIOCISE}(s) \vdots G_{IIOCISE_0}(s) \vdots G_{IIOCISR_0}(s) \right], \tag{6.55}$$

is the *IIO* controller (6.51), (6.52) *input to state (IS) full transfer function matrix*, the inverse Laplace transform of which is the controller *IS full fundamental matrix* $\Psi_{IIOCIS}(t)$ [170],

$$\Psi_{IIOCIS}(t) = \mathcal{L}^{-1}\{F_{IIOCIS}(s)\}, \tag{6.56}$$

and the inverse Laplace transform of $\left(A_C^{(\alpha)} S_\rho^{(\alpha)}(s) \right)^{-1}$ is *the IIO controller IS fundamental matrix* $\Phi_{IIOCIS}(t)$, $\Phi_{IIOCIS}(t) = \mathcal{L}^{-1}\left\{ \left(A_C^{(\alpha)} S_\rho^{(\alpha)}(s) \right)^{-1} \right\}$ [170],
- $G_{IIOCISE}(s)$,

$$G_{IIOCISE}(s) = \left(A_C^{(\alpha)} S_\rho^{(\alpha)}(s) \right)^{-1} J^{(\mu)} S_N^{(\mu)}(s), \tag{6.57}$$

is the *IIO* controller (6.51), (6.52) *input to state (IS) transfer function matrix*,
- $G_{IIOCISE_0}(s)$,

$$G_{IIOCISE_0}(s) = -\left(A_C^{(\alpha)} S_\rho^{(\alpha)}(s) \right)^{-1} J^{(\mu)} Z_N^{(\mu-1)}(s), \tag{6.58}$$

is the *IIO* controller (6.51), (6.52) *initial error to state (IS) transfer function matrix*,
- $G_{IIOCISR_{C0}}(s)$,

$$G_{IIOCISR_{C0}}(s) = \left(A_C^{(\alpha)} S_\rho^{(\alpha)}(s) \right)^{-1} A_C^{(\alpha)} Z_\rho^{(\alpha-1)}(s), \tag{6.59}$$

is the *IIO* controller (6.51), (6.52) *initial internal state to state (IS) transfer function matrix,*

- $F_{IIOC}(s)$,

$$F_{IIOC}(s) =$$

$$\left[G_{IIOCE}(s) \vdots G_{IIOCE_0}(s) \vdots G_{IIOCR_0}(s) \vdots G_{IIOCU_0}(s) \right] \quad (6.60)$$

is the *IIO* controller (6.51), (6.52) *input to output (IO) full transfer function matrix,* the inverse Laplace transform of which is the controller *IO full fundamental matrix* $\Psi_{IIOC}(t)$ [170],

$$\Psi_{IIOC}(t) = \mathcal{L}^{-1}\left\{ F_{IIOC}(s) \right\}, \quad (6.61)$$

- $p_{cIIO}(s)$,

$$p_{cIIO}(s) = \det\left(A_C^{(\alpha)} S_\rho^{(\alpha)}(s) \right) \det\left(E^{(\nu)} S_N^{(\nu)}(s) \right), \quad (6.62)$$

is the characteristic polynomial of the *IIO* controller (6.51), (6.52) and the denominator polynomial of all its transfer function matrices,

- $L_{cIIO}(s)$,

$$L_{cIIO}(s) =$$

$$= \left[\begin{array}{c} adj\left(E^{(\nu)} S_N^{(\nu)}(s) \right) R_C^{(\alpha-1)} S_\rho^{(\alpha-1)}(s) \bullet \\ \bullet adj\left(A_C^{(\alpha)} S_\rho^{(\alpha)}(s) \right) J^{(\mu)} S_N^{(\mu)}(s) + \\ + \left[\det\left(A_C^{(\alpha)} S_\rho^{(\alpha)}(s) \right) \right] adj\left(E^{(\nu)} S_N^{(\nu)}(s) \right) W^{(\mu)} S_N^{(\mu)}(s) \end{array} \right], \quad (6.63)$$

is the numerator matrix polynomial of $G_{IIOCE}(s)$,

- $G_{IIOCE}(s)$,

$$G_{IIOCE}(s) =$$

$$= \left[\begin{array}{c} \left(E^{(\nu)} S_N^{(\nu)}(s) \right)^{-1} R_C^{(\alpha-1)} S_\rho^{(\alpha-1)}(s) \left(A_C^{(\alpha)} S_\rho^{(\alpha)}(s) \right)^{-1} J^{(\mu)} S_N^{(\mu)}(s) \\ + \left(E^{(\nu)} S_N^{(\nu)}(s) \right)^{-1} W^{(\mu)} S_N^{(\mu)}(s) \end{array} \right] =$$

$$= p_{cIIO}^{-1}(s) \bullet$$

$$\bullet \left[\begin{array}{c} adj\left(E^{(\nu)} S_N^{(\nu)}(s) \right) R_C^{(\alpha-1)} S_\rho^{(\alpha-1)}(s) \bullet \\ \bullet adj\left(A_C^{(\alpha)} S_\rho^{(\alpha)}(s) \right) J^{(\mu)} S_N^{(\mu)}(s) + \\ + \left[\det\left(A_C^{(\alpha)} S_\rho^{(\alpha)}(s) \right) \right] adj\left(E^{(\nu)} S_N^{(\nu)}(s) \right) W^{(\mu)} S_N^{(\mu)}(s) \end{array} \right], \quad (6.64)$$

is the *IIO* controller (6.51), (6.52) *transfer function matrix relative to the error* **e**,

- $G_{IIOCEo}(s)$,

$$G_{IIOCEo}(s) =$$

$$= \left[\begin{array}{c} -\left(E^{(\nu)} S_N^{(\nu)}(s)\right)^{-1} R_C^{(\alpha-1)} S_\rho^{(\alpha-1)}(s) \left(A_C^{(\alpha)} S_\rho^{(\alpha)}(s)\right)^{-1} J^{(\mu)} Z_N^{(\mu-1)}(s) \\ -\left(E^{(\nu)} S_N^{(\nu)}(s)\right)^{-1} W^{(\mu)} Z_N^{(\mu-1)}(s) \end{array} \right] =$$

$$= p_{cIIO}^{-1}(s) \bullet$$

$$\bullet \left[\begin{array}{c} -adj\left(E^{(\nu)} S_N^{(\nu)}(s)\right) R_C^{(\alpha-1)} S_\rho^{(\alpha-1)}(s) \bullet \\ \bullet adj\left(A_C^{(\alpha)} S_\rho^{(\alpha)}(s)\right) J^{(\mu)} Z_N^{(\mu-1)}(s) \\ -\left[\det\left(A_C^{(\alpha)} S_\rho^{(\alpha)}(s)\right)\right] adj\left(E^{(\nu)} S_N^{(\nu)}(s)\right) W^{(\mu)} Z_N^{(\mu-1)}(s) \end{array} \right] =$$

$$= \frac{L_{cIIO}(s)}{p_{cIIO}(s)}, \tag{6.65}$$

is the *IIO* controller (6.51), (6.52) *transfer function matrix relative to the extended error vector* $\mathbf{e}_0^{\mu-1}$,

- $G_{IIOCR_{C0}}(s)$,

$$G_{IIOCR_{C0}}(s) =$$

$$= \left[\begin{array}{c} -\left(E^{(\nu)} S_N^{(\nu)}(s)\right)^{-1} R_C^{(\alpha)} S_\rho^{(\alpha)}(s) \bullet \\ \bullet \left(A_C^{(\alpha)} S_\rho^{(\alpha)}(s)\right)^{-1} A_C^{(\alpha)} Z_\rho^{(\alpha-1)}(s) - \\ -\left(E^{(\nu)} S_N^{(\nu)}(s)\right)^{-1} R_C^{(\alpha)} Z_\rho^{(\alpha-1)}(s) \end{array} \right] =$$

$$= p_{cIIO}^{-1}(s) \bullet$$

$$\bullet \left[\begin{array}{c} -adj\left(E^{(\nu)} S_N^{(\nu)}(s)\right) R_C^{(\alpha)} S_\rho^{(\alpha)}(s) \bullet \\ \bullet adj\left(A_C^{(\alpha)} S_\rho^{(\alpha)}(s)\right) A_C^{(\alpha)} Z_\rho^{(\alpha-1)}(s) - \\ -\left[\det\left(A_C^{(\alpha)} S_\rho^{(\alpha)}(s)\right)\right] adj\left(E^{(\nu)} S_N^{(\nu)}(s)\right) R_C^{(\alpha)} Z_\rho^{(\alpha-1)}(s) \end{array} \right] \tag{6.66}$$

is the *IIO* controller (6.51), (6.52) *transfer function matrix relative to the initial internal state vector* $\mathbf{R}_{C0}^{\alpha-1}$,

- $G_{IIOCU_0}(s)$,

$$G_{IIOCU_0}(s) = \left(E^{(\nu)} S_N^{(\nu)}(s)\right)^{-1} E^{(\nu)} Z_N^{(\nu-1)}(s) =$$

$$= p_{cIIO}^{-1}(s) \left(\det\left(A_C^{(\alpha)} S_\rho^{(\alpha)}(s)\right)\right) adj\left(E^{(\nu)} S_N^{(\nu)}(s)\right) E^{(\nu)} Z_N^{(\nu-1)}(s) \tag{6.67}$$

is the *IIO* controller (6.51), (6.52) *transfer function matrix relative to the initial control vector* $\mathbf{U}_0^{\nu-1}$,

   - $\mathbf{V}_{IIOC}(s)$ and $\mathbf{C}_{IIOC0}$,

$$\mathbf{V}_{IIOC}(s) = \begin{bmatrix} \mathbf{I}_C^{\mp}(s) \\ \mathbf{C}_{IIOC0} \end{bmatrix}, \ \mathbf{I}_{IIOC}(s) = \mathbf{E}(s), \ \mathbf{C}_{IIOC0} = \begin{bmatrix} \mathbf{e}_0^{\mu-1} \\ \mathbf{R}_{C0}^{\alpha-1} \\ \mathbf{U}_0^{\nu-1} \end{bmatrix}, \quad (6.68)$$

are the Laplace transform of the action vector $\mathbf{V}_{IIOC}(t)$ and the vector $\mathbf{C}_{IIOC0}$ of all initial conditions, respectively.

6.4   Exercises

Exercise 100 *1. Select a physical IIO plant.*

   2. Determine its time domain IIO mathematical model.

   3. Determine its complex domain IIO mathematical model: its full transfer function matrix and all its transfer function matrices, as well as the vectors $\mathbf{V}_{IIO}(s)$ and $\mathbf{C}_{IIO0}$.

Exercise 101 *1. Select an IIO controller.*

   2. Determine its time domain IIO mathematical model.

   3. Determine its complex domain IIO mathematical model: its full transfer function matrix and all its transfer function matrices, as well as the vectors $\mathbf{V}_{IIOC}(s)$ and $\mathbf{C}_{IIOC0}$.

Exercise 102 *1. Determine the time domain IIO mathematical model of the control system composed of the chosen IIO plant and IIO controller.*

   2. Determine the complex domain IIO mathematical model of the control system composed of the chosen IIO plant and IIO controller: its full transfer function matrix and all its transfer function matrices, as well as the vectors $\mathbf{V}_{IIOCS}(s)$ and $\mathbf{C}_{IIOCS0}$. Hint: Section 1.7 and Section 1.8.1.

Exercise 103 *Test all Lyapunov and BI stability properties of the chosen IIO controller, IIO controller and of the control system composed of them. Hint: [170, Part III.]*

Part II

TRACKING

Chapter 7

Fundamental control principle

7.1 Control axiom

The *fundamental control principle* explains what mutually, i.e., independent r control variables U_i, $i = 1, 2, \ldots, r$, i.e., what the control vector $\mathbf{U}$ (1.46), $\mathbf{U} \in \mathfrak{R}^r$, can achieve at most and what is necessary for them to satisfy in order to govern directly K mutually independent variables $Z_1, Z_2, \ldots$ $, Z_K$, i.e., to govern directly their vector $\mathbf{Z} \in \mathfrak{R}^K$, over some *time* interval $(t_1, t_2) \subseteq \mathfrak{T}$, $t_2 > t_1$, i.e., at every moment $t \in (t_1, t_2)$.

Axiom 104 *The fundamental control principle [175, Axiom 87, pp. 38, 39]*

The scalar form:

*a) In order for r control variables $U_i(.)$, $i = 1, 2, \ldots, r$, **to control simultaneously** K **independent variables** $Z_j(.)$, $j = 1, 2, \ldots, K$, **at every moment** $t \in (t_1, t_2) \subseteq \mathfrak{T}$, $t_2 > t_1$, it is necessary that: $r \geq K$.*

*b) In order for r control variables $U_i(.)$, $i = 1, 2, \ldots, r$, **to control** K **variables** $Z_j(.)$, $j = 1, 2, \ldots, K$, **at every moment** $(t_1, t_2) \subseteq \mathfrak{T}$, $t_2 > t_1$, **by controlling simultaneously** m **independent functions** $v_k(.)$, $v_k(.) :$ $\mathfrak{T} \times \mathfrak{R}^K \longrightarrow \mathfrak{R}$ at every $t \in (t_1, t_2)$, $k = 1, 2, \ldots, m$, which depend on the variables $Z_j(.)$, it is necessary that: $r \geq m$.*

*c) If $r < K$ then for r **control variables** $U_i(.)$, $i = 1, 2, \ldots, r$, **to control** K **variables** $Z_j(.)$, $j = 1, 2, \ldots, K$, at every moment $(t_1, t_2) \subseteq \mathfrak{T}$, $t_2 > t_1$, **by controlling simultaneously** m **independent functions** $v_k(.)$, $v_k(.) : \mathfrak{T} \times \mathfrak{R}^K \longrightarrow \mathfrak{R}$ at every $t \in (t_1, t_2)$, $k = 1, 2, \ldots, m$, which depend on the variables $Z_j(.)$, **it is necessary that** $m \leq r$.*

The vector form:

A) In order for the control vector $\mathbf{U} \in \mathfrak{R}^r$, $\mathbf{U} = [U_1\ U_2...U_r]^{\mathfrak{T}}$, to control elementwise the vector variable $\mathbf{Z} \in \mathfrak{R}^K$, $\mathbf{Z} = [Z_1\ Z_2...Z_K]^{\mathfrak{T}}$ with the independent entries, at every moment $t \in (t_1, t_2) \subseteq \mathfrak{T}$, $t_2 > t_1$, it is necessary that: $r \geq K$.

B) In order for the control vector $\mathbf{U} \in \mathfrak{R}^r$, $\mathbf{U} = [U_1\ U_2...U_r]^{\mathfrak{T}}$, to control the vector variable $Z \in R^K$, $Z = [Z_1\ Z_2...Z_K]^{\mathfrak{T}}$, at every moment $(t_1, t_2) \subseteq \mathfrak{T}$, $t_2 > t_1$, by controlling elementwise the vector function $v = [v_1\ v_2...v_m]^{\mathfrak{T}}$ with independent entries, at every moment (t_1, t_2), $v(.) : \mathfrak{T} \times \mathfrak{R}^K \longrightarrow \mathfrak{R}^m$, which depends on the vector variable Z, it is necessary $r \geq m$.

C) If $r < K$ then for the control vector $\mathbf{U} \in \mathfrak{R}^r$, $\mathbf{U} = [U_1\ U_2...U_r]^{\mathfrak{T}}$, to control K vector variable $Z \in R^K$, $Z = [Z_1\ Z_2...Z_K]^{\mathfrak{T}}$, at every moment $(t_1, t_2) \subseteq \mathfrak{T}$, $t_2 > t_1$, by controlling elementwise m vector function $v = [v_1\ v_2...v_m]^{\mathfrak{T}}$ with independent entries, $v(.) : \mathfrak{T} \times \mathfrak{R}^K \longrightarrow \mathfrak{R}^m$, $m \leq r$, which depends on the vector variable $\mathbf{Z}(.)$, it is necessary that : $r \geq m$.

7.2   Control perpetuum mobile

Corollary 105 *Control "perpetuum mobile" is impossible.*

The scalar form: The control "perpetuum mobile" means that r control variables $U_i(.)$, $i = 1, 2, \ldots, r$, control simultaneously K independent variables $Z_j(.)$, $j = 1, 2, \ldots, K > r$, at every moment $t \in (t_1, t_2)$. It is not possible for $K > r$.

The vector form: The control "perpetuum mobile" means that r control vector variable $\mathbf{U} \in \mathfrak{R}^r$, controls elementwise K vector variable $\mathbf{Z} \in \mathfrak{R}^K$, $K > r$, at every moment $t \in (t_1, t_2)$. It is not possible for $K > r$.

For these reasons the fundamental control principle 104 is the control axiom.

Chapter 8

Tracking fundamentals

8.1 Control goal and tracking concepts

8.1.1 Control purpose and tracking

The system desired behavior reflects its required, its aimed, dynamical behavior. The desired *time* evolution $\mathbf{Y}_d(t)$ of the system real output vector $\mathbf{Y}$ defines mathematically its desired dynamical output behavior.

The basic and primary purpose of control of a dynamical system (which can be *plant*, Definition 18, Section 1.5, or its whole control system, Definition 31, Section 1.5) *is to force the system (the plant, the control system) to behave sufficiently closely (to track sufficiently accurately) any its desired output behavior from a certain functional family over some, usually prespecified, time interval under real (usually unpredictable and unknown, hence arbitrary) both external (i.e., input) actions from another functional family and under unknown, arbitrary, initial conditions influence* [175], [188], [221, pp. 121-127].

Goal 106 *The main control goal [175], [188]*

The very, the primary, the essential, goal of control is to assure that the controlled plant, equivalently, the control system, exhibits a requested kind of **output tracking** *that we call, for short,* **tracking.**

Comment 107 *Control goal and control action*

In order to achieve the control goal the controller should realize such control to compensate both disturbance actions and initial errors influence on the system real behavior relative to its desired behavior (see Note 30, Section 1.6), whatever are the disturbances, initial errors and desired system behavior from the prespecified corresponding functional families or sets.

The term *tracking* in the wide sense concerns all kinds of the plant real output vector $\mathbf{Y}(t)$ *tracking* (i.e., *following*) its desired output vector $\mathbf{Y}_d(t)$. In the specific, but widely and commonly accepted, sense *tracking* signifies *asymptotic output tracking as time* $t \longrightarrow \infty$.

The control that forces the plant to exhibit an adequate type of **tracking** is **tracking control (TC)**.

Tracking and **tracking control synthesis** are the fundamental and original control issues. They have the same importance for control systems as stability for dynamical systems in general. For the more detailed comparison between stability and tracking and for a short historical review see [188, Subsection 8.1, pp. 119-122, Subsection 8.2, pp. 122-124], [175, Subsection 13.2, pp. 102, 103, Subsection 14.2, pp. 106, 107].

8.1.2   Basic tracking meaning

In the control literature tracking has been mainly and largely studied as the zero steady state error problem. It was considered as the stability problem .This means that the control synthesis should assure for the control to force the plant real state to approach asymptotically the plant desired state only as *time* t tends to infinity.

By the definition (Definition 22, Section 1.5) we are substantially interested in the system real output behavior expressed by the *time* evolution of the real output vector $\mathbf{Y}(t)$ relative to the desired output behavior defined by the *time* evolution of the system desired output vector $\mathbf{Y}_d(t)$. The simplest suitable their relationship has been expressed by the demand that the former asymptotically converges to the latter as *time* t escapes to infinity, i.e., that the *output error* (equivalently, *output deviation*) vector $\mathbf{e}(t)$ ($\mathbf{y}(t)$), respectively, approaches the zero vector as *time* t escapes to infinity:

$$\mathbf{Y}(t) \longrightarrow \mathbf{Y}_d(t), \ \ i.e., \ \ \mathbf{e}(t) = -\mathbf{y}(t) \longrightarrow \mathbf{0}_N, \ as \ t \longrightarrow \infty. \qquad (8.1)$$

The beginning of the tracking studies in this sense, might be considered as the beginning of the control studies. They have been known under different names such as studies of servomechanisms/servosystems, or of regulation systems, or of control systems in general comprising the preceding ones. The study of the zero steady *state* (rather than *output) error* has been commonly incorporated in stability and stabilization studies, which might be a reason for which the control field was lacking the common and general tracking theory in its own right probably until it was recognized 1980 [157], [158], [271], [272]. (For more details see [175]).

The definition of any tracking property should clarify the following [175], [188]:

- *The characterization of the plant behavior we are interested in*, whether we are interested in the internal dynamical (i.e., the state) behavior of the plant, or in the plant output dynamical behavior; or in both;

- *The space in which the demanded closeness is to be achieved*, which means that, although originally tracking concerns the output behavior, we can consider the output tracking either via *the output space* or via *the state space* or via their space product;

- *The definition of the distance between the real behavior and the desired behavior* of the plant;

- *The definition of the demanded closeness of the real behavior to the desired behavior* of the plant;

- *The nonempty sets of the initial conditions of all plant variables* under which the demanded closeness is to be achieved;

- *The nonempty set $\mathfrak{D}^{(\cdot)}$ of permitted external disturbances* acting on the plant, under which the demanded closeness is to be realized;

- *The nonempty set $\mathfrak{Y}_d^{(\cdot)}$ of realizable desired plant behaviors* that can be demanded;

- *The time interval over which the demanded closeness is to be guaranteed*; and

- *The requested quality with which the real behavior is to track the desired behavior* of the plant.

The notion, the sense, and the meaning of **tracking** signify in the sequel that the real plant output *tracks, follows*, every plant desired output that belongs to a functional family $\mathfrak{Y}_d^k$, $k \in \{0, 1, ..., \alpha - 1\}$

- *Either the desired output is constant* (in a part of the control literature this is linked exclusively with the regulation systems) *or time-varying* (in another part of the control literature it is associated only with the servomechanisms/servosystems),

- *Under the actions of arbitrary external disturbances* that belong to a family $\mathfrak{D}^j$, $j \in \{0, 1, ...\}$

and

- Under arbitrary (input, state, and output) initial conditions, [175], [188]

Tracking theory incorporates both the servomechanism/servosystem theory and the regulation theory.

We can differently specify the preceding requirements. Their different specifications lead to numerous various tracking concepts, each containing a number of different tracking properties.

For a review of different tracking concepts and the related literature see [175], [188].

Claim 108 *Space and tracking*

Relationship (8.1) explains that the appropriate space for studying the tracking is the system output space $\mathfrak{R}^N$, and that the fully adequate space is the system integral output space $\mathcal{I}$ defined in Equation (1.29) (Section 1.5), i.e.,

$$\mathcal{I} = \mathfrak{T} \times \mathfrak{R}^N. \tag{8.2}$$

It is the set product $\mathcal{I}$ of the time set $\mathfrak{T}$ and the output space $\mathfrak{R}^N$.

Comment 109 *The tracking and the state tracking*

*If and only if the system state vector $\mathbf{S}$ is simultaneously (accepted for) the system output vector $\mathbf{Y}$, $\mathbf{S} = \mathbf{Y}$, then the system output response becomes simultaneously the system motion. Then the system tracking becomes the system **state tracking** and ensures the (global) attraction of any system desired motion that belongs to $\mathfrak{Y}_d^k$ for every $\mathbf{D}(.) \in \mathfrak{D}^k$ and every $\mathbf{Y}_d^k(.) \in \mathfrak{Y}_d^k$ rather than only for a single nominal disturbance $\mathbf{D}_N(.)$ and for a single desired motion. For this reason and due to the page limitation we will not treat separately state tracking in this book, which is done in [175] and in [188].*

This illustrates that the general tracking theory established in [175] incorporates the Lyapunov stability theory.

Comment 110 *The state tracking to guarantee the tracking*

The system desired output behavior $\mathbf{Y}_d(t)$ determines the system desired state behavior $\mathbf{S}_d(t)$. If the system desired state behavior $\mathbf{S}_d(t)$ is calculated for every $\mathbf{Y}_d^k(.) \in \mathfrak{Y}_d^k$ then it is reasonable to accept the state vector $\mathbf{S}$ for the new output vector $\mathbf{Y}^*$, $\mathbf{S} = \mathbf{Y}^*$, so that the system desired state behavior $\mathbf{S}_d(t)$ becomes the new desired output behavior $\mathbf{Y}_d^*(t)$. The state tracking of $\mathbf{S}_d(t)$ becomes the tracking of the new desired output behavior $\mathbf{Y}_d^*(t)$. All what is valid for the tracking of $\mathbf{Y}_d(t)$ is to be applied to the tracking of $\mathbf{Y}_d^*(t)$. With this in mind we continue to deal with the tracking of the desired output $\mathbf{Y}_d(t)$.

Tracking is *perfect* (*ideal*) if and only if the plant real output behavior $\mathbf{Y}(t)$ is always equal to the plant desired output behavior $\mathbf{Y}_d(t)$, $\mathbf{Y}(t) \equiv \mathbf{Y}_d(t)$. If the initial real output $\mathbf{Y}_0$ is different from the initial desired output $\mathbf{Y}_{d0}$, $\mathbf{Y}_0 \neq \mathbf{Y}_{d0}$, then tracking can be only *imperfect*. We will consider both perfect and imperfect tracking.

8.2 Perfect tracking

The perfect tracking means the ideal tracking. It discovers to us tracking that is (even only theoretically) the best possible. The definition of its exact meaning follows [188, Definition 156, p. 124], [175, Definition 195, p. 109]:

Definition 111 *The k-th-order perfect tracking of the system on $\mathfrak{T}_0$*
*The system exhibits **the k-th-order perfect tracking on** $\mathfrak{T}_0$, $\mathfrak{T}_0 \subseteq \mathfrak{T}$, of its desired k-th-order extended output vector response $\mathbf{Y}_d^k(t)$ if and only if its real k-th-order output vector response $\mathbf{Y}^k(t)$ is always equal to its desired k-th-order output vector response $\mathbf{Y}_d^k(t)$,*

$$\mathbf{Y}^k(t) = \mathbf{Y}_d^k(t), \ \forall t \in \mathfrak{T}_0, \ \mathfrak{T}_0 \subseteq \mathfrak{T}, \ k \in \{0, 1, 2, ...\}, \tag{8.3}$$

$$equivalently: \ \mathbf{e}^k(t) = \mathbf{0}_{(k+1)N}, \ \forall t \in \mathfrak{T}_0, \ \mathfrak{T}_0 \subseteq \mathfrak{T}, \tag{8.4}$$

$$equivalently: \ \mathbf{y}^k(t) = \mathbf{0}_{(k+1)N}, \ \forall t \in \mathfrak{T}_0, \ \mathfrak{T}_0 \subseteq \mathfrak{T}. \tag{8.5}$$

*If and only if $k = 0$, then the zero-order perfect tracking on $\mathfrak{T}_0$ is simply called **perfect tracking on** $\mathfrak{T}_0$.*
The expression "on $\mathfrak{T}_0$" is to be omitted if and only if $\mathfrak{T}_0 = \mathfrak{T}$.

This definition implies directly the following obvious simple, but useful, result:

Theorem 112 *The necessary condition for the perfect tracking*
In order for a dynamical system to exhibit the k-th order perfect tracking on $\mathfrak{T}_0$ it is necessary (but not sufficient) that the initial extended real output vector $\mathbf{Y}^k(t_0)$ is equal to the initial extended desired output vector $\mathbf{Y}_d^k(t_0)$, $k \in \{0, 1, 2, ...\}$:

$$\left(\forall t \in \mathfrak{T}_0: \ \mathbf{Y}^k(t) = \mathbf{Y}_d^k(t) \right) \Longrightarrow \mathbf{Y}^k(t_0) = \mathbf{Y}_d^k(t_0) \Longleftrightarrow \mathbf{e}^k(t_0) = \mathbf{0}_{(k+1)N}.$$

Corollary 113 *Possibility for the perfect tracking*
If the initial extended real output vector $\mathbf{Y}^k(t_0)$ is not equal to the initial extended desired output vector $\mathbf{Y}_d^k(t_0)$, $k \in \{0, 1, 2, ...\}$, then the system cannot exhibit the k-th order perfect tracking on $\mathfrak{T}_0$.

Note 114 Tracking reality: a suitable imperfect tracking

It is rare for the initial extended real output vector $\mathbf{Y}^k(t_0)$ to be equal to the initial extended desired output vector $\mathbf{Y}_d^k(t_0)$, $k \in \{0, 1, 2, ...\}$. The k-th order perfect tracking on $\mathfrak{T}_0$ is rarely possible. Facing this fact the only possibility is to look for a most suitable the k-th order imperfect tracking on $\mathfrak{T}_0$.

Comment 115 On the influence of the initial moment t_0

The choice of the initial moment t_0 does not influence the properties of time-invariant dynamical systems. For them, we can accept $\mathfrak{T}_0 = \mathfrak{T}$ and from this point of view we may omit the expression "on $\mathfrak{T}_0$". However, the characteristics of the disturbance variables and of the desired output variables may depend on the initial moment, which is a reason to preserve the term "on $\mathfrak{T}_0$" if t_0 is not fixed. In order to simplify the presentation, we continue with the fixed initial moment to be the zero moment so that $\mathfrak{T}_0$ is also fixed:

$$t_0 = 0 \ and \ \mathfrak{T}_0 = \{t : t \in \mathfrak{T}, \ t \geq 0\} = [0, \infty[. \tag{8.6}$$

With this in mind the expression "on $\mathfrak{T}_0$" may be omitted in the sequel.

The following theorem discovers the link between the k-th order perfect tracking and the perfect tracking:

Theorem 116 Perfect tracking and the k-th order perfect tracking
[188, Theorem 157, p. 125], [175, Theorem 196, p. 109]

If the real output vector function $\mathbf{Y}(.)$ and the desired output vector function $\mathbf{Y}_d(.)$ are k-times continuously differentiable on $\mathfrak{T}_0$ then for the validity of (8.3) it is necessary and sufficient that

$$\mathbf{Y}(t) = \mathbf{Y}_d(t), \ \forall t \in \mathfrak{T}_0, \ i.e., \ \mathbf{e}(t) = \mathbf{0}_N, \ \forall t \in \mathfrak{T}_0, \tag{8.7}$$

holds, i.e.,

$$\mathbf{Y}(t) \in \mathfrak{C}^k(\mathfrak{T}_0) \ \ and \ \ \mathbf{Y}_d(t) \in \mathfrak{C}^k(\mathfrak{T}_0) \Longrightarrow$$

$$\left\langle \mathbf{Y}^k(t) = \mathbf{Y}_d^k(t), \ \forall t \in \mathfrak{T}_0 \right\rangle \Longleftrightarrow \left\langle \mathbf{Y}(t) = \mathbf{Y}_d(t), \forall t \in \mathfrak{T}_0 \right\rangle \Longleftrightarrow$$

$$\Longleftrightarrow \left\langle \mathbf{e}^k(t) = \mathbf{0}_{(k+1)N}, \ \forall t \in \mathfrak{T}_0 \right\rangle \Longleftrightarrow \left\langle \mathbf{e}(t) = \mathbf{0}_N, \forall t \in \mathfrak{T}_0 \right\rangle. \tag{8.8}$$

Equivalently, for the system to exhibit the k-th order perfect tracking it is necessary and sufficient to exhibit the perfect tracking.

Conclusion 117 *This is general theorem on the perfect tracking. Its proof is elaborated in [188, Theorem 157, p. 125]. It holds whatever is the form of the system mathematical model. It can be either linear or nonlinear, time-invariant or time-varying. It permits us to reduce the study of the k-th order perfect tracking to the perfect tracking.*

8.3 Imperfect infinite-*time* tracking

8.3.1 Introduction

The book [175, Part V, pp. 113-141] characterizes the imperfect tracking in the general framework of *time*-varying nonlinear systems. In the framework of the *time*-invariant continuous-*time* linear systems the imperfect tracking characterizations follow.

The extended output error vector $\mathbf{e}^k$, is equal to the zero vector $\mathbf{0}_{(k+1)N}$ if and only if the extended real output vector $\mathbf{Y}^k$ is equal to the extended desired output vector $\mathbf{Y}_d^k$ by its definition, $\mathbf{e}$ is the output error vector, Equation (1.53), Section 1.5. The same holds for the output error deviation vector $\mathbf{y}^k$ due to (2.56), Section 2.2. This means that the difference $\mathbf{Y}_d^k - \mathbf{Y}^k$ can be replaced by $\mathbf{e}^k$ and $\mathbf{Y}^k - \mathbf{Y}_d^k$ can be replaced by $\mathbf{y}^k$, where $\mathbf{y}^k = -\mathbf{e}^k$. This consideration explains that the definitions of various tracking properties can be expressed in terms of the total real output vector $\mathbf{Y}$ and the total desired output vector $\mathbf{Y}_d$, or equivalently in terms of the output error vector $\mathbf{e}$ or the output deviation vector $\mathbf{y}$.

$\|.\| : \mathfrak{R}^N \to \mathfrak{R}_+$ is *any accepted norm on* $\mathfrak{R}^N$, which is to be the *Euclidean norm* $\|.\|_2$ *on* $\mathfrak{R}^N$ if and only if not stated otherwise:

$$\|\mathbf{Y}\| = \|\mathbf{Y}\|_2 = \sqrt{\mathbf{Y}^T\mathbf{Y}} = \sqrt{\sum_{i=1}^{i=N} Y_i^2}. \tag{8.9}$$

8.3.2 Tracking in the Lyapunov sense

If the reader is familiar with the Lyapunov stability concept [195], [230] then it will be very easy to understand that what follows represents *the tracking in the Lyapunov sense*, for short *Lyapunov tracking.*, It generalizes and crucially extends the Lyapunov stability concept in the framework of control systems. The former has been becoming a fundamental concept of the control theory and engineering, the latter has been the fundamental concept of the dynamical systems theory. The control systems form a subfamily of the dynamical systems.

We define in the output space $\mathfrak{R}^N$, or more precisely, in the integral output space $\mathcal{I} = \mathfrak{T} \times \mathfrak{R}^N$, several typical asymptotic tracking properties in *the Lyapunov sense*. We call them *Lyapunov tracking properties* or for short *L-tracking properties*. This means the following:

    1.   The system desired output behavior $\mathbf{Y}_d(.)$ can be any from the functional family $\mathfrak{Y}_d^k$ of realizable desired system output behaviors, where $k \in \{0, 1, ...\}$.

    2.   The vector disturbance function $\mathbf{D}(.)$ can be any from another functional family $\mathfrak{D}^k$ of permitted disturbance vector functions.

    3.   The initial conditions can be arbitrary and unknown.

    4.   If and only if every real system output behavior $\mathbf{Y}^k(t)$ starting initially from $\mathbf{Y}_0^k$ in some $\Delta-$connected neighborhood of the system desired initial output behavior $\mathbf{Y}_{d0}^k$ at the initial moment $t_0 = 0$ converges asymptotically to the system desired output behavior $\mathbf{Y}_d^k(t)$ as time t goes to infinity for every disturbance $\mathbf{D}(.) \in \mathfrak{D}^k$ and for every $\mathbf{Y}_d^k(.) \in \mathfrak{Y}_d^k$,

$$\exists \Delta \in \mathfrak{R}^+, \ \|\mathbf{Y}_0^k - \mathbf{Y}_{d0}^k\| < \Delta \Longrightarrow \mathbf{Y}^k(t) \longrightarrow \mathbf{Y}_d^k(t) \ as \ t \longrightarrow \infty,$$

$$\forall \mathbf{D}(.) \in \mathfrak{D}^k, \ \forall \mathbf{Y}_d^k(.) \in \mathfrak{Y}_d^k, \tag{8.10}$$

then and only then **the system exhibits the asymptotic tracking on** $\mathfrak{D}^k \times \mathfrak{Y}_d^k$, for short ***tracking***, *where the set product* $\mathfrak{D}^k \times \mathfrak{Y}_d^k$ *is fixed. If this closeness holds for every vector* $\mathbf{Y}_0^k \in \mathfrak{R}^{(k+1)N}$, *then and only then the system exhibits its global tracking on* $\mathfrak{D}^k \times \mathfrak{Y}_d^k$.

    5.   The Lyapunov closeness of vector functions $\mathbf{Y}^k(.)$ and $\mathbf{Y}_d^k(.)$ means that for every $\varepsilon \in \mathfrak{R}^+$ there exists $\delta \in \mathfrak{R}^+$ such that for every system real initial extended output vector $\mathbf{Y}_0^k$ in the $\delta-$ connected neighborhood of $\mathbf{Y}_{d0}^k$ the system real extended output behavior $\mathbf{Y}^k(t)$ rests in the $\varepsilon-$ connected neighborhood of $\mathbf{Y}_d^k(t)$ forever, i.e., for all $t \in \mathfrak{T}_0$,

$$\forall \varepsilon \in \mathfrak{R}^+, \ \exists \delta \in \mathfrak{R}^+,$$

$$\|\mathbf{Y}_{d0}^k - \mathbf{Y}_0^k\| = \left\|\mathbf{e}_0^k\right\| < \delta \Longrightarrow \|\mathbf{Y}_d^k(t) - \mathbf{Y}^k(t)\| = \left\|\mathbf{e}^k(t)\right\| < \varepsilon, \ \forall t \in \mathfrak{T}_0. \tag{8.11}$$

If this Lyapunov closeness holds then and only then it ensures also the Lyapunov stability to the desired output system behavior $\mathbf{Y}_d^k(.)$ but not to the system desired motion (desired state behavior) $\mathcal{S}_d(.)$ in general.

    6.   A real behavior $\mathbf{Y}^k(t)$ of a system should track any system desired behavior $\mathbf{Y}_d^k(.)$, $\mathbf{Y}_d(.) \in \mathfrak{Y}_d^k$ for any $\mathbf{D}(.) \in \mathfrak{D}^k$ so that the Lyapunov closeness among them holds on $\mathfrak{T}_0$.

7. When all above requirements under 1. to 6. hold on $\mathfrak{D}^k \times \mathfrak{Y}_d^k$ then and only then the desired output behavior $\mathbf{Y}_d^k(.)$ is asymptotically stable for every $\left(\mathbf{D}(.), \mathbf{Y}_d^k(.)\right) \in \mathfrak{D}^k \times \mathfrak{Y}_d^k$ and **the system exhibits *stablewise tracking on* $\mathfrak{D}^k \times \mathfrak{Y}_d^k$.**

Tracking in the Lyapunov sense demands and ensures more than the Lyapunov stability for the following reasons:

The requested closeness is to be realized under the following conditions:

i) The k-th order *output behavior* $\mathbf{Y}^k(.) = \mathcal{Y}^k(.; \mathbf{Y}^k; \mathbf{D}; \mathbf{U})$ for some $k \in \{0, 1, 2, \ldots\}$ should obey the demanded closeness (rather than only that the whole real system motion $\mathcal{S}(.)$ should obey the Lyapunov closeness, which is strictly demanded in the Lyapunov stability theory).

ii) The k-th order real *output behavior* $\mathbf{Y}^k(.) = \mathcal{Y}^k(.; \mathbf{Y}^k; \mathbf{D}; \mathbf{U})$ should track *any* k-th order *desired system behavior* $\mathbf{Y}_d^k(.)$ *from a (given, or to be determined) family* $\mathfrak{Y}_d^k$ of possibly system demanded realizable desired behaviors $\mathbf{Y}_d(.)$, i.e., *tracking should hold over the desired output family* $\mathfrak{Y}_d^k$.

iii) The condition ii) should be fulfilled *for every external disturbance* $\mathbf{D}(.)$ *from a (given, or to be determined) family* $\mathfrak{D}^k$ of permitted external disturbances, rather than only for the nominal $\mathbf{D}_N(.)$. The Lyapunov stability theory does not permit nonnominal disturbances. Therefore, the Lyapunov stability properties represent special cases of the corresponding asymptotic tracking properties in the Lyapunov sense.

8. The reachability *time* is infinite. This concept does not demand that the real output vector and its derivatives composing $\mathbf{Y}^k(t)$ and the desired output vector and its derivatives forming $\mathbf{Y}_d^k(t)$ become equal in a finite *time*. It ensures the asymptotic convergence of the former to the latter only as *time t* escapes to infinity: $t \longrightarrow \infty$,

$$t \longrightarrow \infty \Longrightarrow \mathbf{Y}^k(t) \longrightarrow \mathbf{Y}_d^k(t), \quad \text{equivalently} \quad \mathbf{e}^k(t) \longrightarrow \mathbf{0}_{(k+1)N}.$$

The tracking concept in the Lyapunov sense is *the infinite time tracking concept*.

8.3.3 Tracking versus stability

A linear system can be stable (its equilibrium vector can be globally asymptotically stable), but it need not exhibit tracking. System stability is not sufficient for system tracking in general [188, Section 8.2, pp. 122-124].

A linear system can exhibit a kind of tracking although it is unstable [188, Section 8.2, Example 152, p.123].

These are some of the reasons and the needs to study tracking both as a self-contained issue and as a phenomenon related to stability.

Conclusion 118 *Stability and tracking are in general mutually independent concepts*

Stability properties do not guarantee tracking properties in general, and tracking properties do not imply stability properties in general. They can be mutually independent. However, system stability appears often necessary for tracking. Besides,tracking can be sufficient for system stability. A good design assures both system stability and requested type of tracking.

Note 119 *Control vector partitioning*

If the plant is not stable, then we can (due to the system linearity) partition the full total control vector $\mathbf{U}_F$ into

-    *The stabilizing total control vector $\mathbf{U}_S$, and*
-    *The tracking total control vector $\mathbf{U}_T$, so that*

$$\mathbf{U}_F = \mathbf{U}_S + \mathbf{U}_T. \tag{8.12}$$

The control synthesis can start by applying any of the stabilization methods to synthesize the stabilizing control vector $\mathbf{U}_S$. The stabilizing controller and so controlled plant constitute then a stable system that will be tretaed as a stable plant for the synthesis of the tracking control $\mathbf{U}_T$. The next step is then to synthesize the tracking control $\mathbf{U}_T$.

We will denote in the sequel $\mathbf{U}_T$ simply by $\mathbf{U}$ regardless of stability or instability of the plant,

$$\mathbf{U}_T = \mathbf{U}. \tag{8.13}$$

The book deals only with synthesis of the tracking control $\mathbf{U}_T$.

8.3.4   Stablewise tracking in general

The natural environment to define tracking properties is the output space $\mathfrak{R}^N$ (Definition 22, Section 1.5), i.e., *the integral output space* $\mathcal{I} = \mathfrak{T} \times \mathfrak{R}^N$, or *the extended output space* $\mathfrak{R}^{(k+1)N}$, i.e., *the extended integral output space* $\mathcal{I}^{(k+1)N} = \mathfrak{T} \times \mathfrak{R}^{(k+1)N}$.

The precise definitions of the tracking properties in the Lyapunov sense follow.

$\mathcal{D}_{(.)} \left(\mathbf{Y}_{d0}^k; \mathbf{D}; \mathbf{U}; \mathbf{Y}_d^k \right)$ is the set of permissible initial extended output vectors $\mathbf{Y}_{d0}^k$. The notation $\mathcal{D}_{(.)} \left(\mathbf{Y}_{d0}^k; \mathbf{D}; \mathbf{U}; \mathbf{Y}_d^k \right)$ emphasizes the dependence of the volume, size and shape of the set $\mathcal{D}_{(.)}$ on:

- The accepted initial desired extended output vector $\mathbf{Y}_{d0}^k$,

- The disturbance vector function $\mathbf{D}\left(.\right)$ and the maximal norm of its instantaneous values on $\mathfrak{T}_0$,

- The control vector function $\mathbf{U}\left(.\right)$ and the maximal vector norm of its instantaneous values on $\mathfrak{T}_0$,

- The accepted desired extended output vector function $\mathbf{Y}_d^k\left(.\right)$.

Note 120 *The influence of the relation between the disturbance and control on the tracking domain* $\mathcal{D}_{(.)}$

It is logical that the greater $\max\left\{\left\|\mathbf{D}\left(t\right)\right\| : t \in \mathfrak{T}_0\right\}$ *demands greater elementwise* $\max\left\{\left|\mathbf{U}\left(t\right)\right| : t \in \mathfrak{T}_0\right\}$ *for a fixed set* $\mathcal{D}_{(.)}$. *If we intend to increase the set* $\mathcal{D}_{(.)}$ *then elementwise* $\max\left\{\left|\mathbf{U}\left(t\right)\right| : t \in \mathfrak{T}_0\right\}$ *should be appropriately increased. Since the system is linear we do not restrict a priory the extended initial output vector* $\mathbf{Y}_{d0}^k$. *However, the relationship between* $\max\left\{\left\|\mathbf{D}\left(t\right)\right\| : t \in \mathfrak{T}_0\right\}$ *and* $\max\left\{\left|\mathbf{U}\left(t\right)\right| : t \in \mathfrak{T}_0\right\}$ *can restrict also permitted initial extended output vector* $\mathbf{Y}_{d0}^k$.

Notice the difference and the following relationship between the neighborhoods **in the output vector space:**

$$\mathfrak{N}_\varepsilon\left[t; \mathbf{Y}_d^k\left(t\right); \mathbf{D}; \mathbf{U}; \mathbf{Y}_d^k\right] \quad and \quad \mathfrak{N}\left(\varepsilon; \mathbf{Y}_{d0}^k; \mathbf{D}; \mathbf{U}; \mathbf{Y}_d^k\right),$$

and **in the output error vector space:**

$$\mathfrak{N}_\varepsilon\left[\mathbf{0}_{(k+1)N}; \mathbf{D}; \mathbf{U}; \mathbf{Y}_d^k\right] \quad and \quad \mathfrak{N}\left(\varepsilon; \mathbf{0}_{(k+1)N}; \mathbf{D}; \mathbf{U}; \mathbf{Y}_d^k\right),$$

$\mathfrak{N}_\varepsilon\left[t; \mathbf{Y}_d^k\left(t\right); \mathbf{D}; \mathbf{U}; \mathbf{Y}_d^k\right]$ is a *time*-varying connected ε-neighborhood of $\mathbf{Y}_d^k\left(t\right)$ at a moment $t \in \mathfrak{T}_0$, $\mathfrak{N}_\varepsilon\left[t; \mathbf{Y}_d^k\left(t\right); \mathbf{D}; \mathbf{U}; \mathbf{Y}_d^k\right] \subseteq \mathfrak{R}^{(k+1)N}$,

$$\mathfrak{N}_\varepsilon\left[t; \mathbf{Y}_d^k\left(t\right); \mathbf{D}; \mathbf{U}; \mathbf{Y}_d^k\right] \equiv \left\{\mathbf{Y}^k \colon \mathbf{Y}^k \in \mathfrak{R}^{(k+1)N}, \ d\left[\mathbf{Y}^k, \mathbf{Y}_d^k\left(t\right)\right] < \varepsilon\right\},$$

where $d\left[\mathbf{Y}^k, \mathbf{Y}_d^k\left(t\right)\right]$ is *the distance* between $\mathbf{Y}^k$ and $\mathbf{Y}_d^k\left(t\right)$ at $t \in \mathfrak{T}_0$,

$$d\left[\mathbf{Y}^k, \mathbf{Y}_d^k\left(t\right)\right] = \left\|\mathbf{Y}^k - \mathbf{Y}_d^k\left(t\right)\right\| = \left\|\mathbf{Y}_d^k\left(t\right) - \mathbf{Y}^k\right\| = \left\|\mathbf{e}^k\left(t\right)\right\|, \ t \in \mathfrak{T}_0,$$

$\mathfrak{N}\left(\varepsilon; \mathbf{Y}_{d0}^k; \mathbf{D}; \mathbf{U}; \mathbf{Y}_d^k\right)$ is a *time*-invariant connected neighborhood of $\mathbf{Y}_{d0}^k$ at the initial moment $t = t_0 = 0$, which is determined by $\varepsilon \in \mathfrak{R}^+, \mathbf{D}\left(.\right) \in \mathfrak{D}^j,$

$\mathbf{U}(.) \in \mathfrak{U}_d^l$, and $\mathbf{Y}_d(.) \in \mathfrak{Y}_d^k$, and the outer radius of which cannot be greater than ε,

$$\forall \mathfrak{N}_\varepsilon \left[t; \mathbf{Y}_d^k(t); \mathbf{D}; \mathbf{U}; \mathbf{Y}_d^k \right] \subseteq \mathfrak{R}^{(k+1)N}, \; \exists \mathfrak{N} \left(\varepsilon; \mathbf{Y}_{d0}^k; \mathbf{D}; \mathbf{U}; \mathbf{Y}_d^k \right) \subseteq \mathfrak{R}^{(k+1)N},$$

$$0 < \varepsilon_1 < \varepsilon_2 \Longrightarrow \mathfrak{N} \left(\varepsilon_1; \mathbf{Y}_{d0}^k; \mathbf{D}; \mathbf{U}; \mathbf{Y}_d^k \right) \subseteq \mathfrak{N} \left(\varepsilon_2; t_0; \mathbf{Y}_{d0}^k; \mathbf{D}; \mathbf{U}; \mathbf{Y}_d^k \right),$$

$$\varepsilon \longrightarrow 0^+ \Longrightarrow \mathfrak{N} \left(\varepsilon; t_0; \mathbf{Y}_{d0}^k; \mathbf{D}; \mathbf{U}; \mathbf{Y}_d^k \right) \longrightarrow \left\{ \mathbf{Y}_{d0}^k \right\},$$

$$\mathfrak{N} \left(\varepsilon; \mathbf{Y}_{d0}^k; \mathbf{D}; \mathbf{U}; \mathbf{Y}_d^k \right) \subseteq \mathfrak{N}_\varepsilon \left[t_0; \mathbf{Y}_d^k(0); \mathbf{D}; \mathbf{U}; \mathbf{Y}_d^k \right]. \tag{8.14}$$

The neighborhood $\mathfrak{N} \left(\varepsilon; \mathbf{Y}_{d0}^k; \mathbf{D}; \mathbf{U}; \mathbf{Y}_d^k \right)$ is usually accepted to be the δ-neighborhood $\mathfrak{N}_\delta \left(\mathbf{Y}_{d0}^k; \mathbf{D}; \mathbf{U}; \mathbf{Y}_d^k \right)$ of $\mathbf{Y}_{d0}^k$ for $0 < \delta = \delta(\varepsilon) \leq \varepsilon$, $\forall \varepsilon \in \mathfrak{R}^+$. In general

$$\mathfrak{N}_{\delta(\varepsilon)} \left(\mathbf{Y}_{d0}^k; \mathbf{D}; \mathbf{U}; \mathbf{Y}_d^k \right) \subseteq \mathfrak{N} \left(\varepsilon; \mathbf{Y}_{d0}^k; \mathbf{D}; \mathbf{U}; \mathbf{Y}_d^k \right) \subseteq \mathfrak{N}_\varepsilon \left(t_0, \mathbf{Y}_{d0}^k; \mathbf{D}; \mathbf{U}; \mathbf{Y}_d^k \right).$$

Definition 121 *Tracking of the extended desired output behavior* $\mathbf{Y}_d^k(t)$ *on the set product* $\mathfrak{D}^j \times \mathfrak{Y}_d^k$ *of the system controlled by a control* $\mathbf{U}(.) \in \mathfrak{U}^l$

a) The system exhibits **the asymptotic output tracking of** $\mathbf{Y}_d^k(t)$, $\mathbf{Y}_d(.) \in \mathfrak{Y}_d^k$, *on* $\mathfrak{D}^j \times \mathfrak{Y}_d^k$, *for short* **the tracking of** $\mathbf{Y}_d^k(t)$ *on* $\mathfrak{D}^j \times \mathfrak{Y}_d^k$, *if and only if for every* $[\mathbf{D}(.), \mathbf{Y}_d(.)] \in \mathfrak{D}^j \times \mathfrak{Y}_d^k$ *there exists a connected neighborhood* $\mathfrak{N} \left(\mathbf{Y}_{d0}^k; \mathbf{D}; \mathbf{U}; \mathbf{Y}_d^k \right) \subseteq \mathfrak{R}^{(k+1)N}$ *of the plant extended desired initial output vector* $\mathbf{Y}_{d0}^k$ *and for every* $\varsigma > 0$ *there exists a nonnegative real number* τ, $\tau = \tau \left(\varsigma, \mathbf{Y}_{d0}^k; \mathbf{D}; \mathbf{U}; \mathbf{Y}_d^k \right) \in \mathfrak{R}_+$, *such that* $\mathbf{Y}_0^k$ *from* $\mathfrak{N} \left(\mathbf{Y}_{d0}^k; \mathbf{D}; \mathbf{U}; \mathbf{Y}_d^k \right)$ *guarantees that the extended output vector* $\mathbf{Y}^k(t; \mathbf{Y}_0^k; \mathbf{D}; \mathbf{U})$ *belongs to the* ς-neighborhood $\mathfrak{N}_\varsigma \left(t; \mathbf{Y}_{d0}^k; \mathbf{D}; \mathbf{U}; \mathbf{Y}_d^k \right)$ *of* $\mathbf{Y}_d^k(t; \mathbf{Y}_{d0}^k)$ *for all time*

$$t \in]\tau \left(\varsigma, \mathbf{Y}_0^k; \mathbf{D}; \mathbf{U}; \mathbf{Y}_d^k \right), \infty[,$$

i.e.,

$$\forall \varsigma > 0, \; \forall [\mathbf{D}(.), \mathbf{Y}_d(.)] \in \mathfrak{D}^j \times \mathfrak{Y}_d^k,$$

$$\exists \mathfrak{N} \left(\mathbf{Y}_{d0}^k; \mathbf{D}; \mathbf{U}; \mathbf{Y}_d^k \right) \subseteq \mathfrak{R}^{(k+1)N}, \; \mathbf{Y}_0^k \in \mathfrak{N} \left(\mathbf{Y}_{d0}^k; \mathbf{D}; \mathbf{U}; \mathbf{Y}_d^k \right) \Longrightarrow$$

$$\left\{ \begin{array}{c} \mathbf{Y}^k(t; \mathbf{Y}_0^k; \mathbf{D}; \mathbf{U}) \in \mathfrak{N}_\varsigma \left(t; \mathbf{Y}_{d0}^k; \mathbf{D}; \mathbf{U}; \mathbf{Y}_d^k \right), \\ \forall t \in]\tau \left(\varsigma, \mathbf{Y}_0^k; \mathbf{D}; \mathbf{U}; \mathbf{Y}_d^k \right), \infty[\end{array} \right\}. \tag{8.15}$$

This is called also **the k-th order asymptotic output tracking of** $\mathbf{Y}_d(t)$, $\mathbf{Y}_d(.) \in \mathfrak{Y}_d^k$, *on* $\mathfrak{D}^j \times \mathfrak{Y}_d^k$, *for short* **the k-th order tracking of** $\mathbf{Y}_d(t)$ *on* $\mathfrak{D}^j \times \mathfrak{Y}_d^k$.

b) The largest connected neighborhood $\mathfrak{N}\left(\mathbf{Y}_{d0}^k; \mathbf{D}; \mathbf{U}; \mathbf{Y}_d^k\right)$ *of* $\mathbf{Y}_{d0}^k$ *that obeys (8.15), is* **the k-th order tracking domain** $\mathcal{D}_T\left(\mathbf{Y}_{d0}^k; \mathbf{D}; \mathbf{U}; \mathbf{Y}_d^k\right)$ *of* $\mathbf{Y}_d^k(t)$ *for every* $[\mathbf{D}(.), \mathbf{Y}_d(.)] \in \mathfrak{D}^j \times \mathfrak{Y}_d^k$ *, i.e., on* $\mathfrak{D}^j \times \mathfrak{Y}_d^k$.
c) The tracking of $\mathbf{Y}_d^k(t)$ *on* $\mathfrak{D}^j \times \mathfrak{Y}_d^k$ *is* **global (in the whole)** *if and only if* $\mathcal{D}_T\left(\mathbf{Y}_{d0}^k; \mathbf{D}; \mathbf{U}; \mathbf{Y}_d^k\right) = \mathfrak{R}^{(k+1)N}$ *for every* $[\mathbf{D}(.), \mathbf{Y}_d(.)] \in \mathfrak{D}^j \times \mathfrak{Y}_d^k$.

This definition determines the tracking properties so that they depend on a particular $[\mathbf{D}(.), \mathbf{Y}_d(.)] \in \mathfrak{D}^j \times \mathfrak{Y}_d^k$. Such tracking properties are nonuniform in $[\mathbf{D}(.), \mathbf{Y}_d(.)] \in \mathfrak{D}^j \times \mathfrak{Y}_d^k$.

The equivalent definition in terms of the output error vector $\mathbf{e}$ is Definition 305 in Appendix B.1.

The boundary of $\mathcal{D}_{(.)}\left(\mathfrak{D}^j; \mathbf{U}; \mathfrak{Y}_d^k\right)$ is denoted by $\partial\mathcal{D}_{(.)}\left(\mathfrak{D}^j; \mathbf{U}; \mathfrak{Y}_d^k\right)$.

Definition 122 *Uniform tracking of the extended desired output behavior* $\mathbf{Y}_d^k(t)$ *on* $\mathfrak{D}^j \times \mathfrak{Y}_d^k$ *of the system controlled by a control* $\mathbf{U}(.) \in \mathfrak{U}^l$

If and only if the system controlled by a control $\mathbf{U}(.) \in \mathfrak{U}^l$ *exhibits tracking of* $\mathbf{Y}_d^k(t)$ *on* $\mathfrak{D}^j \times \mathfrak{Y}_d^k$ *and both the intersection* $\mathcal{D}_{TU}\left(\mathfrak{D}^j; \mathbf{U}; \mathfrak{Y}_d^k\right)$ *of the tracking domains* $\mathcal{D}_T\left(\mathbf{Y}_{d0}^k; \mathbf{D}; \mathbf{U}; \mathbf{Y}_d^k\right)$ *in* $\left[\mathbf{D}(.), \mathbf{Y}_d^k(.)\right] \in \mathfrak{D}^j \times \mathfrak{Y}_d^k$ *is a connected neighborhood of* $\mathbf{Y}_{d0}^k$ *of every* $\mathbf{Y}_d(.) \in \mathfrak{Y}_d^k$:

$$(8.15) \text{ holds and } \exists \xi \in \mathfrak{R}^+ \implies \mathcal{D}_{TU}\left(\mathfrak{D}^j; \mathbf{U}; \mathfrak{Y}_d^k\right) =$$

$$= \cap \left[\mathcal{D}_T\left(\mathbf{Y}_{d0}^k; \mathbf{D}; \mathbf{U}; \mathbf{Y}_d^k\right) : \left[\mathbf{D}(.), \mathbf{Y}_d^k(.)\right] \in \mathfrak{D}^j \times \mathfrak{Y}_d^k \right] \supset \mathfrak{N}_\xi\left(\mathfrak{D}^j; \mathbf{U}; \mathfrak{Y}_d^k\right),$$

$$\partial\mathcal{D}_{TU}\left(\mathfrak{D}^j; \mathbf{U}; \mathfrak{Y}_d^k\right) \cap \partial\mathfrak{N}_\xi\left(\mathfrak{D}^j; \mathbf{U}; \mathfrak{Y}_d^k\right) = \phi, \qquad (8.16)$$

and the minimal $\tau\left(\varsigma, \mathbf{Y}_0^k; \mathbf{D}; \mathbf{U}; \mathbf{Y}_d^k\right)$ *denoted by* $\tau_m\left(\varsigma, \mathbf{Y}_0^k; \mathbf{D}; \mathbf{U}; \mathbf{Y}_d^k\right)$, *which obeys Definition 121, satisfies (8.17):*

$$\forall\left(\varsigma, \mathbf{Y}_0^k\right) \in \mathfrak{R}^+ \times \mathcal{D}_{TU}\left(\mathfrak{D}^j; \mathbf{U}; \mathfrak{Y}_d^k\right) \implies \tau\left(\varsigma, \mathbf{Y}_0^k; \mathfrak{D}^j; \mathbf{U}; \mathfrak{Y}_d^k\right) =$$

$$= \sup\left[\tau_m\left(\varsigma, \mathbf{Y}_0^k; \mathbf{D}; \mathbf{U}; \mathbf{Y}_d^k\right) : [\mathbf{D}(.), \mathbf{Y}_d(.)] \in \mathfrak{D}^j \times \mathfrak{Y}_d^k\right] \in \mathfrak{R}_+ \qquad (8.17)$$

then the tracking of $\mathbf{Y}_d^k(t)$ *is* **uniform in** $[\mathbf{D}(.), \mathbf{Y}_d(.)] \in \mathfrak{D}^j \times \mathfrak{Y}_d^k$ *on* $\mathfrak{D}^j \times \mathfrak{Y}_d^k$ *and the set* $\mathcal{D}_{TU}\left(\mathfrak{D}^j; \mathbf{U}; \mathfrak{Y}_d^k\right)$ *is* **the $(\mathbf{D}, \mathbf{Y}_d^k)$-uniform k-th order tracking domain of** $\mathbf{Y}_d^k(t)$ *on* $\mathfrak{D}^j \times \mathfrak{Y}_d^k$.

Comment 123 *The global tracking of* $\mathbf{Y}_d^k(t)$ *on* $\mathfrak{D}^j \times \mathfrak{Y}_d^k$ *is uniform in* $[\mathbf{D}(.), \mathbf{Y}_d(.)]$ *over* $\mathfrak{D}^j \times \mathfrak{Y}_d^k$ *if and only if the condition (8.17) is satisfied.*

Note 124 *Tracking and realizability of* $\mathbf{Y}_d^k(t)$

The desired plant output $\mathbf{Y}_d^k(t)$ can be unrealizable although the plant can exhibit tracking. It is due to the asymptotic convergence of the real output behavior to the desired one only as $t \longrightarrow \infty$. This essentially means that $\mathbf{Y}^k(t)$ converges to $\mathbf{Y}_d^k(t)$ only as $t \longrightarrow \infty$.

The preceding definitions do not require for tracking of $\mathbf{Y}_d^k(t)$ any closeness of $\mathbf{Y}^k(t)$ to $\mathbf{Y}_d^k(t)$ at any finite $t < \infty$, $t \in \mathfrak{T}_0$.

Tracking does not provide any information about the real output behavior relative to the desired one at any finite moment $t \in \mathfrak{T}_0$, after the initial one, $t > t_0$. The error of the former from the latter can be arbitrarily large at any finite instant. What follows eliminates this drawback.

Definition 125 *Stablewise tracking of the extended desired output behavior* $\mathbf{Y}_d^k(t)$ *on* $\mathfrak{D}^j \times \mathfrak{Y}_d^k$ *of the system controlled by a control* $\mathbf{U}(.) \in \mathfrak{U}^l$

a) *The system exhibits* **stablewise output tracking** *of* $\mathbf{Y}_d^k(t)$ *on* $\mathfrak{D}^j \times \mathfrak{Y}_d^k$, *for short* **the stablewise tracking** *of* $\mathbf{Y}_d^k(t)$ *on* $\mathfrak{D}^j \times \mathfrak{Y}_d^k$, *if and only if it exhibits tracking of* $\mathbf{Y}_d^k(t)$ *on* $\mathfrak{D}^j \times \mathfrak{Y}_d^k$, *and for every connected neighborhood* $\mathfrak{N}_\varepsilon\left[t; \mathbf{Y}_d^k(t)\right]$, $\mathfrak{N}_\varepsilon\left[t; \mathbf{Y}_d^k(t)\right] \subseteq \mathfrak{R}^{(k+1)N}$, *of* $\mathbf{Y}_d^k(t)$ *at any* $t \in \mathfrak{T}_0$, *there is a connected neighborhood* $\mathfrak{N}\left(\varepsilon; \mathbf{Y}_{d0}^k; \mathbf{D}; \mathbf{U}; \mathbf{Y}_d^k\right)$, *(8.14), of the plant desired initial output vector* $\mathbf{Y}_{d0}^k$ *at the initial moment* $t_0 = 0$ *such that for an initial* $\mathbf{Y}_0^k \in \mathfrak{N}\left(\varepsilon; \mathbf{Y}_{d0}^k; \mathbf{D}; \mathbf{U}; \mathbf{Y}_d^k\right)$ *the instantaneous* $\mathbf{Y}^k(t)$ *stays in* $\mathfrak{N}_\varepsilon\left[t; \mathbf{Y}_d^k(t)\right]$ *for all* $t \in \mathfrak{T}_0$; *i.e.*,

$$\forall \mathfrak{N}_\varepsilon\left[t; \mathbf{Y}_d^k(t)\right] \subseteq \mathfrak{R}^{(k+1)N}, \quad \forall t \in \mathfrak{T}_0,$$

$$\forall \left[\mathbf{D}(.), \mathbf{Y}_d(.)\right] \in \mathfrak{D}^j \times \mathfrak{Y}_d^k, \ \exists \mathfrak{N}\left(\varepsilon; \mathbf{Y}_{d0}^k; \mathbf{D}; \mathbf{U}; \mathbf{Y}_d^k\right) \subseteq \mathfrak{R}^{(k+1)N},$$

$$\mathfrak{N}\left(\varepsilon; \mathbf{Y}_{d0}^k; \mathbf{D}; \mathbf{U}; \mathbf{Y}_d^k\right) \subseteq \mathcal{D}_T\left(\mathbf{Y}_{d0}^k; \mathbf{D}; \mathbf{U}; \mathbf{Y}_d^k\right) \cap \mathfrak{N}_\varepsilon\left(t_0; \mathbf{Y}_{d0}^k\right) \Longrightarrow$$

$$\mathbf{Y}_0^k \in \mathfrak{N}\left(\varepsilon; \mathbf{Y}_{d0}^k; \mathbf{D}; \mathbf{U}; \mathbf{Y}_d^k\right) \implies$$

$$\mathbf{Y}^k(t; \mathbf{Y}_0^k; \mathbf{D}; \mathbf{U}) \in \mathfrak{N}_\varepsilon\left[t; \mathbf{Y}_d^k(t)\right], \ \forall t \in \mathfrak{T}_0. \qquad (8.18)$$

b) *The largest connected neighborhood* $\mathfrak{N}_L\left(\varepsilon; \mathbf{Y}_{d0}^k; \mathbf{D}; \mathbf{U}; \mathbf{Y}_d^k\right)$ *of the desired initial output vector* $\mathbf{Y}_{d0}^k$, *(8.18), is the* ε-**tracking domain** *denoted by* $\mathcal{D}_{ST}\left(\varepsilon; \mathbf{Y}_{d0}^k; \mathbf{D}; \mathbf{U}; \mathbf{Y}_d^k\right)$ *of the stablewise tracking of* $\mathbf{Y}_d^k(t)$ *on* $\mathfrak{D}^j \times \mathfrak{Y}_d^k$.

The **domain** $\mathcal{D}_{ST}\left(\mathbf{Y}_{d0}^k; \mathbf{D}; \mathbf{U}; \mathbf{Y}_d^k\right)$ *of the stablewise tracking of* $\mathbf{Y}_d^k(t)$ *on* $\mathfrak{D}^j \times \mathfrak{Y}_d^k$ *is the union of all* $\mathcal{D}_{ST}\left(\varepsilon; \mathbf{Y}_{d0}^k; \mathbf{D}; \mathbf{U}; \mathbf{Y}_d^k\right)$ *over* $\varepsilon \in \mathfrak{R}^+$,

$$\mathcal{D}_{ST}\left(\mathbf{Y}_{d0}^k; \mathbf{D}; \mathbf{U}; \mathbf{Y}_d^k\right) = \cup\left[\mathcal{D}_{ST}\left(\varepsilon; \mathbf{Y}_{d0}^k; \mathbf{D}; \mathbf{U}; \mathbf{Y}_d^k\right) : \varepsilon \in \mathfrak{R}^+\right]. \quad (8.19)$$

Let $[0, \varepsilon_M)$ be the maximal interval over which $\mathcal{D}_{ST}\left(\varepsilon; \mathbf{Y}_{d0}^k; \mathbf{D}; \mathbf{U}; \mathbf{Y}_d^k\right)$ is continuous in $\varepsilon \in \mathfrak{R}_+$,

$$\mathcal{D}_{ST}\left(\varepsilon; \mathbf{Y}_{d0}^k; \mathbf{D}; \mathbf{U}; \mathbf{Y}_d^k\right) \in \mathfrak{C}\left([0, \varepsilon_M)\right), \quad [\mathbf{D}(.), \mathbf{Y}_d(.)] \in \mathfrak{D}^j \times \mathfrak{Y}_d^k.$$

The strict domain $\mathcal{D}_{SST}\left(\mathbf{Y}_{d0}^k; \mathbf{D}; \mathbf{U}; \mathbf{Y}_d^k\right)$ of the stablewise k-th order tracking of $\mathbf{Y}_d(t)$ on $\mathfrak{D}^j \times \mathfrak{Y}_d^k$ is the union of all stablewise tracking domains

$$\mathcal{D}_{ST}\left(\varepsilon; \mathbf{Y}_{d0}^k; \mathbf{D}; \mathbf{U}; \mathbf{Y}_d^k\right):$$

over $\varepsilon \in [0, \varepsilon_M)$,

$$\mathcal{D}_{SST}\left(\mathbf{Y}_{d0}^k; \mathbf{D}; \mathbf{U}; \mathbf{Y}_d^k\right) = \cup\left\{\mathcal{D}_{ST}\left(\varepsilon; \mathbf{Y}_{d0}^k; \mathbf{D}; \mathbf{U}; \mathbf{Y}_d^k\right) : \varepsilon \in [0, \varepsilon_M)\right\},$$

$$[\mathbf{D}(.), \mathbf{Y}_d(.)] \in \mathfrak{D}^j \times \mathfrak{Y}_d^k. \tag{8.20}$$

c) The stablewise tracking of $\mathbf{Y}_d^k(t)$ on $\mathfrak{D}^j \times \mathfrak{Y}_d^k$ is **global (in the whole)** if and only if it is both the global tracking of $\mathbf{Y}_d^k(t)$ on $\mathfrak{D}^j \times \mathfrak{Y}_d^k$ and the stablewise tracking of $\mathbf{Y}_d^k(t)$ on $\mathfrak{D}^j \times \mathfrak{Y}_d^k$ with $\mathcal{D}_{ST}\left(\mathbf{Y}_{d0}^k; \mathbf{D}; \mathbf{U}; \mathbf{Y}_d^k\right) = \mathfrak{R}^{(k+1)N}$ for every $[\mathbf{D}(.), \mathbf{Y}_d(.)] \in \mathfrak{D}^j \times \mathfrak{Y}_d^k$.

The equivalent definition of the stablewise tracking of $\mathbf{Y}_d^k(t)$ on $\mathfrak{D}^j \times \mathfrak{Y}_d^k$ and its domain in terms of the output error vector $\mathbf{e}^k$ is Definition 306 in Appendix B.1.

Definition 126 **The uniform stablewise tracking of the desired output behavior $\mathbf{Y}_d^k(t)$ on $\mathfrak{D}^j \times \mathfrak{Y}_d^k$ of the system controlled by a control $\mathbf{U}(.) \in \mathfrak{U}^l$**

If and only if the system controlled by a control $\mathbf{U}(.) \in \mathfrak{U}^l$ exhibits the stablewise tracking of the desired output behavior $\mathbf{Y}_d^k(t)$ on $\mathfrak{D}^j \times \mathfrak{Y}_d^k$ together with the uniform tracking in $\left[\mathbf{D}(.), \mathbf{Y}_d^k(.)\right] \in \mathfrak{D}^j \times \mathfrak{Y}_d^k$ and the intersection

$$\mathcal{D}_{STU}\left(\mathbf{Y}_{d0}^k; \mathfrak{D}^j; \mathbf{U}; \mathfrak{Y}_d^k\right)$$

of all

$$\mathcal{D}_{ST}\left(\mathbf{Y}_{d0}^k; \mathbf{D}; \mathbf{U}; \mathbf{Y}_d^k\right)$$

[the intersection $\mathcal{D}_{SSTU}\left(\mathbf{Y}_{d0}^k; \mathfrak{D}^j; \mathbf{U}; \mathfrak{Y}_d^k\right)$ of all $\mathcal{D}_{SST}\left(\mathbf{Y}_{d0}^k; \mathbf{D}; \mathbf{U}; \mathbf{Y}_d^k\right)$] over $\mathfrak{D}^j \times \mathfrak{Y}_d^k$ is a connected neighborhood of $\mathbf{Y}_{d0}^k$ then and only then it is the *[strictly] stablewise tracking domain of* $\mathbf{Y}_d^k(t)$ *on* $\mathfrak{D}^j \times \mathfrak{Y}_d^k$ *uniform in the pair* $[\mathbf{D}(.), \mathbf{Y}_d(.)] \in \mathfrak{D}^j \times \mathfrak{Y}_d^k$, *respectively,*

$$\exists \xi \in \mathfrak{R}^+ \Longrightarrow \mathcal{D}_{STU}\left(\mathbf{Y}_{d0}^k; \mathfrak{D}^j; \mathbf{U}; \mathfrak{Y}_d^k\right) =$$
$$= \cap \left\{ \mathcal{D}_{ST}\left(\mathbf{Y}_{d0}^k; \mathbf{D}; \mathbf{U}; \mathbf{Y}_d^k\right) : [\mathbf{D}(.), \mathbf{Y}_d(.)] \in \mathfrak{D}^j \times \mathfrak{Y}_d^k \right\} \supset$$
$$\supset \mathfrak{N}_\xi\left(\mathbf{Y}_{d0}^k; \mathfrak{D}^j; \mathbf{U}; \mathfrak{Y}_d^k\right), \tag{8.21}$$

$$\left[\begin{array}{c} \exists \xi \in \mathfrak{R}^+ \Longrightarrow \mathcal{D}_{SSTU}\left(\mathbf{Y}_{d0}^k; \mathfrak{D}^j; \mathbf{U}; \mathfrak{Y}_d^k\right) = \\ = \cap \left\{ \mathcal{D}_{SST}\left(\mathbf{Y}_{d0}^k; \mathbf{D}; \mathbf{U}; \mathbf{Y}_d^k\right) : [\mathbf{D}(.), \mathbf{Y}_d(.)] \in \mathfrak{D}^j \times \mathfrak{Y}_d^k \right\} \supset \\ \supset \mathfrak{N}_\xi\left(\mathbf{Y}_{d0}^k; \mathfrak{D}^j; \mathbf{U}; \mathfrak{Y}_d^k\right), \end{array} \right]. \tag{8.22}$$

The books [175, Section 41, Definitions 104, 107, 108 and 111, pp. 44-46], [188, Section 3.3, Definitions 50, 52, 54 and 56, p. 37] introduced the concept of the $\mathbf{Y}_d(t)$ realizability, defined its various types and proved [175, Section 41, Lemmae 105, 110 and 112, pp. 44-46], [188, Section 3.3, Theorems 65, 66, 86, 87, 90 and 91, pp. 41, 42, 46-51] the necessary and sufficient conditions for them in general and in the framework of linear systems, respectively (see Sections 2.2, 3.2, 4.2, 5.2 and 6.2). We summarize them as follows:

Theorem 127 *Stablewise tracking and realizability of* $\mathbf{Y}_d^k(t)$
If the plant exhibits stablewise tracking of $\mathbf{Y}_d^k(t)$ *on* $\mathfrak{D}^j \times \mathfrak{Y}_d^k$, *then* $\mathbf{Y}_d^k(t)$ *is realizable on* $\mathfrak{D}^j \times \mathfrak{Y}_d^k$ *for the desired output initial conditions; i.e., the stablewise tracking on* $\mathfrak{D}^j \times \mathfrak{Y}_d^k$ *guarantees that*

$$\left[\mathbf{Y}^k(0) = \mathbf{Y}_d^k(0)\right] \Longrightarrow \left[\mathbf{Y}^k(t) = \mathbf{Y}_d^k(t), \ \forall t \in \mathfrak{T}_0\right].$$

This theorem is proved in Appendix D.3.

Comment 128 *The stablewise tracking of the desired output* $\mathbf{Y}_d^k(t)$ *on* $\mathfrak{D}^j \times \mathfrak{Y}_d^k$ *is sufficient for the realizability of* $\mathbf{Y}_d^k(t)$ *(Definition 24, Section 1.5) on* $\mathfrak{D}^j \times \mathfrak{Y}_d^k$. *This means that the system can realize the perfect tracking on* $\mathfrak{D}^j \times \mathfrak{Y}_d^k$ *(Definition 111, Section 8.2) it exhibits the stablewise tracking of the desired output* $\mathbf{Y}_d^k(t)$ *on* $\mathfrak{D}^j \times \mathfrak{Y}_d^k$..
However, $\mathbf{Y}_d^k(t)$ *can be realizable but the plant need not exhibit either its stablewise tracking or tracking. Realizability of* $\mathbf{Y}_d^k(t)$ *is not sufficient either for stablewise tracking or for tracking in general.*

8.3.5 Exponential tracking

The stablewise tracking expresses stability of the desired output behavior $\mathbf{Y}_d^k(t)$, in addition to its tracking. It does not allow arbitrarily large output error for appropriately bounded initial conditions and for the bounded input vector function. However, it does not show the rate of the convergence of the real output behavior to the desired one.

Definition 129 *Exponential output tracking of* $\mathbf{Y}_d^k(t)$ *on* $\mathfrak{D}^j \times \mathfrak{Y}_d^k$
of the system controlled by a control $\mathbf{U}(.) \in \mathfrak{U}^l$

a) The system exhibits exponential asymptotic output tracking of $\mathbf{Y}_d^k(t)$, $\mathbf{Y}_d^k(.) \in \mathfrak{Y}_d^k$, *on* $\mathfrak{D}^j \times \mathfrak{Y}_d^k$, *for short the exponential tracking of* $\mathbf{Y}_d^k(t)$ *on* $\mathfrak{D}^j \times \mathfrak{Y}_d^k$ *if and only if for every* $[\mathbf{D}(.), \mathbf{Y}_d(.)] \in \mathfrak{D}^j \times \mathfrak{Y}_d^k$ *there exist positive real numbers* $a \geq 1$ *and* $b > 0$, *and a connected neighborhood* $\mathfrak{N}\left(\mathbf{Y}_{d0}^k; a, b; \mathbf{D}; \mathbf{U}; \mathbf{Y}_d^k\right)$ *of* $\mathbf{Y}_{d0}^k$, $a = a(\mathbf{D}, \mathbf{U}, \mathbf{Y}_d^k)$ *and* $b = b(\mathbf{D}, \mathbf{U}, \mathbf{Y}_d^k)$, *such that* $\mathbf{Y}_0^k \in \mathfrak{N}\left(\mathbf{Y}_{d0}^k; a, b; \mathbf{D}; \mathbf{U}; \mathbf{Y}_d^k\right)$ *guarantees that* $\mathbf{Y}_d^k(t)$ *approaches exponentially* $\mathbf{Y}_d^k(t)$ *all the time; i.e.,*

$$\forall [\mathbf{D}(.), \mathbf{Y}_d(.)] \in \mathfrak{D}^j \times \mathfrak{Y}_d^k, \ \exists a \in [1, \infty[, \ \exists b \in \mathfrak{R}^+, \ \exists \xi \in \mathfrak{R}^+,$$

$$a = a(\mathbf{D}, \mathbf{U}, \mathbf{Y}_d^k), \quad b = b(\mathbf{D}, \mathbf{U}, \mathbf{Y}_d^k),$$

$$\exists \mathfrak{N}\left(\mathbf{Y}_{d0}^k; a, b; \mathbf{D}; \mathbf{U}; \mathbf{Y}_d^k\right) \supset In\mathfrak{N}_\xi\left(\mathbf{Y}_{d0}^k; a, b; \mathbf{D}; \mathbf{U}; \mathbf{Y}_d^k\right) \supset \left\{\mathbf{Y}_{d0}^k\right\},$$

$$\partial\mathfrak{N}\left(\mathbf{Y}_{d0}^k; a, b; \mathbf{D}; \mathbf{U}; \mathbf{Y}_d^k\right) \cap \partial\mathfrak{N}_\xi\left(\mathbf{Y}_{d0}^k; a, b; \mathbf{D}; \mathbf{U}; \mathbf{Y}_d^k\right) = \phi,$$

$$\mathbf{Y}_0^k \in \mathfrak{N}\left(\mathbf{Y}_{d0}^k; a, b; \mathbf{D}; \mathbf{U}; \mathbf{Y}_d^k\right) \implies$$

$$\left\|\mathbf{Y}_d^k(t; \mathbf{Y}_{d0}^k) - \mathbf{Y}^k(t; \mathbf{Y}_0^k; \mathbf{D}; \mathbf{U})\right\| \leq a \left\|\mathbf{Y}_{d0}^k - \mathbf{Y}_0^k\right\| exp(-bt), \ \forall t \in \mathfrak{T}_0. \quad (8.23)$$

b) The largest connected neighborhood $\mathfrak{N}\left(\mathbf{Y}_{d0}^k; a, b; \mathbf{D}; \mathbf{U}; \mathbf{Y}_d^k\right)$ *of* $\mathbf{Y}_{d0}^k$ *is the domain* $\mathcal{D}_E\left(\mathbf{Y}_{d0}^k; a, b; \mathbf{D}; \mathbf{U}; \mathbf{Y}_d^k\right)$ *of the exponential tracking of* $\mathbf{Y}_d^k(t)$ *on* $\mathfrak{D}^j \times \mathfrak{Y}_d^k$ *relative to* a *and* b. *When* a *and* b *are fixed then they can be omitted,*

$$\mathcal{D}_E\left(\mathbf{Y}_{d0}^k; a, b; \mathbf{D}; \mathbf{U}; \mathbf{Y}_d^k\right) = \mathcal{D}_E\left(\mathbf{Y}_{d0}^k; \mathbf{D}; \mathbf{U}; \mathbf{Y}_d^k\right).$$

c) The exponential tracking of $\mathbf{Y}_d^k(t)$ *on* $\mathfrak{D}^j \times \mathfrak{Y}_d^k$ *is global (in the whole) if and only if the domain* $\mathcal{D}_E\left(\mathbf{Y}_{d0}^k; \mathbf{D}; \mathbf{U}; \mathbf{Y}_d^k\right) = \mathfrak{R}^{(k+1)N}$ *for every* $[\mathbf{D}(.), \mathbf{Y}_d(.)] \in \mathfrak{D}^j \times \mathfrak{Y}_d^k$.

This definition permits for the exponential tracking parameters $a(.)$ and $b()$, and for the domain $\mathcal{D}_E(.)$, to depend on $\mathbf{D}(.)$ and $\mathbf{Y}_d(.)$. The exponential tracking is nonuniform.

The equivalent definition of the exponential tracking of $\mathbf{Y}_d^k(t)$ on $\mathfrak{D}^j \times \mathfrak{Y}_d^k$ and its domain in terms of the output error vector $\mathbf{e}^k$ is Definition 307 in Appendix B.1.

Definition 130 *The uniform exponential output tracking of* $\mathbf{Y}_d^k(t)$, *on* $\mathfrak{D}^j \times \mathfrak{Y}_d^k$ *of the system controlled by a control* $\mathbf{U}(.) \in \mathfrak{U}^l$

If and only if the values of $a = a\left(\mathbf{D}, \mathbf{U}, \mathbf{Y}_d^k\right)$ *and* $b = b(\mathbf{D}, \mathbf{U}, \mathbf{Y}_d^k)$, *as well as the domain* $\mathcal{D}_E\left(\mathbf{Y}_{d0}^k; a, b; \mathbf{D}; \mathbf{U}; \mathbf{Y}_d^k\right)$, *depend at most on the set product* $\mathfrak{D}^j \times \mathfrak{Y}_d^k$,

$$a = a\left(\mathfrak{D}^j, \mathbf{U}, \mathfrak{Y}_d^k\right), \quad b = b(\mathfrak{D}^j, \mathbf{U}, \mathfrak{Y}_d^k), \tag{8.24}$$

but not on a particular choice of $[\mathbf{D}(.), \mathbf{Y}_d(.)]$ *from* $\mathfrak{D}^j \times \mathfrak{Y}_d^k$, *then the exponential tracking of* $\mathbf{Y}_d^k(t)$ *on* $\mathfrak{D}^j \times \mathfrak{Y}_d^k$ *is* **uniform in** $[\mathbf{D}(.), \mathbf{Y}_d(.)]$ *relative to* (a, b). *Its* **domain** $\mathcal{D}_E\left(a, b; \mathfrak{D}^j; \mathbf{U}; \mathfrak{Y}_d^k\right)$ *is the intersection of all* $\mathcal{D}_T\left(\mathbf{Y}_{d0}^k; a, b; \mathbf{D}; \mathbf{U}; \mathbf{Y}_d^k\right)$ *in* $[\mathbf{D}(.), \mathbf{Y}_d(.)] \in \mathfrak{D}^j \times \mathfrak{Y}_d^k$ *and the connected neighborhood of* $\mathbf{Y}_{d0}^k$ *for every* $[\mathbf{D}(.), \mathbf{Y}_d(.)] \in \mathfrak{D}^j \times \mathfrak{Y}_d^k$,

$$\exists \xi > 0 \Longrightarrow \mathcal{D}_E\left(a, b; \mathfrak{D}^j; \mathbf{U}; \mathfrak{Y}_d^k\right) =$$

$$= \cap \left\{ \begin{array}{c} \mathcal{D}_E\left[\mathbf{Y}_{d0}^k; a\left(\mathbf{D}, \mathbf{U}, \mathbf{Y}_d^k\right), b\left(\mathbf{D}, \mathbf{U}, \mathbf{Y}_d^k\right); \mathbf{D}; \mathbf{U}; \mathbf{Y}_d^k\right] : \\ : [\mathbf{D}(.), \mathbf{Y}_d(.)] \in \mathfrak{D}^j \times \mathfrak{Y}_d^k \end{array} \right\} \supset$$

$$\supset \mathfrak{N}_\xi\left(\mathbf{Y}_{d0}^k; \mathfrak{D}^j; \mathbf{U}; \mathfrak{Y}_d^k\right),$$

$$\partial \mathfrak{D}_E^k\left(a, b; \mathfrak{D}^j; \mathbf{U}; \mathfrak{Y}_d^k\right) \cap \mathfrak{N}_\xi\left(\mathbf{Y}_{d0}^k; \mathfrak{D}^j; \mathbf{U}; \mathfrak{Y}_d^k\right) = \phi. \tag{8.25}$$

The expression **"relative to** (a, b)**"** *can be omitted if and only if a and b are fixed and known.*

The general Theorem 226 presented and proved in [175, pp. 124,125] takes the following form in the framework of the linear systems:

Theorem 131 *Exponential tracking and stablewise tracking*

If the system exhibits exponential tracking of $\mathbf{Y}_d^k(t)$ *on the product set* $\mathfrak{D}^j \times \mathfrak{Y}_d^k$ *then the tracking is also stablewise tracking on* $\mathfrak{D}^j \times \mathfrak{Y}_d^k$.

Theorem 127 and Theorem 131 imply directly the following results [175, pp. 125, 126].

Corollary 132 *Exponential tracking and realizability of* $\mathbf{Y}_d^k(t)$

If the plant exhibits exponential tracking of the extended desired output $\mathbf{Y}_d^k(t)$, *then* $\mathbf{Y}_d^k(t)$ *is realizable for the extended desired output initial conditions* $\mathbf{Y}_{d0}^k$.

Corollary 133 *Necessity of realizability of* $\mathbf{Y}_d^k(t)$

Realizability of $\mathbf{Y}_d^k(t)$ *is necessary, but not sufficient, for the stablewise tracking, hence also for the exponential tracking.*

Note 134 *Tracking allows arbitrary big error overshoot for arbitrary small initial output error. Stablewise tracking eliminates this drawback. Both tracking and stablewise tracking permit very slow error convergence to the zero error. Exponential tracking eliminates this drawback.*

Note 135 *Lyapunov tracking and Lyapunov stability*

*The concept of Lyapunov (the infinite time) tracking (**Ly-tracking**) preserves the sense of the Lyapunov stability concept; i.e., retains the Lyapunov sense [175], [188]. The former broadens the Lyapunov theory to disturbed and to (be) controlled plants and generalizes the latter. The latter can be considered as a special case of the former.*

The above explanations under 1. through 3. at the beginning of Subsection 8.3.2, together with the above definitions, show that, and why, the infinite-time tracking properties and the Lyapunov stability properties are mutually different. For more details see [188, Note 190, p. 137], [175, Note 230, pp. 125,126].

Note 136 *Tracking is necessary for all other, above-defined, tracking properties.*

8.4 Tracking with finite reachability *time*

8.5 Finite scalar reachability *time*

This section slightly refines [175, Section 26, pp. 221-246]

Let *the final moment* t_F of the system operation obey

$$t_F \in \mathfrak{T}_0 \ or \ t_F = \infty, \quad \text{and} \ t_F > t_0 = 0, \tag{8.26}$$

i.e., the final moment t_F can be finite ($t_F \in \mathfrak{T}_0$) or infinite ($t_F = \infty$). It determines, together with the initial instant $t_0 = 0$, the *time set* $\mathfrak{T}_{0F}$ over which the system should work properly,

$$\mathfrak{T}_{0F} = \left\{ t : \ t \in \mathfrak{T}, \left\langle \begin{array}{l} 0 \leq t \leq t_F \ iff \ t_F < \infty, \\ 0 \leq t < t_F \ iff \ t_F = \infty \end{array} \right\rangle \right\} \subseteq \mathfrak{T}_0, \tag{8.27}$$

so that its *closure* $Cl \ \mathfrak{T}_{0F}$ is compact if $t_F < \infty$,

$$t_F < \infty \Longrightarrow Cl \ \mathfrak{T}_{0F} = \{ t : \ t \in Cl \ \mathfrak{T}, \ 0 \leq t \leq t_F \} \subset \mathfrak{T}_0. \tag{8.28}$$

This enables us to study *the finite-time tracking* that demands $t_F < \infty$.

The Lyapunov tracking properties guarantee asymptotic convergence of the real output response to the desired one only as $t \to \infty$, (8.10). They do not ensure that the real output response reaches the desired one in a finite *time* and that they stay equal since then until the final moment t_F of the plant work. In order to overcome this essential drawback from the engineering and control system customer points of view, we present definitions of some tracking properties with the *finite reachability time (FRT)*. It can be *finite scalar reachability time (FSRT) (8.31)*, or *finite vector reachability time (FVRT)*.

Let

$$[t_R, t_F) = \left\{ \begin{array}{l} [t_R, t_F] \iff t_F < \infty, \\ [t_R, t_F[\iff t_F = \infty. \end{array} \right\} \tag{8.29}$$

Definition 137 *Reachability time*

A moment denoted by t_R,

$$t_R \in In\mathfrak{T}_0, \tag{8.30}$$

*and called **the (output) finite scalar reachability time (FSRT)**, is the first moment when the plant real output variable Y, or the plant real output vector $\mathbf{Y}$, becomes equal to its desired output variable Y_d, or to its desired output vector $\mathbf{Y}_d$, respectively, and since then they rest always equal ,*

$$Y(t) \left\{ \begin{array}{l} \neq Y_d(t), \ \exists t \in]t_0, t_R[, \\ = Y_d(t), \ \forall t \in [t_R, t_F) \end{array} \right\}, \quad \mathbf{Y}(t) \left\{ \begin{array}{l} \neq \mathbf{Y}_d(t), \ \exists t \in]t_0, t_R[, \\ = \mathbf{Y}_d(t), \ \forall t \in [t_R, t_F), \end{array} \right\}. \tag{8.31}$$

*Equivalently, a moment denoted by t_R, $t_R \in In\mathfrak{T}_0$, and called **the (error) finite scalar reachability time (FSRT)**, is the first moment when the plant real output error e, or the plant real output error vector $\mathbf{e}$, becomes equal to zero, or to the zero vector $\mathbf{0}_N$, respectively, and since then they rest equal until the final moment t_F,*

$$e(t) \left\{ \begin{array}{l} \neq 0, \ \exists t \in]t_0, t_R[, \\ = 0, \ \forall t \in [t_R, t_F), \end{array} \right\}, \quad \mathbf{e}(t) \left\{ \begin{array}{l} \neq \mathbf{0}_N, \ \exists t \in]t_0, t_R[, \\ = \mathbf{0}_N, \ \forall t \in [t_R, t_F), \end{array} \right\}. \tag{8.32}$$

This means that the (finite scalar) reachability *time* t_R is the same for all output variables, i.e., for the whole output vector. We can extend this to hold also for their derivatives,

$$t_R \leq t \leq t_F \implies \mathbf{Y}^k(t) = \mathbf{Y}_d^k(t). \tag{8.33}$$

The finite scalar reachability *time* t_R induces the *time* sets $\mathfrak{T}_R$ and $\mathfrak{T}_{RF}$ as the subsets of $\mathfrak{T}_0$,

$$\mathfrak{T}_R = \{t \in \mathfrak{T}_0 : t_0 \leq t \leq t_R,\} \subset \mathfrak{T}_{0F},$$

$$\mathfrak{T}_{RF} = \left\{ t \in Cl\ \mathfrak{T}_0 : \left\langle \begin{array}{c} t_R \leq t \leq t_F < \infty, \\ t_R \leq t < t_F = \infty \end{array} \right\rangle \right\} \subset \mathfrak{T}_{0F},$$

$$\mathfrak{T}_R \cup \mathfrak{T}_{RF} = \mathfrak{T}_{0F},\ t_R = \infty \Longrightarrow \mathfrak{T}_{R\infty} = \{\infty\}. \tag{8.34}$$

$\mathfrak{T}_R$ is *the reachability time set, and* $\mathfrak{T}_{RF}$ *and* $\mathfrak{T}_{R\infty}$ *are the post reachability time sets.*

We continue to use the abbreviated notation:

$$\mathbf{Y}_d^k(t) \equiv \mathbf{Y}_d^k(t; t_0; \mathbf{Y}_{d0}^k), \quad \mathbf{Y}^k(t) \equiv \mathbf{Y}^k(t; t_0; \mathbf{Y}_0^k; \mathbf{D}; \mathbf{U}), \tag{8.35}$$

The following definitions determine **in the (extended) output space** $\mathfrak{R}^{(k+1)N}$ various types of *the k-th order tracking of* $\mathbf{Y}_d(t)$ *with the finite scalar reachability time (FSRT)*, i.e., *various types of the tracking of* $\mathbf{Y}_d^k(t)$ *with the finite scalar reachability time (FSRT)*. The engineering demand is that the system should possess a required tracking with the finite reachability *time* over the *time* set $\mathfrak{T}_{0F}$. We omit the expression "over the *time* set $\mathfrak{T}_{0F}$" from the following definitions.

The concept of **the (scalar or vector) reachability time tracking** is essentially different from the Lyapunov tracking concept.

Definition 138 *Tracking with finite scalar reachability time (FSRT) of the desired output* $\mathbf{Y}_d^k(t)$ *of the system controlled by a control* $\mathbf{U}(.) \in \mathfrak{U}^l$

a) The system exhibits **the output tracking of** $\mathbf{Y}_d^k(t)$ **with the finite scalar reachability time (FSRT)** t_R *on* $\mathfrak{T}_{0F} \times \mathfrak{D}^i \times \mathfrak{Y}_d^k$, *for short* **the tracking with the finite reachability time** t_R *of* $\mathbf{Y}_d^k(t)$ *on* $\mathfrak{T}_{0F} \times \mathfrak{D}^i \times \mathfrak{Y}_d^k$ *if and only if for every* $(\mathbf{D}(.), \mathbf{Y}_d(.)) \in \mathfrak{D}^i \times \mathfrak{Y}_d^k$ *there exists a* (t_R, t_F)-*dependent connected neighborhood*

$$\mathfrak{N}\left(t_R; t_F; \mathbf{Y}_{d0}^k; \mathbf{D}; \mathbf{U}; \mathbf{Y}_d\right) \subseteq \mathfrak{R}^{(k+1)N},$$

of the system desired initial output vector $\mathbf{Y}_{d0}^k$ *at the initial moment* $t_0 = 0$ *such that* $\mathbf{Y}_0^k$ *from the neighborhood* $\mathfrak{N}\left(t_R; t_F; \mathbf{Y}_{d0}^k; \mathbf{D}; \mathbf{U}; \mathbf{Y}_d\right)$ *guarantees both that* $\mathbf{Y}^k(t; \mathbf{Y}_0^k; \mathbf{D}; \mathbf{U})$ *becomes equal to* $\mathbf{Y}_d^k(t; \mathbf{Y}_{d0}^k)$ *at the moment* $t_R \in$

In $\mathfrak{T}_{0F}$, and that they stay equal on $\mathfrak{T}_{RF}$, i.e.,

$$\forall \, [\mathbf{D}(.), \mathbf{Y}_d(.)] \in \mathfrak{D}^i \times \mathfrak{Y}_d^k,$$

$$\exists \mathfrak{N} \left(t_R; t_F; \mathbf{Y}_{d0}^k; \mathbf{D}; \mathbf{U}; \mathbf{Y}_d \right), \; \mathfrak{N} \left(t_R; t_F; \mathbf{Y}_{d0}^k; \mathbf{D}; \mathbf{U}; \mathbf{Y}_d \right) \subseteq \mathfrak{R}^{(k+1)N},$$

$$\mathbf{Y}_0^k \in \mathfrak{N} \left(t_R; t_F; \mathbf{Y}_{d0}^k; \mathbf{D}; \mathbf{U}; \mathbf{Y}_d \right) \implies$$

$$\mathbf{Y}^k(t; \mathbf{Y}_0^k; \mathbf{D}; \mathbf{U}) = \mathbf{Y}_d^k(t; \mathbf{Y}_{d0}^k), \; \forall t \in \mathfrak{T}_{RF}. \tag{8.36}$$

*b) The largest connected neighborhood $\mathfrak{N} \left(t_R; t_F; \mathbf{Y}_{d0}^k; \mathbf{D}; \mathbf{U}; \mathbf{Y}_d \right)$ of $\mathbf{Y}_{d0}^k$, (8.36), is the **FSRT tracking domain** $\mathcal{D}_T \left(t_R; t_F; \mathbf{Y}_{d0}^k; \mathbf{D}; \mathbf{U}; \mathbf{Y}_d \right)$ of $\mathbf{Y}_d^k(t)$ on $\mathfrak{T}_{0F} \times \mathfrak{D}^i \times \mathfrak{Y}_d^k$.*

The FSRT tracking domain $\mathcal{D}_T \left(t_R; t_F; \mathbf{Y}_{d0}^k; \mathbf{D}; \mathbf{U}; \mathbf{Y}_d \right)$ of $\mathbf{Y}_d^k(t)$ on $\mathfrak{T}_{0F} \times \mathfrak{D}^i \times \mathfrak{Y}_d^k$ and the FSRT tracking domain $\mathcal{D}_T \left(t_R; t_F; \mathbf{0}_{(k+1)N}; \mathbf{D}; \mathbf{U}; \mathbf{Y}_d \right)$ of $\mathbf{e}^k = \mathbf{0}_{(k+1)N}$ on $\mathfrak{T}_{0F} \times \mathfrak{D}^i \times \mathfrak{Y}_d^k$ obey:

$$\mathcal{D}_T \left(t_R; t_F; \mathbf{Y}_{d0}^k; \mathbf{D}; \mathbf{U}; \mathbf{Y}_d \right) =$$

$$= \left\{ \mathbf{Y}^k : \mathbf{Y}^k = \mathbf{Y}_{d0}^k + \mathbf{y}^k, \; \mathbf{y}^k \in \mathcal{D}_T \left(t_R; t_F; \mathbf{0}_{(k+1)N}; \mathbf{D}; \mathbf{U}; \mathbf{Y}_d \right) \right\}.$$

*c) The FSRT tracking of $\mathbf{Y}_d^k(t)$ on $\mathfrak{T}_{0F} \times \mathfrak{D}^i \times \mathfrak{Y}_d^k$ is **global** (in the whole) if and only if $\mathcal{D}_T \left(t_R; t_F; \mathbf{Y}_{d0}^k; \mathbf{D}; \mathbf{U}; \mathbf{Y}_d \right) = \mathfrak{R}^{(k+1)N}$ for every $[\mathbf{D}(.), \mathbf{Y}_d(.)] \in \mathfrak{D}^i \times \mathfrak{Y}_d^k$.*

The preceding definition allows nonuniformity of the tracking on $\mathfrak{D}^i \times \mathfrak{Y}_d^k$.

Exercise 139 *Define the FSRT tracking of $\mathbf{Y}_d^k(t)$ on $\mathfrak{D}^j \times \mathfrak{Y}_d^k$ and its domain in terms of the output error vector $\mathbf{e}^k$.*

Definition 140 ***The uniform tracking with the finite scalar reachability time of the desired output behavior $\mathbf{Y}_d^k(t)$ on $\mathfrak{T}_{0F} \times \mathfrak{D}^i \times \mathfrak{Y}_d^k$ of the system controlled by a control $\mathbf{U}(.) \in \mathfrak{U}^l$***

If and only if the intersection $\mathcal{D}_{TU} \left(t_R; t_F; \mathfrak{D}^i; \mathbf{U}; \mathfrak{Y}_d^k \right)$ of the tracking domains $\mathcal{D}_T \left(t_R; t_F; \mathbf{Y}_{d0}^k; \mathbf{D}; \mathbf{U}; \mathbf{Y}_d \right)$ in $[\mathbf{D}(.), \mathbf{Y}_d(.)] \in \mathfrak{D}^i \times \mathfrak{Y}_d^k$ is a connected neighborhood of $\mathbf{Y}_{d0}^k$ for every $\mathbf{Y}_d(.) \in \mathfrak{Y}_d^k$,

$$\exists \xi \in \mathfrak{R}^+ \implies \mathcal{D}_{TU} \left(t_R; t_F; \mathfrak{D}^i; \mathbf{U}; \mathfrak{Y}_d^k \right)$$

$$= \cap \left[\begin{array}{c} \mathcal{D}_T \left(t_R; t_F; \mathbf{Y}_{d0}^k; \mathbf{D}; \mathbf{U}; \mathbf{Y}_d \right) : \\ [\mathbf{D}(.), \mathbf{Y}_d(.)] \in \mathfrak{D}^i \times \mathfrak{Y}_d^k \end{array} \right] \supset \mathfrak{N}_\xi \left(t_R; t_F; \mathfrak{D}^i; \mathbf{U}; \mathfrak{Y}_d^k \right),$$

$$\partial \mathcal{D}_{TU} \left(t_R; t_F; \mathfrak{D}^i; \mathbf{U}; \mathfrak{Y}_d^k \right) \cap \partial \mathfrak{N}_\xi \left(t_R; t_F; \mathfrak{D}^i; \mathbf{U}; \mathfrak{Y}_d^k \right) = \phi, \tag{8.37}$$

then the FSRT tracking of $\mathbf{Y}_d^k(t)$ *is **uniform in** $[\mathbf{D}(.), \mathbf{Y}_d(.)] \in \mathfrak{D}^i \times \mathfrak{Y}_d^k$ on* $\mathfrak{T}_{0F} \times \mathfrak{D}^i \times \mathfrak{Y}_d^k$ *and the set* $\mathcal{D}_{TU}\left(t_R; t_F; \mathfrak{D}^i; \mathbf{U}; \mathfrak{Y}_d^k\right)$ *is the $(\mathbf{D}, \mathbf{Y}_d)$-uniform FSRT tracking domain of* $\mathbf{Y}_d^k(t)$ *on* $\mathfrak{T}_{0F} \times \mathfrak{D}^i \times \mathfrak{Y}_d^k$.

Comment 141 *The global FSRT tracking of* $\mathbf{Y}_d^k(t)$ *on* $\mathfrak{T}_{0F} \times \mathfrak{D}^i \times \mathfrak{Y}_d^k$ *is uniform in the functional pair* $[\mathbf{D}(.), \mathbf{Y}_d(.)]$ *over* $\mathfrak{D}^i \times \mathfrak{Y}_d^k$.

In order to avoid the *FSRT* tracking with a big overshoot we introduce the following.

Definition 142 **The stablewise FSRT tracking of the desired output behavior** $\mathbf{Y}_d^k(t)$ **of the system controlled by a control** $\mathbf{U}(.) \in \mathfrak{U}^l$
*a) The system exhibits **the stablewise output FSRT tracking of*** $\mathbf{Y}_d^k(t)$ **on** $\mathfrak{T}_{0F} \times \mathfrak{D}^i \times \mathfrak{Y}_d^k$, *for short **the stablewise FSRT tracking of** $\mathbf{Y}_d^k(t)$ **on** $\mathfrak{T}_{0F} \times \mathfrak{D}^i \times \mathfrak{Y}_d^k$, if and only if it exhibits the FSRT tracking of* $\mathbf{Y}_d^k(t)$ *on* $\mathfrak{T}_{0F} \times \mathfrak{D}^i \times \mathfrak{Y}_d^k$, *and for every connected neighborhood* $\mathfrak{N}_\varepsilon\left[t; \mathbf{Y}_d^k(t)\right]$ *of* $\mathbf{Y}_d^k(t)$ *at any* $t \in \mathfrak{T}_R$, *there is a connected neighborhood* $\mathfrak{N}\left(\varepsilon; t_R; t_F; \mathbf{Y}_{d0}^k; \mathbf{D}; \mathbf{U}; \mathbf{Y}_d\right)$, (8.14) (Section 8.3), of the plant desired initial output vector $\mathbf{Y}_{d0}^k$ at the initial moment $t_0 = 0$ such that it is subset of $\mathcal{D}_T\left(t_R; t_F; \mathbf{Y}_{d0}^k; \mathbf{D}; \mathbf{U}; \mathbf{Y}_d\right)$ and for initial $\mathbf{Y}_0^k \in \mathfrak{N}\left(\varepsilon; t_R; t_F; \mathbf{Y}_{d0}^k; \mathbf{D}; \mathbf{U}; \mathbf{Y}_d\right)$ the instantaneous $\mathbf{Y}^k(t)$ stays in the neighborhood $\mathfrak{N}_\varepsilon\left[t; \mathbf{Y}_d^k(t)\right]$ for all $t \in \mathfrak{T}_R$; i.e.,

$$\forall \mathfrak{N}_\varepsilon\left[t; \mathbf{Y}_d^k(t)\right] \subseteq \mathfrak{R}^{(k+1)N}, \forall t \in \mathfrak{T}_R,$$

$$\forall [\mathbf{D}(.), \mathbf{Y}_d(.)] \in \mathfrak{D}^i \times \mathfrak{Y}_d^k, \exists \mathfrak{N}\left(\varepsilon; t_R; t_F; \mathbf{Y}_{d0}^k; \mathbf{D}; \mathbf{U}; \mathbf{Y}_d\right) \subseteq \mathfrak{R}^{(k+1)N},$$

$$\mathfrak{N}\left(\varepsilon; t_R; t_F; \mathbf{Y}_{d0}^k; \mathbf{D}; \mathbf{U}; \mathbf{Y}_d\right) \subseteq \mathcal{D}_T\left(t_R; t_F; \mathbf{Y}_{d0}^k; \mathbf{D}; \mathbf{U}; \mathbf{Y}_d\right),$$

$$\mathbf{Y}_0^k \in \mathfrak{N}\left(\varepsilon; t_R; t_F; \mathbf{Y}_{d0}^k; \mathbf{D}; \mathbf{U}; \mathbf{Y}_d\right) \implies$$

$$\mathbf{Y}^k(t; \mathbf{Y}_0^k; \mathbf{D}; \mathbf{U}) \in \mathfrak{N}_\varepsilon\left[t; \mathbf{Y}_d^k(t)\right], \forall t \in \mathfrak{T}_R. \tag{8.38}$$

b) The largest connected neighborhood $\mathfrak{N}\left(\varepsilon; t_R; t_F; \mathbf{Y}_{d0}^k; \mathbf{D}; \mathbf{U}; \mathbf{Y}_d\right)$ *of* $\mathbf{Y}_{d0}^k$, *(8.38), is the ε-tracking domain*

$$\mathcal{D}_{ST}\left(\varepsilon; t_R; t_F; \mathbf{Y}_{d0}^k; \mathbf{D}; \mathbf{U}; \mathbf{Y}_d\right)$$

of the stablewise FSRT tracking of $\mathbf{Y}_d^k(t)$ *on* $\mathfrak{T}_{0F} \times \mathfrak{D}^i \times \mathfrak{Y}_d^k$.
The domain $\mathcal{D}_{ST}\left(t_R; t_F; \mathbf{Y}_{d0}^k; \mathbf{D}; \mathbf{U}; \mathbf{Y}_d\right)$ *of the stablewise FSRT tracking of* $\mathbf{Y}_d^k(t)$ *on* $\mathfrak{T}_{0F} \times \mathfrak{D}^i \times \mathfrak{Y}_d^k$ *is the union of all*

$$\mathcal{D}_{ST}\left(\varepsilon; t_R; t_F; \mathbf{Y}_{d0}^k; \mathbf{D}; \mathbf{U}; \mathbf{Y}_d\right)$$

over $\varepsilon \in \Re^+$,

$$\mathcal{D}_{ST}\left(t_R; t_F; \mathbf{Y}_{d0}^k; \mathbf{D}; \mathbf{U}; \mathbf{Y}_d\right) = \cup \left[\begin{array}{c} \mathcal{D}_{ST}\left(\varepsilon; t_R; t_F; \mathbf{Y}_{d0}^k; \mathbf{D}; \mathbf{U}; \mathbf{Y}_d\right) : \\ : \varepsilon \in \Re^+ \end{array}\right].$$
(8.39)

Let $[0, \varepsilon_M)$ be the maximal interval over which

$$\mathcal{D}_{ST}\left(\varepsilon; t_R; t_F; \mathbf{Y}_{d0}^k; \mathbf{D}; \mathbf{U}; \mathbf{Y}_d\right)$$

is continuous in $\varepsilon \in \Re_+$,

$$\mathcal{D}_{ST}\left(\varepsilon; t_R; t_F; \mathbf{Y}_{d0}^k; \mathbf{D}; \mathbf{U}; \mathbf{Y}_d\right) \in \mathfrak{C}\left([0, \varepsilon_M)\right),$$

$$\forall [\mathbf{D}(.), \mathbf{Y}_d(.)] \in \mathfrak{D}^i \times \mathfrak{Y}_d^k.$$

The domain $\mathcal{D}_{SST}\left(t_R; t_F; \mathbf{Y}_{d0}^k; \mathbf{D}; \mathbf{U}; \mathbf{Y}_d\right)$ **of the strict stablewise FSRT tracking of** $\mathbf{Y}_d^k(t)$ **on** $\mathfrak{T}_{0F} \times \mathfrak{D}^i \times \mathfrak{Y}_d^k$ *is the union of all stable FSRT tracking domains* $\mathcal{D}_{ST}\left(\varepsilon; t_R; t_F; \mathbf{Y}_{d0}^k; \mathbf{D}; \mathbf{U}; \mathbf{Y}_d\right)$ *in* $\varepsilon \in [0, \varepsilon_M)$,

$$\mathcal{D}_{SST}\left(t_R; t_F; \mathbf{Y}_{d0}^k; \mathbf{D}; \mathbf{U}; \mathbf{Y}_d\right)$$

$$= \cup \left\{\begin{array}{c} \mathcal{D}_{ST}\left(\varepsilon; t_R; t_F; \mathbf{Y}_{d0}^k; \mathbf{D}; \mathbf{U}; \mathbf{Y}_d\right) : \\ : \varepsilon \in [0, \varepsilon_M) \end{array}\right\},$$

$$\forall [\mathbf{D}(.), \mathbf{Y}_d(.)] \in \mathfrak{D}^i \times \mathfrak{Y}_d^k.$$
(8.40)

c) The stable FSRT tracking of $\mathbf{Y}_d^k(t)$ *on* $\mathfrak{T}_{0F} \times \mathfrak{D}^i \times \mathfrak{Y}_d^k$ *is* **global** *(in* **the whole)** *if and only if it is the global FSRT tracking of* $\mathbf{Y}_d^k(t)$ *on* $\mathfrak{T}_{0F} \times \mathfrak{D}^i \times \mathfrak{Y}_d^k$ *and the stable FSRT tracking of* $\mathbf{Y}_d^k(t)$ *on* $\mathfrak{T}_{0F} \times \mathfrak{D}^i \times \mathfrak{Y}_d^k$ *is with*

$$\mathcal{D}_{ST}\left(t_R; t_F; \mathbf{Y}_{d0}^k; \mathbf{D}; \mathbf{U}; \mathbf{Y}_d\right) = \Re^{(k+1)N},$$

or equivalently with

$$\mathcal{D}_{ST}\left(t_R; t_F; \mathbf{0}_{(k+1)N}; \mathbf{D}; \mathbf{U}; \mathbf{Y}_d\right) = \Re^{(k+1)N},$$

for every $[\mathbf{D}(.), \mathbf{Y}_d(.)] \in \mathfrak{D}^i \times \mathfrak{Y}_d^k$.

This definition permits the nonuniformity of the stable $FSRT$ tracking.

Exercise 143 *Define the stable FSRT tracking of* $\mathbf{Y}_d^k(t)$ *on* $\mathfrak{D}^j \times \mathfrak{Y}_d^k$ *and its domain in terms of the output error vector* $\mathbf{e}^k$.

Definition 144 *The uniform stablewise FSRT tracking of the de-sired output behavior* $\mathbf{Y}_d^k(t)$ *of the system controlled by a control* $\mathbf{U}(.) \in \mathfrak{U}^l$

If and only if the intersection $\mathcal{D}_{STU}\left(t_R; t_F; \mathfrak{D}^i; \mathbf{U}; \mathfrak{Y}_d^k\right)$ *of all*

$$\mathcal{D}_{ST}\left(t_R; t_F; \mathbf{Y}_{d0}^k; \mathbf{D}; \mathbf{U}; \mathbf{Y}_d\right)$$

[$\mathcal{D}_{SSTU}\left(t_R; t_F; \mathfrak{D}^i; \mathbf{U}; \mathfrak{Y}_d^k\right)$ *of all* $\mathcal{D}_{SST}\left(t_R; t_F; \mathbf{Y}_{d0}^k; \mathbf{D}; \mathbf{U}; \mathbf{Y}_d\right)$*] over* $\mathfrak{D}^i \times \mathfrak{Y}_d^k$ *is a connected neighborhood of* $\mathbf{Y}_{d0}^k$ *then and only then it is the (strictly) stablewise FSRT tracking domain of* $\mathbf{Y}_d^k(t)$ *on* $\mathfrak{T}_{0F} \times \mathfrak{D}^i \times \mathfrak{Y}_d^k$ *uni-form in* $[\mathbf{D}(.), \mathbf{Y}_d(.)] \in \mathfrak{D}^i \times \mathfrak{Y}_d^k$, *respectively,*

$$\exists \xi \in \mathfrak{R}^+ \Longrightarrow \mathcal{D}_{STU}\left(t_R; t_F; \mathfrak{D}^i; \mathbf{U}; \mathfrak{Y}_d^k\right) =$$
$$= \cap \left\{\mathcal{D}_{ST}\left(t_R; t_F; \mathbf{Y}_{d0}^k; \mathbf{D}; \mathbf{U}; \mathbf{Y}_d\right) : [\mathbf{D}(.), \mathbf{Y}_d(.)] \in \mathfrak{D}^i \times \mathfrak{Y}_d^k\right\}$$
$$\supset \mathfrak{N}_\xi\left(\mathfrak{D}^i; \mathbf{U}; \mathfrak{Y}_d^k\right), \tag{8.41}$$

$$\left[\begin{array}{c} \exists \xi \in \mathfrak{R}^+ \Longrightarrow \mathcal{D}_{SSTU}\left(t_R; t_F; \mathfrak{D}^i; \mathbf{U}; \mathfrak{Y}_d^k\right) = \\ = \cap \left\{\begin{array}{c} \mathcal{D}_{SST}\left(t_R; t_F; \mathbf{Y}_{d0}^k; \mathbf{D}; \mathbf{U}; \mathbf{Y}_d\right) : \\ : [\mathbf{D}(.), \mathbf{Y}_d(.)] \in \mathfrak{D}^i \times \mathfrak{Y}_d^k \end{array}\right\} \\ \supset \mathfrak{N}_\xi\left(\mathfrak{D}^i; \mathbf{U}; \mathfrak{Y}_d^k\right) \end{array}\right]. \tag{8.42}$$

8.6 Finite vector reachability *time*

We define **in the output space** various types of *tracking with the finite vector reachability time (FVRT)*.

Elementwise tracking with the finite vector reachability time represents better tracking than the preceding tracking types. It allows different *FSRTs* to be associated, mutually independently, to different output variables.

By referring to Equations (1.13), (Section 1.3), we introduce also *the elementwise* $(k+1)N$- *zero vector* $\mathbf{0}_{(k+1)N}$, all elements of which are equal to zero, and $(k+1)N$- *unit vector* $\mathbf{1}_{(k+1)N}$, all elements of which are equal to one,

$$\mathbf{0}_{(k+1)N} = [0\ 0...0]^T \in \mathfrak{R}^{(k+1)N}, \quad \mathbf{1}_{(k+1)N} = [1\ 1...1]^T \in \mathfrak{R}^{(k+1)N},$$
$$k \in \{0, 1, 2, ..\}, \tag{8.43}$$

Let us introduce the $(k+1)N$-*time vector* $\mathbf{t}^{(k+1)N}$, [183, p. 387], [175] all $(k+1)N$ elements of which are the same temporal variable, *time t*,

$$\mathbf{t}^{(k+1)N} = t\mathbf{1}_{(k+1)N} = [t\ t...t]^T \in \left(\mathfrak{T}_0^{(k+1)N} \cup \{\infty\}^{(k+1)N}\right),\ \ k \in \{0,1,2,...\},$$

$$\mathbf{t} = \mathbf{t}^N = t\mathbf{1}_N = [t\ t...t]^T \in \mathfrak{T}_0^N \cup \{\infty\}^N,$$

$$\mathbf{t}_0^{(k+1)N} = t_0\mathbf{1}_{(k+1)N} = 0\mathbf{1}_{(k+1)N} = [0\ 0...0]^T = \mathbf{0}_{(k+1)N} \in In\ \mathfrak{T}^{(k+1)N}. \tag{8.44}$$

where

$$\mathfrak{T}_0^{(k+1)N} = \underbrace{\mathfrak{T}_0 \times \mathfrak{T}_0 \times ... \times \mathfrak{T}_0}_{(k+1)N-times}, \tag{8.45}$$

$$Cl\ \mathfrak{T}_0^{(k+1)N} = \underbrace{Cl\ \mathfrak{T}_0 \times Cl\ \mathfrak{T}_0 \times .. \times\ Cl\ \mathfrak{T}_0}_{(k+1)N-times},$$

$$In\ \mathfrak{T}_0^{(k+1)N} = \underbrace{In\ \mathfrak{T}_0 \times In\ \mathfrak{T}_0 \times .. \times In\ \mathfrak{T}_0}_{(k+1)N-times}, \tag{8.46}$$

We can associate with every output variable Y_i its own *scalar reachability time* $t_{Ri} \in In\ \mathfrak{T}_{0F}$, $i = 0, 1, 2, ..., N$, respectively. They compose the following *vector reachability time* with N entries being possibly mutually different instants:

$$\mathbf{t}_R^N = \mathbf{t}_{R(0)}^N = \begin{bmatrix} t_{R1} \\ t_{R2} \\ ... \\ t_{RN} \end{bmatrix} = \begin{bmatrix} t_{R1,(0)} \\ t_{R2,(0)} \\ ... \\ t_{RN,(0)} \end{bmatrix} \in In\ \mathfrak{T}_{0F}^N. \tag{8.47}$$

We associate also with the j-th derivative $Y_1^{(j)}$, $Y_2^{(j)}$, ..., $Y_N^{(j)}$ of the scalar variables Y_i, $i = 0, 1, 2, ..., N$, their own scalar reachability *times* $t_{R1(j)} \in \mathfrak{T}_0 \cup \{\infty\}$ $t_{R2(j)} \in \mathfrak{T}_0 \cup \{\infty\}$, ..., $t_{RN(j)} \in \mathfrak{T}_0 \cup \{\infty\}$, respectively. They induce the *time* sets $\mathfrak{T}_{Ri(j)}$ and $\mathfrak{T}_{Ri(j)F}$:

$$0 \leq t_{Ri(j)} < \infty \Longrightarrow \left\{ \begin{array}{c} \mathfrak{T}_{Ri(j)} = \{t : 0 \leq t \leq t_{Ri(j)}\} \subset \mathfrak{T}_{0F}, \\ \mathfrak{T}_{Ri(j)F} = \left\{ \begin{array}{l} t:\ t_{Ri(j)} \leq t \leq t_F < \infty, \\ t:\ t_{Ri(j)} \leq t < t_F = \infty, \end{array} \right\} \subset \mathfrak{T}_{0F}, \\ \mathfrak{T}_{Ri(j)} \cup \mathfrak{T}_{Ri(j)F} = \mathfrak{T}_{0F}, \end{array} \right\},$$

$$t_{Ri(j)} = t_F = \infty \Longrightarrow \mathfrak{T}_{Ri(j)\infty} = \{\infty\},\ i = 1, 2, ..., N,\ j \in \{0, 1, .., k\}. \tag{8.48}$$

They also compose the following *vector reachability time* with N entries being possibly mutually different instants. It is the generalization of the *time vector* $\mathbf{t}_R^N$ (8.47) to the j-th derivatives $Y_1^{(j)}$, $Y_2^{(j)}$, ..., $Y_{Ni}^{(j)}$ of the output variables Y_1, Y_2, ..., Y_{Ni} reads

$$\mathbf{t}_{R(j)}^N = \begin{bmatrix} t_{R1,(j)} \\ t_{R2,(j)} \\ \cdots \\ t_{RN,(j)} \end{bmatrix} \in In \ \mathfrak{T}_{0F}^N, \ j \in \{0,1,2,..\}. \tag{8.49}$$

In order to treat mathematically effectively and simply such cases, we define *the finite vector reachability time (FVRT)* $t_R^{(k+1)N} \in (In \ \mathfrak{T}_0)^{(k+1)N}$, which is related to the output vector and its derivatives up to the order k:

$$\mathbf{t}_R^{(k+1)N} = \begin{bmatrix} \mathbf{t}_R^N \\ \mathbf{t}_{R(1)}^N \\ \mathbf{t}_{R(2)}^N \\ \cdots \\ \mathbf{t}_{R(k)}^N \end{bmatrix} = \begin{bmatrix} \mathbf{t}_{R(0)}^N \\ \mathbf{t}_{R(1)}^N \\ \mathbf{t}_{R(2)}^N \\ \cdots \\ \mathbf{t}_{R(k)}^N \end{bmatrix} \in In \ \mathfrak{T}_{0F}^{(k+1)N}. \tag{8.50}$$

In the scalar form:

$$\mathbf{t}_R^{(k+1)N} = [t_{R1,(0)} \ \cdots \ t_{RN,(0)} \ \ t_{R1,(1)} \cdots \ t_{RN,(1)} \ \cdots \ t_{R1,(k)} \ \cdots \ t_{RN,(k)}]^T,$$
$$\forall k = 0, \ 1, ... \tag{8.51}$$

Notice that

$$\mathbf{t}_R^{(k+1)N} = \mathbf{t}_F^{(k+1)N} = \infty \mathbf{1}_{k+1)N} \Longleftrightarrow \mathfrak{T}_{RF}^{(k+1)N} = \{\infty\}^{(k+1)N} = \{\infty \mathbf{1}_{(k+1)N}\}. \tag{8.52}$$

Note 145 *The value t_R is infinite, $t_R = t_F = \infty$; hence, $\mathfrak{T}_{RF} = \mathfrak{T}_{R\infty} = \{\infty\}$, i.e., $\mathbf{t}_R^{(k+1)N} = \mathbf{t}_F^{(k+1)N} = \infty \mathbf{1}_{(k+1)N}$ and $\mathfrak{T}_{RF}^{(k+1)N} = \mathfrak{T}_{R\infty}^{(k+1)N} = \{\infty\}^{(k+1)N}$, if and only if tracking is, or should be, asymptotic. Otherwise $t_R \in In\mathfrak{T}_0$; hence, $\mathfrak{T}_{RF} \subset \mathfrak{T}_0$, and $\mathbf{0}_{k+1)N} \leq \mathbf{t}_R^{(k+1)N} < \infty \mathbf{1}_{(k+1)N}$ so that*

$$\mathfrak{T}_{RF}^{(k+1)N} = \left\{ \mathbf{t}^{(k+1)N} : \ \mathbf{t}_R^{(k+1)N} \leq \mathbf{t}^{(k+1)N} < \infty \mathbf{1}_{(k+1)N} \right\} \subset \mathfrak{T}_0^{(k+1)N}.$$

Let

$$\mathbf{Y}^k(\mathbf{t}^{(k+1)N}) = \left[\mathbf{Y}^T(t) \ \ \mathbf{Y}^{(1)T}(t) \ \dots \ \mathbf{Y}^{(k)T}(t)\right]^T \in \mathfrak{R}^{(k+1)N},$$

$$\mathbf{Y}^{(j)}(\mathbf{t}^N_{R(j)}) = \left[Y_1^{(j)T}(t_{R1,(j)}) \ \ Y_2^{(j)T}(t_{R2,(j)}) \ \dots \ Y_N^{(j)T}(t_{RN,(j)})\right]^T \in \mathfrak{R}^N,$$

$$j \in \{0, 1, 2, ..., k\},$$

$$\mathbf{Y}^k(\mathbf{t}^{(k+1)N}_R) = \left[\mathbf{Y}^T(\mathbf{t}^N_{R(0)}) \ \ \mathbf{Y}^{(1)T}(\mathbf{t}^N_{R(1)}) \ \dots \ \mathbf{Y}^{(k)T}(\mathbf{t}^N_{R(k)})\right]^T \in \mathfrak{R}^{(k+1)N}.$$

We can now summarize the above presentation about *the finite vector reachability time (FVRT)* $t^{(k+1)N}_R \in In \ \mathfrak{T}^{(k+1)N}_{0F}$. It is the first vector instant $t^{(k+1)N}$ at which and after which the real output vector $\mathbf{Y}^k(\mathbf{t}^{(k+1)N})$ becomes and stays elementwise equal to the desired output vector $\mathbf{Y}^k_d(\mathbf{t}^{(k+1)N})$ until the final vector instant $t^{(k+1)N}_F$:

$$\mathbf{Y}^k(\mathbf{t}^{(k+1)N}) = \mathbf{Y}^k_d(\mathbf{t}^{(k+1)N}), \quad \forall \mathbf{t}^{(k+1)N} \in \mathfrak{T}^{(k+1)N}_{RF}. \tag{8.53}$$

We relate $t^{(k+1)N}_R$ to the tracking treated via *the extended output space* $\mathfrak{R}^{(k+1)N}$, which for $k = \nu - 1$ becomes also *the state space* $\mathfrak{R}^{\nu N}$ if the plant is the *IO* plant. However, $\mathfrak{R}^{(k+1)N}$ becomes the ordinary output space $\mathfrak{R}^N$ for $k = 0$ if the plant is the *ISO* plant or the *EISO* plant because then formally $\nu = 1$. For $k = \alpha - 1$ the space $\mathfrak{R}^{(k+1)N}$ rests *the extended output space* $\mathfrak{R}^{\alpha N}$ if the plant is the *HISO* plant and for $k = \alpha + \nu - 2$ it preserves its meaning of the extended output space $\mathfrak{R}^{(\alpha+\nu-1)N}$ of the *IIO* plant.

The above notation leads to

$$\mathfrak{T}^{(k+1)N} = \left\{\mathbf{t}^{(k+1)N} : \ -\infty\mathbf{1}_{(k+1)N} < \mathbf{t}^{(k+1)N} < \infty\mathbf{1}_{(k+1)N}\right\}, \tag{8.54}$$

to

$$\mathfrak{T}^{(k+1)N}_R = \left\{\mathbf{t}^{(k+1)N} : \ \mathbf{t}^{(k+1)N}_0 = \mathbf{0}_{(k+1)N} \le \mathbf{t}^{(k+1)N} \le \mathbf{t}^{(k+1)N}_R < \infty\mathbf{1}_{(k+1)N}\right\}$$

$$\mathbf{t}^{(k+1)N}_R = \infty\mathbf{1}_{(k+1)N} \Longrightarrow \mathfrak{T}^{(k+1)N}_{R\infty} = \left\{\infty\mathbf{1}_{(k+1)N}\right\}, \tag{8.55}$$

and to

$$\mathfrak{T}^{(k+1)N}_{RF} = \left\{\mathbf{t}^{(k+1)N} : \left\langle \begin{array}{l} \mathbf{t}^{(k+1)N}_R \le \mathbf{t}^{(k+1)N} \le \mathbf{t}^{(k+1)N}_F < \infty\mathbf{1}_{(k+1)N}, \ or \\ \mathbf{t}^{(k+1)N}_R \le \mathbf{t}^{(k+1)N} < \mathbf{t}^{(k+1)N}_F = \infty\mathbf{1}_{(k+1)N} \end{array} \right\rangle \right\}. \tag{8.56}$$

The symbolic vector notation

$$\mathbf{Y}^k(\mathbf{t}^{(k+1)N}) = \mathbf{Y}^k_d(\mathbf{t}^{(k+1)N}), \ \forall \mathbf{t}^{(k+1)N} \in [\mathbf{t}^{(k+1)N}_R, \ \infty\mathbf{1}_{(k+1)N}[,$$

$$k \in \{0, 1, 2, ...\}$$

means in the scalar form

$$Y_i^{(j)}(t) = Y_{di}^{(j)}(t), \ \forall t \in [t_{Ri(j)}, \ \infty[, \ \forall i = 1, 2, ..., N, \ \forall j \in \{0, 1, 2, .., k\}.$$

Besides

$$\left| \mathbf{Y}^{(j)}(\mathbf{t}^N) - \mathbf{Y}_d^{(j)}(\mathbf{t}^N) \right| = \begin{vmatrix} Y_1^{(j)}(t) - Y_{d1}^{(j)}(t) \\ Y_2^{(j)}(t) - Y_{d2}^{(j)}(t) \\ \cdots \\ Y_N^{(j)}(t) - Y_{dN}^{(j)}(t) \end{vmatrix} \in \mathfrak{R}_+^N, \ \forall j = 0, 1, 2, ..., k,$$

and

$$\left| \mathbf{Y}^k(\mathbf{t}^{(k+1)N}) - \mathbf{Y}_d^k(\mathbf{t}^{(k+1)N}) \right| = \begin{vmatrix} \mathbf{Y}(\mathbf{t}^N) - \mathbf{Y}_d(\mathbf{t}^N) \\ \mathbf{Y}^{(1)}(\mathbf{t}^N) - \mathbf{Y}_d^{(1)}(\mathbf{t}^N) \\ \cdots \\ \mathbf{Y}^{(k)}(\mathbf{t}^N) - \mathbf{Y}_d^{(k)}(\mathbf{t}^N) \end{vmatrix} \in \mathfrak{R}_+^{(k+1)N},$$

$$k \in \{0, 1, 2, ...\}.$$

Let a positive real number $\varepsilon_{i(j)}$, or $\varepsilon_{i(j)} = \infty$, be associated with the j-th derivative of Y_i and of Y_{di}, and be taken for the i-th entry of the positive N vector $\varepsilon_{(j)}^N$, i.e., of the positive $(k+1)N-$ vector $\varepsilon^{(k+1)N}$, respectively,

$$\varepsilon_{(j)}^N = \begin{bmatrix} \varepsilon_{1,(j)} \\ \varepsilon_{2,(j)} \\ \cdots \\ \varepsilon_{N,(j)} \end{bmatrix} \in \mathfrak{R}^{+N} \cup \{\infty\}^N, \ \forall j = 0, 1, 2, ..., k, \ \varepsilon_{i,(0)} \equiv \varepsilon_i, \varepsilon_{(0)}^N \equiv \varepsilon^N,$$

(8.57)

$$\varepsilon^{(k+1)N} = \begin{bmatrix} \varepsilon_{(0)}^N \\ \varepsilon_{(1)}^N \\ \cdots \\ \varepsilon_{(k)}^N \end{bmatrix} = \begin{bmatrix} \varepsilon^N \\ \varepsilon_{(1)}^N \\ \cdots \\ \varepsilon_{(k)}^N \end{bmatrix} \in \mathfrak{R}^{+(k+1)N} \cup \{\infty\}^{(k+1)N}, \ k \in \{1, 2, ..\},$$

(8.58)

so that

$$\left| \mathbf{Y}_{d0}^k - \mathbf{Y}_0^k \right| = \left| \mathbf{Y}_0^k - \mathbf{Y}_{d0}^k \right| < \varepsilon^{(k+1)N}, \ \forall k = 0, 1, 2, ..., \alpha - 1, \tag{8.59}$$

signifies that the relationship holds element by element, i.e., **elementwise**,

$$\left| Y_{di0}^{(j)} - Y_{i0}^{(j)} \right| = \left| Y_{i0}^{(j)} - Y_{di0}^{(j)} \right| < \varepsilon_{i,(j)}, \ \forall i = 1, 2, ..., N, \ \forall j = 0, 1, 2, ..., k.$$

(8.60)

We use the above simplified notation in the sequel,

$$\mathbf{Y}_d^k(\mathbf{t}^{(k+1)N}) \equiv \mathbf{Y}_d^k(\mathbf{t}^{(k+1)N}; \mathbf{Y}_{d0}^k),$$
$$\mathbf{Y}^k(\mathbf{t}^{(k+1)N}) \equiv \mathbf{Y}^k(\mathbf{t}^{(k+1)N}; \mathbf{Y}_0^k; \mathbf{D}; \mathbf{U}), \tag{8.61}$$

The following definition generalizes Definition 142.

Definition 146 *Elementwise tracking with the finite vector reachability time $t_R^{(k+1)N}$ of the desired output $\mathbf{Y}_d^k(t^{(k+1)N})$ of the system
controlled by a control $\mathbf{U}(.) \in \mathfrak{U}^l$*

*a) The system exhibits **the elementwise output tracking of the desired output response** $\mathbf{Y}_d^k(t^{(k+1)N})$ with the finite vector reachabil-
ity time (FVRT) $t_R^{(k+1)N}$ on $\mathfrak{T}_{0F}^{(k+1)N} \times \mathfrak{D}^i \times \mathfrak{Y}_d^k$; i.e., the elementwise
tracking of $\mathbf{Y}_d^k(t^{(k+1)N})$ with FVRT $t_R^{(k+1)N}$ on $\mathfrak{T}_{0F}^{(k+1)N} \times \mathfrak{D}^i \times \mathfrak{Y}_d^k$ if
and only if for every $\left[\mathbf{D}(.), \mathbf{Y}_d^k(.)\right] \in \mathfrak{D}^i \times \mathfrak{Y}_d^k$ there exists a connected neighborhood $\mathfrak{N}\left(\mathbf{t}_R^{(k+1)N}; \mathbf{t}_F^{(k+1)N}; \mathbf{Y}_{d0}^k; \mathbf{D}; \mathbf{U}; \mathbf{Y}_d^k\right)$, which is dependent on the pair
$\left(\mathbf{t}_R^{(k+1)N}, \mathbf{t}_F^{(k+1)N}\right)$,*

$$\mathfrak{N}\left(\mathbf{t}_R^{(k+1)N}; \mathbf{t}_F^{(k+1)N}; \mathbf{Y}_{d0}^k; \mathbf{D}; \mathbf{U}; \mathbf{Y}_d^k\right) \subseteq \mathfrak{R}^{(k+1)N},$$

*of the plant desired initial output vector $\mathbf{Y}_{d0}^k$ at the initial vector moment
$\mathbf{t}_0^{(k+1)N} = \mathbf{0}_{(k+1)N}$ such that $\mathbf{Y}_0^k$ from the neighborhood*

$$\mathfrak{N}\left(\mathbf{t}_R^{(k+1)N}; \mathbf{t}_F^{(k+1)N}; \mathbf{Y}_{d0}^k; \mathbf{D}; \mathbf{U}; \mathbf{Y}_d^k\right)$$

*guarantees both that $\mathbf{Y}^k(t^{(k+1)N})$ becomes equal to $\mathbf{Y}_d^k(t^{(k+1)N})$ at the finite
vector reachability moment $t_R^{(k+1)N} \in In \ \mathfrak{T}_{0F}^{(k+1)N}$, and that they stay equal
since then on $\mathfrak{T}_{RF}^{(k+1)N}$; i.e.,*

$$\forall \left[\mathbf{D}(.), \mathbf{Y}_d^k(.)\right] \in \mathfrak{D}^i \times \mathfrak{Y}_d^k,$$
$$\exists \mathfrak{N}\left(\mathbf{t}_R^{(k+1)N}; \mathbf{t}_F^{(k+1)N}; \mathbf{Y}_{d0}^k; \mathbf{D}; \mathbf{U}; \mathbf{Y}_d^k\right),$$
$$\mathfrak{N}\left(\mathbf{t}_R^{(k+1)N}; \mathbf{t}_F^{(k+1)N}; \mathbf{Y}_{d0}^k; \mathbf{D}; \mathbf{U}; \mathbf{Y}_d^k\right) \subseteq \mathfrak{R}^{(k+1)N},$$
$$\mathbf{Y}_0^k \in \mathfrak{N}\left(\mathbf{t}_R^{(k+1)N}; \mathbf{t}_F^{(k+1)N}; \mathbf{Y}_{d0}^k; \mathbf{D}; \mathbf{U}; \mathbf{Y}_d^k\right) \implies$$
$$\mathbf{Y}^k(\mathbf{t}^{(k+1)N}) = \mathbf{Y}_d^k(\mathbf{t}^{(k+1)N}), \ \forall \mathbf{t}^{(k+1)N} \in \mathfrak{T}_{RF}^{(k+1)N}. \tag{8.62}$$

b) The largest connected neighborhood

$$\mathfrak{N}_L\left(\mathbf{t}_R^{(k+1)N};\mathbf{t}_F^{(k+1)N};\mathbf{Y}_{d0}^k;\mathbf{D};\mathbf{U};\mathbf{Y}_d^k\right)$$

of $\mathbf{Y}_{d0}^k$, *which obeys (8.62), is the **FVRT elementwise tracking domain***

$$\mathcal{D}_T\left(\mathbf{t}_R^{(k+1)N};\mathbf{t}_F^{(k+1)N};\mathbf{Y}_{d0}^k;\mathbf{D};\mathbf{U};\mathbf{Y}_d^k\right)$$

of $\mathbf{Y}_d^k(t^{(k+1)N})$ *on* $\mathfrak{T}_{0F}^{(k+1)N}\times\mathfrak{D}^i\times\mathfrak{Y}_d^k$.

c) FVRT elementwise tracking of $\mathbf{Y}_d^k(t^{(k+1)N})$ *on* $\mathfrak{T}_{0F}^{(k+1)N}\times\mathfrak{D}^i\times\mathfrak{Y}_d^k$ *is **global (in the whole)** if and only if*

$$\mathcal{D}_T\left(\mathbf{t}_R^{(k+1)N};\mathbf{t}_F^{(k+1)N};\mathbf{Y}_{d0}^k;\mathbf{D};\mathbf{U};\mathbf{Y}_d^k\right)=\mathfrak{R}^{(k+1)N}$$

for every $\left[\mathbf{D}(.),\mathbf{Y}_d^k(.)\right]\in\mathfrak{D}^i\times\mathfrak{Y}_d^k$.

FVRT elementwise tracking of $\mathbf{Y}_d^k(t^{(k+1)N})$ *on* $\mathfrak{T}_{0F}^{(k+1)N}\times\mathfrak{D}^i\times\mathfrak{Y}_d^k$ *is then uniform in* $\left[\mathbf{D}(.),\mathbf{Y}_d^k(.)\right]\in\mathfrak{D}^i\times\mathfrak{Y}_d^k$.

Exercise 147 *Define the elementwise FVRT tracking of* $\mathbf{Y}_d^k(t)$ *on* $\mathfrak{D}^j\times\mathfrak{Y}_d^k$ *and its domain in terms of the output error vector* $\mathbf{e}^k$.

We present the vector generalization of Definition 140.

Definition 148 *The **uniform elementwise tracking with the finite vector reachability time** of the desired output behavior* $\mathbf{Y}_d^k(t^{(k+1)N})$ *on* $\mathfrak{T}_{0F}^{(k+1)N}\times\mathfrak{D}^i\times\mathfrak{Y}_d^k$ *of the system controlled by a control* $\mathbf{U}(.)\in\mathfrak{U}^l$ *If and only if the intersection*

$$\mathcal{D}_{TU}\left(\mathbf{t}_R^{(k+1)N};\mathbf{t}_F^{(k+1)N};\mathfrak{D}^i;\mathbf{U};\mathfrak{Y}_d^k\right)$$

of the elementwise tracking domains

$$\mathcal{D}_T\left(\mathbf{t}_R^{(k+1)N};\mathbf{t}_F^{(k+1)N};\mathbf{Y}_{d0}^k;\mathbf{D};\mathbf{U};\mathbf{Y}_d^k\right)$$

in $\left[\mathbf{D}(.),\mathbf{Y}_d^k(.)\right]\in\mathfrak{D}^i\times\mathfrak{Y}_d^k$ *is a connected neighborhood of* $\mathbf{Y}_{d0}^k$ *of every*

$\mathbf{Y}_d^k(.) \in \mathfrak{Y}_d^k,$

$$\exists \xi \in \mathfrak{R}^+ \Longrightarrow \mathcal{D}_{TU}\left(\mathbf{t}_R^{(k+1)N}; \mathbf{t}_F^{(k+1)N}; \mathfrak{D}^i; \mathbf{U}; \mathfrak{Y}_d^k\right) =$$

$$= \cap \left[\begin{array}{c} \mathcal{D}_T\left(\mathbf{t}_R^{(k+1)N}; \mathbf{t}_F^{(k+1)N}; \mathbf{Y}_{d0}^k; \mathbf{D}; \mathbf{U}; \mathbf{Y}_d^k\right) : \\ : [\mathbf{D}(.), \mathbf{Y}_d^k(.)] \in \mathfrak{D}^i \times \mathfrak{Y}_d^k \end{array} \right] \supset$$

$$\supset \mathfrak{N}_\xi\left(\mathbf{t}_R^{(k+1)N}; \mathbf{t}_F^{(k+1)N}; \mathfrak{D}^i; \mathbf{U}; \mathfrak{Y}_d^k\right),$$

$$\partial \mathcal{D}_{TU}\left(\mathbf{t}_R^{(k+1)N}; \mathbf{t}_F^{(k+1)N}; \mathfrak{D}^i; \mathbf{U}; \mathfrak{Y}_d^k\right) \cap$$

$$\cap \partial \mathfrak{N}_\xi\left(\mathbf{t}_R^{(k+1)N}; \mathbf{t}_F^{(k+1)N}; \mathfrak{D}^i; \mathbf{U}; \mathfrak{Y}_d^k\right) = \phi, \tag{8.63}$$

then the FVRT elementwise tracking of $\mathbf{Y}_d^k(t^{(k+1)N})$ *is **uniform in the pair*** $[\mathbf{D}(.), \mathbf{Y}_d^k(.)] \in \mathfrak{D}^i \times \mathfrak{Y}_d^k$ *on* $\mathfrak{T}_{0F}^{(k+1)N} \times \mathfrak{D}^i \times \mathfrak{Y}_d^k$ *and the set*

$$\mathcal{D}_{TU}\left(\mathbf{t}_R^{(k+1)N}; \mathbf{t}_F^{(k+1)N}; \mathfrak{D}^i; \mathbf{U}; \mathfrak{Y}_d^k\right)$$

is the $(\mathbf{D}, \mathbf{Y}_d^k)$**-uniform FVRT elementwise tracking domain of the extended desired output vector** $\mathbf{Y}_d^k(\mathbf{t}^{(k+1)N})$ **on** $\mathfrak{T}_{0F}^{(k+1)N} \times \mathfrak{D}^i \times \mathfrak{Y}_d^k.$

Exercise 149 *Define the uniform elementwise FVRT tracking of* $\mathbf{Y}_d^k(t)$ *on* $\mathfrak{D}^j \times \mathfrak{Y}_d^k$ *and its domain in terms of the output error vector* $\mathbf{e}^k.$

Note 150 *The global FVRT elementwise tracking of* $\mathbf{Y}_d^k(t^{(k+1)N})$ *on the product set* $\mathfrak{T}_{0F}^{(k+1)N} \times \mathfrak{D}^i \times \mathfrak{Y}_d^k$ *is uniform in* $\left[\mathbf{D}(.), \mathbf{Y}_d^k(.)\right]$ *over* $\mathfrak{D}^i \times \mathfrak{Y}_d^k.$

In order to assure a stability property of the tracking with *FVRT* we introduce:

Definition 151 *The stablewise elementwise tracking with the finite vector reachability time* $t_R^{(k+1)N}$ *of the desired output* $\mathbf{Y}_d^k(t^{(k+1)N})$ *of the system controlled by a control* $\mathbf{U}(.) \in \mathfrak{U}^l$
a) The system exhibits the stablewise elementwise tracking of the desired output $\mathbf{Y}_d^k(t^{(k+1)N})$ *with the finite vector reachability time* $t_R^{(k+1)N}$ *on* $\mathfrak{T}_{0F}^{(k+1)N} \times \mathfrak{D}^i \times \mathfrak{Y}_d^k,$ *i.e., the stablewise elementwise tracking of* $\mathbf{Y}_d^k(t^{(k+1)N})$ *with the finite vector reachability time* $t_R^{(k+1)N}$ *on* $\mathfrak{T}_{0F}^{(k+1)N} \times \mathfrak{D}^i \times \mathfrak{Y}_d^k$ *if and only if it exhibits the elementwise tracking with the finite vector reachability time* $t_R^{(k+1)N}$ *on* $\mathfrak{T}_{0F}^{(k+1)N} \times \mathfrak{D}^i \times \mathfrak{Y}_d^k,$ *and for*

every connected neighborhood $\mathfrak{N}_\varepsilon\left[\mathbf{t}^{(k+1)N};\mathbf{Y}_d^k(\mathbf{t}^{(k+1)N})\right]$ *of* $\mathbf{Y}_d^k(t^{(k+1)N})$ *at any* $t^{(k+1)N} \in \mathfrak{T}_{0F}^{(k+1)N}$, *there is a connected neighborhood*

$$\mathfrak{N}\left(\varepsilon;\mathbf{t}_R^{(k+1)N};\mathbf{t}_F^{(k+1)N};\mathbf{Y}_{d0}^k;\mathbf{D};\mathbf{U};\mathbf{Y}_d^k\right),\ (8.14),$$

of the plant desired initial output vector $\mathbf{Y}_{d0}^k$ *at the initial vector moment* $t_0^{(k+1)N}=\mathbf{0}_{(k+1)N}$ *such that it is subset of*

$$\mathcal{D}_T\left(\mathbf{t}_R^{(k+1)N};\mathbf{t}_F^{(k+1)N};\mathbf{Y}_{d0}^k;\mathbf{D};\mathbf{U};\mathbf{Y}_d^k\right)$$

and for the initial vector

$$\mathbf{Y}_0^k \in \mathfrak{N}\left(\varepsilon;\mathbf{t}_R^{(k+1)N};\mathbf{t}_F^{(k+1)N};\mathbf{Y}_{d0}^k;\mathbf{D};\mathbf{U};\mathbf{Y}_d^k\right)$$

the instantaneous $\mathbf{Y}^k(t^{(k+1)N})$ *stays in the neighborhood*

$$\mathfrak{N}_\varepsilon\left[\mathbf{t}^{(k+1)N};\mathbf{Y}_d^k(\mathbf{t}^{(k+1)N})\right]$$

for all $t^{(k+1)N} \in \mathfrak{T}_R^{(k+1)N}$; *i.e.*,

$$\forall\mathfrak{N}_\varepsilon\left[\mathbf{t}^{(k+1)N};\mathbf{Y}_d^k(\mathbf{t}^{(k+1)N})\right] \subseteq \mathfrak{R}^{(k+1)N},\ \forall\mathbf{t}^{(k+1)N} \in \mathfrak{T}_R^{(k+1)N},$$

$$\forall\left[\mathbf{D}(.),\mathbf{Y}_d^k(.)\right] \in \mathfrak{D}^i \times \mathfrak{Y}_d^k,$$

$$\exists\mathfrak{N}\left(\varepsilon;\mathbf{t}_R^{(k+1)N};\mathbf{t}_F^{(k+1)N};\mathbf{Y}_{d0}^k;\mathbf{D};\mathbf{U};\mathbf{Y}_d^k\right) \subseteq \mathfrak{R}^{(k+1)N},$$

$$\mathfrak{N}\left(\varepsilon;\mathbf{t}_R^{(k+1)N};\mathbf{t}_F^{(k+1)N};\mathbf{Y}_{d0}^k;\mathbf{D};\mathbf{U};\mathbf{Y}_d^k\right) \subseteq$$

$$\subseteq \mathcal{D}_T\left(\mathbf{t}_R^{(k+1)N};\mathbf{t}_F^{(k+1)N};\mathbf{Y}_{d0}^k;\mathbf{D};\mathbf{U};\mathbf{Y}_d^k\right),$$

$$\mathbf{Y}_0^k \in \mathfrak{N}\left(\varepsilon;\mathbf{t}_R^{(k+1)N};\mathbf{t}_F^{(k+1)N};\mathbf{Y}_{d0}^k;\mathbf{D};\mathbf{U};\mathbf{Y}_d^k\right) \Longrightarrow$$

$$\mathbf{Y}^k(\mathbf{t}^{(k+1)N}) \in \mathfrak{N}_\varepsilon\left[\mathbf{t}^{(k+1)N};\mathbf{Y}_d^k(\mathbf{t}^{(k+1)N})\right],\ \forall\mathbf{t}^{(k+1)N} \in \mathfrak{T}_R^{(k+1)N}.\quad (8.64)$$

b) The largest connected neighborhood

$$\mathfrak{N}_L\left(\varepsilon;\mathbf{t}_R^{(k+1)N};\mathbf{t}_F^{(k+1)N};\mathbf{Y}_{d0}^k;\mathbf{D};\mathbf{U};\mathbf{Y}_d^k\right)$$

of $\mathbf{Y}_{d0}^k$, *(8.38), is **the elementwise ε-tracking domain***

$$\mathcal{D}_{ST}\left(\varepsilon;\mathbf{t}_R^{(k+1)N};\mathbf{t}_F^{(k+1)N};\mathbf{Y}_{d0}^k;\mathbf{D};\mathbf{U};\mathbf{Y}_d^k\right)$$

at $t_0^{(k+1)N} = \mathbf{0}_{(k+1)N}$ of the stablewise elementwise tracking with FVRT of $\mathbf{Y}_d^k(t^{(k+1)N})$ on $\mathfrak{T}_{0F}^{(k+1)N} \times \mathfrak{D}^i \times \times \mathfrak{Y}_d^k$.

The domain $\mathcal{D}_{ST}\left(\mathbf{t}_R^{(k+1)N}; \mathbf{t}_F^{(k+1)N}; \mathbf{Y}_{d0}^k; \mathbf{D}; \mathbf{U}; \mathbf{Y}_d^k\right)$ of the stablewise elementwise tracking with FVRT of $\mathbf{Y}_d^k(t^{(k+1)N})$ on $\mathfrak{T}_{0F}^{(k+1)N} \times \mathfrak{D}^i \times \mathfrak{Y}_d^k$ is the union of all

$$\mathcal{D}_{ST}\left(\varepsilon; \mathbf{t}_R^{(k+1)N}; \mathbf{t}_F^{(k+1)N}; \mathbf{Y}_{d0}^k; \mathbf{D}; \mathbf{U}; \mathbf{Y}_d^k\right)$$

over $\varepsilon \in \mathfrak{R}^+$,

$$\mathcal{D}_{ST}\left(\mathbf{t}_R^{(k+1)N}; \mathbf{t}_F^{(k+1)N}; \mathbf{Y}_{d0}^k; \mathbf{D}; \mathbf{U}; \mathbf{Y}_d^k\right) =$$
$$= \cup\left[\mathcal{D}_{ST}\left(\varepsilon; \mathbf{t}_R^{(k+1)N}; \mathbf{t}_F^{(k+1)N}; \mathbf{Y}_{d0}^k; \mathbf{D}; \mathbf{U}; \mathbf{Y}_d^k\right) : \varepsilon \in \mathfrak{R}^+\right]. \qquad (8.65)$$

Let $[0, \varepsilon_M)$ be the maximal interval over which

$$\mathcal{D}_{ST}\left(\varepsilon; \mathbf{t}_R^{(k+1)N}; \mathbf{t}_F^{(k+1)N}; \mathbf{Y}_{d0}^k; \mathbf{D}; \mathbf{U}; \mathbf{Y}_d^k\right)$$

is continuous in $\varepsilon \in \mathfrak{R}_+$,

$$\mathcal{D}_{ST}\left(\varepsilon; \mathbf{t}_R^{(k+1)N}; \mathbf{t}_F^{(k+1)N}; \mathbf{Y}_{d0}^k; \mathbf{D}; \mathbf{U}; \mathbf{Y}_d^k\right) \in \mathfrak{C}\left([0, \varepsilon_M)\right),$$
$$\forall\left[\mathbf{D}(.), \mathbf{Y}_d^k(.)\right] \in \mathfrak{D}^i \times \mathfrak{Y}_d^k.$$

The strict domain $\mathcal{D}_{SST}\left(\mathbf{t}_R^{(k+1)N}; \mathbf{t}_F^{(k+1)N}; \mathbf{Y}_{d0}^k; \mathbf{D}; \mathbf{U}; \mathbf{Y}_d^k\right)$ of the stablewise tracking with FVRT of $\mathbf{Y}_d^k(t^{(k+1)N})$ on $\mathfrak{T}_{0F}^{(k+1)N} \times \mathfrak{D}^i \times \mathfrak{Y}_d^k$ is the union of all stable $\varepsilon-$tracking domains

$$\mathcal{D}_{ST}\left(\varepsilon; \mathbf{t}_R^{(k+1)N}; \mathbf{t}_F^{(k+1)N}; \mathbf{Y}_{d0}^k; \mathbf{D}; \mathbf{U}; \mathbf{Y}_d^k\right)$$

over $\varepsilon \in [0, \varepsilon_M)$,

$$\mathcal{D}_{SST}\left(\mathbf{t}_R^{(k+1)N}; \mathbf{t}_F^{(k+1)N}; \mathbf{Y}_{d0}^k; \mathbf{D}; \mathbf{U}; \mathbf{Y}_d^k\right) =$$
$$= \cup\left\{\mathcal{D}_{ST}\left(\varepsilon; \mathbf{t}_R^{(k+1)N}; \mathbf{t}_F^{(k+1)N}; \mathbf{Y}_{d0}^k; \mathbf{D}; \mathbf{U}; \mathbf{Y}_d^k\right) : \varepsilon \in [0, \varepsilon_M)\right\},$$
$$\forall\left[\mathbf{D}(.), \mathbf{Y}_d^k(.)\right] \in \mathfrak{D}^i \times \mathfrak{Y}_d^k. \qquad (8.66)$$

c) The stablewise elementwise tracking with FVRT of $\mathbf{Y}_d^k(t^{(k+1)N})$ on $\mathfrak{T}_{0F}^{(k+1)N} \times \mathfrak{D}^i \times \mathfrak{Y}_d^k$ is global (in the whole) if and only if it is the FVRT *global tracking of* $\mathbf{Y}_d^k(t^{(k+1)N})$ *on* $\mathfrak{T}_{0F}^{(k+1)N} \times \mathfrak{D}^i \times \mathfrak{Y}_d^k$, *and the FVRT stablewise tracking of* $\mathbf{Y}_d^k(t^{(k+1)N})$ *on* $\mathfrak{T}_{0F}^{(k+1)N} \times \mathfrak{D}^i \times \mathfrak{Y}_d^k$ *is with*

$$\mathcal{D}_{ST}\left(\mathbf{t}_R^{(k+1)N}; \mathbf{t}_F^{(k+1)N}; \mathbf{Y}_{d0}^k; \mathbf{D}; \mathbf{U}; \mathbf{Y}_d^k\right) = \mathfrak{R}^{(k+1)N}$$

for every $\left[\mathbf{D}(.), \mathbf{Y}_d^k(.)\right] \in \mathfrak{D}^i \times \mathfrak{Y}_d^k$.

Exercise 152 *Define the stable elementwise FVRT tracking of* $\mathbf{Y}_d^k(t)$ *on* $\mathfrak{D}^j \times \mathfrak{Y}_d^k$ *and its domain in terms of the output error vector* $\mathbf{e}^k$.

Definition 153 *The uniform stablewise elementwise tracking with FVRT of the desired output behavior* $\mathbf{Y}_d^k(t^{(k+1)N})$ *on* $\mathfrak{T}_{0F}^{(k+1)N} \times \mathfrak{D}^i \times \mathfrak{Y}_d^k$ *of the system controlled by a control* $\mathbf{U}(.) \in \mathfrak{U}^l$
If and only if the intersection

$$\mathcal{D}_{STU}\left(\mathbf{t}_R^{(k+1)N}; \mathbf{t}_F^{(k+1)N}; \mathfrak{D}^i; \mathbf{U}; \mathfrak{Y}_d^k\right)$$

of all

$$\mathcal{D}_{ST}\left(\mathbf{t}_R^{(k+1)N}; \mathbf{t}_F^{(k+1)N}; \mathbf{Y}_{d0}^k; \mathbf{D}; \mathbf{U}; \mathbf{Y}_d^k\right)$$

[the intersection $\mathcal{D}_{SSTU}\left(\mathbf{t}_R^{(k+1)N}; \mathbf{t}_F^{(k+1)N}; \mathfrak{D}^i; \mathbf{U}; \mathfrak{Y}_d^k\right)$ *of all domains*

$$\mathcal{D}_{SST}\left(\mathbf{t}_R^{(k+1)N}; \mathbf{t}_F^{(k+1)N}; \mathbf{Y}_{d0}^k; \mathbf{D}; \mathbf{U}; \mathbf{Y}_d^k\right)]$$

over $\mathfrak{D}^i \times \mathfrak{Y}_d^k$ *is a connected neighborhood of* $\mathbf{Y}_{d0}^k$ *then and only then it is the* **FVRT (strictly) stablewise elementwise tracking domain of** $\mathbf{Y}_d^k(t^{(k+1)N})$ *on* $\mathfrak{T}_{0F}^{(k+1)N} \times \mathfrak{D}^i \times \mathfrak{Y}_d^k$ *uniform in* $\left[\mathbf{D}(.), \mathbf{Y}_d^k(.)\right] \in \mathfrak{D}^i \times \mathfrak{Y}_d^k$, *respectively*,

$$\exists \xi \in \mathfrak{R}^+ \Longrightarrow \mathcal{D}_{STU}\left(\mathbf{t}_R^{(k+1)N}; \mathbf{t}_F^{(k+1)N}; \mathfrak{D}^i; \mathbf{U}; \mathfrak{Y}_d^k\right) =$$

$$= \cap \left\{ \begin{array}{l} \mathcal{D}_{ST}\left(\mathbf{t}_R^{(k+1)N}; \mathbf{t}_F^{(k+1)N}; \mathbf{Y}_{d0}^k; \mathbf{D}; \mathbf{U}; \mathbf{Y}_d^k\right) : \\ : \left[\mathbf{D}(.), \mathbf{Y}_d^k(.)\right] \in \mathfrak{D}^i \times \mathfrak{Y}_d^k \end{array} \right\} \supset$$

$$\supset \mathfrak{N}_\xi\left(\mathfrak{D}^i; \mathbf{U}; \mathfrak{Y}_d^k\right), \tag{8.67}$$

$$\left[\begin{array}{l} \exists \xi \in \mathfrak{R}^+ \Longrightarrow \mathcal{D}_{SSTU}\left(\mathbf{t}_R^{(k+1)N}; \mathbf{t}_F^{(k+1)N}; \mathfrak{D}^i; \mathbf{U}; \mathfrak{Y}_d^k\right) = \\ = \cap \left\{ \begin{array}{l} \mathcal{D}_{SST}\left(\mathbf{t}_R^{(k+1)N}; \mathbf{t}_F^{(k+1)N}; \mathbf{Y}_{d0}^k; \mathbf{D}; \mathbf{U}; \mathbf{Y}_d^k\right) : \\ : \left[\mathbf{D}(.), \mathbf{Y}_d^k(.)\right] \in \mathfrak{D}^i \times \mathfrak{Y}_d^k \end{array} \right\} \\ \qquad \supset \mathfrak{N}_\xi\left(\mathfrak{D}^i; \mathbf{U}; \mathfrak{Y}_d^k\right). \end{array} \right]. \tag{8.68}$$

Exercise 154 *Define the uniform stablewise elementwise FVRT tracking of* $\mathbf{Y}_d^k(t)$ *on* $\mathfrak{D}^j \times \mathfrak{Y}_d^k$ *and its domain in terms of the output error vector* $\mathbf{e}^k$.

Comment 155 *Every tracking with the finite (scalar or vector) reachability time implies the perfect tracking that begins at the (scalar or vector) reachability instant* t_R *or* $t_R^{(k+1)N}$ *and continues until the final (scalar or vector) moment* t_F *or* $t_F^{(k+1)N}$, *respectively. It expresses a high tracking quality.*

Part III

TRACKABILITY

Chapter 9

Trackability fundamentals

This chapter is essentially a slightly refined extract of the corresponding parts of the recently published books [175], [188].

9.1   Trackability of *a* plant and its regime

Throughout this text the following notation means:

$i \in \{0, 1, ..., \eta, \mu\}$ the highest derivative of the disturbance vector **D** acting on the system,

$k \in \{0, 1, ..., m - 1\}$ the highest derivative $\mathbf{Y}_d^{(k)}$ of the plant desired output vector $\mathbf{Y}_d$ to be tracked,

$l \in \{0, 1, ..., k\}$ the highest order of the tracking,

$m \in \{1, \alpha, \nu, \alpha + \nu\}$ the system order.

The dynamical systems, in general, and plants, hence their control systems, in particular, are exposed in reality to actions of unpredictable external perturbations (called usually *disturbances*) and initial conditions. Any their study aimed to be complete should treat them in the forced regime under arbitrary initial conditions.

The controllability problems and the disturbance compensation problems [175, Section 6.3, p. 62], [188, Remark 234, pp. 169, 170] have been mainly studied separately. However, each of them does not satisfy the basic control goal that is **to force the plant subjected to disturbance actions and to arbitrary initial conditions** *to follow, i.e., to track*, **its desired behavior**. Since this is the very goal of the control to be realized in the real plant environment and under real operating conditions, it led to the introduction of a new control concept called ***trackability*** ([117]–[121], [147]–[156], [175, Section 6.3, p. 62], [188, Remark 234, pp. 169, 170], [190]–[194],

[247]–[257]).

The trackability concept clarifies whether *the plant itself has a property to enable the existence* of a control that can guarantee **tracking under arbitrary initial conditions (globally or from a domain) and under external perturbations** belonging to a set $\mathfrak{D}^i$ of permitted disturbances, and all that for every plant desired output response from a given functional family $\mathfrak{Y}_d^k$. The trackability is the plant property. It is independent of the controller and control. It incorporates and unites the controllability concept and the disturbance compensation concept as special cases.

Another plant property called **natural trackability** is a type of the plant *trackability* that permits control synthesis and implementation *without using information about the real values and forms of the disturbances and about the mathematical model of the plant internal dynamics, i.e., about the plant state.* Such control that is (almost always) continuous in *time* is **Natural Tracking Control (NTC)**.

9.2   Trackability versus controllability

Controllability concepts assume the nonexistence of any external perturbation acting on the dynamical plant. The only external influences on the dynamical plant are control actions. However, trackability treats the *simultaneous influence of both* disturbances and initial conditions on the dynamical plant behavior.

Controllability ensures only that the dynamical plant state or output will become at *some* moment a prespecified state or output but without any concern relative to the dynamical plant state or output before and after that moment. Trackability takes care about the dynamical plant behavior at every moment from the initial moment until the final moment.

Controllability is not related to tracking. Trackability is necessary for tracking.

Trackability is more complex phenomenon than controllability.

Controllability is well established and effectively studied only in the framework of the linear systems [7, page 313], [27], [39], [48, page 216], [99], [197], [206]–[211]. For the generalization of the controllability concept and its extension to *IO*, *EISO*, *HISO*, and *IIO* systems see the book [171].

Trackability has been recently well established and completed in the general framework of *time*-varying nonlinear systems [175]. This book presents it in the framework of the *time*-invariant continuous-*time* linear systems, which extends and completes that of [188].

9.3 Tracking demands trackability

Attacking the tracking and the tracking control synthesis problems we discover another fundamental control problem:

Problem 156 *The fundamental control problem [175, Problem 9, p. xxii], [188, Problem 1, pp. xvii]*
Do the properties of the plant enable the existence of a tracking control for all initial conditions from a neighborhood of $\mathbf{Y}_{d0}^{m-1}$, *[i.e., of* $\mathbf{Y}_d(t)$ *at the initial moment* $t = 0$*], for all permitted disturbances* $\mathbf{D}(.) \in \mathfrak{D}^i$ *and for every plant desired output behavior* $\mathbf{Y}_d(.) \in \mathfrak{Y}_d^k$? *If and only if they do, then the plant is* **trackable** *over* $\mathfrak{D}^i \times \mathfrak{Y}_d^k$.

The prerequisite for the control synthesis is the test of the plant **trackability**.

We will consider **tracking** and **trackability** of *time*-invariant continuous-*time* linear systems, as well as **tracking control synthesis** for them by synthesizing and extending the results of [175] and [188].

Various tracking properties and trackability kinds defined in the sequel will illustrate richness of the tracking and trackability phenomena as well as their greater complexity than that of the related stability or controllability properties, respectively, [175], [188].

The task of the tracking control synthesis is meaningful if the plant is able to exhibit tracking under an appropriate action of control, i.e., if the plant is **trackable**. The book defines various trackability properties and establishes conditions on the plant to possess the corresponding trackability property. This permits us to continue with **tracking control synthesis**, which completes the main body of the book.

9.4 Perfect trackability: various types

9.4.1 Perfect and elementwise perfect trackability

Definitions of perfect trackability follow in terms of the system total desired output $\mathbf{Y}_d(t)$. The equivalent perfect trackability definitions in terms of the output error $\mathbf{e}(t)$ are given in Appendix B.2.

We present the following due to [175, Chapter 6, pp. 65-68], [188, Definition 235, p. 171]:

Definition 157 *Definition of the l-th order perfect trackability and elementwise perfect trackability of* $\mathbf{Y}_d(.)$ *for the given* $\mathbf{D}(.)$

a) The desired output vector function $\mathbf{Y}_d(.)$ of the m-th order plant is **the l-th order perfect trackable under the action of the given** $\mathbf{D}(.)$ if and only if there exists a control vector function $\mathbf{U}(.)$ such that the plant real output vector $\mathbf{Y}(t)$ and its first l derivatives are always equal to the desired plant output vector $\mathbf{Y}_d(t)$ and its first l derivatives, respectively, as soon as $\mathbf{Y}^{m-1}(0) = \mathbf{Y}_d^{m-1}(0)$,

$$given\ \mathbf{D}(.),\ \exists \mathbf{U}(.)\ and\ \mathbf{Y}^{m-1}(0) = \mathbf{Y}_d^{m-1}(0) \Longrightarrow \mathbf{Y}^l(t) = \mathbf{Y}_d^l(t),\ \forall t \in \mathfrak{T}_0.$$
(9.1)

The zero order $(l = 0)$ perfect trackability is called simply **perfect trackability**.

b) **The perfect trackability is elementwise** if and only if the control vector $\mathbf{U}$ can act simultaneously on every entry Y_i, $\forall i = 1, 2, \dots, N$, of $\mathbf{Y}$ mutually independently at every $t \in \mathfrak{T}_0$.

Lemma 158 *The right-hand functional identity [175, Lemma 137, p. 66], [188, Lemma 236, pp. 171, 172]*

If two functions $\mathbf{Y}(.)$ and $\mathbf{Y}_d(.)$ are defined, k-times continuously differentiable on $]\sigma, \infty[$, $\sigma \in \mathfrak{T}_0$, $]\sigma, \infty[\subseteq$ In $\mathfrak{T}_0$, as well as at $t = \sigma$ from the right-hand side, i.e., at $t = \sigma^+$, and identical on $[\sigma, \infty[$, then all their derivatives up to the order k included are also identical on $]\sigma, \infty[$ and at $t = \sigma^+$.

Note 159 *This lemma allows different vector values of the derivatives of* $\mathbf{Y}(t)$ *from the left-hand side and the right-hand side of the moment* $t = \sigma$, $\sigma \in \mathfrak{T}_0$.

Definition 24 (Section 1.5), Definition 157 and Lemma 158 imply the following:

Lemma 160 *If the desired output vector function* $\mathbf{Y}_d(.)$ *is differentiable at least up to the order* $k \geq l$, $\mathbf{Y}_d(t) \in \mathfrak{C}^k$, *then for it to be:*

i)   the l-th-order perfect trackable under the perturbation of $\mathbf{D}(.) \in \mathfrak{D}^i$ it is necessary and sufficient to be realizable for $\mathbf{D}(.) \in \mathfrak{D}^i$, equivalently, to be perfect trackable under the action of $\mathbf{D}(.) \in \mathfrak{D}^i$,

ii)   the l-th order elementwise perfect trackable under the perturbation of $\mathbf{D}(.) \in \mathfrak{D}^i$ it is necessary and sufficient to be elementwise realizable for $\mathbf{D}(.) \in \mathfrak{D}^i$, equivalently, to be elementwise perfect trackable under the action of $\mathbf{D}(.) \in \mathfrak{D}^i$.

We are interested also in perfect trackability of every plant desired output $\mathbf{Y}_d(.)$ from $\mathfrak{Y}_d^k$ rather than only in perfect trackability of a single plant desired output.

Definition 161 *The l-th order perfect trackability and elementwise perfect trackability of the plant on* $\mathfrak{D}^i \times \mathfrak{Y}_d^k$

a) The m-th order dynamical plant is the l-th order perfect trackable on $\mathfrak{D}^i \times \mathfrak{Y}_d^k$ *if and only if for every* $[\mathbf{D}(.), \mathbf{Y}_d(.)] \in \mathfrak{D}^i \times \mathfrak{Y}_d^k$ *there exists a control vector function* $\mathbf{U}(.)$ *such that the plant real output and its first l derivatives are always equal to the plant desired output and its first l derivatives, respectively, as soon as* $\mathbf{Y}^{m-1}(0) = \mathbf{Y}_d^{m-1}(0)$,

$$\mathbf{Y}^{m-1}(0) = \mathbf{Y}_d^{m-1}(0), \ \forall [\mathbf{D}(.), \mathbf{Y}_d(.)] \in \mathfrak{D}^i \times \mathfrak{Y}_d^k,$$
$$\exists \mathbf{U}(.) \Longrightarrow \ \mathbf{Y}^l(t) = \mathbf{Y}_d^l(t), \ \forall t \in \mathfrak{T}_0, \tag{9.2}$$

The zero order (l=0) perfect trackability on $\mathfrak{D}^i \times \mathfrak{Y}_d^k$ *is called simply* **perfect trackability on** $\mathfrak{D}^i \times \mathfrak{Y}_d^k$.

b) **The perfect trackability is elementwise** *if and only if the control vector* $\mathbf{U}$ *can act on every entry* Y_i, $\forall i = 1, 2, ..., N$, *of* $\mathbf{Y}$ *mutually independently.*

From Definition 24, Lemma 158 and Definition 161 imply the following.

Lemma 162 *The l-th order perfect trackability of the m-th order dynamical plant on* $\mathfrak{D}^i \times \mathfrak{Y}_d^k$ *and its perfect trackability on* $\mathfrak{D}^i \times \mathfrak{Y}_d^k$

i) For the m-th order dynamical plant to be the l-th order perfect track-able on $\mathfrak{D}^i \times \mathfrak{Y}_d^k$ *it is necessary and sufficient that every* $\mathbf{Y}_d(.) \in \mathfrak{Y}_d^k$ *is realizable on* $\mathfrak{D}^i$, *equivalently, to be perfect trackable on* $\mathfrak{D}^i \times \mathfrak{Y}_d^k$.

ii) For the m-th order dynamical plant to be the l-th order elementwise perfect trackable on $\mathfrak{D}^i \times \mathfrak{Y}_d^k$ *it is necessary and sufficient that every* $\mathbf{Y}_d(.) \in \mathfrak{Y}_d^k$ *is elementwise realizable on* $\mathfrak{D}^i$, *equivalently, to be elementwise perfect trackable on* $\mathfrak{D}^i \times \mathfrak{Y}_d^k$.

The above lemmas discover the equivalence between the realizability of the plant desired output and the plant perfect trackability. The type of the plant, i.e., the form of its mathematical model, governs the form of the realizability conditions [175, Chapter 4, pp. 41-51], [188, Section 3.3, pp. 35-51.].

Except for the existence requirement, the preceding definitions do not impose any other condition on the control vector function $\mathbf{U}(.)$. Its existence means that its instantaneous vector value $\mathbf{U}(t)$ is defined at every moment

$t \in \mathfrak{T}_0$. This permits   piecewise continuity of $\mathbf{U}(t)$; i.e., it allows $\mathbf{U}(t) \in \mathfrak{C}^-(\mathfrak{T}_0)$. A piecewise continuous variable can only be a mathematical, but not a physical variable. It is not exactly physically realizable, which is explained by $PCUP$ (Principles 7 and 8). In order to be physically realizable, control variable $\mathbf{U}(.)$ should obey $PCUP$ (Principle 7, Section 1.2), equivalently $TCUP$ (Principle 10, Section 1.2).

The preceding definitions determine the control vector function $\mathbf{U}(.)$ in terms of the disturbance vector function $\mathbf{D}(.)$.

The vector form and the instantaneous value of the disturbance variable $\mathbf{D}(.)$ are most often unknown, unpredictable, and their values can be unmeasurable. These disturbance features cause the problem of the control realization if control is determined in terms of $\mathbf{D}(.)$.

9.4.2   Trackability and nature

Every being belongs to the nature. The being is alive due to the appropriate control of all its organs. It is the nature of the being that creates such control without any consciousness of the being. The control is natural and automatically natural, biologically natural.

Problem 163 *Disturbance and the control synthesis problem*

Do the plant properties enable a control synthesis without using information about the real form and the value of the disturbance vector $\mathbf{D}(t)$ *at any* $t \in \mathfrak{T}_0$*? Do they enable that for every* $\mathbf{D}(.) \in \mathfrak{D}^i$*?*

Mathematical models of plants, which are usually the starting point for the control synthesis, are approximative both qualitatively (due to their nonlinear nature, their forms and the dynamical complexity) and quantitatively (due to their order, their dimensionality and parameter values).

Problem 164 *System state and the control synthesis problem*

Is it possible to determine control without knowing the mathematical model of the dynamical plant state? Do the properties of the dynamical plant permit the existence of such control?

Comment 165 *Nature (e.g., the brain as a biological natural controller) does not use any information about the mathematical model of the dynamical plant (of any organ) in order to create very effective time-continuous control (of the organ). Moreover, nature (the brain) often does not need precise, or any, information about the forms and/or the values of disturbances. Such control exists. It exists in ourselves. Nature (brain) creates such control. We* call it in general **natural control** (*NC*) *regardless of the controller physical nature and regardless of the creator of the controller.*

Natural control is the **Natural Tracking Control** (*NTC*) *if and only if it ensures a kind of tracking and its implementation does not need any information about the form and the value of any* $\mathbf{D}(.) \in \mathfrak{D}^i$ *and about the mathematical model of the plant state.*

This comment leads to the definition of the natural tracking control.

Definition 166 *Natural Tracking Control*

Control that forces a plant to exhibit a requested tracking and for synthesis and implementation of which there is not any use of information about the real values and form of the variation either of the plant state or of the disturbances acting on the plant or on the mathematical model of the plant is **the Natural Tracking Control** (*NTC*) *of the plant.*

The scientific and engineering concept of **the** *natural trackability* was established and developed by discovering algorithms for synthesis of **Natural Tracking Control** in [160], [117]–[121], [148]–[154], [166], [169]–[179], [188], [190]–[194], [247]–[257].

PCUP (Principles 8 and 7, Section 1.2) and *TCUP* (*Time Continuity and Uniqueness Principle* 10, Section 1.2) jointly express the crucial properties of *time* [161], [163], [177], [175], [181]–[184] that enable effective *Natural Tracking Control synthesis* for linear [188] and nonlinear [175] dynamical plants.

9.4.3 Perfect and elementwise perfect natural trackability

In order to reply to the preceding questions we accept the following definition of the ideal, i.e., perfect natural trackability [175, Definition 146, p. 68], [188, Definition 248, p. 175]:

Definition 167 *The l-th order perfect natural trackability and elementwise perfect natural trackability on* $\mathfrak{D}^i \times \mathfrak{Y}_d^k$

a) The m-th order dynamical plant is **the l-th order perfect natural trackable on** $\mathfrak{D}^i \times \mathfrak{Y}_d^k$ *if and only if for every pair* $[\mathbf{D}(.), \mathbf{Y}_d(.)] \in \mathfrak{D}^i \times \mathfrak{Y}_d^k$ *there exists a control vector function* $\mathbf{U}(.)$ *obeying TCUP on* $\mathfrak{T}_0$, *which can be synthesized without using information about the form and the value of any* $\mathbf{D}(.) \in \mathfrak{D}^i$ *and about the plant state, such that the plant real output and its* *first l derivatives are always equal to the desired plant output and its first l derivatives, respectively, i.e., that (9.3) holds,*

$$\mathbf{Y}_0^{m-1} = \mathbf{Y}_{d0}^{m-1}, \ \forall [\mathbf{D}(.), \mathbf{Y}_d(.)] \in \mathfrak{D}^i \times \mathfrak{Y}_d^k \Longrightarrow$$
$$\exists \mathbf{U}(.) \in \mathfrak{C}(\mathfrak{T}_0) \Longrightarrow \mathbf{Y}^l(t) = \mathbf{Y}_d^l(t), \ \forall t \in \mathfrak{T}_0. \qquad (9.3)$$

The zero order ($l = 0$) perfect natural trackability on $\mathfrak{D}^i \times \mathfrak{Y}_d^k$ is simply called **the perfect natural trackability on $\mathfrak{D}^i \times \mathfrak{Y}_d^k$.**

b) **The perfect natural trackability is elementwise** *if and only if the control vector* $\mathbf{U}$ *can act simultaneously on every entry Y_i, $\forall i = 1, 2, ..., N$, of* $\mathbf{Y}$ *mutually independently.*

Comment 168 *[188, Comment 249, p. 175] Definition 161 and Definition 310 imply that the l-th order perfect trackability on $\mathfrak{D}^i \times \mathfrak{Y}_d^k$ is necessary for the l-th order perfect natural trackability on $\mathfrak{D}^i \times \mathfrak{Y}_d^k$, and that the l-th order perfect natural trackability on $\mathfrak{D}^i \times \mathfrak{Y}_d^k$ is sufficient for the l-th order perfect trackability on $\mathfrak{D}^i \times \mathfrak{Y}_d^k$.*

We deduce the following directly from Lemma 158 and Definition 310.

Lemma 169 *[188, Lemma 250, p. 175]* **The l-th order perfect natural trackability on $\mathfrak{D}^i \times \mathfrak{Y}_d^k$ and the perfect natural trackability on $\mathfrak{D}^i \times \mathfrak{Y}_d^k$**
For the m-th order dynamical plant to be the l-th order perfect natural trackable on $\mathfrak{D}^i \times \mathfrak{Y}_d^k$ it is necessary and sufficient to be perfect trackable on $\mathfrak{D}^i \times \mathfrak{Y}_d^k$.

Comment 170 *Perfect trackability and the fundamental control principle*

Every perfect trackability property demands that all output variables are simultaneously mutually independently controlled. Therefore, the number r of control variables should not be less than the number N of the output variables, i.e., the dimension r of the control vector should not be less than the dimension N of the output vector,

$$\dim \mathbf{Y} = N \le \dim \mathbf{U} = r.$$

This is due to the definitions of the perfect trackability properties and the fundamental control principle 104, Section 7.1.

9.5 Imperfect trackability: various types

Definitions of imperfect trackability follow in terms of the system total desired output $\mathbf{Y}_d(t)$. The equivalent imperfect trackability definitions in terms of the output error $\mathbf{e}(t)$ are presented in Appendix B.2.

9.5.1 Imperfect trackability

We slightly refine the definition of functionally interrelated variables [175, Definition 135, p. 65].

Let variables $Z_1, Z_2, \dots, Z_K$ form the vector $\mathbf{Z}$ that induces the extended vector $\mathbf{Z}^k$,

$$\left[Z_1 \vdots Z_2 \vdots \dots \vdots Z_K \right]^T = \mathbf{Z} \in \mathfrak{R}^K,$$

$$\mathbf{Z}^k = \left[\mathbf{Z} \vdots \mathbf{Z}^{(1)} \vdots \dots \vdots \mathbf{Z}^{(k)} \right]^T \in \mathfrak{R}^{(k+1)K}.$$

Definition 171 *Functionally interrelated variables*
*Variables $Z_1, Z_2, \dots, Z_K$ are **functionally interrelated** if and only if there exist a natural number P, instant $\sigma \in \mathfrak{T}_0$ and a vector function $\mathbf{v}(.) : \mathfrak{T} \times \mathfrak{R}^{(k+1)K} \times \mathfrak{T}_0 \longrightarrow \mathfrak{R}^P$ such that*

$$\mathbf{v}\left(t, \mathbf{Z}^k; \sigma \right) = \mathbf{0}_P, \quad \forall t \in \mathfrak{T}_\sigma = \{ t \in \mathfrak{T}_0 : t \geq \sigma \}. \tag{9.4}$$

*If and only if, additionally, the moment σ is fixed then the variables are **functionally interrelated on** $\mathfrak{T}_\sigma$*

Definitions introduced in [175] for *time*-varying systems reduce to the following definitions valid for the *time*-invariant systems [188]:

Definition 172 *The l-th order trackability on $\mathfrak{D}^i \times \mathfrak{Y}_d^k$ [175, Definition 168, pp. 81,82], [188, Definition 253, pp. 176, 177]*
*a) The m-th order dynamical plant is **the l-th order trackable on** $\mathfrak{D}^i \times \mathfrak{Y}_d^k$ if and only if there is $\Delta \in \mathfrak{R}^+$, or $\Delta = \infty$, such that for every disturbance vector function $\mathbf{D}(.) \in \mathfrak{D}^i$, for every plant output desired response $\mathbf{Y}_d(.) \in \mathfrak{Y}_d^k$, and for every instant $\sigma \in Int\ \mathfrak{T}_0$, there is a control vector function $\mathbf{U}(.)$ such that for every plant initial output vector $\mathbf{Y}_0^{m-1}$ in the Δ neighborhood of the plant initial desired output vector $\mathbf{Y}_{d0}^{m-1}$, the extended*

real output vector $\mathbf{Y}^l(t)$ *becomes equal to the extended desired output vector* $\mathbf{Y}^l_d(t)$ *at latest at the moment* σ, *after which they rest equal forever, i.e.,*

$$\exists \Delta \in]0,\ \infty], \ \forall [\mathbf{D}(.), \mathbf{Y}_d(.)] \in \mathfrak{D}^i \times \mathfrak{Y}^k_d, \ \forall \sigma \in Int\ \mathfrak{T}_0,$$

$$\exists \mathbf{U}(.), \ \mathbf{U}(t) = \mathbf{U}(t; \sigma; \mathbf{D}; \mathbf{Y}_d) \implies$$

$$\left\| \mathbf{Y}^{m-1}_0 - \mathbf{Y}^{m-1}_{d0} \right\| < \Delta \implies \mathbf{Y}^l(t) = \mathbf{Y}^l_d(t), \ \forall\, (t \geq \sigma) \in \mathfrak{T}_0. \tag{9.5}$$

*Such control is **the l-th order tracking control on** $\mathfrak{D}^i \times \mathfrak{Y}^k_d$, for short,* **the l-th order tracking control.**

The zero, (l = 0), order trackability on $\mathfrak{D}^i \times \mathfrak{Y}^k_d$ *is simply called **trackability on** $\mathfrak{D}^i \times \mathfrak{Y}^k_d$.*

The zero, (l = 0), order tracking control on $\mathfrak{D}^i \times \mathfrak{Y}^k_d$ *is simply called **the tracking control on** $\mathfrak{D}^i \times \mathfrak{Y}^k_d$, for short, **the tracking control.***

b) The l-th order trackability on $\mathfrak{D}^i \times \mathfrak{Y}^k_d$ *is **global (in the whole)** if and only if $\Delta = \infty$.*

c) The l-th order trackability on $\mathfrak{D}^i \times \mathfrak{Y}^k_d$ *is **uniform** over* $\mathfrak{D}^i \times \mathfrak{Y}^k_d$ *if and only if* $\mathbf{U}(.)$ *depends on* $\mathfrak{D}^i \times \mathfrak{Y}^k_d$ *but not on an individual pair* $[\mathbf{D}(.), \mathbf{Y}_d(.)]$ *from* $\mathfrak{D}^i \times \mathfrak{Y}^k_d$, $\mathbf{U}(t) = \mathbf{U}(t; \sigma; \mathfrak{D}^k; \mathfrak{Y}^k_d)$.

*d) If and only if, additionally to a), the output variables are **mutually functionally interrelated** by P functional constraints, P < N, then the l-th order trackability on* $\mathfrak{D}^i \times \mathfrak{Y}^k_d$ *of* $\mathbf{Y}_d(.)$ *is **incomplete with** f = N − P **degrees of freedom.***

If and only if, additionally to a), all output variables can be controlled simultaneously mutually independently (P = 0), then the l-th order trackability on $\mathfrak{D}^i \times \mathfrak{Y}^k_d$ *of* $\mathbf{Y}_d(.)$ *is **with complete, i.e., with N, degrees of freedom**, for short, it is **complete.***

Comment 173 *The trackability of the m-th order dynamical plant can be incomplete because its output variables need not be controlled simultaneously mutually independently at every moment until the moment* σ, *while its perfect trackability is complete.*

This explains the following theorem on imperfect trackability properties. It clarifies the conditions of Theorem 254 of [188, pp. 177, 178], which are correct for perfect trackability related to the complete trackability or to the elementwise trackability.

Theorem 174 *The perfect versus the imperfect trackability on the product set* $\mathfrak{D}^i \times \mathfrak{Y}^k_d$ *[188, Theorem 254, pp. 177, 178]*

For the m-th order dynamical plant to be the l-th order perfect trackable on $\mathfrak{D}^i \times \mathfrak{Y}^k_d$ *it is necessary and sufficient to be the l-th order complete trackable on* $\mathfrak{D}^i \times \mathfrak{Y}^k_d$.

Lemma 158 and Definition 172 directly imply the following.

Lemma 175 *The l-th order trackability and the trackability*
For the m-th order dynamical plant to be the l-th order trackable on $\mathfrak{D}^i$
$\times \mathfrak{Y}_d^k$ it is necessary and sufficient to be trackable on $\mathfrak{D}^i \times \mathfrak{Y}_d^k$.

9.5.2 Imperfect natural trackability

Perfect natural trackability properties demand that $\mathbf{Y}_0^{m-1} = \mathbf{Y}_{d0}^{m-1}$, Definition 310. Let us analyze the cases when this initial condition is not satisfied, for which the perfection of the natural trackability is impossible. This leads us to introduce imperfect natural trackability properties.

Definition 176 *The l-th order natural trackability on $\mathfrak{D}^i \times \mathfrak{Y}_d^k$ [175,*
Deinition 171, pp.83-85], [188, Definition 259, p. 180]
*a) The m-th order dynamical plant is **the l-th order natural trackable***
on $\mathfrak{D}^i \times \mathfrak{Y}_d^k$ if and only if there is $\Delta \in \mathfrak{R}^+$, or $\Delta = \infty$, such that for
every disturbance vector function $\mathbf{D}(.) \in \mathfrak{D}^i$, for every plant output desired
response $\mathbf{Y}_d(.) \in \mathfrak{Y}_d^k$, and for every instant $\sigma \in Int\ \mathfrak{T}_0$, there is a control
vector function $\mathbf{U}(.)$ obeying $TCUP$ on $\mathfrak{T}_0$, which can be synthesized without
using information about the form and the value of $\mathbf{D}(.) \in \mathfrak{D}^i$ and about the
plant state, such that for every plant initial output vector $\mathbf{Y}_0^{m-1}$ in the Δ
neighborhood of the plant initial desired output vector $\mathbf{Y}_{d0}^{m-1}$, the extended
real output vector $\mathbf{Y}^l(t)$ becomes equal to the extended desired output vector
$\mathbf{Y}_d^l(t)$ at latest at the moment σ, after which they rest equal forever, i.e.,

$$\exists \Delta \in]0,\ \infty], \ \forall [\mathbf{D}(.), \mathbf{Y}_d(.)] \in \mathfrak{D}^i \times \mathfrak{Y}_d^k, \ \forall \sigma \in Int\ \mathfrak{T}_0, \ \sigma \longrightarrow 0^+,$$
$$\exists \mathbf{U}(.), \ \mathbf{U}(t) = \mathbf{U}(t; \sigma; \mathbf{Y}_d) \in \mathfrak{C}(\mathfrak{T}_0) \Longrightarrow$$
$$\left\| \mathbf{Y}_0^{m-1} - \mathbf{Y}_{d0}^{m-1} \right\| < \Delta \Longrightarrow \mathbf{Y}^l(t) = \mathbf{Y}_d^l(t), \ \forall (t \geq \sigma) \in \mathfrak{T}_0. \qquad (9.6)$$

*Such control is **the l-th order natural tracking control on $\mathfrak{D}^i \times \mathfrak{Y}_d^k$**, for*
*short **the l-th order natural tracking control**.*
*The zero, (l = 0), order natural trackability on $\mathfrak{D}^i \times \mathfrak{Y}_d^k$ is called **natural***
***trackability on $\mathfrak{D}^i \times \mathfrak{Y}_d^k$**.*
The zero, (l = 0), order natural tracking control on $\mathfrak{D}^i \times \mathfrak{Y}_d^k$ is called for
*short **natural tracking control on $\mathfrak{D}^i \times \mathfrak{Y}_d^k$**, or shorter **natural tracking***
***control (NTC)**.*
*b) The l-th order natural trackability on $\mathfrak{D}^i \times \mathfrak{Y}_d^k$ is **global (in the whole)***
if and only if $\Delta = \infty$.

*c) The l-th order natural trackability on $\mathfrak{D}^i \times \mathfrak{Y}_d^k$ is **uniform** over $\mathfrak{Y}_d^k$ if and only if control $\mathbf{U}(.)$ depends on $\mathfrak{Y}_d^k$ but not on an individual $\mathbf{Y}_d(.)$ from $\mathfrak{Y}_d^k$, $\mathbf{U}(t) = \mathbf{U}(t; \sigma; \mathfrak{Y}_d^k)$.*

*d) If and only if, additionally to a), the output variables are **mutually functionally interrelated** by P functional constraints then the l-th-order natural trackability on $\mathfrak{D}^i \times \mathfrak{Y}_d^k$ of $\mathbf{Y}_d(.)$ is **incomplete with** $f = N - P$ **degrees of freedom,**.*

*If and only if, additionally to a), all output variables can be controlled simultaneously mutually independently, $(P = 0)$, then the l-th-order natural trackability on $\mathfrak{D}^i \times \mathfrak{Y}_d^k$ of $\mathbf{Y}_d(.)$ is **complete**.*

*The term **incomplete** means **incomplete ($f < N$) degrees of freedom**.*

*The term **complete** means **complete**, i.e., **full, ($f = N$), degrees of freedom**.*

Comment 177 *Definition 172 and Definition 312 show that the l-th order trackability on $\mathfrak{D}^i \times \mathfrak{Y}_d^k$ is necessary for the l-th order natural trackability on $\mathfrak{D}^i \times \mathfrak{Y}_d^k$, and the l-th order natural trackability on $\mathfrak{D}^i \times \mathfrak{Y}_d^k$ is sufficient for the l-th order trackability on $\mathfrak{D}^i \times \mathfrak{Y}_d^k$.*

Comment 178 *The natural trackability of the m-th order dynamical plant need not be complete because its output variables need not be controlled mutually independently, while its perfect natural trackability is complete.*

This explains the following theorem on imperfect natural trackability properties. It refines Theorem 295 of [188, p. 198], which is correct for the complete trackability and for the elementwise natural trackability. Otherwise, their conditions are sufficient but not necessary for non-elementwise natural trackability.

Theorem 179 *Perfect natural trackability versus natural trackability on $\mathfrak{D}^i \times \mathfrak{Y}_d^k$*

In order for the m-th order dynamical plant to be the l-th order perfect natural trackable on $\mathfrak{D}^i \times \mathfrak{Y}_d^k$ it is necessary and sufficient to be the l-th order complete natural trackable on $\mathfrak{D}^i \times \mathfrak{Y}_d^k$.

Lemma 158 and Definition 312 result in the following.

Lemma 180 *Natural trackability and the l-th order natural trackability*

For the m-th order dynamical plant to be the l-th order natural trackable on $\mathfrak{D}^i \times \mathfrak{Y}_d^k$ it is necessary and sufficient to be natural trackable on $\mathfrak{D}^i \times \mathfrak{Y}_d^k$.

9.5.3 Elementwise trackability

The elementwise tracking permits us to associate different tracking requirements with different output variables of the plant. The concept of the elementwise trackability generalizes in that sense the preceding concept of the imperfect trackability.

Definition 181 *The l-th order elementwise trackability on* $\mathfrak{D}^i \times \mathfrak{Y}_d^k$
[175, Definition 175, pp. 86, 87], [188, Definition 263, pp 181, 182]
a) The m-th order dynamical plant is **the l-th order elementwise trackable on** $\mathfrak{D}^i \times \mathfrak{Y}_d^k$ *if and only if there is* $\mathbf{\Delta}^{mN} \in \mathfrak{R}^{+mN}$, *or* $\mathbf{\Delta}^{mN} = \infty \mathbf{1}_{mN}$, *such that for every disturbance vector function* $\mathbf{D}(.) \in \mathfrak{D}^i$, *for every plant output desired response* $\mathbf{Y}_d(.) \in \mathfrak{Y}_d^k$, *and for every vector instant* $\sigma \in (Int \ \mathfrak{T}_0)^{(l+1)N}$, *there is a control vector function* $\mathbf{U}(.)$ *such that for every plant initial output vector* $\mathbf{Y}_0^{m-1}$ *in the* $\mathbf{\Delta}^{mN}$ *elementwise neighborhood of the plant initial desired output vector* $\mathbf{Y}_{d0}^{m-1}$, *the plant real extended output response* $\mathbf{Y}^l(t)$ *becomes elementwise equal to* $\mathbf{Y}_d^l(t)$ *at latest at the vector moment* σ, *after which they rest equal forever, i.e.,*

$$\exists \mathbf{\Delta}^{mN} \in \left]\mathbf{0}_{mN}, \ \infty \mathbf{1}_{mN}\right], \ \forall \left[\mathbf{D}(.), \mathbf{Y}_d(.)\right] \in \mathfrak{D}^i \times \mathfrak{Y}_d^k,$$

$$\forall \sigma \in (Int \ \mathfrak{T}_0)^{(l+1)N}, \ \exists \mathbf{U}(.),$$

$$\mathbf{U}(t) = \mathbf{U}(t; \sigma; \mathbf{D}; \mathbf{Y}) \in \mathfrak{C}\left(\mathfrak{T}_0\right) \implies \left|\mathbf{Y}_0^{m-1} - \mathbf{Y}_{d0}^{m-1}\right| < \mathbf{\Delta}^{mN} \implies$$

$$\mathbf{Y}^l(\mathbf{t}^{(l+1)N}) = \mathbf{Y}_d^l(\mathbf{t}^{(l+1)N}), \ \forall \left(\mathbf{t}^{(l+1)N} \geq \sigma\right) \in \mathfrak{T}_0^{(l+1)N}. \quad (9.7)$$

Such control is **the l-th order elementwise tracking control on** $\mathfrak{D}^i \times \mathfrak{Y}_d^k$.
The zero, (l = 0), order elementwise trackability on $\mathfrak{D}^i \times \mathfrak{Y}_d^k$ *is called* **elementwise trackability on** $\mathfrak{D}^i \times \mathfrak{Y}_d^k$.
The zero, (l = 0), order elementwise tracking control on $\mathfrak{D}^i \times \mathfrak{Y}_d^k$ *is called for short* **elementwise tracking control on** $\mathfrak{D}^i \times \mathfrak{Y}_d^k$, *or shorter,* **elementwise tracking control.**
b) The l-th order elementwise trackability on $\mathfrak{D}^i \times \mathfrak{Y}_d^k$ *is* **global (in the whole)** *if and only if* $\mathbf{\Delta}^{mN} = \infty \mathbf{1}_{mN}$.
The l-th order elementwise trackability on $\mathfrak{D}^i \times \mathfrak{Y}_d^k$ *is* **uniform** *over* $\mathfrak{D}^i \times \mathfrak{Y}_d^k$ *if and only if* $\mathbf{U}(.)$ *depends on* $\mathfrak{D}^i \times \mathfrak{Y}_d^k$ *but not on an individual pair* $[\mathbf{D}(.), \mathbf{Y}_d(.)]$ *from* $\mathfrak{D}^i \times \mathfrak{Y}_d^k$, $\mathbf{U}(t) = \mathbf{U}(t; \sigma; \mathfrak{D}^i; \mathfrak{Y}_d^k)$.

Note 182 Complete trackability and elementwise trackability
The elementwise trackability is simultaneously the complete trackability.
The complete trackability is necessary for the elementwise trackability, and the elementwise trackability is sufficient for the complete trackability.

We use the following result from [188, Lemma 265, pp. 182, 183].

Lemma 183 *If two functions* $\mathbf{Y}(.)$ *and* $\mathbf{Y}_d(.)$ *are defined, l-times contin-
uously differentiable on* $]\sigma, \infty \mathbf{1}_N[$, $\sigma \in \mathfrak{T}_0^N$, $]\sigma, \infty \mathbf{1}_N[\subseteq (In\ \mathfrak{T}_0)^N$, *as well
as at* $\mathbf{t}^N = \sigma$ *from the right-hand side, i.e., at* $\mathbf{t}^N = \sigma^+$, *and identical on*
$[\sigma, \infty \mathbf{1}_N[$ *then all their derivatives up to the order l included are also identical
on* $]\sigma, \infty \mathbf{1}_N[$ *and at* $\mathbf{t}^N = \sigma^+$.

Note 184 *This lemma is the vector generalization of Lemma 158.*

Lemma 183 and Definition 313 imply the following.

Lemma 185 *The l-th order elementwise trackability and elementwise trackability [188, Lemma 265, pp. 182, 183]*
*For the m-th order dynamical plant to be the l-th order (global) elemen-
twise trackable on* $\mathfrak{D}^i \times \mathfrak{Y}_d^k$ *it is necessary and sufficient to be (global) elementwise trackable on* $\mathfrak{D}^i \times \mathfrak{Y}_d^k$.

9.5.4   Elementwise natural trackability

The natural trackability concept can also satisfy the demand for different
reachability *times* to be associated with different output variables.

Definition 186 *The l-th order elementwise natural trackability on*
$\mathfrak{D}^i \times \mathfrak{Y}_d^k$
*a) The m-th order dynamical plant is **the l-th order elementwise
natural trackable on** $\mathfrak{D}^i \times \mathfrak{Y}_d^k$ if and only if there is* $\boldsymbol{\Delta}^{mN} \in \mathfrak{R}^{+mN}$, *or*
$\boldsymbol{\Delta}^{mN} = \infty \mathbf{1}_{mN}$, *such that for every disturbance vector function* $\mathbf{D}(.) \in \mathfrak{D}^i$,
for every plant desired output response $\mathbf{Y}_d(.) \in \mathfrak{Y}_d^k$, *and for every vector
instant* $\sigma \in (Int\ \mathfrak{T}_0)^{(l+1)N}$, *there is control vector function* $\mathbf{U}(.)$ *obeying
$TCUP$ on* $\mathfrak{T}_0$, *which can be synthesized without using information about the
form and value of* $\mathbf{D}(.) \in \mathfrak{D}^k$ *and about the plant state, such that for every
plant initial output vector* $\mathbf{Y}_0^{m-1}$ *in the* $\boldsymbol{\Delta}^{mN}$-*elementwise neighborhood of
the plant initial desired output vector* $\mathbf{Y}_{d0}^{m-1}$, *the plant extended real output
response* $\mathbf{Y}^l(t)$ *becomes elementwise equal to* $\mathbf{Y}_d^l(t)$ *at latest at the vector
moment σ, after which they rest equal forever, i.e.,*

$$\exists \boldsymbol{\Delta}^{mN} \in]\mathbf{0}_{mN},\ \infty \mathbf{1}_{mN}],\ \forall [\mathbf{D}(.), \mathbf{Y}_d(.)] \in \mathfrak{D}^i \times \mathfrak{Y}_d^k,$$

$$\forall \sigma \in (Int\ \mathfrak{T}_0)^{(l+1)N},\ \exists \mathbf{U}(.),$$

$$\mathbf{U}(t) = \mathbf{U}(t; \sigma; \mathbf{Y}_d) \in \mathfrak{C}(\mathfrak{T}_0)\ \ and\ \ \left|\mathbf{Y}_0^{m-1} - \mathbf{Y}_{d0}^{m-1}\right| < \boldsymbol{\Delta}^{mN} \implies$$

$$\mathbf{Y}^l(\mathbf{t}^{(l+1)N}) = \mathbf{Y}_d^l(\mathbf{t}^{(l+1)N})\ \forall \left(\mathbf{t}^{(l+1)N} \geq \sigma\right) \in \mathfrak{T}_0^{(l+1)N}. \tag{9.8}$$

Such control is **the l-th order elementwise natural tracking control on** $\mathfrak{D}^i \times \mathfrak{Y}_d^k$, *for short,* **the l-th order elementwise natural tracking control.**

The zero, (l = 0), order elementwise natural trackability on $\mathfrak{D}^i \times \mathfrak{Y}_d^k$ *is called* **elementwise natural trackability on** $\mathfrak{D}^i \times \mathfrak{Y}_d^k$. *The zero, (l = 0), order elementwise natural tracking control on* $\mathfrak{D}^i \times \mathfrak{Y}_d^k$ *is called* **elementwise natural tracking control on** $\mathfrak{D}^i \times \mathfrak{Y}_d^k$, *for short,* **elementwise natural tracking control.**

b) The l-th order elementwise natural trackability on $\mathfrak{D}^i \times \mathfrak{Y}_d^k$ *is* **global** *(in the whole) if and only if* $\mathbf{\Delta}^{mN} = \infty \mathbf{1}_{mN}$.

c) The l-th order elementwise natural trackability on $\mathfrak{D}^i \times \mathfrak{Y}_d^k$ *is* **uniform** *over* $\mathfrak{Y}_d^k$ *if and only if* $\mathbf{U}(.)$ *depends on* $\mathfrak{Y}_d^k$ *but not on an individual* $\mathbf{Y}_d(.)$ *from* $\mathfrak{Y}_d^k$, $\mathbf{U}(t) = \mathbf{U}(t; \sigma; \mathfrak{Y}_d^k)$.

Comment 187 *Definition 313 and Definition 314 show the difference between the l-th order elementwise trackability and the l-th order elementwise natural trackability. The former is necessary for the latter, and the latter is sufficient for the former.*

Note 188 *Complete natural trackability and elementwise natural trackability*

The elementwise natural trackability is simultaneously the complete natural trackability.

Comment 189 *The l-th order elementwise trackability is necessary for the l-th order elementwise natural trackability. The latter is sufficient for the former.*

Lemma 183 and Definition 314 induce the following.

Lemma 190 *The l-th order elementwise natural trackability on the product set* $\mathfrak{D}^i \times \mathfrak{Y}_d^k$ *and the elementwise natural trackability on the same product set* $\mathfrak{D}^i \times \mathfrak{Y}_d^k$ *[175, Lemma 184, p.90], [188, Lemma 270, p184]*

For the m-th order dynamical plant to be the l-th order (global) elementwise natural trackable on the product set $\mathfrak{D}^i \times \mathfrak{Y}_d^k$ *it is necessary and sufficient to be (global) elementwise natural trackable on the product set* $\mathfrak{D}^i \times \mathfrak{Y}_d^k$, *respectively.*

Note 191 *General importance of the trackability*

The trackability concept represents the inherent bridge between the engineer who designs the plant and the engineer who synthesizes the control, and/or designs the controller, for the plant.

Trackability is the fundamental link between the manufacturer of the plant and the manufacturer of the controller for the plant.

Trackability is also the link between the dynamics and mathematical modeling of the plant, and the controller. It is crucial for the tracking control synthesis.

Comment 192 *Minimal tracking control*

For a relevant trackability of $\mathbf{Y}_d(.)$ the tracking control $\mathbf{U} \in \mathfrak{R}^r$ is minimal if and only if its dimension r is its minimal dimension with which it satisfies the corresponding trackability conditions.

Chapter 10

Various systems trackability

10.1 *IO* system trackability

What follows continues, refines and generalizes the results of the book [188, Section 9.3, pp. 185-190, Section 9.4, pp. 197-201] on the *IO* systems trackability.

10.1.1 Perfect trackability criteria

Section 2.1 exposes the *IO* plant characteristics. Its Subsection 2.1.1 determines the the *IO* plant *time* domain description, Equation (2.15), in terms of the total values of all variables:

$$A^{(\nu)}\mathbf{Y}^{\nu}(t) = D^{(\eta)}\mathbf{D}^{\eta}(t) + B^{(\mu)}\mathbf{U}^{\mu}(t), \ \forall t \in \mathfrak{T}_0. \tag{10.1}$$

The Laplace transform of Equation (10.1) is Equation (2.23) (Section 2.1), i.e.,

$$A^{(\nu)}S_N^{(\nu)}(s)\mathbf{Y}(s) - A^{(\nu)}Z_N^{(\nu-1)}(s)\mathbf{Y}_0^{\nu-1} = D^{(\eta)}S_d^{(\eta)}(s)\mathbf{D}(s) -$$
$$-D^{(\eta)}Z_d^{(\eta-1)}(s)\mathbf{D}_0^{\eta-1} + B^{(\mu)}S_r^{(\mu)}(s)\mathbf{U}(s) - B^{(\mu)}Z_r^{(\mu-1)}(s)\mathbf{U}_0^{\mu-1}. \tag{10.2}$$

Which properties of the *IO* plant enable the existence of a control that can force the plant to exhibit perfect tracking as soon as the initial real output vector is equal to the initial desired output vector? This means, which properties of the *IO* plant ensure its perfect trackability?

We present, at first, the *time* domain conditions for the perfect trackability properties of the *IO* plant (10.1).

Theorem 193 *Time-domain condition for the perfect trackability of the IO plant (10.1) on $\mathfrak{D}^\eta \times \mathfrak{Y}_d^\nu$*

For the IO plant (10.1) to be perfect trackable on $\mathfrak{D}^\eta \times \mathfrak{Y}_d^\nu$ it is necessary and sufficient that for every $[\mathbf{D}(.), \mathbf{Y}_d(.)] \in \mathfrak{D}^\eta \times \mathfrak{Y}_d^\nu$ there is a control vector function $\mathbf{U}(.)$ that obeys the differential equation (10.3),

$$B^{(\mu)}\mathbf{U}^\mu(t) = A^{(\nu)}\mathbf{Y}_d^\nu(t) - D^{(\eta)}\mathbf{D}^\eta(t), \ \forall t \in \mathfrak{T}_0,$$

$$\text{under the condition } \mathbf{Y}_0^{\nu-1} = \mathbf{Y}_{d0}^{\nu-1}. \tag{10.3}$$

for which it is necessary and sufficient that $B^{(\mu)}$ has the full rank N,

$$rank B^{(\mu)} = full \ rank B^{(\mu)} = N \leq r. \tag{10.4}$$

Proof. *Necessity.* Let the *IO* plant (10.1) be perfect trackable on $\mathfrak{D}^\eta \times \mathfrak{Y}_d^\nu$. Definition 161 holds. The control vector $\mathbf{U}(t)$ and the desired initial conditions: $\mathbf{Y}_0^{\nu-1} = \mathbf{Y}_{d0}^{\nu-1}$ guarantee $\mathbf{Y}^\nu(t) \equiv \mathbf{Y}_d^\nu(t)$, which transforms Equation (10.1) into Equation (10.3). The existence of the unique solution $\mathbf{U}(t)$ to this linear differential vector equation implies

$$rank B^{(\mu)} = full \ rank B^{(\mu)} = N \leq r,$$

which is the condition (10.4) and proves its necessity.

Sufficiency. Let the condition 10.4) be satisfied. Then $det\left(B^{(\mu)}B^{(\mu)T}\right) \neq 0$, which permits us to define the control vector $\mathbf{U}(t)$ so that it is the unique solution of the linear differential equation (10.3). Let us eliminate $B^{(\mu)}\mathbf{U}^\mu(t)$ from Equation (10.1) and Equation (10.3), i.e., let us replace $B^{(\mu)}\mathbf{U}^\mu(t)$ by the right hand side of Equation (10.3) into Equation (10.1):

$$A^{(\nu)}\mathbf{Y}^\nu(t) = D^{(\eta)}\mathbf{D}^\eta(t) + A^{(\nu)}\mathbf{Y}_d^\nu(t) - D^{(\eta)}\mathbf{D}^\eta(t) = A^{(\nu)}\mathbf{Y}_d^\nu(t), \ \forall t \in \mathfrak{T}_0,$$

which yields

$$A^{(\nu)}\mathbf{Y}^\nu(t) = A^{(\nu)}\mathbf{Y}_d^\nu(t), \forall t \in \mathfrak{T}_0.$$

Condition 39, equivalently Equation (2.3) : $det A_\nu \neq 0$, (Section 2.1), ensures $rank A^{(\nu)} = N$ and the nonsingularity of $\left(A^{(\nu)}A^{(\nu)T}\right)$. If $\mathbf{Y}^\nu(t) = \mathbf{Y}_d^\nu(t)$, $\forall t \in \mathfrak{T}_0$, were not true then there would be $\mathbf{W}(t) \in \mathfrak{R}^N$ such that

$$A^{(\nu)T}\left(A^{(\nu)}A^{(\nu)T}\right)^{-1}\mathbf{W}(t) = \mathbf{Y}^\nu(t) - \mathbf{Y}_d^\nu(t) \neq \mathbf{0}_{(\nu+1)N},$$

i.e.,

$$\mathbf{W} = A^{(\nu)}\mathbf{Y}^\nu(t) - A^{(\nu)}\mathbf{Y}_d^\nu(t) \neq \mathbf{0}_{(\nu+1)N},$$

which contradicts the obtained $A^{(\nu)}\mathbf{Y}^{\nu}(t) = A^{(\nu)}\mathbf{Y}_d^{\nu}(t)$. The contradiction implies $\mathbf{Y}^{\nu}(t) = \mathbf{Y}_d^{\nu}(t)$, $\forall t \in \mathfrak{T}_0$. The IO plant (10.1) is perfect trackable on $\mathfrak{D}^{\eta} \times \mathfrak{Y}_d^{\nu}$ in view of Definition 161. ∎

A matrix function

$$P(.) : \mathbb{C} \longrightarrow \mathbb{C}^{N \times r}, \ P(s) = [p_{j,k}(s)] \in \mathbb{C}^{N \times r}, \qquad (10.5)$$

is *polynomial matrix if and only if every its entry* $p_{j,k}(s)$ is a polynomial in the complex variable $s \in \mathbb{C}$,

$$p_{j,k}(s) = \sum_{i=0}^{i=\mu} p_{j,k}^i s^i, \ p_{j,k}^i \in \mathfrak{R}, \ \forall j = 1, 2, .., N, \ \forall k = 1, 2, ..., r. \qquad (10.6)$$

The polynomial matrix $P(s)$ can be set in the form of *the matrix polynomial*

$$P(s) = \sum_{i=0}^{i=\mu} P_i s^i, \ P_i \in \mathfrak{R}^{N \times r}, \ \forall i = 0, 1, ..., \mu, \qquad (10.7)$$

or in the compact form by applying $P^{(\mu)}$,

$$P^{(\mu)} = \left[P_0 \vdots P_1 \vdots ... \vdots P_{\mu} \right] \in \mathfrak{R}^{N \times (\mu+1)r}, \qquad (10.8)$$

by using $S_r^{(\mu)}(s)$ and the identity matrix $I_r \in \mathfrak{R}^{r \times r}$, so that

$$P(s) = \sum_{i=0}^{i=\mu} P_i s^i = P^{(\mu)} S_r^{(\mu)}(s). \qquad (10.9)$$

The matrix $P^{(\mu)}$ (10.8) is *the generating matrix* of both the matrix polynomial (10.7) and the polynomial matrix $P(s)$ (10.5).

Theorem 194 Rank of a polynomial matrix [171]

Let $N \leq r$.

1) In order for the polynomial matrix $P(s)$ *(10.5),* $P(s) \in \mathbb{C}^{N \times r}$, *to have the full rank* $\rho = \min(N, r) = N$ *it is necessary and sufficient that there is* $s^* \in \mathbb{C}$ *such that* $rank P(s^*) = N$:

$$full \ rank P(s) = N \ on \ \mathbb{C} \Longleftrightarrow \exists s^* \in \mathbb{C}, \ rank P(s^*) = N, \qquad (10.10)$$

2) If the polynomial matrix $P(s)$ *(10.5),* $P(s) \in \mathbb{C}^{N \times r}$, *has the full rank* $\rho = \min(N, r) = N$ *then its generating matrix* $P^{(\mu)} \in \mathfrak{R}^{N \times (\mu+1)r}$ *has also the full rank* $\rho = N$,

$$full \ rank P(s) = N \Longrightarrow rank P^{(\mu)} = full \ rank P^{(\mu)} = N. \qquad (10.11)$$

Let the rational matrix $R(s)$ be defined by

$$R(s) = \left[C(sI - A)^{-1} B + U \right] \in \mathbb{C}^{N \times r}. \qquad (10.12)$$

Let

$$p(s) = \det(sI - A), \qquad (10.13)$$

and

$$L(s) = C\, adj(sI - A)\, B + p(s)\, U, \ \ L(s) \in \mathbb{C}^{N \times r}. \qquad (10.14)$$

Equation (10.12) simplifies the definition of $R(s)$:

$$R(s) = \frac{L(s)}{p(s)} = p^{-1}(s)\, L(s). \qquad (10.15)$$

Theorem 195 *Rank of a rational matrix [171]*
   The rank of the rational matrix $R(s)$ (10.12) is the rank of its numerator polynomial matrix $L(s)$ (10.14),

$$rank F(s) = rank \left[C(sI - A)^{-1} B + U \right] =$$
$$= rank \left[p^{-1}(s)\, L(s) \right] = rank L(s) \leq \min(N, r). \qquad (10.16)$$

We can express the trackability conditions in the complex domain, as well. The crucial condition is the rank condition on the generic polynomial $B^{(\mu)}$ of the numerator matrix polynomial $B^{(\mu)} S_r^{(\mu)}(s)$ of the plant transfer function matrix $G_{IOU}(s)$ (Equation 2.30, Section 2.1) relative to the control vector:

$$G_{IOU}(s) = \left(A^{(\nu)} S_N^{(\nu)}(s) \right)^{-1} B^{(\mu)} S_r^{(\mu)}(s) \in \mathbb{C}^{N \times r}. \qquad (10.17)$$

Theorem 196 *Complex domain conditions for the perfect tracka-bility of the IO plant (10.1) on $\mathfrak{D}^\eta \times \mathfrak{Y}_d^\nu$*
   For the IO plant (10.1) to be (the $(\nu\text{-}1)$ th order) perfect trackable on $\mathfrak{D}^\eta \times \mathfrak{Y}_d^\nu$ it is necessary and sufficient that both
   1) $N \leq r,$
   and
   2) $rank G_{IOU}(s) = full\ rank G_{IOU}(s) = full\ rank B^{(\mu)} = N.$

Proof. *Necessity.* Let the *IO* plant (10.1) be $(\nu\text{-}1)$ th order perfect trackable on $\mathfrak{D}^\eta \times \mathfrak{Y}_d^\nu$. Let $\mathbf{Y}_{d0}^{\nu-1}$ be arbitrary and let us accept $\mathbf{Y}_0^{\nu-1} = \mathbf{Y}_{d0}^{\nu-1}$ (Definition 161, Section 9.4). The *IO* plant (10.1) is also perfect trackable on

$\mathfrak{D}^\eta \times \mathfrak{Y}_d^\nu$ due to Definition 161, Section 9.4. There exists control vector $\mathbf{U}(t)$ such that its Laplace transform obeys Equation (10.2) that has the following equivalent form for $\mathbf{Y}(t) \equiv \mathbf{Y}_d(t)$, i.e., for $\mathbf{Y}(s) \equiv \mathbf{Y}_d(s)$:

$$B^{(\mu)} S_r^{(\mu)}(s)\mathbf{U}(s) = A^{(\nu)} S_N^{(\nu)}(s)\mathbf{Y}_d(s) - D^{(\eta)} S_d^{(\eta)}(s)\mathbf{D}(s) -$$
$$- A^{(\nu)} Z_N^{(\nu-1)}(s)\mathbf{Y}_{d0}^{\nu-1} + D^{(\eta)} Z_d^{(\eta-1)}(s)\mathbf{D}_0^{\eta-1} + B^{(\mu)} Z_r^{(\mu-1)}(s)\mathbf{U}_0^{\mu-1}. \quad (10.18)$$

The existence of the solution $\mathbf{U}(s)$ of this equation implies

$$rank\left(B^{(\mu)} S_r^{(\mu)}(s)\right) = full\ rank\left(B^{(\mu)} S_r^{(\mu)}(s)\right) = N \le r.$$

This, Theorem 194 and Theorem 195 prove the necessity of the conditions 1) and 2) since $\left(A^{(\nu)} S_N^{(\nu)}(s)\right)^{-1}$ is nonsingular matrix so that the full rank N of the polynomial numerator $B^{(\mu)} S_r^{(\mu)}(s)$ and the full rank r of the matrix $S_r^{(\mu)}(s)$, imply $N \le r$, $rank B^{(\mu)} = full\ rank B^{(\mu)} = N$, and guarantee the full rank N of $G_{IOU}(s)$, i.e.,

$$full\ rank\left(B^{(\mu)} S_r^{(\mu)}(s)\right) = full\ rank B^{(\mu)} = full\ rank G_{IOU}(s) = N \le r.$$

This proves necessity of the conditions 1) and 2).

Sufficiency. Let the conditions 1) and 2) be satisfied. They, Equations (2.30) (Section 2.1), the nonsingularity of $det\left(A^{(\nu)} S_N^{(\nu)}(s)\right)$ almost everywhere on $\mathbb{C}$, the nonsingularity of $\left(A^{(\nu)} S_N^{(\nu)}(s)\right)^{-1}$ and $adj\left(A^{(\nu)} S_N^{(\nu)}(s)\right)$ almost everywhere on $\mathbb{C}$, together with Theorem 194 and Theorem 195, prove the validity of

$$r \ge N = full\ rank G_{IOU}(s) = full\ rank B^{(\mu)} S_r^{(\mu)}(s).$$

The full rank N of $G_{IOU}(s)$ guarantees the nonsingularity of

$$G_{IOU}(s) G_{IOU}^T(s) \in \mathbb{C}^{N \times N}$$

so that $\Gamma_{IOU}(s)$ introduced by

$$\Gamma_{IOU}(s) = G_{IOU}^T(s)\left[G_{IOU}(s) G_{IOU}^T(s)\right]^{-1}. \quad (10.19)$$

is fully defined as well as $\mathbf{U}(s)$ defined by

$$\mathbf{U}(s) = \Gamma_{IOU}(s)\left\{ \begin{array}{c} \mathbf{Y}_d(s) - G_{IOD}(s)\mathbf{D}(s) - G_{IOY_0}(s)\mathbf{Y}_{d0}^{\nu-1} - \\ + G_{IOD_0}(s)\mathbf{D}_0^{\eta-1} + G_{IOU_0}(s)\mathbf{U}_0^{\mu-1} \end{array} \right\}. \quad (10.20)$$

The multiplication of this equation on the left by $G_{IOU}(s)$ gives

$$G_{IOU}(s)\,\mathbf{U}(s) = \left\{ \begin{array}{c} \mathbf{Y}_d(s) - G_{IOD}(s)\,\mathbf{D}(s) - G_{IOY_0}(s)\,\mathbf{Y}_{d0}^{\nu-1} + \\ + G_{IOD_0}(s)\,\mathbf{D}_0^{\eta-1} + G_{IOU_0}(s)\,\mathbf{U}_0^{\mu-1} \end{array} \right\}.$$
$$(10.21)$$

Let $\left(A^{(\nu)}S_N^{(\nu)}(s)\right)^{-1}$ multiply Equation (10.2) on the left. The result takes the following form in view of Equations (2.31)–(2.34), Section 2.1:

$$\mathbf{Y}(s) = G_{IOU}(s)\,\mathbf{U}(s) + G_{IOD}(s)\,\mathbf{D}(s) -$$
$$-G_{IOD_0}(s)\,\mathbf{D}_0^{\eta-1} - G_{IOU_0}(s)\,\mathbf{U}_0^{\mu-1} + G_{IOY_0}(s)\,\mathbf{Y}_0^{\nu-1}.$$

We eliminate $\mathbf{U}(s)$ from this equation by replacing it by the right-hand side of Equation (10.21):

$$\mathbf{Y}(s) = \mathbf{Y}_d(s) - G_{IOD}(s)\,\mathbf{D}(s) - G_{IOY_0}(s)\,\mathbf{Y}_{d0}^{\nu-1} -$$
$$+G_{IOD_0}(s)\,\mathbf{D}_0^{\eta-1} + G_{IOU_0}(s)\,\mathbf{U}_0^{\mu-1} + G_{IOD}(s)\,\mathbf{D}(s) -$$
$$-G_{IOD_0}(s)\,\mathbf{D}_0^{\eta-1} - G_{IOU_0}(s)\,\mathbf{U}_0^{\mu-1} + G_{IOY_0}(s)\,\mathbf{Y}_{d0}^{\nu-1} = \mathbf{Y}_d(s),$$

which in the *time* domain reads

$$\mathbf{Y}(t) \equiv \mathbf{Y}_d(t), \ \forall\,[D\,(.)\,,\mathbf{Y}_d(.)] \in \mathfrak{D}^\eta \times \mathfrak{Y}_d^\nu.$$

This proves the perfect trackability of the *IO* plant (10.1) in view of Definition 161. ∎

Note 197 *This theorem slightly refines Note 277 of [188, p 188]. The conditions for the perfect trackability on $\mathfrak{D}^\eta \times \mathfrak{Y}_d^\nu$ of the IO plant (10.1) do not impose any requirement on the internal dynamics, i.e., on the state, of the object. Besides, the conditions do not impose any demand on the disturbance. They are independent of the disturbance. They are purely algebraic and simple. They are in terms of the rank of the generic matrix $B^{(\mu)}$ of the numerator matrix polynomial $B^{(\mu)}S_r^{(\mu)}(s)$ of the plant transfer function matrix $G_{IOU}(s)$ relative to control, which, due to Theorem 195 and Theorem 194, imply:*

$$full\ rank G_{IOU}(s) = full\ rank\left(B^{(\mu)}S_r^{(\mu)}(s)\right) = N \Longrightarrow$$
$$full\ rank B^{(\mu)} = N.$$
$$(10.22)$$

Comment 198 *[188, Comment 278, p 188] The perfect trackability incorporates the output function controllability. The perfect trackability of the IO plant (10.1) on $\mathfrak{D}^\eta \times \mathfrak{Y}_d^\nu$ takes into account the influence of all disturbances $\mathbf{D}(.) \in \mathfrak{D}^\mu$, while the output function controllability is defined only for the unperturbed systems (e.g. [48, p. 216]), i.e., that $\mathbf{D}(t) = \mathbf{0}_d, \forall t \in \mathfrak{T}_0$.*

10.1.2   Conditions for perfect natural trackability

Theorem 199 *Conditions for the perfect natural trackability of the IO plant (10.1) on $\mathfrak{D}^\eta \times \mathfrak{Y}_d^\nu$*

For the IO plant (10.1) to be perfect natural trackable on $\mathfrak{D}^\eta \times \mathfrak{Y}_d^\nu$ it is necessary and sufficient that both

1) $N \leq r$,

and

2) $rank G_{IOU}(s) = full \ rank G_{IOU}(s) = full \ rank B^{(\mu)} = N$.

Proof. Let the *IO* plant (10.1) be perfect natural trackable on $\mathfrak{D}^\eta \times \mathfrak{Y}_d^\nu$. Definition 310, Section 9.4, is fulfilled. It and Definition 161 guarantee that the plant is also perfect trackable on $\mathfrak{D}^\eta \times \mathfrak{Y}_d^\nu$ (which results also from Comment 168, Section 9.4). The necessity part of Theorem 196 is valid, which is expressed by the necessity of the conditions 1) and 2).

Sufficiency. Let the conditions 1) and 2) hold. The *IO* plant (10.1) is perfect trackable on $\mathfrak{D}^\eta \times \mathfrak{Y}_d^\nu$, Theorem 196. We should show that the perfect tracking control can be synthesized without using information about the plant state and about the disturbance $\mathbf{D}(.) \in \mathfrak{D}^\mu$, i.e., that it is natural tracking control (Definition 166, Section 9.4). We recall Equation (2.23) (Section 2.1) and set it into the following form:

$$A^{(\nu)} S_N^{(\nu)}(s) \mathbf{Y}(s) = B^{(\mu)} S_r^{(\mu)}(s) \mathbf{U}(s) -$$
$$- B^{(\mu)} Z_r^{(\mu-1)}(s) \mathbf{U}_0^{(\mu-1)} + D^{(\eta)} S_d^{(\eta)}(s) \mathbf{D}(s) -$$
$$- D^{(\eta)} Z_d^{(\eta-1)}(s) \mathbf{D}_0^{(\mu-1)} + A^{(\nu)} Z_N^{(\nu-1)}(s) \mathbf{Y}_0^{(\nu-1)}. \qquad (10.23)$$

The Laplace transform of the error vector $\mathbf{e}(t) = \mathbf{Y}_d(t) - \mathbf{Y}(t)$ is $\mathbf{E}(s)$, $\mathbf{E}(s) = \mathbf{Y}_d(s) - \mathbf{Y}(s)$. Let $\sigma \in \mathfrak{R}^+$ be arbitrarily small, i.e., $\sigma \longrightarrow 0^+$ and in the ideal case $\sigma = 0^+$. Let

$$\phi(.) : \mathfrak{T}_0 \longrightarrow \mathfrak{R}^N, \ \phi(t) \in \mathfrak{C}(\mathfrak{T}),$$
$$\phi(t) = \mathbf{0}_N, \ \forall t \in [\sigma, \infty[, \ \phi(0) = -\mathbf{e}(0), \qquad (10.24)$$

The matrix $\Gamma_{IOU}(s)$ is defined in Equation (10.19). Let the control be governed by:

$$\mathbf{U}(s) = (1 - e^{-\sigma s})^{-1} \Gamma_{IOU}(s) [\mathbf{\Phi}(s) + \mathbf{E}(s)], \ \mathbf{\Phi}(s) = \mathcal{L}\{\phi(t)\} \Rightarrow$$
$$\mathbf{U}(s) = e^{-\sigma s} \mathbf{U}(s) + \Gamma_{IOU}(s) [\mathbf{\Phi}(s) + \mathbf{E}(s)], \qquad (10.25)$$

equivalently,

$$\mathbf{U}(t) = \mathbf{U}(t^-) + \Gamma_{IOU}\left[\phi(t) + \mathbf{e}\,(t)\right],$$

$$\Gamma_{IOU} = \left(B^{(\mu)}\right)^T \left[\left(B^{(\mu)}\right)\left(B^{(\mu)}\right)^T\right]^{-1}. \tag{10.26}$$

The control $\mathbf{U}(.)$ (10.25) is independent of the plant state and of the disturbance $\mathbf{D}(.)$. The control $\mathbf{U}(.)$ is natural control (Definition 166, Section 9.4). We replace $\mathbf{U}(s)$ with the right-hand side of (10.25) into (10.23):

$$A^{(\nu)}S_N^{(\nu)}(s)\mathbf{Y}(s) = B^{(\mu)}S_r^{(\mu)}(s)\left\{e^{-\sigma s}\mathbf{U}(s) + \Gamma_{IOU}(s)\left[\mathbf{\Phi}(s) + \mathbf{E}\,(s)\right]\right\} -$$
$$-B^{(\mu)}Z_r^{(\mu-1)}(s)\mathbf{U}_0^{(\mu_{Pu}-1)} + D^{(\eta)}S_d^{(\eta)}(s)\mathbf{D}(s) -$$
$$-D^{(\eta)}Z_d^{(\eta-1)}(s)\mathbf{D}_0^{(\eta-1)} + A^{(\nu)}Z_N^{(\nu-1)}(s)\mathbf{Y}_0^{(\nu-1)}.$$

We subtract this equation from (10.23). The result is

$$\mathbf{0}_N = B^{(\mu)}S_r^{(\mu)}(s)\left\{\left(1 - e^{-\sigma s}\right)\mathbf{U}(s) - \Gamma_{IOU}(s)\left[\mathbf{\Phi}(s) - \mathbf{E}\,(s)\right]\right\}.$$

For $\sigma \longrightarrow 0^+$, or for $\sigma = 0^+$ in the ideal case of the signal transmission through the local feedback of the controller, the preceding equation takes the following form:

$$\mathbf{0}_N = \underbrace{B^{(\mu)}S_r^{(\mu)}(s)\Gamma_{IOU}(s)}_{I_N}\left[\mathbf{\Phi}(s) + \mathbf{E}\,(s)\right], \; i.e., \; \mathbf{0}_N = \mathbf{\Phi}(s) + \mathbf{E}\,(s)$$

due to (10.19). The last equation yields in the *time* domain:

$$\mathbf{e}(t) = -\phi(t) = \mathbf{0}_N, \; \forall t \in \mathfrak{T}_0,$$

due to the equation in (10.24), i.e., equivalently:

$$\mathbf{e}(t) = \mathbf{Y}_d(t) - \mathbf{Y}(t) = \mathbf{0}_N, \; \forall t \in \mathfrak{T}_0.$$

The natural control $\mathbf{U}(.)$ (10.25) satisfies Definition 310. The *IO* plant (10.1) is perfect natural trackable on $\mathfrak{D}^\eta \times \mathfrak{Y}_d^\nu$. $\blacksquare$

Comment 200 *[188, Comment 280, p. 190] Theorem 196 and Theorem 199 show that the IO plant (10.1) is perfect natural trackable on $\mathfrak{D}^\eta \times \mathfrak{Y}_d^\nu$ if and only if it is perfect trackable on $\mathfrak{D}^\eta \times \mathfrak{Y}_d^\nu$. This completes Comment 168, (Section 9.4).*

Note 201 *[188, Note 281, p. 190] Theorem 199 enables the controller to use only*

- *information about the output error* $\mathbf{e}(t) = \mathbf{Y}_d(t)$-$\mathbf{Y}(t)$, *which is expressed through*

$$\mathbf{U}(s) = e^{-\sigma s}\mathbf{U}(s) + \Gamma_{IOU}(s)[\mathbf{\Phi}(s) + \mathbf{E}(s)] \qquad (10.27)$$

with

$$\Gamma_{IOU}(s) = \left[B^{(\mu)}S_r^{(\mu)}(s)\right]^T \left\{\left[B^{(\mu)}S_r^{(\mu)}(s)\right]\left[B^{(\mu)}S_r^{(\mu)}(s)\right]^T\right\}^{-1},$$

$$(10.28)$$

$$\phi(t) \in \mathfrak{C}(\mathfrak{T}),\ \phi(0) = -\mathbf{e}(0),\ \phi(t) = 0,\ \forall t \in [\sigma, \infty[,\ 0 < \sigma <<< 1,$$
$$\mathbf{\Phi}(s) = \mathcal{L}\{\phi(t)\}, \qquad (10.29)$$

for which there is the classical global negative feedback loop in the control system from the plant output to the controller input, which transfers the signal of the plant real output vector $\mathbf{Y}$,

- *and the control* $\mathbf{U}(t^-) = \mathbf{U}(t - \sigma)_{\sigma \longrightarrow 0+}$, *with which the controller has just acted on the plant at the moment* t^- *and which is then an input to the controller itself at the instant t, for which there is local, internal, unit positive feedback in the controller. The value of* $\sigma \longrightarrow 0^+$ *expresses the infinitesimal duration of the time interval during which the controller output* $\mathbf{U}(t^-) = \mathbf{U}(t - \sigma)$ *becomes its input* $\mathbf{U}(t)$. *The positive unit controller local feedback does not allow the controller to be ON when it is disconnected. It should be connected with the plant in the global negative feedback.*

10.1.3   Imperfect trackability criteria

The imperfect trackability, for short: *trackability*, reflects the plant ability to enable the existence of control that can steer the output vector from its arbitrary initial value to its desired value in a finite *time*, after which the real output vector stays always equal to the desired output vector.

Theorem 202 *Criteria for the global complete trackability of the IO plant (10.1) on* $\mathfrak{D}^\eta \times \mathfrak{Y}_d^\nu$

For the IO plant (10.1) to be global complete trackable on $\mathfrak{D}^\eta \times \mathfrak{Y}_d^\nu$ *it is necessary and sufficient that both*

1) $N \leq r$,

and

2) $rankG_{IOU}(s) = full \ rankG_{IOU}(s) = full \ rankB^{(\mu)} = N$.

Proof. *Necessity.* Let the *IO* plant (10.1) be global complete trackable on $\mathfrak{D}^\eta \times \mathfrak{Y}_d^\nu$. Definition 172 (Section 9.5) is satisfied. Then the plant is also perfect trackable on $\mathfrak{D}^\eta \times \mathfrak{Y}_d^\nu$ (due to Definition 172 and Theorem 174, Section 9.5, and Theorem 196), which proves (Theorem 196) the necessity of the conditions 1) and 2) for the global complete trackability on $\mathfrak{D}^\eta \times \mathfrak{Y}_d^\nu$.

Sufficiency. The sufficiency of the conditions results also from Definition 172, Theorem 174 and Theorem 196. Let us show the proof in details. It will be useful for the subsequent study. Let the conditions 1) and 2) hold. Let $[\mathbf{D}(.), \mathbf{Y}_d(.)] \in \mathfrak{D}^\eta \times \mathfrak{Y}_d^\nu$, the instant $\sigma \in Int \ \mathfrak{T}_0$, and $\mathbf{Y}_0^{\nu-1} \in \mathfrak{R}^{\nu N}$ be arbitrarily chosen (see Definition 172). Let the control be defined by

$$\mathbf{U}(t) = \Gamma_{IO} \left[-D^{(\eta)}\mathbf{D}^\eta(t) + A^{(\nu)}\mathbf{Y}^\nu(t) - \mathbf{Z}(t) + |\mathbf{e}(t)| \right] \in \mathfrak{C}(\mathfrak{T}_0),$$

$$\Gamma_{IO} = \left(B^{(\mu)} \right)^T \left[B^{(\mu)} \left(B^{(\mu)} \right)^T \right]^{-1},$$

together with

$$\mathbf{Z}(t) = \left\{ \begin{array}{c} |\mathbf{e}(0)| \left(1 - \frac{t}{\sigma} \right), \ t \in [0, \sigma], \\ \mathbf{0}_N, \ \forall (t \geq \sigma) \in \mathfrak{T}_0 \end{array} \right\} \in \mathfrak{C}(\mathfrak{T}_0), \qquad (10.30)$$

where we used the elementwise absolute value of the error vector by recalling Equation (1.14), Section 1.3:

$$|\mathbf{e}| = [|e_1| \quad |e_2| \quad \dots \quad |e_N|]^T.$$

For such control the plant mathematical model (10.1) becomes

$$A^{(\nu)}\mathbf{Y}^\nu(t) = D^{(\eta)}\mathbf{D}^\eta(t) + B^{(\mu)} \left(B^{(\mu)} \right)^T \left[B^{(\mu)} \left(B^{(\mu)} \right)^T \right]^{-1} \bullet$$

$$\bullet \left[-D^{(\eta)}\mathbf{D}^\eta(t) + A^{(\nu)}\mathbf{Y}^\nu(t) - \mathbf{Z}(t) + |\mathbf{e}(t)| \right] =$$

$$= D^{(\eta)}\mathbf{D}^\eta(t) - D^{(\eta)}\mathbf{D}^\eta(t) + A^{(\nu)}\mathbf{Y}^\nu(t) - \mathbf{Z}(t) + |\mathbf{e}(t)|,$$

i.e.,

$$|\mathbf{e}(t)| = \mathbf{Z}(t), \ \forall t \in \mathfrak{T}_0.$$

This and (10.30) imply

$$\mathbf{e}(t) = \mathbf{0}_N, \ \forall (t \geq \sigma) \in \mathfrak{T}_0,$$

i.e.,

$$\mathbf{Y}(t) = \mathbf{Y}_d(t), \ \forall (t \geq \sigma) \in \mathfrak{T}_0,$$

which proves the global complete trackability on $\mathfrak{D}^\eta \times \mathfrak{Y}_d^\nu$ of the *IO* plant (10.1) due to Definition 172. ∎

The complete trackability and the elementwise trackability are interrelated as explained in Note 182, Section 9.5.

Theorem 203 *Conditions for the $(\nu-1)$th order global elementwise trackability of the IO plant (10.1) on $\mathfrak{D}^\eta \times \mathfrak{Y}_d^\nu$*

For the IO plant (10.1) to be $(\nu-1)$th order global elementwise trackable on $\mathfrak{D}^\eta \times \mathfrak{Y}_d^\nu$ it is necessary and sufficient that both

1) $N \leq r$,

and

2) $rank G_{IOU}(s) = full \ rank G_{IOU}(s) = full \ rank B^{(\mu)} = N$.

Proof. *Necessity.* Let the *IO* plant (10.1) be $(\nu-1)th$ order global elemntwise trackable on $\mathfrak{D}^\eta \times \mathfrak{Y}_d^\nu$. Definition 313 (Section 9.5) is satisfied. Then the plant is also complete trackable on $\mathfrak{D}^\eta \times \mathfrak{Y}_d^\nu$ (due to Definition 172 and Definition 313), which proves (due to Theorem 202) the necessity of the conditions 1) and 2) for the $(\nu - 1)th$ order global elementwise trackability on $\mathfrak{D}^\eta \times \mathfrak{Y}_d^\nu$.

Sufficiency. Let the conditions 1) and 2) hold. Let $[\mathbf{D}(.), \mathbf{Y}_d(.)] \in \mathfrak{D}^\eta \times \mathfrak{Y}_d^\nu$, the vector instant $\sigma \in (Int \ \mathfrak{T}_0)^\nu$, $\sigma = \begin{bmatrix} \sigma_0 \vdots \sigma_1 \vdots ... \vdots \sigma_{\nu-1} \end{bmatrix}^T$, and $\mathbf{Y}_0^{\nu-1} \in \mathfrak{R}^{\nu N}$ be arbitrarily chosen (in view of Definition 313). Let the control be defined by

$$\mathbf{U}(t) = \Gamma_{=IO} \left[-D^{(\eta)} \mathbf{D}^\eta(t) + A^{(\nu)} \mathbf{Y}^\nu(t) - \mathbf{Z}(t) + \sum_{j=0}^{j=\nu-1} \left| \mathbf{e}^{(j)}(t) \right| \right] \in \mathfrak{C}(\mathfrak{T}_0),$$

together with

$$\mathbf{Z}(t) = \left\{ \begin{array}{l} \sum_{j=0}^{j=\nu-1} \left| \mathbf{e}^{(j)}(0) \right| \left(1 - \frac{t}{\sigma_m} \right), \ \forall t \in [0, \sigma_m], \\ \mathbf{0}_N, \ \forall (t \geq \sigma_m) \in \mathfrak{T}_0, \end{array} \right\} \in \mathfrak{C}(\mathfrak{T}_0)$$

$$0 < \sigma_m = \min(\sigma_0, \sigma_1, ... , \sigma_{\nu-1}) <<< 0^+, \ \sigma_m \longrightarrow 0^+. \quad (10.31)$$

For such control the plant mathematical model (10.1) becomes

$$A^{(\nu)}\mathbf{Y}^{\nu}(t) = D^{(\eta)}\mathbf{D}^{\eta}(t) + B^{(\mu)}\left(B^{(\mu)}\right)^{T}\left[B^{(\mu)}\left(B^{(\mu)}\right)^{T}\right]^{-1} \bullet$$

$$\bullet\left[-D^{(\eta)}\mathbf{D}^{\eta}(t) + A^{(\nu)}\mathbf{Y}^{\nu}(t) - \mathbf{Z}(t) + \sum_{j=0}^{j=\nu-1}\left|\mathbf{e}^{(j)}(t)\right|\right] =$$

$$= D^{(\eta)}\mathbf{D}^{\eta}(t) - D^{(\eta)}\mathbf{D}^{\eta}(t) + A^{(\nu)}\mathbf{Y}^{\nu}(t) - \mathbf{Z}(t) + \sum_{j=0}^{j=\nu-1}\left|\mathbf{e}^{(j)}(t)\right|,$$

i.e.,

$$\sum_{j=0}^{j=\nu-1}\left|\mathbf{e}^{(j)}(t)\right| = \mathbf{Z}(t), \ \forall t \in \mathfrak{T}_{0}.$$

This and (10.31) imply

$$\mathbf{e}^{(j)}(t) = \mathbf{0}_{N}, \ \forall (t \geq \sigma_{j}) \in \mathfrak{T}_{0}, \ \forall j = 0, 1, ..., \nu - 1,$$

i.e.,

$$\mathbf{Y}^{(j)}(t) = \mathbf{Y}_{d}^{(j)}(t), \ \forall (t \geq \sigma_{j}) \in \mathfrak{T}_{0}, \ \forall j = 0, 1, ..., \nu - 1,$$

which proves the $(\nu - 1)th$ order global elementwise trackability on $\mathfrak{D}^{\eta} \times \mathfrak{Y}_{d}^{\nu}$ of the IO plant (10.1) due to Definition 313. ∎

10.1.4   Conditions for natural trackability

A refinement and generalization of Theorem 295 in [188, pp. 198-200] has the following form:

Theorem 204 *Conditions for the global complete natural trackability of the IO plant (10.1) on $\mathfrak{D}^{\eta} \times \mathfrak{Y}_{d}^{\nu}$*
    For the IO plant (10.1) to be global complete natural trackable on $\mathfrak{D}^{\eta} \times \mathfrak{Y}_{d}^{\nu}$ it is necessary and sufficient that both
    1) $N \leq r$,
    and
    2) $rankG_{IOU}(s) = full \ rankG_{IOU}(s) = full \ rankB^{(\mu)} = N.$

Proof. *Necessity.* Let the IO plant (10.1) be global complete natural trackable on $\mathfrak{D}^{\eta} \times \mathfrak{Y}_{d}^{\nu}$. Definition 312, (Section 9.5), holds. It and Definition 172, (Section 9.5), confirm that the plant is complete trackable on $\mathfrak{D}^{\eta} \times \mathfrak{Y}_{d}^{\nu}$ so that the conditions 1) and 2) are necessary due to Theorem 202.

Sufficiency. Let the conditions 1) and 2) be valid. The *IO* plant (10.1) is global complete trackable on $\mathfrak{D}^\eta \times \mathfrak{Y}_d^\nu$ (Theorem 202). Let the Laplace transform of the control vector be defined by Equations (10.24), (10.25). By repeating from Equation (10.25) the sufficiency part of the proof of Theorem 199 we prove that the so defined control is natural tracking control that guarantees

$$\mathbf{e}(t) = \mathbf{Y}_d(t) - \mathbf{Y}(t) = \mathbf{0}_N, \ \forall t \in \mathfrak{T}_0.$$

This proves the global complete natural trackability of the *IO* plant (10.1) on $\mathfrak{D}^\eta \times \mathfrak{Y}_d^\nu$ in view of Definition 312. ■

The following theorem slightly refines Theorem 295 of [188, pp. 198-200].

Theorem 205 *Conditions for the $(\nu-1)$th order global elementwise natural trackability of the IO plant (10.1) on $\mathfrak{D}^\eta \times \mathfrak{Y}_d^\nu$*

For the *IO* plant (10.1) to be the $(\nu-1)$th order global elementwise natural trackable on $\mathfrak{D}^\eta \times \mathfrak{Y}_d^\nu$ it is necessary and sufficient that both

1) $N \leq r$,

and

2) $rankG_{IOU}(s) = full \ rankG_{IOU}(s) = full \ rankB^{(\mu)} = N.$

Proof. *Necessity.* Let the *IO* plant (10.1) be the $(\nu - 1)th$ order global elementwise natural trackable on $\mathfrak{D}^\eta \times \mathfrak{Y}_d^\nu$. Definition 314 (Section 9.5) is fulfilled. This means that the plant and control satisfy also Definition 313 (Section 9.5). It and Theorem 203 prove the necessity of the conditions 1) and 2).

Sufficiency. Let the conditions 1) and 2) hold. Let $[\mathbf{D}(.), \mathbf{Y}_d(.)] \in \mathfrak{D}^\eta \times \mathfrak{Y}_d^\nu$, $\sigma \in (Int \ \mathfrak{T}_0)^\nu$, $\sigma = [\sigma_0 \ \sigma_1 \ ... \ \sigma_{\nu-1}]^T$, and $\mathbf{Y}_0^{\nu-1} \in \mathfrak{R}^{\nu N}$ be arbitrarily chosen (by following Definition 314). Let the control be defined by

$$B^{(\mu)}\mathbf{U}^\mu(t) = B^{(\mu)}\mathbf{U}^\mu(t^-) + \sum_{j=0}^{j=\nu-1} \left|\mathbf{e}^{(j)}(t)\right| - \mathbf{Z}(t) \in \mathfrak{C}(\mathfrak{T}_0), \qquad (10.32)$$

for $\mathbf{Z}(t)$ determined by (10.31). For such control the plant mathematical model (10.1), becomes

$$A^{(\nu)}\mathbf{Y}^\nu(t) = B^{(\mu)}\mathbf{U}^\mu(t^-) + \sum_{j=0}^{j=\nu-1} \left|\mathbf{e}^{(j)}(t)\right| - \mathbf{Z}(t) + D^{(\eta)}\mathbf{D}^\eta(t). \qquad (10.33)$$

The plant mathematical model at $t^- \in \mathfrak{T}_0$ reads

$$A^{(\nu)}\mathbf{Y}^\nu(t^-) = B^{(\mu)}\mathbf{U}^\mu(t^-) + D^{(\eta)}\mathbf{D}^\eta(t^-).$$

We solve this for $B^{(\mu)}\mathbf{U}^{\mu}(t^{-})$,

$$B^{(\mu)}\mathbf{U}^{\mu}(t^{-}) = A^{(\nu)}\mathbf{Y}^{\nu}(t^{-}) - D^{(\eta)}\mathbf{D}^{\eta}(t^{-})$$

so that (10.33) becomes

$$A^{(\nu)}\mathbf{Y}^{\nu}(t) = A^{(\nu)}\mathbf{Y}^{\nu}(t^{-}) - D^{(\eta)}\mathbf{D}^{\eta}(t^{-}) -$$
$$-\mathbf{Z}(t) + \sum_{j=0}^{j=\nu-1} \left|\mathbf{e}^{(j)}(t)\right| + D^{(\eta)}\mathbf{D}^{\eta}(t).$$

Linearity of (10.1) and $[\mathbf{D}(.), \mathbf{Y}_d(.)] \in \mathfrak{D}^{\eta} \times \mathfrak{Y}_d^{\nu}$, imply, (also due to Principle 10, Section 1.2), continuity of all variables in (10.1) so that

$$\mathbf{Y}^{\nu}(t) = \mathbf{Y}^{\nu}(t^{-}), \ \mathbf{D}^{\mu Pd}(t) = \mathbf{D}^{\mu Pd}(t^{-}).$$

Hence,

$$\sum_{j=0}^{j=\nu-1} \left|\mathbf{e}^{(j)}(t)\right| = \mathbf{Z}(t) = \mathbf{0}_N, \ \forall(t \geq \sigma_m) \in \mathfrak{T}_0,$$

i.e.,

$$\mathbf{Y}^{(j)}(t) = \mathbf{Y}_d^{(j)}(t), \ \forall(t \geq \sigma_m) \in \mathfrak{T}_0, \ \sigma_m < \sigma_j, \ \forall j = 0, 1, ..., \nu - 1,$$

which proves the $(\nu\text{-}1)th$ order global elementwise natural trackability on $\mathfrak{D}^{\eta} \times \mathfrak{Y}_d^{\nu}$ of the *IO* plant (10.1). ∎

Comment 206 *Theorem 193 - Theorem 205 prove that the necessary and sufficient conditions are the same for:*
-       *The perfect trackability,*
-       *The perfect natural trackability,*
-       *The global complete trackability,*
-       *The $(\nu - 1)$th order global elementwise trackability,*
-       *The $(\nu - 1)$th order global natural trackability,*
-       *The $(\nu - 1)$th order global elementwise natural trackability on $\mathfrak{D}^{\eta} \times \mathfrak{Y}_d^{\nu}$ of the IO plant (10.1).*

Comment 207 *Theorem 193–Theorem 205 prove that the any trackability property guarantees the output controllability of the IO plant (10.1) (see also the book [171]).*

Summary 208 *Trackability conditions, IO plant properties and external actions*

   The above obtained results discover the following very important IO plant (10.1) characteristics:

   a) The trackability properties of the IO plant (10.1):

   - Are independent of the external (control and disturbance) actions on the plant,

   - Depend on the plant parameters, more precisely, depend on the rank of the plant transfer function matrix relative to control, or equivalently, depend on the rank of the extended matrix $B^{(\mu)}$,

   b) The trackability conditions are algebraic and simple,

   c) The identity of the trackability conditions for all trackability properties shows that the conditions are independent of the chosen trackability type. The conditions express that they depend exclusively on the plant control characteristic that is formulated in the conditions 1) and 2) of all above theorems, i.e.,

$$rank G_{IOU}(s) = rank B^{(\mu)} = N \leq r.$$

d) The trackability conditions are independent of the stability property of the plant that can be either stable, critically stable or unstable and simultaneously can be trackable,

   e) The necessary and sufficient condition $rank B^{(\mu)} = N \leq r$ is common to both the controllability and trackability and it agrees with the Fundamental control principle (Axiom 104),

   f) The plant trackability is sufficient (but not necessary) for its output controllability. The output controllability of the IO plant (10.1) is necessary, but not sufficient, for its trackability.

   g) The common trackability condition $rank B^{(\mu)} = N \leq r$ agrees with Fundamental control principle (Axiom 104).

   The designer of the IO plant (10.1) should guarantee its trackability that will ensure simultaneously its controllability.

10.2 *ISO* system trackability

This section continues, refines and generalizes the results of the book [188, Section 9.3, pp. 191-196, Section 9.4, pp. 201-205].

10.2.1 Perfect trackability criteria

The characteristics of the *ISO* plant are presented in Section 3.1:

- In the *time* domain (Subsection 3.1.1),
- In the complex domain (Subsection 3.1.2).

Equation (3.1) and Equation (3.2) determine the *ISO* plant *time* domain description in terms of the total coordinates (Section 3.1). Equation (3.10) and Equation (3.11), respectively, or, equivalently, Equation (3.12) and Equation(3.13), respectively, specify their Laplace transforms (in the same section).

Equations (3.23)–(3.30) permit us to set Equation (3.13) in a compact form in terms of the plant transfer function matrices:

$$\mathbf{Y}(s) = G_{ISOD}(s)\, \mathbf{D}(s) + G_{ISOU}(s)\, \mathbf{U}(s) + G_{ISOX_0}(s)\, \mathbf{X}_0. \qquad (10.34)$$

In order to discover conditions for any trackability property of the *ISO* plant (3.1), (3.2) we should show the existence of the control that obeys the definition of the corresponding trackability property. The following result offers the solution in general.

We recall the characteristic polynomial $p_{ISO}(s)$, Equation (3.22), of the matrix A, the numerator matrix polynomial $L_{ISO}(s)$, Equation (3.25)–Equation (3.28) and its generic matrix L_{ISO}, Equation (3.29), of $G_{ISOU}(s)$ (3.24),

$$G_{ISOU}(s) = C(sI - A)^{-1} B + U, \qquad (10.35)$$

all from Section 3.1.

Theorem 194 (Subsection 10.1.1) and Theorem 195 (Subsection 10.1.1) are crucial for the proofs of the next theorems. We repeat the former jointly accommodated to the *ISO* plant:

Theorem 209 *Full rank of the transfer function matrix relative to the control*

The full rank ρ of the transfer function matrix $G_{ISOU}(s)$ (10.35) is also the full rank of both its polynomial matrix $L_{ISO}(s)$ (3.26) and its generating matrix L_{ISO} (3.29),

$$full\ rank G_{ISOU}(s) = full\ rank\left[p_{ISO}^{-1}(s)\, L_{ISO}(s)\right] =$$
$$= full\ rank L_{ISO}(s) = \rho = \min(N, r) \Longrightarrow$$
$$rank L_{ISO} = full\ rank L_{ISO} = \rho = \min(N, r). \qquad (10.36)$$

However, the full rank of L_{ISO} does not necessarily implies either the full rank of $L_{ISO}(s)$ or the full rank of $G_{ISOU}(s)$.

The full rank of L_{ISO} is the necessary, but not sufficient, condition for the full rank of $L_{ISO}(s)$ and for the full rank of $G_{ISOU}(s)$.

We wish to investigate which properties of the ISO plant enable the existence of a control that can force the plant to exhibit perfect tracking as soon as the initial real output vector is equal to the initial desired output vector, i.e., which properties of the ISO plant guarantee its perfect trackability?

Definition 161 (Section 9.4) determines the l-th order perfect trackability of the plant on $\mathfrak{D}^1 \times \mathfrak{Y}_d^1$. Lemma 162 (Section 9) reduces the study of the l-th order perfect trackability of the plant on $\mathfrak{D}^1 \times \mathfrak{Y}_d^1$ to the study of the perfect trackability of the plant on $\mathfrak{D}^1 \times \mathfrak{Y}_d^1$.

The following theorem extends and generalizes Theorem 285 of [188, p. 193].

Theorem 210 *Conditions for the perfect trackability of the ISO plant (3.1), (3.2) on $\mathfrak{D}^1 \times \mathfrak{Y}_d^1$*

In order for the ISO plant (3.1), (3.2) to be perfect trackable on $\mathfrak{D}^1 \times \mathfrak{Y}_d^1$ it is necessary and sufficient that there is $s^* \in \mathbb{C}$ such that the plant transfer function matrix $G_{ISOU}(s)$ relative to control has the full row rank N for $s = s^*$, $rankG_{ISOU}(s^*) = N$, so that the dimension r of the control vector $\mathbf{U}$, which is then the nominal control $\mathbf{U}_N$, is not less than the dimension N of the output vector $\mathbf{Y}$, i.e.,

$$\exists s^* \in \mathbb{C} \Longrightarrow rankG_{ISOU}(s^*) = full \ rankG_{ISOU}(s^*) =$$
$$= full \ rankL_{ISO}(s^*) = N \leq r. \tag{10.37}$$

The Laplace transform $\mathbf{U}_N(s)$ of the nominal control $\mathbf{U}_N(t)$ relative to the pair $(\mathbf{D}(.), \mathbf{Y}_d(.))$ is determined by

$$\mathbf{U}_N(s) = G_{ISOU}^T(s) \left\{ G_{ISOU}(s)G_{ISOU}^T(s) \right\}^{-1} \bullet$$
$$\bullet \left\{ \mathbf{Y}_d(s) - G_{ISOD}(s)\mathbf{D}(s) - G_{ISOX_o}(s)\mathbf{X}_{d0} \right\}. \tag{10.38}$$

Proof. *Necessity.* Let the ISO plant (3.1), (3.2) be perfect trackable on $\mathfrak{D}^1 \times \mathfrak{Y}_d^1$. Definition 161, (Section 9.4), is satisfied. There exists control vector $\mathbf{U}(t)$ such that its Laplace transform obeys Equation (10.34) that has the following equivalent form for $\mathbf{Y}(t) \equiv \mathbf{Y}_d(t)$, i.e., for $\mathbf{Y}(s) \equiv \mathbf{Y}_d(s)$ that implies $\mathbf{U}(s) = \mathbf{U}_N(s)$, and for $\mathbf{Y}_0 = \mathbf{Y}_{d0}$ due to Definition 161 and Definition 60 (Section 3.2), which implies $\mathbf{X}_0 = \mathbf{X}_{d0}$:

$$\mathbf{Y}_d(s) = G_{ISOD}(s)\mathbf{D}(s) + G_{ISOU}(s)\mathbf{U}(s) + G_{ISOX_0}(s)\mathbf{X}_{d0}. \tag{10.39}$$

The existence of the solution $\mathbf{U}(s)$ of this equation implies

$$rankG_{ISOU}(s) = N \leq r.$$

Since $N \leq r$ then N is the full rank of $G_{ISOU}(s)$,

$$rank G_{ISOU}(s) = full\ rank G_{ISOU}(s) = N \leq r$$

This, the statement under 1) of Theorem 194 and Theorem 209 prove the necessity of (10.37) that guarantees the nonsingularity of $G_{ISOU}(s)G_{ISOU}^T(s)$ so that the solution $\mathbf{U}(s) = \mathbf{U}_N(s)$ of Equation (10.39) is given by Equation (10.38).

   Sufficiency. Let the conditions (10.37) hold so that $G_{ISOU}(s)$ has the full rank (due to the statement under 1) of Theorem 194 and Theorem 209. The full rank of $G_{ISOU}(s)$ ensures that $\Gamma_{ISOU}(s)$ introduced by

$$\Gamma_{ISOU}(s) = G_{ISOU}^T(s)\left[G_{ISOU}(s)G_{ISOU}^T(s)\right]^{-1}. \tag{10.40}$$

is fully defined as well as $\mathbf{U}(s)$ determined by

$$\mathbf{U}(s) = \Gamma_{ISOU}(s)\left\{\mathbf{Y}_d(s) - G_{ISOD}(s)\mathbf{D}(s) - G_{ISOX_0}(s)\mathbf{X}_{d0}\right\}. \tag{10.41}$$

We eliminate $\mathbf{U}(s)$ from Equation (10.34) by replacing it by the right-hand side of Equation (10.41) and by exploiting $G_{ISOU}(s)\Gamma_{ISOU}(s) = I_N$ and $\mathbf{X}_0 = \mathbf{X}_{d0}$ due to Definition 161 and Definition 60:

$$\mathbf{Y}(s) = G_{ISOD}(s)\mathbf{D}(s) + G_{ISOU}(s)\mathbf{U}(s) + G_{ISOX_0}(s)\mathbf{X}_{d0} =$$
$$= G_{ISOD}(s)\mathbf{D}(s) + \mathbf{Y}_d(s) - G_{ISOD}(s)\mathbf{D}(s) - G_{ISOX_0}(s)\mathbf{X}_{d0}+$$
$$+ G_{ISOX_0}(s)\mathbf{X}_{d0} = \mathbf{Y}_d(s),$$

which in the *time* domain reads

$$\mathbf{Y}(t) = \mathbf{Y}_d(t), \ \forall[t, D(.), \mathbf{Y}_d(.)] \in \mathfrak{T}_0 \times \mathfrak{D}^\mu \times \mathfrak{Y}_d^\nu.$$

This proves the perfect trackability of the *ISO* plant (3.1), (3.2) on $\mathfrak{D}^1 \times \mathfrak{Y}_d^1$ in view of Definition 161 and that the control defined by Equation (10.38) is the nominal control *relative to the functional pair* $[\mathbf{D}(.), \mathbf{Y}_d(.)]$ (Definition 60). ∎

Note 211 *Notice that the condition (10.37) of Theorem 210 is the necessary and sufficient for the output function controllability [7, p. 313]; [48, p. 216, Theorem 5-23]; [312, p. 164, Theorem 5.5.7], which assumes* $\mathbf{D}(t) \equiv \mathbf{0}_d$.

Comment 212 *[188, Comment 286, p. 193] Equation (10.38) determines the nominal control in terms of the disturbance vector. This means that the disturbance vector should be measurable, which is rarely satisfied.*

10.2.2 Conditions for perfect natural trackability

Definition 310 (Section 9.4) determines the perfect natural trackability of the *ISO* plant (3.1), (3.2) on $\mathfrak{D}^1 \times \mathfrak{Y}_d^1$.

Let us refine and extend Theorem 287 of [188, p. 194] as follows:

Theorem 213 *Conditions for the perfect natural trackability of the ISO plant (3.1), (3.2) on $\mathfrak{D}^1 \times \mathfrak{Y}_d^1$*

In order for the ISO plant (3.1), (3.2) to be perfect natural trackable on $\mathfrak{D}^1 \times \mathfrak{Y}_d^1$ it is necessary and sufficient that the plant transfer function matrix $G_{ISOU}(s)$ relative to control has the full row rank N, so that the dimension r of the control vector $\mathbf{U}$ is not less than the dimension N of the output vector $\mathbf{Y}$, i.e., that any of the conditions (10.37) holds.

Proof. *Necessity.* Let the *ISO* plant (3.1), (3.2) be perfect natural trackable on $\mathfrak{D}^1 \times \mathfrak{Y}_d^1$. Definitions 310 and 161, (Section 9.4), guarantee that the plant is also perfect trackable on $\mathfrak{D}^1 \times \mathfrak{Y}_d^1$. Theorem 210 implies the necessity of the conditions (10.37).

Sufficiency. Let the conditions (10.37) be fulfilled. The transfer function matrix $G_{ISOU}(s)$ has the full rank N (due to the statement under 1) of Theorem 194 and Theorem 209) and $\Gamma_{ISOU}(s)$ (10.40) is well defined. The *ISO* plant (3.1), (3.2) is perfect trackable on $\mathfrak{D}^1 \times \mathfrak{Y}_d^1$, Theorem 210. We should show that the perfect tracking control can be synthesized without using information about the plant state and about the disturbance $\mathbf{D}(.) \in \mathfrak{D}^1$, i.e., that it is natural tracking control (Definition 166, Section 9.4). We recall Equation (10.34). Let $\sigma \in \mathfrak{R}^+$ be arbitrarily small, i.e., $\sigma \longrightarrow 0^+$ and in the ideal case $\sigma = 0^+$. Let

$$\phi(.) : \mathfrak{T} \longrightarrow \mathfrak{R}^N, \ \phi(t) \in \mathfrak{C}(\mathfrak{T}),$$
$$\phi(t) = \mathbf{0}_N, \ \forall t \in [\sigma, \infty[, \ \phi(0) = -\mathbf{e}(0), \tag{10.42}$$

Let Equation (10.40) define $\Gamma_{ISOU}(s)$ and the control be governed by:

$$\mathbf{U}(s) = (1 - e^{-\sigma s})^{-1} \Gamma_{ISOU}(s) [\mathbf{\Phi}(s) + \mathbf{E}(s)], \ \mathbf{\Phi}(s) = \mathcal{L}\{\phi(t)\}. \tag{10.43}$$

The control $\mathbf{U}(.)$ (10.43) is independent of the plant internal dynamics, i.e., of its state, and of the disturbance $\mathbf{D}(.)$. The control $\mathbf{U}(.)$ (10.43) is natural control (Definition 166, Section 9.4). We multiply Equation (10.43) on the left by $G_{ISOU}(s)$ and use $G_{ISOU}(s)\Gamma_{ISOU}(s) \equiv I_N$:

$$G_{ISOU}(s)\mathbf{U}(s) = (1 - e^{-\sigma s})^{-1} [\mathbf{\Phi}(s) + \mathbf{E}(s)] \tag{10.44}$$

Let us replace $G_{ISOU}(s)\mathbf{U}(s)$ with the right-hand side of (10.44) into Equation (10.34):

$$\mathbf{Y}(s) = G_{ISOD}(s)\,\mathbf{D}(s) + G_{ISOU}(s)\,\mathbf{U}(s) + G_{ISOX_0}(s)\,\mathbf{X}_0 =$$
$$= G_{ISOD}(s)\,\mathbf{D}(s)\, + (1 - e^{-\sigma s})^{-1}\,[\Phi(s) + \mathbf{E}(s)] + G_{ISOX_0}(s)\,\mathbf{X}_0 \Longrightarrow$$

$$(1 - e^{-\sigma s})\,\{\mathbf{Y}(s) - G_{ISOD}(s)\,\mathbf{D}(s)\, - G_{ISOX_0}(s)\,\mathbf{X}_0\} = \Phi(s) + \mathbf{E}(s)\,.$$

For $\sigma \longrightarrow 0^+$, or for $\sigma = 0^+$ in the ideal case, the preceding equation reduces to

$$\mathbf{0}_N = \Phi(s) + \mathbf{E}(s)\,.$$

This equation yields in the *time* domain:

$$\mathbf{e}(t) = -\phi(t) = \mathbf{0}_N,\ \forall t \in \mathfrak{T}_0$$

due to (10.42) and $\mathbf{Y}_0 = \mathbf{Y}_{d0}$ due to Definition 161 and Definition 60, or equivalently:

$$\mathbf{e}(t) = \mathbf{Y}_d(t) - \mathbf{Y}(t) = \mathbf{0}_N,\ \forall t \in \mathfrak{T}_0.$$

The natural control $\mathbf{U}(.)$ defined by (10.43) satisfies Definition 310. The *ISO* plant (3.1), (3.2) is perfect natural trackable on $\mathfrak{D}^1 \times \mathfrak{Y}_d^1$. ∎

Comment 214 *[188, Comment 288, p. 196] Theorem 210 and Theorem 213 show that for the ISO plant (3.1), (3.2) to be perfect natural trackable on $\mathfrak{D}^1 \times \mathfrak{Y}_d^1$ it is necessary and sufficient that the plant is perfect trackable on $\mathfrak{D}^1 \times \mathfrak{Y}_d^1$. This confirms Comment 168 (Section 9.4).*

Comment 215 *[188, Comment 289, p. 196] The proof of Theorem 213 shows that the control implementation does not need any information about the disturbance vector. This means that the disturbance vector need not be measurable, which corresponds to the reality.*

Note 201, Section 10.1, holds in this settings if $\Gamma_{ISOU}(s)$ replaces $\Gamma_{IOU}(s)$.

10.2.3 Imperfect trackability criteria

The necessary and sufficient conditions for various imperfect trackability properties of the *ISO* plant (3.1), (3.2) represent the topic of what follows.

We refine and generalize Theorem 293 of [188, pp. 197, 198] as follows:

Theorem 216 *Conditions for the global complete trackability of the ISO plant (3.1), (3.2) on $\mathfrak{D}^1 \times \mathfrak{Y}_d^1$*

In order for the ISO plant (3.1), (3.2) to be global complete trackable on $\mathfrak{D}^1 \times \mathfrak{Y}_d^1$ it is necessary and sufficient that the plant transfer function matrix $G_{ISOU}(s)$ relative to control has the full row rank N, so that the dimension r of the control vector $\mathbf{U}$ is not less than the dimension N of the output vector $\mathbf{Y}$, i.e., that all conditions (10.37) hold.

Proof. *Necessity.* Let the ISO plant (3.1), (3.2) be global complete trackable on $\mathfrak{D}^1 \times \mathfrak{Y}_d^1$. Definition 172 (Section 9.5) is fulfilled. The plant is also perfect trackable on $\mathfrak{D}^1 \times \mathfrak{Y}_d^1$ (due to Definition 172 and Theorem 174, Section 9.5). Theorem 210 proves the necessity of the conditions (10.37) for the global complete trackability on $\mathfrak{D}^1 \times \mathfrak{Y}_d^1$.

Sufficiency. Let the conditions (10.37) hold, which ensure that $G_{ISOU}(s)$ has the full rank (due to the statement under 1) of Theorem 194 and Theorem 209). Let $[\mathbf{D}(.), \mathbf{Y}_d(.)] \in \mathfrak{D}^1 \times \mathfrak{Y}_d^1$, the instant $\sigma \in Int \; \mathfrak{T}_0$, and $\mathbf{Y}_0^{\nu-1} \in \mathfrak{R}^{\nu N}$ be arbitrarily chosen (see Definition 172). Let the control be defined in the complex domain by

$$\mathbf{U}(s) = \Gamma_{ISOU}(s) \begin{bmatrix} \mathbf{Y}(s) - G_{ISOD}(s)\mathbf{D}(s) - \\ -\mathbf{Z}(s) + |\mathbf{E}(s)| - G_{ISOX_0}(s)\mathbf{X}_0 \end{bmatrix},$$

together with

$$\mathbf{Z}(t) = \left\{ \begin{array}{c} |\mathbf{e}(0)| \left(1 - \frac{t}{\sigma}\right), \; t \in [0, \sigma], \\ \mathbf{0}_N, \; \forall (t \geq \sigma) \in \mathfrak{T}_0 \end{array} \right\} \in \mathfrak{C}(\mathfrak{T}_0),$$

$$\mathbf{Z}(s) = \mathcal{L}\{\mathbf{Z}(t)\}. \tag{10.45}$$

For such control Equation (10.34) becomes

$$\mathbf{Y}(s) = G_{ISOD}(s)\mathbf{D}(s) + \mathbf{Y}(s) - G_{ISOD}(s)\mathbf{D}(s) - \\ -\mathbf{Z}(s) + |\mathbf{E}(s)| - G_{ISOX_0}(s)\mathbf{X}_0 + G_{ISOX_0}(s)\mathbf{X}_0 = \\ = \mathbf{Y}(s) - \mathbf{Z}(s) + |\mathbf{E}(s)|$$

due to $G_{ISOU}(s)\Gamma_{ISOU}(s) \equiv I_N$, i.e.,

$$-\mathbf{Z}(s) + |\mathbf{E}(s)| = \mathbf{0}_N \Longrightarrow |\mathbf{e}(t)| = \mathbf{Z}(t), \; \forall t \in \mathfrak{T}_0.$$

This and (10.45) imply

$$\mathbf{e}(t) = \mathbf{0}_N, \; \forall (t \geq \sigma) \in \mathfrak{T}_0,$$

i.e.,

$$\mathbf{Y}(t) = \mathbf{Y}_d(t), \ \forall (t \geq \sigma) \in \mathfrak{T}_0,$$

which proves the global complete trackability on $\mathfrak{D}^1 \times \mathfrak{Y}_d^1$ of the *ISO* plant (3.1), (3.2) due to Definition 172. ∎

Theorem 216 continues, refines and generalizes Theorem 298 of [188, pp. 201, 202], while the following theorem does the same with Theorem 300 of [188, pp. 202-204].

Theorem 217 *Conditions for the first order global elementwise trackability of the ISO plant (3.1), (3.2) on $\mathfrak{D}^1 \times \mathfrak{Y}_d^1$*

In order for the ISO plant (3.1), (3.2) to be the first order global elementwise trackable on $\mathfrak{D}^1 \times \mathfrak{Y}_d^1$ it is necessary and sufficient that the plant transfer function matrix $G_{ISOU}(s)$ relative to control has the full row rank N so that the dimension r of the control vector $\mathbf{U}$ is not less than the dimension N of the output vector $\mathbf{Y}$, i.e., that any of the conditions (10.37) holds.

Proof. *Necessity.* Let the *ISO* plant (3.1), (3.2) be the first order global elementwise trackable on $\mathfrak{D}^1 \times \mathfrak{Y}_d^1$. Definition 313 (Section 9.5) is fulfilled. The plant and control satisfy also Definition 172. It and Theorem 174 prove the necessity of the conditions (10.37).

Sufficiency. Let the conditions (10.37) be satisfied. They guarantee that $G_{ISOU}(s)$ has the full rank (due to the statement under 1) of Theorem 194 and Theorem 209). Let $[\mathbf{D}(.), \mathbf{Y}_d(.)] \in \mathfrak{D}^1 \times \mathfrak{Y}_d^1$, $\sigma \in Int\ \mathfrak{T}_0 \times Int\mathfrak{T}_0$, $\sigma = [\sigma_1 \ \sigma_2]^T$, and $\mathbf{Y}_0^{\nu-1} \in \mathfrak{R}^{\nu N}$ be arbitrarily chosen (in view of Definition 313). We select the control vector $\mathbf{U}(t)$ to satisfy the following complex domain equation:

$$\mathbf{U}(s) = \Gamma_{ISOU}(s) \left[\begin{array}{c} \mathbf{Y}(s) - G_{ISOD}(s)\mathbf{D}(s)- \\ -\mathbf{Z}(s) + \mathcal{L}\left\{\sum_{j=0}^{j=1} \left|\mathbf{e}^{(j)}(t)\right|\right\} - G_{ISOX_0}(s)\mathbf{X}_0 \end{array} \right]$$

$$(10.46)$$

for

$$0 < \sigma_M = \max(\sigma_1, \sigma_2) <<< 0^+, \ i.e., \ \sigma_M \longrightarrow 0^+, \tag{10.47}$$

and for the Laplace transform $\mathbf{Z}(s)$ of $\mathbf{Z}(t)$ determined by

$$\mathbf{Z}(t) = \left\{ \begin{array}{c} \sum_{j=0}^{j=1} \left(\left|\mathbf{e}^{(j)}(0)\right| \left(1 - \frac{t}{\sigma_m}\right), \ \forall t \in [0, \sigma_m]\right), \\ \mathbf{0}_N, \ \forall (t \geq \sigma_m) \in \mathfrak{T}_0 \end{array} \right\} \in \mathfrak{C}(\mathfrak{T}_0),$$

$$0 < \sigma_m = \min(\sigma_1, \sigma_2) \leq \sigma_M, \ \mathbf{Z}(s) = \mathcal{L}\{\mathbf{Z}(t)\}. \tag{10.48}$$

For the control defined in Equation (10.46) the plant complex domain mathematical model (10.34), becomes, in view of $G_{ISOU}(s)\Gamma_{ISOU}(s) \equiv I_N$,

$$\mathbf{Y}(s) = G_{ISOD}(s)\,\mathbf{D}(s) + \mathbf{Y}(s) - G_{ISOD}(s)\mathbf{D}(s) - \mathbf{Z}(s) +$$

$$+\mathcal{L}\left\{ \sum_{j=0}^{j=1} \left| \mathbf{e}^{(j)}(t) \right| \right\} - G_{ISOX_0}(s)\,\mathbf{X}_0 + G_{ISOX_0}(s)\,\dot{\mathbf{X}}_0 \Longrightarrow$$

$$\mathbf{0}_N = . - \mathbf{Z}(s) + \mathcal{L}\left\{ \sum_{j=0}^{j=1} \left| \mathbf{e}^{(j)}(t) \right| \right\},$$

or in the *time* domain :

$$\sum_{j=0}^{j=1} \left| \mathbf{e}^{(j)}(t) \right| = \mathbf{Z}(t) = \mathbf{0}_N, \ \forall (t \geq \sigma_j) \in \mathfrak{T}_0, \ \forall j = 0, 1,$$

i.e.,

$$\mathbf{Y}^{(j)}(t) = \mathbf{Y}_d^{(j)}(t), \ \forall (t \geq \sigma_j) \in \mathfrak{T}_0, \ \forall j = 0, 1,$$

which proves the first order global elementwise trackability of the *ISO* plant (3.1), (3.2) on $\mathfrak{D}^\eta \times \mathfrak{Y}_d^\nu$ (Definition 313). ∎

The control algorithm (10.46) demands the full information about the disturbance vector $\mathbf{D}(t)$.

10.2.4 Conditions for imperfect natural trackability

A slight generalization of Theorem 302 [188, pp. 204, 205] reads:

Theorem 218 *Conditions for global complete natural trackability of the ISO plant (3.1), (3.2) on $\mathfrak{D}^1 \times \mathfrak{Y}_d^1$*

In order for the ISO plant (3.1), (3.2) to be global complete natural trackable on $\mathfrak{D}^1 \times \mathfrak{Y}_d^1$ it is necessary and sufficient that the plant transfer function matrix $G_{ISOU}(s)$ relative to control has the full row rank N so that the dimension r of the control vector $\mathbf{U}$ is not less than the dimension N of the output vector $\mathbf{Y}$, i.e., that any of the conditions (10.37) holds.

Proof. *Necessity.* Let the *ISO* plant (3.1), (3.2) be global complete natural trackable on $\mathfrak{D}^1 \times \mathfrak{Y}_d^1$. Definition 312 is applicable. It and Definition 172 confirm that the plant is complete trackable on $\mathfrak{D}^1 \times \mathfrak{Y}_d^1$ so that the conditions (10.37) are necessary due to Theorem 216.

Sufficiency. Let the conditions (10.37) be valid. The *ISO* plant (3.1), (3.2) is global complete trackable on $\mathfrak{D}^1 \times \mathfrak{Y}_d^1$ (Theorem 216). Let the Laplace transform of the control vector be defined by Equations (10.42)–(10.43). By repeating from Equation (10.43) the sufficiency part of the proof of Theorem 213 we prove that the so defined natural tracking control implies

$$\mathbf{e}(t) = \mathbf{Y}_d(t) - \mathbf{Y}(t) = \mathbf{0}_N, \ \forall (t \geq \sigma) \in \mathfrak{T}_0.$$

This proves the global complete natural trackability of the *ISO* plant (3.1), (3.2) on $\mathfrak{D}^1 \times \mathfrak{Y}_d^1$ in view of Definition 312. ∎

A further refinement and generalization of Theorem 302 [188, pp. 204, 205] follows:

Theorem 219 *Conditions for the first order global elementwise natural trackability of the ISO plant (3.1), (3.2) on $\mathfrak{D}^1 \times \mathfrak{Y}_d^1$*

In order for the ISO plant (3.1), (3.2) to be first order global elementwise natural trackable on $\mathfrak{D}^1 \times \mathfrak{Y}_d^1$ it is necessary and sufficient that the plant transfer function matrix $G_{ISOU}(s)$ relative to control has the full row rank N so that the dimension r of the control vector $\mathbf{U}$ is not less than the dimension N of the output vector $\mathbf{Y}$, i.e., that any of the conditions (10.37) holds.

Proof. *Necessity.* Let the *ISO* plant (3.1), (3.2) be the first order global elementwise natural trackable on $\mathfrak{D}^1 \times \mathfrak{Y}_d^1$. Definition 314 (Section 9.5) is fulfilled. The plant and control satisfy also Definition 172. It and Theorem 174 (Section 9.5) prove the necessity of the conditions (10.37).

Sufficiency. Let the conditions (10.37) be satisfied so that $G_{ISOU}(s)$ has the full rank (due to the statement under 1) of Theorem 194 and Theorem 209). Let $[\mathbf{D}(.), \mathbf{Y}_d(.)] \in \mathfrak{D}^1 \times \mathfrak{Y}_d^1$, $\sigma \in (Int \ \mathfrak{T}_0)^2$, $\sigma = [\sigma_1 \ \sigma_2]^T$, and $\mathbf{Y}_0^{\nu-1} \in \mathfrak{R}^{\nu N}$ be arbitrarily chosen (in view of Definition 314). We select the control vector $\mathbf{U}(t)$ to satisfy the following complex domain equation:

$$G_{ISOU}(s)\mathbf{U}(s) = e^{-\sigma_M s}G_{ISOU}(s)\mathbf{U}(s) + \mathcal{L}\left\{\sum_{j=0}^{j=1}\left|\mathbf{e}^{(j)}(t)\right|\right\} - \mathbf{Z}(s), \quad (10.49)$$

for σ_M defined by (10.47) and for the Laplace transform $\mathbf{Z}(s)$ of $\mathbf{Z}(t)$ determined by (10.48). For such control the plant complex domain mathematical model (10.34), becomes

$$\mathbf{Y}(s) = G_{ISOD}(s)\mathbf{D}(s) + G_{ISOU}(s)\left[e^{-\sigma_M s}\mathbf{U}(s)\right] +$$

$$+ \mathcal{L}\left\{\sum_{j=0}^{j=1}\left|\mathbf{e}^{(j)}(t)\right|\right\} - \mathbf{Z}(s) + G_{ISOX_0}(s)\mathbf{X}_0. \quad (10.50)$$

Let us subtract Equation (10.50) from Equation (10.34):

$$\mathbf{0}_N = .G_{ISOU}\left(s\right)\left(1 - e^{-\sigma_M s}\right)\mathbf{U}(s) + \mathbf{Z}(s) - \mathcal{L}\left\{\sum_{j=0}^{j=\nu-1}\left|\mathbf{e}^{(j)}(t)\right|\right\}.$$

In the limes as $\sigma_M \longrightarrow 0^+$ the preceding equation reduces to:

$$\mathbf{0}_N = \mathbf{Z}(s) - \mathcal{L}\left\{\sum_{j=0}^{j=1}\left|\mathbf{e}^{(j)}(t)\right|\right\},$$

or in the *time* domain :

$$\sum_{j=0}^{j=1}\left|\mathbf{e}^{(j)}(t)\right| = \mathbf{Z}(t) = \mathbf{0}_N, \ \forall(t \geq \sigma_m) \in \mathfrak{T}_0, \ \sigma_m \leq \sigma_j, \ \forall j = 0, 1,$$

i.e.,

$$\mathbf{Y}^{(j)}(t) = \mathbf{Y}_d^{(j)}(t), \ \forall(t \geq \sigma_j) \in \mathfrak{T}_0, \ \forall j = 0, 1,$$

which proves the first order global elementwise natural trackability of the *ISO* plant (3.1), (3.2) on $\mathfrak{D}^\eta \times \mathfrak{Y}_d^\nu$ (Definition 314). ∎

Summary 220 *Trackability conditions, ISO plant properties and external actions*

The trackability properties of the ISO plant (3.1), (3.2):

- Are independent of the external (control and disturbance) actions on the plant,

- Depend on the plant parameters, i.e., on the rank of the corresponding plant transfer function matrix relative to control, or equivalently, depend on the rank of the polynomial matrix $L_{ISO}(s)$.

Besides, the following statements hold:

- The trackability conditions are purely algebraic and simple for the implementation,

- The identity of the trackability conditions for all trackability properties discovers that the conditions do not depend on the selected trackability type. The conditions reflect that they depend only on the plant control characteristic that is formulated in the conditions (10.37) of all above theorems, i.e.,

$$rankG_{ISOU}(s) = full \ rankG_{ISOU}(s) = full \ rankL_{ISO}(s) = N \leq r.$$

- *The trackability conditions do not depend on the stability property of the plant or on the treated trackability property or on disturbance,*

- *The trackability conditions are independent of the controllability conditions,*

- *The characteristics of the plant part that transmits the control action on the plant processing part, i.e., on the plant process, are those plant characteristics that determine whether the plant is trackable or not; such plant characteristics are mathematically expressed by the plant transfer function matrix relative to control, or equivalently by the polynomial matrix $L_{ISO}(s)$,*

- *The trackability conditions guarantee the output controllability,*

- *The common trackability conditions (10.37) agree with the Fundamental control principle (Axiom 104).*

The designer of the ISO plant (3.1), (3.2) should ensure its trackability.

10.3   *EISO* system trackability

The book [188, Chapter 9, pp. 169-205] contains the initial investigation of the trackability of the linear *time*-invariant continuous-*time IO* and *ISO* dynamical plants. The book [175, Part III, p. 59-97] presents the trackability theory in the general framework of the *time*-varying continuous-*time* nonlinear systems. The study of the trackability related specifically to the linear *time*-invariant continuous-*time EISO* plants starts in this section.

10.3.1   Perfect trackability criteria

Section 4.1 determines the characteristics of the *EISO* plant:

- In the *time* domain by the state Equation (4.1), and by the output Equation (4.2),

- And in the complex domain by their Laplace transforms, Equation (4.22) and Equation (4.23).

Equations (4.32), (4.35)–(4.44) simplify the form of Equation (4.23), (Section 4.1) to:

$$\mathbf{Y}(s) = G_D(s)\,\mathbf{D}(s) + G_U(s)\,\mathbf{U}(s) + G_{D_0}(s)\,\mathbf{D}_0^{\mu-1} +$$
$$+ G_{U_0}(s)\,\mathbf{U}_0^{\mu-1} + G_{X_0}(s)\,\mathbf{X}_0. \qquad (10.51)$$

Theorem 194 (Subsection 10.1.1) and Theorem 195 (Subsection 10.1.1) are inherent for the proofs of the results on the trackability of the *EISO* plant (4.1), (4.2). We state them jointly in this framework:

Theorem 221 *Full rank of the transfer function matrix relative to the control*

a) The full rank ρ of the transfer function matrix $G_U(s)$ (4.36), (4.50) is also the full rank of its numerator matrix polynomial $L_U(s)$ (4.37), of the transfer function matrix $G_{U^\mu}(s)$ (4.50), of its numerator matrix polynomial $N_U(s)$ (4.54) and implies the full ranks of their generic matrices L_U (4.41) and N_U (4.57),

$$full\ rank G_U(s) = full\ rank\left\{ C(sI - A)^{-1} B^{(\mu)} S_r^{(\mu)}(s) + U \right\} =$$
$$= full\ rank\left\{ C(sI - A)^{-1} B^{(\mu)} S_r^{(\mu)}(s) + U \right\} =$$
$$= full\ rank L_U(s) = \rho = \min(N, r) \Longrightarrow$$
$$rank L_U = full\ rank L_U = \rho = \min(N, r), \tag{10.52}$$

$$full\ rank G_U(s) = full\ rank G_{U\mu}(s) =$$
$$= full\ rank\left\{ C(sI - A)^{-1} B^{(\mu)} + U^{(\mu)} \right\} =$$
$$= full\ rank\left[p^{-1}(s) N_U(s) \right] = full\ rank N_U(s) \Longrightarrow$$
$$rank N_U = full\ rank N_U = \rho = \min(N, r). \tag{10.53}$$

b) The full rank of L_U is necessary, but not sufficient, for the full rank either of $L_U(s)$ or of $G_U(s)$.

The full rank of N_U is necessary, but not sufficient, for the full rank either of $N_U(s)$ or of $G_{U\mu}(s)$.

10.3.2 Conditions for perfect trackability

The general Definition 161 (Section 9.4) specifies the *l-th* order perfect trackability of the plant on $\mathfrak{D}^k \times \mathfrak{Y}_d^k$. In view of Lemma 162 (Section 9.4), which simplifies the study of the *l-th* order perfect trackability of the plant on $\mathfrak{D}^k \times \mathfrak{Y}_d^k$ to the study of the perfect trackability of the plant on $\mathfrak{D}^k \times \mathfrak{Y}_d^k$, we will explore the conditions for the perfect trackability of the *EISO* plant (4.1), (4.2) on $\mathfrak{D}^k \times \mathfrak{Y}_d^k$.

Theorem 222 *Conditions for the perfect trackability on $\mathfrak{D}^k \times \mathfrak{Y}_d^k$*

In order for the EISO plant (4.1), (4.2) to be perfect trackable on $\mathfrak{D}^k \times \mathfrak{Y}_d^k$ it is necessary and sufficient that there is $s^* \in \mathbb{C}$ such that the plant transfer function matrix $G_U(s)$ relative to control has the full row rank N for $s = s^*$, $rank G_U(s^*) = N$, so that the dimension r of the control vector $\mathbf{U}$, which is *then the nominal control* $\mathbf{U}_N$, *is not less than the dimension* N *of the output vector* $\mathbf{Y}$, *i.e.,*

$$\exists s^* \in \mathbb{C} \Longrightarrow rankG_U(s^*) = rank\left[C(s^*I - A)^{-1}\mathbf{B}^{(\mu)}S_r^{(\mu)}(s^*) + U\right] =$$

$$= rankL_U(s^*) = N \leq r,$$

$$\exists s^* \in \mathbb{C} \Longrightarrow rankG_{U\mu}(s^*) = rank\left[C(s^*I - A)^{-1}\mathbf{B}^{(\mu)} + U^{(\mu)}\right] =$$

$$= rankN_U(s^*) = N \leq r. \tag{10.54}$$

The Laplace transform $\mathbf{U}_N(s)$ *of the nominal control* $\mathbf{U}_N(t)$ *is then determined by*

$$\mathbf{U}_N(s) = G_U(s)^T \left\{G_U(s) G_U(s)^T\right\}^{-1} \bullet$$

$$\bullet \left\{\begin{array}{c} \mathbf{Y}_d(s) - G_D(s)\mathbf{D}_N(s) - G_{D_0}(s)\mathbf{D}_{N0}^{\mu-1} \\ -G_{U_0}(s)\mathbf{U}_{N0}^{\mu-1} - G_{X_0}(s)\mathbf{X}_{d0} \end{array}\right\}. \tag{10.55}$$

Proof. *Necessity.* Let the *EISO* plant (4.1), (4.2) be perfect complete trackable on $\mathfrak{D}^k \times \mathfrak{Y}_d^k$. We use Definition 161, Section 9.4. It specifies that there is control vector $\mathbf{U}(t)$ such that its Laplace transform obeys Equation (10.51) that has the following equivalent form for $\mathbf{Y}(t) \equiv \mathbf{Y}_d(t)$, i.e., for $\mathbf{Y}(s) \equiv \mathbf{Y}_d(s)$ that implies $\mathbf{U}(s) = \mathbf{U}_N(s)$:

$$\mathbf{Y}_d(s) = \left\{\begin{array}{c} G_D(s)\mathbf{D}(s) + G_U(s)\mathbf{U}(s)+ \\ +G_{D_0}(s)\mathbf{D}_0^{\mu-1} + G_{U_0}(s)\mathbf{U}_0^{\mu-1}+ \\ +G_{X_0}(s)\mathbf{X}_{d0} \end{array}\right\}. \tag{10.56}$$

The existence of the solution $\mathbf{U}(s)$ of this equation implies

$$rankG_U(s) = full \; rankG_U(s) = N \leq r.$$

This, Theorem 194 (Subsection 10.1.1), and Theorem 221 prove the necessity of the conditions (10.54). They guarantee the nonsingularity of $G_U(s) G_U^T(s)$ so that the solution $\mathbf{U}(s) = \mathbf{U}_N(s)$ of Equation (13.52) is given by Equation (10.55).

Sufficiency. Let the conditions (10.54) hold so that $G_U(s)$ has the full rank N (due to Theorem 194 and Theorem 221) and $\Gamma_{EISOU}(s)$ introduced by

$$\Gamma_{EISOU}(s) = G_U(s) [G_U(s) G_U(s)]^{-1} \tag{10.57}$$

is fully defined as well as $\mathbf{U}(s)$ determined by

$$\mathbf{U}(s) = \Gamma_{EISOU}(s)\left\{\begin{array}{c} \mathbf{Y}_d(s) - G_D(s)\mathbf{D}(s) - G_{D_0}(s)\mathbf{D}_0^{\mu-1} \\ -G_{U_0}(s)\mathbf{U}_0^{\mu-1} - G_{X_0}(s)\mathbf{X}_{d0} \end{array}\right\}. \tag{10.58}$$

We eliminate $\mathbf{U}(s)$ from Equation (10.51) by replacing it by the right hand side of Equation (10.58) and by exploiting $G_U(s)\,\Gamma_{EISOU}(s) \equiv I_N$ and $\mathbf{X}_0 = \mathbf{X}_{d0}$ due $\mathbf{Y}_0 = \mathbf{Y}_{d0}$ (Definition 161):

$$\mathbf{Y}(s) = \left\{ \begin{array}{c} G_D(s)\,\mathbf{D}(s) + G_U(s)\,\mathbf{U}(s)+ \\ +G_{D_0}(s)\,\mathbf{D}_0^{\mu-1} + G_{U_0}(s)\,\mathbf{U}_0^{\mu-1} + G_{X_0}(s)\,\mathbf{X}_{d0} \end{array} \right\} \Longrightarrow$$

$$\mathbf{Y}(s) = G_D(s)\,\mathbf{D}(s) + \left\{ \begin{array}{c} \mathbf{Y}_d(s) - G_D(s)\,\mathbf{D}(s)- \\ -G_{D_0}(s)\,\mathbf{D}_0^{\mu-1} - G_{U_0}(s)\,\mathbf{U}_0^{\mu-1} \\ -G_{X_0}(s)\,\mathbf{X}_{d0} \end{array} \right\} +$$

$$+ \left\{ \begin{array}{c} G_{D_0}(s)\,\mathbf{D}_0^{\mu-1} + G_{U_0}(s)\,\mathbf{U}_0^{\mu-1}+ \\ +G_{X_0}(s)\,\mathbf{X}_0 \end{array} \right\} = \mathbf{Y}_d(s).$$

This result, $\mathbf{Y}(s) = \mathbf{Y}_d(s)$, in the *time* domain reads

$$\mathbf{Y}(t) = \mathbf{Y}_d(t), \ \forall\,[t, D\,(.)\,, \mathbf{Y}_d(.)] \in \mathfrak{T}_0 \times \mathfrak{D}^k \times \mathfrak{Y}_d^k.$$

This proves the perfect trackability of the *EISO* plant (4.1), (4.2) on $\mathfrak{D}^k \times \mathfrak{Y}_d^k$ in view of Definition 161 and that the control $\mathbf{U}(s)$ (10.58) is the plant nominal control relative to the functional vector pair $[\mathbf{D}(.), \mathbf{Y}_d(.)]$ in view of Definition 75, Section 4.2. ■

Comment 223 *Equation (10.55) defines the nominal control $\mathbf{U}_N(t)$ in terms of the disturbance vector $D\,(t)$. This requires the measurability of the disturbance vector, which is rarely possible.*

10.3.3 Conditions for perfect natural trackability

Definition 310 (Section 9.4) determines the perfect natural trackability of the *EISO* plant (4.1), (4.2) on $\mathfrak{D}^k \times \mathfrak{Y}_d^k$.

Theorem 224 *Conditions for the perfect natural trackability of the EISO plant (4.1), (4.2) on $\mathfrak{D}^k \times \mathfrak{Y}_d^k$*
In order for the EISO plant (4.1), (4.2) to be perfect natural trackable on $\mathfrak{D}^k \times \mathfrak{Y}_d^k$ it is necessary and sufficient that the plant transfer function matrix relative to control $G_U(s)$ has the full row rank N, so that the dimension r of the control vector $\mathbf{U}$ is not less than the dimension N of the output vector $\mathbf{Y}$, i.e., that all conditions (10.54) hold.

Proof. *Necessity.* Let the *EISO* plant (4.1), (4.2) be perfect natural trackable on $\mathfrak{D}^k \times \mathfrak{Y}_d^k$. Definitions 310 and 161 guarantee that the plant is also perfect trackable on $\mathfrak{D}^k \times \mathfrak{Y}_d^k$. Theorem 222 implies the necessity of the conditions (10.54).

Sufficiency. Let the conditions (10.54) hold. The system transfer function matrix $G_U(s)$ has the full rank $N \leq r$ (due to Theorem 194 and Theorem 221). The *EISO* plant (4.1), (4.2) is perfect trackable on $\mathfrak{D}^k \times \mathfrak{Y}_d^k$, Theorem 210. We should show that the perfect tracking control can be synthesized without using information about the plant state and about the disturbance $\mathbf{D}(.) \in \mathfrak{D}^k$, i.e., that it is natural tracking control (Definition 166, Section 9.4). We recall Equation (10.51). Let $\sigma \in \mathfrak{R}^+$ be arbitrarily small, i.e., $\sigma \longrightarrow 0^+$ and in the ideal case $\sigma = 0^+$. Let

$$\phi(.) : \mathfrak{T}_0 \longrightarrow \mathfrak{R}^N, \ \phi(t) \in \mathfrak{C},$$
$$\phi(t) = \mathbf{0}_N, \ \forall t \in [\sigma, \infty[, \ \phi(0) = -\mathbf{e}(0), \tag{10.59}$$

Let Equation (10.57) define $\Gamma_{EISOU}(s)$ and let the control be defined in the complex domain by:

$$\mathbf{U}(s) = (1 - e^{-\sigma s})^{-1} \Gamma_{EISOU}(s) \left[\Phi(s) + \mathbf{E}(s) \right], \ \Phi(s) = \mathcal{L}\{\phi(t)\}. \tag{10.60}$$

The control $\mathbf{U}(.)$ (10.60) does not depend on the plant state or on the disturbance $\mathbf{D}(.)$. The control $\mathbf{U}(.)$ (10.60) is natural control (Definition 166, Section 9.4). We multiply Equation (10.60) on the left by $G_U(s)$ and use $G_U(s)\Gamma_{EISOU}(s) \equiv I_N$:

$$G_U(s)\mathbf{U}(s) = (1 - e^{-\sigma s})^{-1} \left[\Phi(s) + \mathbf{E}(s) \right] \tag{10.61}$$

After replacing $\mathbf{U}(s)$ with the right-hand side of Equation (10.61) into Equation (10.51) the result reads:

$$(1 - e^{-\sigma s}) \left\{ \begin{array}{c} \mathbf{Y}(s) - G_D(s)\mathbf{D}(s) - G_{D_0}(s)\mathbf{D}_0^{\mu-1} - \\ -G_{U_0}(s)\mathbf{U}_0^{\mu-1} - G_{X_0}(s)\mathbf{X}_0 \end{array} \right\} = $$
$$= \Phi(s) + \mathbf{E}(s).$$

Let $\sigma \longrightarrow 0^+$, or $\sigma = 0^+$ in the ideal case. Then the preceding equation becomes

$$\mathbf{0}_N = \Phi(s) + \mathbf{E}(s) \Longrightarrow \mathbf{e}(t) = -\phi(t) = \mathbf{0}_N, \ \forall t \in \mathfrak{T}_0$$

due to $\mathbf{Y}_0 = \mathbf{Y}_{d0}$ (Definition 161) and (10.59), or equivalently:

$$\mathbf{e}(t) = \mathbf{Y}_d(t) - \mathbf{Y}(t) = \mathbf{0}_N, \ \forall t \in \mathfrak{T}_0.$$

The natural control $\mathbf{U}(.)$ defined by (10.60) satisfies Definition 310. The *EISO* plant (4.1), (4.2) is perfect natural trackable on $\mathfrak{D}^k \times \mathfrak{Y}_d^k$. ∎

Comment 225 *Theorem 222 and Theorem 224 show that for the EISO plant (4.1), (4.2) to be perfect natural trackable on $\mathfrak{D}^k \times \mathfrak{Y}_d^k$ it is necessary and sufficient that the plant is perfect trackable on $\mathfrak{D}^k \times \mathfrak{Y}_d^k$. This confirms Comment 168 (Section 9.4).*

Comment 226 *The proof of Theorem 224 demonstrates that the control implementation can be successful without using any information about the disturbance vector.*

The disturbance vector can be unmeasurable, which corresponds to the reality. That is the great advantage of the natural perfect trackability over the perfect trackability.

If $\Gamma_{EISOU}(s)$ replaces $\Gamma_{IOU}(s)$ then Note 201, (Section 10.1), holds in this settings.

10.3.4   Imperfect trackability criteria

This section treats the problem of the necessary and sufficient conditions for imperfect trackability properties of the *EISO* plant (4.1), (4.2). They are for the first *time* studied in this book.

Theorem 227 *Conditions for the global complete trackability of the EISO plant (4.1), (4.2) on $\mathfrak{D}^k \times \mathfrak{Y}_d^k$*
   In order for the EISO plant (4.1), (4.2) to be global complete trackable on $\mathfrak{D}^k \times \mathfrak{Y}_d^k$ it is necessary and sufficient that the plant transfer function matrix $G_U(s)$ relative to control has the full row rank N, so that the dimension r of the control vector $\mathbf{U}$ is not less than the dimension N of the output vector $\mathbf{Y}$, i.e., that all conditions (10.54) hold.

Proof. *Necessity.* Let the *EISO* plant (4.1), (4.2) be global complete trackable on $\mathfrak{D}^k \times \mathfrak{Y}_d^k$. Definition 172 (Section 9.5) is fulfilled. The plant is also perfect trackable on $\mathfrak{D}^k \times \mathfrak{Y}_d^k$ (due to Definition 172 and Theorem 174, Section 9.5). Theorem 222 establishes the necessity of the conditions (10.54) for the global complete trackability on $\mathfrak{D}^k \times \mathfrak{Y}_d^k$.
   Sufficiency. Let the conditions (10.54) hold. The system transfer function matrix $G_U(s)$ has the full rank N (due to Theorem 194 and Theorem 221). Let $[\mathbf{D}(.), \mathbf{Y}_d(.)] \in \mathfrak{D}^k \times \mathfrak{Y}_d^k$, the instant $\sigma \in Int\ \mathfrak{T}_0$, and $\mathbf{Y}_0^{\nu-1} \in \mathfrak{R}^{\nu N}$

be arbitrarily chosen (see Definition 172). Let the control be defined in the complex domain by

$$\mathbf{U}(s) = \Gamma_{EISOU}(s) \bullet$$

$$\bullet \left[\begin{array}{c} \mathbf{Y}(s) - G_D(s)\, \mathbf{D}(s) - G_{D_0}(s)\, \mathbf{D}_0^{\mu-1} - \\ -G_{U_0}(s)\, \mathbf{U}_0^{\mu-1} - G_{X_0}(s)\, \mathbf{X}_0 - \mathbf{Z}(s) + |\mathbf{E}(s)| \end{array} \right],$$

together with

$$\mathbf{Z}(t) = \left\{ \begin{array}{c} |\mathbf{e}(0)| \left(1 - \frac{t}{\sigma}\right),\ t \in [0,\, \sigma], \\ \mathbf{0}_N,\ \forall\, (t \geq \sigma) \in \mathfrak{T}_0 \end{array} \right\} \in \mathfrak{C}(\mathfrak{T}_0),$$

$$\mathbf{Z}(s) = \mathcal{L}\{\mathbf{Z}(t)\}. \tag{10.62}$$

For such control Equation (10.51) becomes

$$\mathbf{Y}(s) = G_D(s)\, \mathbf{D}(s) + \mathbf{Y}(s) - G_D(s)\, \mathbf{D}(s) -$$

$$-\mathbf{Z}(s) + |\mathbf{E}(s)| - G_{X_0}(s)\, \mathbf{X}_0 + G_{X_0}(s)\, \mathbf{X}_0 = \mathbf{Y}(s) - \mathbf{Z}(s) + |\mathbf{E}(s)|$$

due to $G_U(s)\, \Gamma_{EISOU}(s) \equiv I_N$, i.e.,

$$-\mathbf{Z}(s) + |\mathbf{E}(s)| = \mathbf{0}_N \Longrightarrow |\mathbf{e}(t)| = \mathbf{Z}(t),\ \forall t \in \mathfrak{T}_0.$$

This and (10.62) imply

$$\mathbf{e}(t) = \mathbf{0}_N,\ \forall\, (t \geq \sigma) \in \mathfrak{T}_0,\ i.e.,\ \mathbf{Y}(t) = \mathbf{Y}_d(t),\ \forall\, (t \geq \sigma) \in \mathfrak{T}_0,$$

which proves the global complete trackability on $\mathfrak{D}^k \times \mathfrak{Y}_d^k$ of the *EISO* plant (4.1), (4.2) due to Definition 172. ∎

Theorem 228 *On the global k-th order elementwise trackability of the EISO plant (4.1), (4.2) on $\mathfrak{D}^k \times \mathfrak{Y}_d^k$*

In order for the EISO plant (4.1), (4.2) to be the k-th order global elementwise trackable on $\mathfrak{D}^k \times \mathfrak{Y}_d^k$ it is necessary and sufficient that the plant transfer function matrix $G_U(s)$ relative to control has the full row rank N so that the dimension r of the control vector $\mathbf{U}$ is not less than the dimension N of the output vector $\mathbf{Y}$, i.e., that all conditions (10.54) hold.

Proof. *Necessity.* Let the *EISO* plant (4.1), (4.2) be the the k-th order global elementwise trackable on $\mathfrak{D}^k \times \mathfrak{Y}_d^k$. Definition 313 (Section 9.5) is fulfilled. The plant and control satisfy also Definition 172. It and Theorem 174 prove the necessity of the conditions (10.54).

Sufficiency. Let the conditions (10.54) be satisfied. They, Theorem 194 and Theorem 221 imply that the system transfer function matrix $G_U(s)$ has the full rank N. Let $[\mathbf{D}(.), \mathbf{Y}_d(.)] \in \mathfrak{D}^k \times \mathfrak{Y}_d^k$, $\sigma \in (Int \ \mathfrak{T}_0)^{k+1}$, $\sigma = \left[\sigma_0 \vdots \sigma_1 \vdots ... \vdots \sigma_k \right]^T$, and $\mathbf{Y}_0^{\nu-1} \in \mathfrak{R}^{\nu N}$ be arbitrarily chosen (in view of Definition 313). We select the control vector $\mathbf{U}(t)$ so that its Laplace transform satisfies the following equation:

$$\mathbf{U}(s) = \Gamma_{EISOU}(s)\bullet$$

$$\bullet \left[\begin{array}{c} \mathbf{Y}(s) - G_D(s)\mathbf{D}(s) - G_{D_0}(s)\mathbf{D}_0^{\mu-1} - G_{U_0}(s)\mathbf{U}_0^{\mu-1} - \\ -G_{X_0}(s)\mathbf{X}_0 - \mathbf{Z}(s) + \mathcal{L}\left\{ \sum_{j=0}^{j=k\leq\mu-1} \left| \mathbf{e}^{(j)}(t) \right| \right\} \end{array} \right] \qquad (10.63)$$

for the Laplace transform $\mathbf{Z}(s)$ of $\mathbf{Z}(t)$ determined by

$$\mathbf{Z}(t) = \left\{ \begin{array}{c} \sum_{j=0}^{j=k\leq\mu-1} \left(\left| \mathbf{e}^{(j)}(0) \right| \left(1 - \frac{t}{\sigma_m} \right), \ t \in [0, \sigma_m] \right), \\ \mathbf{0}_N, \ \forall (t \geq \sigma_m) \in \mathfrak{T}_0 \end{array} \right\} \in \mathfrak{C}(\mathfrak{T}_0),$$

$$0 < \sigma_m = \min(\sigma_1, \sigma_2, ..., \sigma_k) \leq \sigma_M, \ \mathbf{Z}(s) = \mathcal{L}\{\mathbf{Z}(t)\}. \qquad (10.64)$$

For the control defined in Equation (10.63) the system complex domain mathematical model (10.51) becomes, in view of $G_U(s)\Gamma_{EISOU}(s) \equiv I_N$,

$$\mathbf{Y}(s) = G_D(s)\mathbf{D}(s) + \mathbf{Y}(s) - G_D(s)\mathbf{D}(s) - G_{D_0}(s)\mathbf{D}_0^{\mu-1} -$$

$$-G_{U_0}(s)\mathbf{U}_0^{\mu-1} - G_{X_0}(s)\mathbf{X}_0 - \mathbf{Z}(s) + \mathcal{L}\left\{ \sum_{j=0}^{j=\leq\mu-1} \left| \mathbf{e}^{(j)}(t) \right| \right\} +$$

$$+G_{D_0}(s)\mathbf{D}_0^{\mu-1} + +G_{U_0}(s)\mathbf{U}_0^{\mu-1} + G_{X_0}(s)\mathbf{X}_0 \Longrightarrow$$

$$\mathbf{0}_N = . - \mathbf{Z}(s) + \mathcal{L}\left\{ \sum_{j=0}^{j=\leq\mu-1} \left| \mathbf{e}^{(j)}(t) \right| \right\},$$

or in the *time* domain:

$$\sum_{j=0}^{j=\leq\mu-1} \left| \mathbf{e}^{(j)}(t) \right| = \mathbf{Z}(t) = \mathbf{0}_N, \ \forall t \in \mathfrak{T}_0,$$

i.e.,

$$\mathbf{Y}^{(j)}(t) = \mathbf{Y}_d^{(j)}(t), \ \forall (t \geq \sigma_j) \in \mathfrak{T}_0, \ \forall j = 0, 1, ..., k \leq \mu - 1,$$

which proves the *k-th* order global elementwise trackability of the *EISO* plant (4.1), (4.2) on $\mathfrak{D}^\eta \times \mathfrak{Y}_d^\nu$ (Definition 313). ∎

10.3.5   Conditions for natural trackability

Theorem 229 *Conditions for global complete natural trackability of the EISO plant (4.1), (4.2) on $\mathfrak{D}^k \times \mathfrak{Y}_d^k$*

In order for the EISO plant (4.1), (4.2) to be global complete natural trackable on $\mathfrak{D}^k \times \mathfrak{Y}_d^k$ it is necessary and sufficient that the plant transfer function matrix $G_U(s)$ relative to control has the full row rank N so that the dimension r of the control vector $\mathbf{U}$ is not less than the dimension N of the output vector $\mathbf{Y}$, i.e., that all conditions (10.54) hold.

Proof. *Necessity.* Let the EISO plant (4.1), (4.2) be global complete natural trackable on $\mathfrak{D}^k \times \mathfrak{Y}_d^k$. Definition 312 is applicable. It and Definition 172 confirm that the plant is complete trackable on $\mathfrak{D}^k \times \mathfrak{Y}_d^k$ so that the conditions (10.54) are necessary due to Theorem 227.

Sufficiency. Let the conditions (10.54)) be valid. The EISO plant (4.1), (4.2) is global complete trackable on $\mathfrak{D}^k \times \mathfrak{Y}_d^k$ (Theorem 227). Let Equations (10.59)-(10.60) define the Laplace transform of the control vector. By repeating from Equation (10.60) the sufficiency part of the proof of Theorem 224 we prove that the so defined natural tracking control implies

$$\mathbf{e}(t) = \mathbf{Y}_d(t) - \mathbf{Y}(t) = \mathbf{0}_N, \ \forall (t \geq \sigma) \in \mathfrak{T}_0.$$

This proves the global complete natural trackability of the EISO plant (4.1), (4.2) on $\mathfrak{D}^k \times \mathfrak{Y}_d^k$ in view of Definition 312. ∎

Theorem 230 *Conditions for the   k-th order global elementwise natural trackability of the EISO plant (4.1), (4.2) on $\mathfrak{D}^k \times \mathfrak{Y}_d^k$*

In order for the EISO plant (4.1), (4.2) to be the k-th order global elementwise natural trackable on $\mathfrak{D}^k \times \mathfrak{Y}_d^k$, $0 \leq k \leq \mu - 1$, it is necessary and sufficient that the plant transfer function matrix $G_U(s)$ relative to control has the full row rank N so that the dimension r of the control vector $\mathbf{U}$ is not less than the dimension N of the output vector $\mathbf{Y}$, i.e., that all conditions (10.54) hold.

Proof. *Necessity.* Let the EISO plant (4.1), (4.2) be the k-th order global elementwise natural trackable on $\mathfrak{D}^k \times \mathfrak{Y}_d^k$, $0 \leq k \leq \mu - 1$. Definition 314 (Section 9.5) is fulfilled. The plant and control satisfy also Definition 172. It and Theorem 227 prove the necessity of the conditions (10.54).

Sufficiency. Let the conditions (10.54) be satisfied. Let $[\mathbf{D}(.), \mathbf{Y}_d(.)] \in \mathfrak{D}^k \times \mathfrak{Y}_d^k$, $\sigma \in (Int\ \mathfrak{T}_0)^{k+1}$, $\sigma = \left[\sigma_0 \vdots \sigma_1 \vdots ... \vdots \sigma_k\right]^T$, and $\mathbf{Y}_0^{\nu-1} \in \mathfrak{R}^{\nu N}$ be arbitrarily chosen (in view of Definition 314). We select the control vector $\mathbf{U}(t)$ to satisfy the following complex domain equation:

$$G_U(s)\mathbf{U}(s) = e^{-\sigma_M s}G_U(s)\mathbf{U}(s) + \mathcal{L}\left\{\sum_{j=0}^{j=k\leq\mu-1}\left|\mathbf{e}^{(j)}(t)\right|\right\} - \mathbf{Z}(s) \quad (10.65)$$

for

$$\sigma_M = \max(\sigma_0, \sigma_1,\dots, \sigma_k) <<< 0^+, \; i.e., \; \sigma_M \longrightarrow 0^+, \quad (10.66)$$

and for the Laplace transform $\mathbf{Z}(s)$ of $\mathbf{Z}(t)$ that is defined by

$$\mathbf{Z}(t) = \left\{\begin{array}{c}\sum_{j=0}^{j=k\leq\mu-1}\left(\left|\mathbf{e}^{(j)}(0)\right|\left(1-\frac{t}{\sigma_m}\right), \; t \in [0, \sigma_m]\right), \\ \mathbf{0}_N, \; \forall(t \geq \sigma_m) \in \mathfrak{T}_0\end{array}\right\} \in \mathfrak{C}(\mathfrak{T}_0),$$

$$0 < \sigma_m = \min(\sigma_0, \sigma_1, \dots, \sigma_k) < \sigma_M, \; \mathbf{Z}(s) = \mathcal{L}\{\mathbf{Z}(t)\}. \quad (10.67)$$

For such control the plant complex domain mathematical model (10.51) becomes

$$\mathbf{Y}(s) = G_D(s)\mathbf{D}(s) + G_U(s)\left[e^{-\sigma_M s}\mathbf{U}(s)\right] + \mathcal{L}\left\{\sum_{j=0}^{j=k\leq\mu-1}\left|\mathbf{e}^{(j)}(t)\right|\right\} -$$

$$-\mathbf{Z}(s) + G_{D_0}(s)\mathbf{D}_0^{\mu-1} + G_{U_0}(s)\mathbf{U}_0^{\mu-1} + G_{X_0}(s)\mathbf{X}_0. \quad (10.68)$$

Let us subtract Equation (10.68) from Equation (10.51). The result reads:

$$\mathbf{0}_N = .G_U(s)\left(1 - e^{-\sigma_M s}\right)\mathbf{U}(s) + \mathbf{Z}(s) - \mathcal{L}\left\{\sum_{j=0}^{j=k\leq\mu-1}\left|\mathbf{e}^{(j)}(t)\right|\right\}.$$

In the limit as $\sigma_M \longrightarrow 0^+$ for σ_M defined by (10.66) the preceding equation reduces to:

$$\mathbf{0}_N = \mathbf{Z}(s) - \mathcal{L}\left\{\sum_{j=0}^{j=k\leq\mu-1}\left|\mathbf{e}^{(j)}(t)\right|\right\},$$

or in the *time* domain due to Equation (10.67):

$$\sum_{j=0}^{j=k\leq\mu-1}\left|\mathbf{e}^{(j)}(t)\right| = \mathbf{Z}(t) = \mathbf{0}_N, \; \forall(t \geq \sigma_j) \in \mathfrak{T}_0,$$

i.e.,

$$\mathbf{Y}^{(j)}(t) = \mathbf{Y}_d^{(j)}(t), \ \forall (t \geq \sigma_j) \in \mathfrak{T}_0, \ \forall j = 0, 1, ..., k \leq \mu - 1,$$

which proves the $k-th$ order global elementwise natural trackability of the *EISO* plant (4.1), (4.2) on $\mathfrak{D}^k \times \mathfrak{Y}_d^k$ (Definition 314). ∎

If in Summary 220, (Section 10.2):

- "The *EISO* plant (4.1), (4.2)" replaces "the *ISO* plant (3.1), (3.2)",
- $L_U(s)$ replaces $L_{pISO}(s)$,
- $G_U(s)$ replaces $G_{ISOPU}(s)$, and
- "(10.54) and 3) of Theorem 222" replace "(10.37) and 3) of Theorem 210"

then Summary 220 becomes valid for the *EISO* plant (4.1), (4.2).

10.4   *HISO* system trackability

This section initiates the trackability study of the *HISO* plant (5.1), (5.2).

10.4.1   Perfect trackability criteria

The *time* domain mathematical model of the *HISO* plant is given in Equations (5.1), (5.2) (Section 5.1).

The Laplace transforms of Equation (5.1) and Equation (5.2) yield the complex domain characteristics of the plant determined in Equations (5.5), (5.14), (5.16-(5.28), (Section 5.1). We summarize them as follows:

$$\mathbf{Y}(s) = G_{HISOD}(s)\,\mathbf{D}(s) + G_{HISOU}(s)\,\mathbf{U}(s) + G_{HISOD_0}(s)\,\mathbf{D}_0^{\mu-1} +$$

$$+G_{HISOR_0}(s)\,\mathbf{R}_0^{\alpha-1} + G_{HISOU_0}(s)\,\mathbf{U}_0^{\mu-1} = F_{HISO}(s)\,\mathbf{V}_{HISO}(s). \quad (10.69)$$

The general Definition 161 (Section 9.4) is basic. It specifies the *l-th* order perfect trackability of the plant on $\mathfrak{D}^k \times \mathfrak{Y}_d^k$. Lemma 162 (Section 9.4) enables us to reduce the study of the *l-th* order perfect trackability of the plant on $\mathfrak{D}^k \times \mathfrak{Y}_d^k$ to the study of the perfect trackability of the plant on $\mathfrak{D}^k \times \mathfrak{Y}_d^k$. We will investigate the perfect trackability of the *HISO* plant (5.1), (5.2) on $\mathfrak{D}^k \times \mathfrak{Y}_d^k$. The result will be valid simultaneously for the *l-th* order perfect trackability of the plant on $\mathfrak{D}^k \times \mathfrak{Y}_d^k$.

Equations (5.16), (5.18)–(5.23) set the expression of $G_{HISOU}(s)$ in the following simple form:

$$G_{HISOU}(s) = \frac{L_{HISO}(s)}{p_{HISO}(s)} = p_{HISO}^{-1}(s)\,L_{HISO}(s). \quad (10.70)$$

Theorem 194 (Subsection 10.1.1) and Theorem 195 (Subsection 10.1.1) simplify effectively the next proves. In this framework they take the following joint form:

Theorem 231 *Full rank of the transfer function matrix relative to the control*

The full rank ρ of the transfer function matrix $G_{HISOU}(s)$ (5.18) is also the full rank of its numerator matrix polynomial $L_{HISO}(s)$ (5.19) and of its generating matrix L_{pHISO} (5.23),

$$full \ rank G_{HISOU}(s) = \rho = \min(N, r) =$$
$$= full \ rank \begin{bmatrix} R_P^{(\alpha)} S_\rho^{(\alpha)}(s) adj \left(A_P^{(\alpha)} S_\rho^{(\alpha)}(s) \right) B^{(\mu)} S_r^{(\mu)}(s) + \\ + p_{HISO}(s) U \end{bmatrix} =$$
$$= full \ rank \left[p_{HISO}^{-1}(s) L_{HISO}(s) \right] = full \ rank L_{HISO}(s) \Longrightarrow$$
$$\Longrightarrow rank L_{HISO} = full \ rank L_{HISO} = \rho = \min(N, r). \qquad (10.71)$$

Theorem 232 *Conditions for the perfect trackability on $\mathfrak{D}^k \times \mathfrak{Y}_d^k$*

In order for the HISO plant (5.1), (5.2) to be perfect trackable on $\mathfrak{D}^k \times \mathfrak{Y}_d^k$ it is necessary and sufficient that there is $s^* \in \mathbb{C}$ for which its transfer function matrix $G_{HISOU}(s)$ relative to control has the full row rank N, $rank G_{HISOU}(s^*) = N$, so that the dimension r of the control vector $\mathbf{U}$, which is then the nominal control $\mathbf{U}_N$, is not less than the dimension N of the output vector $\mathbf{Y}$, i.e.,

$$\exists s^* \in \mathbb{C} \Longrightarrow full \ rank G_{HISOU}(s^*) =$$
$$= full \ rank \left[R_P^{(\alpha)} S_\rho^{(\alpha)}(s^*) \left(A_P^{(\alpha)} S_\rho^{(\alpha)}(s^*) \right)^{-1} B^{(\mu)} S_r^{(\mu)}(s^*) + U \right] =$$
$$= full \ rank L_{HISO}(s^*) = N \leq r. \qquad (10.72)$$

The Laplace transform $\mathbf{U}_N(s)$ of the nominal control $\mathbf{U}_N(.)$ is then determined by

$$\mathbf{U}_N(s) = G_{HISOU}(s)^T \left\{ G_{HISOU}(s) G_{HISOU}(s)^T \right\}^{-1} \bullet$$
$$\bullet \left\{ \begin{array}{c} \mathbf{Y}_d(s) - G_{HISOD}(s) \mathbf{D}(s) - G_{HISOD_0}(s) \mathbf{D}_0^{\mu-1} \\ - G_{HISOR_0}(s) \mathbf{R}_{d0}^{\alpha-1} - G_{HISOU_0}(s) \mathbf{U}_0^{\mu-1} \end{array} \right\}. \qquad (10.73)$$

Proof. *Necessity.* Let the *HISO* plant (5.1), (5.2) be perfect trackable on $\mathfrak{D}^k \times \mathfrak{Y}_d^k$. We recall Definition 161, Section 9.4. Its satisfaction guarantees that there is a control vector $\mathbf{U}(t)$, the Laplace transform of which obeys Equation (10.69). The perfect trackability is characterized by $\mathbf{Y}(t) \equiv \mathbf{Y}_d(t)$ that yields $\mathbf{U}(t) \equiv \mathbf{U}_N(t)$, i.e., by $\mathbf{Y}(s) \equiv \mathbf{Y}_d(s)$ that implies $\mathbf{U}(s) \equiv \mathbf{U}_N(s)$:

$$\mathbf{Y}_d(s) = \left\{ \begin{array}{c} G_{HISOD}(s)\,\mathbf{D}(s) + G_{HISOU}(s)\,\mathbf{U}_N(s) + \\ + G_{HISOD_0}(s)\,\mathbf{D}_0^{\mu-1} + G_{HISOR_0}(s)\,\mathbf{R}_{d0}^{\alpha-1} + \\ + G_{HISOU_0}(s)\,\mathbf{U}_{0N}^{\mu-1} \end{array} \right\}. \qquad (10.74)$$

The existence of the solution $\mathbf{U}_N(s)$ of this equation yields

$$rank G_{HISOU}(s) = full\ rank G_{HISOU}(s) = N \le r.$$

This, Theorem 194 (Subsection 10.1.1), and Theorem 231 prove the necessity of the conditions (10.72). They guarantee that $G_{HISOU}(s)\,G_{HISOU}^T(s)$ is nonsingular so that the solution $\mathbf{U}(s) = \mathbf{U}_N(s)$ of Equation (10.74) is given by Equation (10.73).

Sufficiency. Let the conditions (10.72) hold. They, together with Theorem 194 and Theorem 231 guarantee that $G_{HISOU}(s)$ has the full rank N so that $\Gamma_{HISOU}(s)$ introduced by

$$\Gamma_{HISOU}(s) = G_{HISOU}^T(s) \left[G_{HISOU}(s)\,G_{HISOU}^T(s) \right]^{-1}. \qquad (10.75)$$

is fully defined as well as $\mathbf{U}(s)$ determined by

$$\mathbf{U}(s) = \Gamma_{HISOU}(s) \bullet$$

$$\bullet \left\{ \begin{array}{c} \mathbf{Y}_d(s) - G_{HISOD}(s)\,\mathbf{D}(s) - G_{HISOD_0}(s)\,\mathbf{D}_0^{\mu-1} - \\ - G_{HISOR_0}(s)\,\mathbf{R}_{d0}^{\alpha-1} - G_{HISOU_0}(s)\,\mathbf{U}_0^{\mu-1} \end{array} \right\}. \qquad (10.76)$$

We eliminate $\mathbf{U}(s)$ from Equation (5.5) by replacing it by the right hand side of Equation (10.76) and by applying $G_{HISOU}(s)\,\Gamma_{HISOU}(s) \equiv I_N$ and $\mathbf{R}_0^{\alpha-1} = \mathbf{R}_{d0}^{\alpha-1}$ due to Definition 161:

$$\mathbf{Y}(s) = G_{HISOD}(s)\,\mathbf{D}(s) +$$

$$+ \left\{ \begin{array}{c} \mathbf{Y}_d(s) - G_{HISOD}(s)\,\mathbf{D}(s) - G_{HISOD_0}(s)\,\mathbf{D}_0^{\mu-1} - \\ - G_{HISOR_0}(s)\,\mathbf{R}_0^{\alpha-1} - G_{HISOU_0}(s)\,\mathbf{U}_0^{\mu-1} \end{array} \right\} +$$

$$+ \left\{ \begin{array}{c} G_{HISOD_0}(s)\,\mathbf{D}_0^{\mu-1} + G_{HISOR_0}(s)\,\mathbf{R}_0^{\alpha-1} + \\ + G_{HISOU_0}(s)\,\mathbf{U}_0^{\mu-1} \end{array} \right\} = \mathbf{Y}_d(s).$$

This result, $\mathbf{Y}(s) = \mathbf{Y}_d(s)$ that implies $\mathbf{U}_N(s) = \mathbf{U}(s)$ (10.76), in the *time* domain reads

$$\mathbf{Y}(t) = \mathbf{Y}_d(t),\ \mathbf{U}_N(t) = \mathbf{U}(t),\ \forall [t, D(.), \mathbf{Y}_d(.)] \in \mathfrak{T}_0 \times \mathfrak{D}^\mu \times \mathfrak{Y}_d^\nu.$$

This proves the perfect trackability of the $HISO$ plant (5.1), (5.2) on $\mathfrak{D}^k \times \mathfrak{Y}_d^k$ in view of Definition 161 and that the control $\mathbf{U}(s)$ (10.58) is the Laplace transform of the plant nominal control relative to the functional vector pair $[\mathbf{D}(.), \mathbf{Y}_d(.)]$ due to Definition 84, Section 5.2. ∎

Comment 233 *Equation (10.73) defines the nominal control* $\mathbf{U}_N(t)$ *in terms of the often unmeasurable disturbance vector* $D(t)$.

10.4.2 Conditions for perfect natural trackability

Definition 310 (Section 9.4) introduced the notion and the sense of the perfect natural trackability valid also for the $HISO$ plant (5.1), (5.2) on $\mathfrak{D}^k \times \mathfrak{Y}_d^k$.

Theorem 234 *Conditions for the perfect natural trackability of the* **HISO plant (5.1), (5.2) on** $\mathfrak{D}^k \times \mathfrak{Y}_d^k$

In order for the HISO plant (5.1), (5.2) to be perfect natural trackable on $\mathfrak{D}^k \times \mathfrak{Y}_d^k$ *it is necessary and sufficient that the plant transfer function matrix* $G_{HISOU}(s)$ *relative to control has the full row rank* N, *so that the dimension* r *of the control vector* $\mathbf{U}$ *is not less than the dimension* N *of the output vector* $\mathbf{Y}$, *i.e., that all conditions (10.72) hold.*

Proof. *Necessity.* Let the $HISO$ plant (5.1), (5.2) be perfect natural trackable on $\mathfrak{D}^k \times \mathfrak{Y}_d^k$. Definitions 310 and 161 guarantee that the plant is also perfect trackable on $\mathfrak{D}^k \times \mathfrak{Y}_d^k$. Theorem 232 implies the necessity of the conditions (10.72).

Sufficiency. Let the conditions (10.72) hold.

The $HISO$ plant (5.1), (5.2) is perfect trackable on $\mathfrak{D}^k \times \mathfrak{Y}_d^k$, Theorem 210. We should show that the perfect tracking control can be synthesized without using information about the plant state and about the disturbance $\mathbf{D}(.) \in \mathfrak{D}^k$, i.e., that it is natural tracking control (Definition 166, Section 9.4). We recall Equation (5.5). Let $\sigma \in \mathfrak{R}^+$ be arbitrarily small, i.e., $\sigma \longrightarrow 0^+$ and in the ideal case $\sigma = 0^+$. Let

$$\phi(.) : \mathfrak{T}_0 \longrightarrow \mathfrak{R}^N, \ \phi(t) \in \mathfrak{C},$$
$$\phi(t) = \mathbf{0}_N, \ \forall t \in [\sigma, \infty[, \ \phi(0) = -\mathbf{e}(0), \qquad (10.77)$$

Let Equation (10.75) define $\Gamma_{HISOU}(s)$ and let the control be defined in the complex domain by:

$$\mathbf{U}(s) = (1 - e^{-\sigma s})^{-1} \Gamma_{HISOU}(s) [\Phi(s) + \mathbf{E}(s)], \ \Phi(s) = \mathcal{L}\{\phi(t)\}. \quad (10.78)$$

The control $\mathbf{U}(.)$ (10.78) does not depend on the plant state and on the disturbance $\mathbf{D}(.)$. The control $\mathbf{U}(.)$ (10.78) is natural control (Definition 166, Section 9.4). We multiply Equation (10.78) on the left by $G_{HISOU}(s)$ and use $G_{HISOU}(s)\,\Gamma_{HISOU}(s) \equiv I_N$. The result is

$$G_{HISOU}(s)\,\mathbf{U}(s) = (1 - e^{-\sigma s})^{-1}\,[\Phi(s) + \mathbf{E}(s)] \qquad (10.79)$$

After replacing $\mathbf{U}(s)$ with the right-hand side of (10.79) into Equation (5.5) we get the following:

$$(1 - e^{-\sigma s})\left\{ \begin{array}{c} \mathbf{Y}(s) - G_{HISOD}(s)\,\mathbf{D}(s) - G_{HISOD_0}(s)\,\mathbf{D}_0^{\mu-1} - \\ -G_{HISOR_0}(s)\,\mathbf{R}_0^{\alpha-1} - G_{HISOU_0}(s)\,\mathbf{U}_0^{\mu-1} \end{array} \right\} =$$
$$= \Phi(s) + \mathbf{E}(s).$$

Let $\sigma \longrightarrow 0^+$, or $\sigma = 0^+$ in the ideal case. Then the preceding equation becomes

$$\mathbf{0}_N = \Phi(s) + \mathbf{E}(s) \Longrightarrow \mathbf{e}(t) = -\phi(t),\ \forall t \in \mathfrak{T}_0,$$

due to $\mathbf{Y}_0 = \mathbf{Y}_{d0}$ and (10.77), or equivalently:

$$\mathbf{e}(t) = \mathbf{Y}_d(t) - \mathbf{Y}(t) = \mathbf{0}_N,\ \forall t \in \mathfrak{T}_0.$$

The natural control $\mathbf{U}(.)$ defined by (10.78) satisfies Definition 310. The $HISO$ plant (5.1), (5.2) is perfect natural trackable on $\mathfrak{D}^k \times \mathfrak{Y}_d^k$. ∎

Comment 235 *Theorem 232 and Theorem 234 show that for the HISO plant (5.1), (5.2) to be perfect natural trackable on $\mathfrak{D}^k \times \mathfrak{Y}_d^i$ it is necessary and sufficient that the plant is perfect trackable on $\mathfrak{D}^k \times \mathfrak{Y}_d^k$. This confirms Comment 168 (Section 9.4).*

Comment 236 *The proof of Theorem 234 demonstrates that the control implementation can be successful without using any information about the disturbance vector. The disturbance vector can be unmeasurable, which corresponds to the reality. That is the great advantage of the natural perfect trackability over the perfect trackability.*

Note 201, (Section 10.1) holds in this settings if $\Gamma_{HISOU}(s)$ replaces $\Gamma_{IOU}(s)$.

10.4.3 Imperfect trackability criteria

This section deals with the necessary and sufficient conditions for imperfect trackability properties of the *HISO* plant (5.1), (5.2). The fact that they are for the first time studied in this book for the *HISO* system (5.1), (5.2) we will present the proofs of the next theorems.

Theorem 237 *Conditions for the global complete trackability of the HISO plant (5.1), (5.2) on* $\mathfrak{D}^k \times \mathfrak{Y}_d^k$

In order for the HISO plant (5.1), (5.2) to be global complete trackable on $\mathfrak{D}^k \times \mathfrak{Y}_d^k$ *it is necessary and sufficient that the plant transfer function matrix* $G_{HISOU}(s)$ *relative to control has the full row rank N, so that the dimension r of the control vector* $\mathbf{U}$ *is not less than the dimension N of the output vector* $\mathbf{Y}$, *i.e., that all conditions (10.72) hold.*

Proof. *Necessity.* Let the *HISO* plant (5.1), (5.2) be global complete trackable on $\mathfrak{D}^k \times \mathfrak{Y}_d^k$. Definition 172 (Section 9.5) is fulfilled. The plant is also perfect trackable on $\mathfrak{D}^k \times \mathfrak{Y}_d^k$ (due to Definition 172 and Theorem 174, Section 9.5). Theorem 232 proves the necessity of the conditions (10.72) for the global complete trackability of the *HISO* plant (5.1), (5.2) on $\mathfrak{D}^k \times \mathfrak{Y}_d^k$.

Sufficiency. Let the conditions (10.72) hold. They, together with Theorem 194 and Theorem 231 guarantee that $G_{HISOU}(s)$ has the full rank N. Let $[\mathbf{D}(.), \mathbf{Y}_d(.)] \in \mathfrak{D}^k \times \mathfrak{Y}_d^k$, the instant $\sigma \in Int\ \mathfrak{T}_0$, and $\mathbf{Y}_0^{\nu-1} \in \mathfrak{R}^{\nu N}$ be arbitrarily chosen (see Definition 172). Let the control be defined in the complex domain by

$$\mathbf{U}(s) = \Gamma_{HISOU}(s)\bullet$$

$$\bullet \left[\begin{array}{c} \mathbf{Y}(s) - G_{HISOD}(s)\mathbf{D}(s) - G_{HISOD_0}(s)\mathbf{D}_0^{\mu-1} - \\ -G_{HISOR_0}(s)\mathbf{R}_0^{\alpha-1} - G_{HISOU_0}(s)\mathbf{U}_0^{\mu-1} - \mathbf{Z}(s) + |\mathbf{E}(s)| \end{array} \right],$$

together with

$$\mathbf{Z}(t) = \left\{ \begin{array}{c} |\mathbf{e}(0)|\left(1 - \frac{t}{\sigma}\right), \ t \in [0, \sigma], \\ \mathbf{0}_N, \ \forall(t \geq \sigma) \in \mathfrak{T}_0 \end{array} \right\} \in \mathfrak{C}(\mathfrak{T}_0),$$

$$\mathbf{Z}(s) = \mathcal{L}\{\mathbf{Z}(t)\}. \tag{10.80}$$

For such control Equation (10.69) becomes

$$\mathbf{Y}(s) = G_{HISOD}(s)\mathbf{D}(s) + \mathbf{Y}(s) - G_{HISOD}(s)\mathbf{D}(s) -$$

$$-\mathbf{Z}(s) + |\mathbf{E}(s)| - G_{HISOD_0}(s)\mathbf{D}_0^{\mu-1} - G_{HISOR_0}(s)\mathbf{R}_0^{\alpha-1} -$$

$$-G_{HISOU_0}(s)\mathbf{U}_0^{\mu-1} + G_{HISOD_0}(s)\mathbf{D}_0^{\mu-1} + G_{HISOR_0}(s)\mathbf{R}_0^{\alpha-1} +$$

$$+G_{HISOU_0}(s)\mathbf{U}_0^{\mu-1} = \mathbf{Y}(s) - \mathbf{Z}(s) + |\mathbf{E}(s)|$$

due to $G_{HISOU}(s)\,\Gamma_{HISOU}(s) = I_N$, i.e.,

$$\mathbf{0}_N = -\mathbf{Z}(s) + |\mathbf{E}(s)|$$

or in the *time* domain:

$$|\mathbf{e}(t)| = \mathbf{Z}(t), \ \forall t \in \mathfrak{T}_0.$$

This and (10.80) imply

$$\mathbf{e}(t) = \mathbf{0}_N, \ \forall\,(t \geq \sigma) \in \mathfrak{T}_0,$$

i.e.,

$$\mathbf{Y}(t) = \mathbf{Y}_d(t), \ \forall\,(t \geq \sigma) \in \mathfrak{T}_0,$$

which proves the global complete trackability of the *HISO* plant (5.1), (5.2) on $\mathfrak{D}^k \times \mathfrak{Y}_d^k$ due to Definition 172. ■

Theorem 238 *On the global k-th order elementwise trackability of the HISO plant (5.1), (5.2) on $\mathfrak{D}^k \times \mathfrak{Y}_d^k$*

In order for the HISO plant (5.1), (5.2) to be the k-th order global elementwise trackable on $\mathfrak{D}^k \times \mathfrak{Y}_d^k$ it is necessary and sufficient that the plant transfer function matrix $G_{HISOU}(s)$ relative to control has the full row rank N so that the dimension r of the control vector $\mathbf{U}$ is not less than the dimension N of the output vector $\mathbf{Y}$, i.e., that all conditions (10.72) hold.

Proof. *Necessity.* Let the *HISO* plant (5.1), (5.2) be the *k-th* order global elementwise trackable on $\mathfrak{D}^k \times \mathfrak{Y}_d^k$. Definition 313 (Section 9.5) is fulfilled. The plant and control satisfy also Definition 172. It, Theorem 232 and Theorem 174 prove the necessity of the conditions (10.72).

Sufficiency. Let the conditions (10.72) be satisfied. They, Theorem 194 and Theorem 231 ensure that $G_{HISOU}(s)$ has the full rank N. Let $[\mathbf{D}(.), \mathbf{Y}_d(.)] \in \mathfrak{D}^k \times \mathfrak{Y}_d^k$, $\sigma \in (Int\ \mathfrak{T}_0)^{k+1}$, $\sigma = \left[\sigma_0 \vdots \sigma_1 \vdots ... \vdots \sigma_k\right]^T$, and $\mathbf{Y}_0^{\nu-1} \in \mathfrak{R}^{\nu N}$ be arbitrarily chosen (in view of Definition 313). We select the control vector $\mathbf{U}(t)$ to satisfy the following complex domain equation:

$$\mathbf{U}(s) =$$

$$= \Gamma_{HISOU}(s) \begin{bmatrix} \mathbf{Y}(s) - G_{HISOD}(s)\mathbf{D}(s) - G_{HISOD_0}(s)\,\mathbf{D}_0^{\mu-1} - \\ -G_{HISOR_0}(s)\,\mathbf{R}_0^{\alpha-1} - G_{HISOU_0}(s)\,\mathbf{U}_0^{\mu-1} - \\ -\mathbf{Z}(s) + \mathcal{L}\left\{\sum_{j=0}^{j=k\leq\mu-1} \left|\mathbf{e}^{(j)}(t)\right|\right\} \end{bmatrix} \quad (10.81)$$

for the Laplace transform $\mathbf{Z}(s)$ of $\mathbf{Z}(t)$ determined by

$$\mathbf{Z}(t) = \left\{ \begin{array}{l} \sum_{j=0}^{j=k\le\mu-1} \left(\left| \mathbf{e}^{(j)}(0) \right| \left(1 - \frac{t}{\sigma_m} \right), \ t \in [0, \sigma_m] \right), \\ \mathbf{0}_N, \ \forall (t \ge \sigma_m) \in \mathfrak{T}_0 \end{array} \right\} \in \mathfrak{C}(\mathfrak{T}_0),$$

$$0 < \sigma_m = \min(\sigma_1, \sigma_2, ..., \sigma_k) \le \sigma_M, \ \mathbf{Z}(s) = \mathcal{L}\{\mathbf{Z}(t)\}. \tag{10.82}$$

For the control defined in Equation (10.81) the plant complex domain mathematical model (10.69) becomes, in view of $G_{HISOU}(s)\Gamma_{HISOU}(s) \equiv I_N$,

$$\mathbf{Y}(s) = G_{HISOD}(s)\mathbf{D}(s) + \mathbf{Y}(s) - G_{HISOD}(s)\mathbf{D}(s) -$$

$$-G_{HISOD_0}(s)\mathbf{D}_0^{\mu-1} - G_{HISOR_0}(s)\mathbf{R}_0^{\alpha-1} - G_{HISOU_0}(s)\mathbf{U}_0^{\mu-1} -$$

$$-\mathbf{Z}(s) + \mathcal{L}\left\{ \sum_{j=0}^{j=\le\mu-1} \left| \mathbf{e}^{(j)}(t) \right| \right\} + G_{HISOD_0}(s)\mathbf{D}_0^{\mu-1} +$$

$$+G_{HISOR_0}(s)\mathbf{R}_0^{\alpha-1} + G_{HISOU_0}(s)\mathbf{U}_0^{\mu-1} \Longrightarrow$$

$$\mathbf{0}_N = . - \mathbf{Z}(s) + \mathcal{L}\left\{ \sum_{j=0}^{j=\le\mu-1} \left| \mathbf{e}^{(j)}(t) \right| \right\}, \ \forall t \in \mathfrak{T}_0,$$

or in the *time* domain:

$$\sum_{j=0}^{j=\le\mu-1} \left| \mathbf{e}^{(j)}(t) \right| = \mathbf{Z}(t) = \mathbf{0}_N, \ \forall (t \ge \sigma_m) \in \mathfrak{T}_0,$$

i.e.,

$$\mathbf{Y}^{(j)}(t) = \mathbf{Y}_d^{(j)}(t), \ \forall (t \ge \sigma_j) \in \mathfrak{T}_0, \ \forall j = 0, 1, ..., k \le \mu - 1,$$

which proves the *k-th* order global elementwise trackability of the *HISO* plant (5.1), (5.2) on $\mathfrak{D}^\eta \times \mathfrak{Y}_d^\nu$ (Definition 313). ∎

10.4.4   Conditions for imperfect natural trackability

Theorem 239 *Conditions for global complete natural trackability of the HISO plant (5.1), (5.2) on $\mathfrak{D}^k \times \mathfrak{Y}_d^k$*

In order for the HISO plant (5.1), (5.2) to be global complete natural trackable on $\mathfrak{D}^k \times \mathfrak{Y}_d^k$ it is necessary and sufficient that the plant transfer function matrix $G_{HISOU}(s)$ relative to control has the full row rank N so that the dimension r of the control vector $\mathbf{U}$ is not less than the dimension N of the output vector $\mathbf{Y}$, i.e., that all conditions (10.72) hold.

Proof. *Necessity.* Let the *HISO* plant (5.1), (5.2) be global complete natural trackable on $\mathfrak{D}^k \times \mathfrak{Y}^k_d$. Definition 312 is applicable. It and Definition 172 confirm that the plant is complete trackable on $\mathfrak{D}^k \times \mathfrak{Y}^k_d$ so that the conditions (10.72) are necessary due to Theorem 237.

Sufficiency. Let the conditions (10.72)) be valid. The *HISO* plant (5.1), (5.2) is global complete trackable on $\mathfrak{D}^k \times \mathfrak{Y}^k_d$ (Theorem 237). Let the Laplace transform of the control vector be defined by Equations (10.77)–10.78). By repeating from Equation (10.78) the sufficiency part of the proof of Theorem 234 we prove that the so defined natural tracking control implies

$$\mathbf{Y}(t) = \mathbf{Y}_d(t), \ \forall\,(t \geq \sigma) \in \mathfrak{T}_0,$$

This proves the global complete natural trackability of the *HISO* plant (5.1), (5.2) on $\mathfrak{D}^k \times \mathfrak{Y}^k_d$ in view of Definition 312. ∎

Theorem 240 *Conditions for the k-th order global elementwise natural trackability of the HISO plant (5.1), (5.2) on $\mathfrak{D}^k \times \mathfrak{Y}^k_d$*

In order for the HISO plant (5.1), (5.2) to be the k-th order global elementwise natural trackable on $\mathfrak{D}^k \times \mathfrak{Y}^k_d$, $0 \leq k \leq \mu - 1$, it is necessary and sufficient that the plant transfer function matrix $G_{HISOU}(s)$ relative to control has the full row rank N so that the dimension r of the control vector $\mathbf{U}$ is not less than the dimension N of the output vector $\mathbf{Y}$, i.e., that all conditions (10.72) hold.

Proof. *Necessity.* Let the *HISO* plant (5.1), (5.2) be the *k-th* order global elementwise natural trackable on $\mathfrak{D}^k \times \mathfrak{Y}^k_d$. Definition 314 (Section 9.5) is fulfilled. The plant and control satisfy also Definition 172. It and Theorem 237 prove the necessity of the conditions (10.72).

Sufficiency. Let the conditions (10.72) be satisfied so that $G_{HISOU}(s)$ has the full rank N due to Theorem 194 and Theorem 231. Let $[\mathbf{D}(.), \mathbf{Y}_d(.)] \in \mathfrak{D}^k \times \mathfrak{Y}^k_d$, $\sigma \in (Int\ \mathfrak{T}_0)^{k+1}$, $\sigma = \begin{bmatrix} \sigma_0 \vdots \sigma_1 \vdots \dots \vdots \sigma_k \end{bmatrix}^T$, and $\mathbf{Y}^{\nu-1}_0 \in \mathfrak{R}^{\nu N}$ be arbitrarily chosen (in view of Definition 314). We select the control vector $\mathbf{U}(t)$ so that its Laplace transform obeys the following complex domain equation:

$$G_{HISOU}(s)\,\mathbf{U}(s) =$$

$$= e^{-\sigma_M s}G_{HISOU}(s)\,\mathbf{U}(s) + \mathcal{L}\left\{ \sum_{j=0}^{j=k\leq\mu-1} \left| \mathbf{e}^{(j)}(t) \right| \right\} - \mathbf{Z}(s) \qquad (10.83)$$

for

$$\sigma_M = \max(\sigma_0, \sigma_1,..., \sigma_k) <<< 0^+, \ 0 \le k \le \mu - 1, \ i.e., \ \sigma_M \longrightarrow 0^+, \tag{10.84}$$

and for the Laplace transform $\mathbf{Z}(s)$ of $\mathbf{Z}(t)$ that is determined by

$$\mathbf{Z}(t) = \left\{ \begin{array}{c} \sum_{j=0}^{j=k\le\mu-1} \left(\left| \mathbf{e}^{(j)}(0) \right| \left(1 - \frac{t}{\sigma_m} \right), \ t \in [0, \sigma_m] \right), \\ \mathbf{0}_N, \ \forall (t \ge \sigma_m) \in \mathfrak{T}_0 \end{array} \right\} \in \mathfrak{C}(\mathfrak{T}_0),$$

$$0 < \sigma_m = \min(\sigma_0, \sigma_1,..., \sigma_k) < \sigma_M, \ \mathbf{Z}(s) = \mathcal{L}\{\mathbf{Z}(t)\}. \tag{10.85}$$

For such control the plant complex domain mathematical model (10.69) becomes

$$\mathbf{Y}(s) = G_{HISOD}(s)\mathbf{D}(s) + G_{HISOU}(s)\left[e^{-\sigma_M s}\mathbf{U}(s) \right] +$$

$$+ \mathcal{L} \left\{ \sum_{j=0}^{j=k\le\mu-1} \left| \mathbf{e}^{(j)}(t) \right| \right\} - \mathbf{Z}(s) +$$

$$+ G_{HISOD_0}(s)\mathbf{D}_0^{\mu-1} + G_{HISOR_0}(s)\mathbf{R}_0^{\alpha-1} + G_{HISOU_0}(s)\mathbf{U}_0^{\mu-1}. \tag{10.86}$$

Let us subtract Equation (10.86) from Equation (10.69):

$$\mathbf{0}_N = .G_{HISOU}(s)\left(1 - e^{-\sigma_M s}\right)\mathbf{U}(s) + \mathbf{Z}(s) - \mathcal{L}\left\{ \sum_{j=0}^{j=k\le\mu-1} \left| \mathbf{e}^{(j)}(t) \right| \right\}.$$

In the limit as $\sigma_M \longrightarrow 0^+$ for σ_M defined by (10.84) the preceding equation reduces to:

$$\mathbf{0}_N = \mathbf{Z}(s) - \mathcal{L}\left\{ \sum_{j=0}^{j=k\le\mu-1} \left| \mathbf{e}^{(j)}(t) \right| \right\},$$

or in the *time* domain due to Equation (10.85):

$$\sum_{j=0}^{j=k\le\mu-1} \left| \mathbf{e}^{(j)}(t) \right| = \mathbf{Z}(t) = \mathbf{0}_N, \ \forall(t \ge \sigma_m) \in \mathfrak{T}_0,$$

i.e.,

$$\mathbf{Y}^{(j)}(t) = \mathbf{Y}_d^{(j)}(t), \ \forall(t \ge \sigma_J) \in \mathfrak{T}_0, \ \forall j = 0, 1, ..., k \le \mu - 1.$$

which proves the *k-th* order global elementwise natural trackability of the *HISO* plant (5.1), (5.2) on $\mathfrak{D}^k \times \mathfrak{Y}_d^k$ (Definition 314). ∎

If in Summary 220, (Section 10.2):

- "The $HISO$ plant (5.1), (5.2)" replaces "the ISO plant (3.1), (3.2)",
- $L_{HISO}(s)$ replaces $L_{pISO}(s)$,
- $G_{HISOU}(s)$ replaces $G_{ISOU}(s)$, and
- "(10.72) and 3) of Theorem 232" replace "(10.37) and 3) of Theorem 210"

then Summary 220 becomes valid for the $HISO$ plant (5.1), (5.2).

10.5 IIO system trackability

This chapter initiates the study of the IIO plant (6.1), (6.2) trackability properties. The complete solutions will follow for the most important of them.

10.5.1   Perfect trackability criteria

Equations (6.1), (6.2) determine the *time* domain mathematical model of the IIO plant (Section 6.1).

Equation (6.35) determines the Laplace transform $\mathbf{Y}(s)$ of the IIO plant output vector $\mathbf{Y}(t)$ and Equations (6.19)–(6.34) present the complex domain characteristics of the IIO plant (6.1), (6.2) in Section 6.1.

The general Definition 161 (Section 9.4) determines the *l-th* order perfect trackability of the plant on $\mathfrak{D}^i \times \mathfrak{Y}_d^k$. Lemma 162 (Section 9.4) reduces the study of the *l-th* order perfect trackability of the plant on $\mathfrak{D}^i \times \mathfrak{Y}_d^k$ to the study of the perfect trackability of the plant on $\mathfrak{D}^i \times \mathfrak{Y}_d^k$. This explains why it is crucial to explore the perfect trackability of the IIO plant (6.1), (6.2) on $\mathfrak{D}^i \times \mathfrak{Y}_d^k$. The result is directly applicable to the *l-th* order perfect trackability of the plant on $\mathfrak{D}^i \times \mathfrak{Y}_d^k$.

Equations (6.21), (6.23)-(6.28), (Section 6.1), lead to

$$G_{IIOU}(s) = \frac{L_{IIO}(s)}{p_{IIO}(s)} = p_{IIO}^{-1}(s)\, L_{IIO}(s). \qquad (10.87)$$

In order to relax the verification of the next proofs we accommodate Theorem 194 (Subsection 10.1.1) and Theorem 195 (Subsection 10.1.1) to the IIO plant (6.1), (6.2):

Theorem 241 *Full rank of the transfer function matrix relative to the control*

The full rank ρ of the transfer function matrix $G_{IIOU}(s)$ (6.23) is also the full rank of both its numerator polynomial matrix $L_{IIO}(s)$ (6.25) and its generating matrix L_{IIO} (6.29),

$$full\ rank G_{IIOU}(s) = \rho = \min(N, r)) =$$

$$= full\ rank \left[\begin{array}{c} \left(E^{(\nu)}S_N^{(\nu)}(s)\right)^{-1} R_y^{(\alpha)} S_\rho^{(\alpha)}(s) \left(A^{(\alpha)}S_\rho^{(\alpha)}(s)\right)^{-1} B^{(\mu)}S_r^{(\mu)}(s) + \\ + \left(E^{(\nu)}S_N^{(\nu)}(s)\right)^{-1} U^{(\mu)}S_r^{(\mu)}(s) \end{array} \right]$$

$$= full\ rank \left[p_{IIO}^{-1}(s)\ L_{IIO}(s) \right] = full\ rank L_{IIO}(s) \Longrightarrow$$

$$rank L_{IIO}^{(n)} = full\ rank L_{IIO}^{(n)} = \rho = \min(N, r). \tag{10.88}$$

Theorem 242 *Conditions for the perfect trackability on $\mathfrak{D}^k \times \mathfrak{Y}_d^k$*
In order for the IIO plant (6.1), (6.2) to be perfect trackable on $\mathfrak{D}^k \times \mathfrak{Y}_d^k$ it is necessary and sufficient that there is $s^* \in \mathbb{C}$ such that the plant transfer function matrix $G_{IIOU}(s)$ relative to control has the full rank N for $s = s^*$, $rank G_{IIOU}(s^*) = N$, so that the dimension r of the control vector $\mathbf{U}$, which is then the nominal control $\mathbf{U}_N$, is not less than the dimension N of the output vector $\mathbf{Y}$, i.e.,

$$\exists s^* \in \mathbb{C} \Longrightarrow rank G_{IIOU}(s^*) =$$

$$- rank \left[\begin{array}{c} \left\langle \begin{array}{c} \left(E^{(\nu)}S_N^{(\nu)}(s^*)\right)^{-1} R_y^{(\alpha)} S_\rho^{(\alpha)}(s^*) \bullet \\ \bullet \left(A^{(\alpha)}S_\rho^{(\alpha)}(s^*)\right)^{-1} B^{(\mu)}S_r^{(\mu)}(s^*) \end{array} \right\rangle + \\ + \left(E^{(\nu)}S_N^{(\nu)}(s^*)\right)^{-1} U^{(\mu)}S_r^{(\mu)}(s^*) \end{array} \right] = N \leq r. \tag{10.89}$$

The Laplace transform $\mathbf{U}_N(s)$ of the nominal control $\mathbf{U}_N(.)$ is then determined by

$$\mathbf{U}_N(s) = G_{IIOU}(s)^T \left\{ G_{IIOU}(s) G_{IIOU}(s)^T \right\}^{-1} \bullet$$

$$\bullet \left\{ \begin{array}{c} \mathbf{Y}_d(s) - G_{IIOD}(s)\mathbf{D}(s) - G_{IIOD_0}(s)\mathbf{D}_0^{\mu-1} \\ -G_{IIOR_0}(s)\mathbf{R}_0^{\alpha-1} - G_{IIOU_0}(s)\mathbf{U}_0^{\mu-1} - G_{IIOY_0}(s)\mathbf{Y}_0^{\nu-1} \end{array} \right\}. \tag{10.90}$$

Proof. *Necessity.* Let the *IIO* plant (6.1), (6.2) be perfect trackable on $\mathfrak{D}^k \times \mathfrak{Y}_d^k$. We recall Definition 161, Section 9.4. Its satisfaction guarantees that there is a control vector $\mathbf{U}(t)$, the Laplace transform of which obeys Equation (6.35). The perfect trackability is characterized by $\mathbf{Y}(t) \equiv \mathbf{Y}_d(t)$

that yields $\mathbf{U}(t) \equiv \mathbf{U}_N(t)$, i.e., by $\mathbf{Y}(s) \equiv \mathbf{Y}_d(s)$ that implies $\mathbf{U}(s) \equiv \mathbf{U}_N(s)$:

$$\mathbf{Y}_d(s) = \left\{ \begin{array}{c} G_{IIOD}(s)\,\mathbf{D}(s) + G_{IIOU}(s)\,\mathbf{U}_N(s) + G_{IIOD_0}(s)\,\mathbf{D}_0^{\mu-1} + \\ + G_{IIOR_{d0}}(s)\,\mathbf{R}_{d0}^{\alpha-1} + G_{IIOU_0}(s)\,\mathbf{U}_{N0}^{\mu-1} + G_{IIOY_0}(s)\,\mathbf{Y}_{d0}^{\nu-1} \end{array} \right\}.$$

$$(10.91)$$

The existence of the solution $\mathbf{U}(s)$ of this equation yields

$$rank G_{IIOU}(s) = full\ rank G_{IIOU}(s) = N \leq r.$$

This, the statement under 1) of Theorem 194 (Subsection 10.1.1) and Theorem 241 prove the necessity of the conditions (10.89). They guarantee the nonsingularity of $G_{IIOU}(s)\,G_{IIOU}^T(s)$ so that the solution $\mathbf{U}(s) = \mathbf{U}_N(s)$ of Equation (10.91) is given by Equation (10.90).

Sufficiency. Let the conditions (10.89) hold so that, in view of Theorem 194 and Theorem 241, *rank* $G_{IIOU}(s) = N$ and $\Gamma_{IIOU}(s)$ introduced by

$$\Gamma_{IIOU}(s) = G_{IIOU}^T(s)\left[G_{IIOU}(s)\,G_{IIOU}^T(s)\right]^{-1}.$$

$$(10.92)$$

is fully defined as well as $\mathbf{U}(s)$ determined by

$$\mathbf{U}(s) = \Gamma_{IIOU}(s)\bullet$$

$$\bullet\left\{ \begin{array}{c} \mathbf{Y}_d(s) - G_{IIOD}(s)\,\mathbf{D}(s) - G_{IIOD_0}(s)\,\mathbf{D}_0^{\mu-1} - \\ - G_{IIOR_0}(s)\,\mathbf{R}_{d0}^{\alpha-1} - G_{IIOU_0}(s)\,\mathbf{U}_0^{\mu-1} - G_{IIOY_0}(s)\,\mathbf{Y}_0^{\nu-1}. \end{array} \right\}$$

$$(10.93)$$

We eliminate $\mathbf{U}(s)$ from Equation (6.35) by replacing it by the right hand side of Equation (10.93) and by applying $G_{IIOU}(s)\,\Gamma_{IIOU}(s) = I_N$ and $\mathbf{R}_0^{\alpha-1} = \mathbf{R}_{d0}^{\alpha-1}$ due to Definition 161:

$$\mathbf{Y}(s) = G_{IIOD}(s)\,\mathbf{D}(s) +$$

$$+\left\{ \begin{array}{c} \mathbf{Y}_d(s) - G_{IIOD}(s)\,\mathbf{D}(s) - G_{IIOD_0}(s)\,\mathbf{D}_0^{\mu-1} - \\ - G_{IIOR_0}(s)\,\mathbf{R}_{d0}^{\alpha-1} - G_{IIOU_0}(s)\,\mathbf{U}_0^{\mu-1} - G_{IIOY_0}(s)\,\mathbf{Y}_0^{\nu-1}. \end{array} \right\} +$$

$$+\left\{ \begin{array}{c} G_{IIOD_0}(s)\,\mathbf{D}_0^{\mu-1} + G_{IIOR_0}(s)\,\mathbf{R}_0^{\alpha-1} + \\ + G_{IIOU_0}(s)\,\mathbf{U}_0^{\mu-1} + G_{IIOY_0}(s)\,\mathbf{Y}_0^{\nu-1} \end{array} \right\} = \mathbf{Y}_d(s).$$

This result, $\mathbf{Y}(s) = \mathbf{Y}_d(s)$ that implies $\mathbf{U}_N(s) = \mathbf{U}(s)$ (10.93), in the *time* domain reads

$$\mathbf{Y}(t) = \mathbf{Y}_d(t),\ \mathbf{U}_N(t) = \mathbf{U}(t),\ \forall\,[t, D\,(.)\,, \mathbf{Y}_d(.)] \in \mathfrak{T}_0 \times \mathfrak{D}^k \times \mathfrak{Y}_d^k.$$

This proves the perfect trackability of the *IIO* plant (6.1), (6.2) on $\mathfrak{D}^k \times \mathfrak{Y}_d^k$ in view of Definition 161. ∎

Comment 243 *Equation (10.90) defines the nominal control* $\mathbf{U}_N(t)$ *in terms of the disturbance vector* $D\,(t)$. *This is disadvantage due to often obstacles to measure the disturbance vector.*

10.5.2   Conditions for perfect natural trackability

Definition 310 (Section 9.4) introduced the notion and the sense of the perfect natural trackability on $\mathfrak{D}^k \times \mathfrak{Y}_d^k$, which is valid for the IIO plant (6.1), (6.2).

Theorem 244 *Conditions for the perfect natural trackability of the IIO plant (6.1), (6.2) on $\mathfrak{D}^k \times \mathfrak{Y}_d^k$*

In order for the IIO plant (6.1), (6.2) to be perfect natural trackable on $\mathfrak{D}^k \times \mathfrak{Y}_d^k$ it is necessary and sufficient that the plant transfer function matrix $G_{IIOU}(s)$ relative to control has the full row rank N, so that the dimension r of the control vector $\mathbf{U}$ is not less than the dimension N of the output vector $\mathbf{Y}$, i.e., that all conditions (10.89) hold.

Proof. *Necessity.* Let the IIO plant (6.1), (6.2) be perfect natural trackable on $\mathfrak{D}^k \times \mathfrak{Y}_d^k$. Definitions 310 and 161 guarantee that the plant is also perfect trackable on $\mathfrak{D}^k \times \mathfrak{Y}_d^k$. Theorem 242 implies the necessity of the conditions (10.89).

Sufficiency. Let the conditions (10.89) hold. The IIO plant (6.1), (6.2) is perfect trackable on $\mathfrak{D}^k \times \mathfrak{Y}_d^k$, Theorem 210. We should show that the perfect tracking control can be synthesized without using information about the plant state and about the disturbance $\mathbf{D}(.) \in \mathfrak{D}^k$, i.e., that it is natural tracking control (Definition 166, Section 9.4). We recall Equation (6.35). Let $\sigma \in \mathfrak{R}^+$ be arbitrarily small, i.e., $\sigma \longrightarrow 0^+$ and in the ideal case $\sigma = 0^+$. Let

$$\phi(.) : \mathfrak{T}_0 \longrightarrow \mathfrak{R}^N, \ \phi(t) \in \mathfrak{C},$$
$$\phi(t) = \mathbf{0}_N, \ \forall t \in [\sigma, \infty[, \ \phi(0) = -\mathbf{e}(0), \tag{10.94}$$

Let Equation (10.92) define $\Gamma_{IIOU}(s)$ and let the control be defined in the complex domain by:

$$\mathbf{U}(s) = (1 - e^{-\sigma s})^{-1} \Gamma_{IIOU}(s) \left[\Phi(s) + \mathbf{E}(s) \right], \ \ \Phi(s) = \mathcal{L}\{\phi(t)\}. \tag{10.95}$$

The control $\mathbf{U}(.)$ (10.95) does not depend on the plant state or on the disturbance $\mathbf{D}(.)$. The control $\mathbf{U}(.)$ (10.95) is natural control (Definition 166, Section 9.4). We multiply Equation (10.95) on the left by $G_{IIOU}(s)$ and use $G_{IIOU}(s)\Gamma_{IIOU}(s) \equiv I_N$,

$$G_{IIOU}(s)\mathbf{U}(s) = (1 - e^{-\sigma s})^{-1} \left[\Phi(s) + \mathbf{E}(s) \right] \tag{10.96}$$

After replacing $\mathbf{U}(s)$ with the right-hand side of (10.96) into Equation (6.35) the result reads:

$$(1 - e^{-\sigma s}) \left\{ \begin{array}{c} \mathbf{Y}(s) - G_{IIOD}(s)\,\mathbf{D}(s) - G_{IIOD_0}(s)\,\mathbf{D}_0^{\mu-1} - \\ -G_{IIOR_0}(s)\,\mathbf{R}_0^{\alpha-1} - G_{IIOU_0}(s)\,\mathbf{U}_0^{\mu-1} - \\ -G_{IIOY_0}(s)\,\mathbf{Y}_0^{\nu-1} \end{array} \right\} =$$

$$= \Phi(s) + \mathbf{E}(s).$$

Let $\sigma \longrightarrow 0^+$, or $\sigma = 0^+$ in the ideal case. Then the preceding equation becomes

$$\mathbf{0}_N = \Phi(s) + \mathbf{E}(s) \Longrightarrow \mathbf{0}_N = \mathbf{e}(t) + \phi(t),\ \forall t \in \mathfrak{T}_0.$$

Hence,

$$\mathbf{e}(t) = -\phi(t) = \mathbf{0}_N,\ \forall t \in \mathfrak{T}_0$$

due to $\mathbf{Y}_0 = \mathbf{Y}_{d0}$ in view of Definition 161 and (10.94), or equivalently:

$$\mathbf{e}(t) = \mathbf{Y}_d(t) - \mathbf{Y}(t) = \mathbf{0}_N,\ \forall t \in \mathfrak{T}_0.$$

The natural control $\mathbf{U}(.)$ defined by (10.95) satisfies Definition 310. The IIO plant (6.1), (6.2) is perfect natural trackable on $\mathfrak{D}^k \times \mathfrak{Y}_d^k$. ∎

Comment 245 *Theorem 242 and Theorem 244 show that for the IIO plant (6.1), (6.2) to be perfect natural trackable on $\mathfrak{D}^k \times \mathfrak{Y}_d^k$ it is necessary and sufficient that the plant is perfect trackable on $\mathfrak{D}^k \times \mathfrak{Y}_d^k$. This confirms Comment 168 (Section 9.4).*

Comment 246 *The proof of Theorem 244 demonstrates that the control implementation can be successful without using any information about the disturbance vector. The disturbance vector can be unmeasurable, which expresses the reality. That is the significant advantage of the natural perfect trackability over the perfect trackability.*

Note 201, (Section 10.1) is valid in this framework if $\Gamma_{IIOU}(s)$ replaces $\Gamma_{IOU}(s)$.

10.5.3   Imperfect trackability criteria

This section discovers the necessary and sufficient conditions for various imperfect trackability properties of the IIO system (6.1), (6.2). Since their exploration begins in the sequel we present the detailed proofs of the next theorems.

Theorem 247 *Conditions for the global complete trackability of the IIO plant (6.1), (6.2) on* $\mathfrak{D}^k \times \mathfrak{Y}_d^k$

In order for the IIO plant (6.1), (6.2) to be global complete trackable on $\mathfrak{D}^k \times \mathfrak{Y}_d^k$ *it is necessary and sufficient that the plant transfer function matrix* $G_{IIOU}(s)$ *relative to control has the full row rank* N, *so that the dimension* r *of the control vector* $\mathbf{U}$ *is not less than the dimension* N *of the output vector* $\mathbf{Y}$, *i.e., that all conditions (10.89) hold.*

Proof. *Necessity.* Let the *IIO* plant (6.1), (6.2) be global complete trackable on $\mathfrak{D}^k \times \mathfrak{Y}_d^k$. Definition 172 (Section 9.5) is fulfilled. The plant is also perfect trackable on $\mathfrak{D}^k \times \mathfrak{Y}_d^k$ (due to Definition 172 and Theorem 174, Section 9.5). Theorem 242 proves the necessity of the conditions (10.89) for the global complete trackability on $\mathfrak{D}^k \times \mathfrak{Y}_d^k$.

Sufficiency. Let the conditions (10.89) hold. Theorem 194 and Theorem 241, The matrix $G_{IIOU}(s)$ has the full rank N (the conditions (10.89) linked with Theorem 194 and Theorem 241). Let $[\mathbf{D}(.), \mathbf{Y}_d(.)] \in \mathfrak{D}^k \times \mathfrak{Y}_d^k$, the instant $\sigma \in Int\ \mathfrak{T}_0$, and $\mathbf{Y}_0^{\nu-1} \in \mathfrak{R}^{\nu N}$ be arbitrarily chosen (see Definition 172). Let the control be defined in the complex domain by

$$\mathbf{U}(s) = \Gamma_{IIOU}(s) \bullet$$

$$\bullet \left[\begin{array}{c} \mathbf{Y}(s) - G_{IIOD}(s)\,\mathbf{D}(s) - G_{IIOD_0}(s)\,\mathbf{D}_0^{\mu-1} - \\ -G_{IIOR_0}(s)\,\mathbf{R}_0^{\alpha-1} - G_{IIOU_0}(s)\,\mathbf{U}_0^{\mu-1} - G_{IIOY_0}(s)\,\mathbf{Y}_0^{\nu-1} - \\ -\mathbf{Z}(s) + \mathbf{E}(s) \end{array} \right],$$

together with

$$\mathbf{Z}(t) = \left\{ \begin{array}{c} |\mathbf{e}(0)|\left(1 - \frac{t}{\sigma}\right),\ t \in [0,\ \sigma], \\ \mathbf{0}_N,\ \forall (t \geq \sigma) \in \mathfrak{T}_0 \end{array} \right\} \in \mathfrak{C}(\mathfrak{T}_0),$$

$$\mathbf{Z}(s) = \mathcal{L}\{\mathbf{Z}(t)\}. \tag{10.97}$$

For such control, Equation (6.35) becomes

$$\mathbf{Y}(s) = G_{IIOD}(s)\,\mathbf{D}(s) + \mathbf{Y}(s) - G_{IIOD}(s)\,\mathbf{D}(s) -$$

$$-\mathbf{Z}(s) + |\mathbf{E}(s)| - G_{IIOD_0}(s)\,\mathbf{D}_0^{\mu-1} - G_{IIOR_0}(s)\,\mathbf{R}_0^{\alpha-1} -$$

$$-G_{IIOU_0}(s)\,\mathbf{U}_0^{\mu-1} - G_{IIOY_0}(s)\,\mathbf{Y}_0^{\nu-1} + G_{IIOD_0}(s)\,\mathbf{D}_0^{\mu-1} +$$

$$+G_{IIOR_0}(s)\,\mathbf{R}_0^{\alpha-1} + G_{IIOU_0}(s)\,\mathbf{U}_0^{\mu-1} + G_{IIOY_0}(s)\,\mathbf{Y}_0^{\nu-1} \Longrightarrow$$

$$\mathbf{Y}(s) = \mathbf{Y}(s) - \mathbf{Z}(s) + \mathbf{E}(s)$$

due to $G_{IIOU}(s)\,\Gamma_{IIOU}(s) \equiv I_N$, i.e.,

$$\mathbf{0}_N = -\mathbf{Z}(s) + \mathbf{E}(s) \Longrightarrow \mathbf{e}(t) = \mathbf{Z}(t),\ \forall t \in \mathfrak{T}_0$$

This and (10.97) imply

$$\mathbf{e}(t) = \mathbf{0}_N, \ \forall (t \geq \sigma) \in \mathfrak{T}_0 \Longrightarrow \mathbf{Y}(t) = \mathbf{Y}_d(t), \ \forall (t \geq \sigma) \in \mathfrak{T}_0,$$

which proves the global complete trackability on $\mathfrak{D}^k \times \mathfrak{Y}_d^k$ of the IIO plant (6.1), (6.2) due to Definition 172. ∎

Theorem 248 *On the global k-th order elementwise trackability of the IIO plant (6.1), (6.2) on $\mathfrak{D}^k \times \mathfrak{Y}_d^k$*
 In order for the IIO plant (6.1), (6.2) to be the k-th order global elementwise trackable on $\mathfrak{D}^k \times \mathfrak{Y}_d^k$ it is necessary and sufficient that the plant transfer function matrix $G_{IIOU}(s)$ relative to control has the full row rank N so that the dimension r of the control vector $\mathbf{U}$ is not less than the dimension N of the output vector $\mathbf{Y}$, i.e., that all conditions (10.89) hold.

Proof. *Necessity.* Let the IIO plant (6.1), (6.2) be the *k-th* order global elementwise trackable on $\mathfrak{D}^k \times \mathfrak{Y}_d^k$. Definition 313 (Section 9.5) is fulfilled. The plant and control satisfy also Definition 172. It and Theorem 174 prove the necessity of the conditions (10.89).
 Sufficiency. Let the conditions (10.89) be satisfied so that $rankG_{IIOU}(s)$
$=N$. Let $[\mathbf{D}(.), \mathbf{Y}_d(.)] \in \mathfrak{D}^k \times \mathfrak{Y}_d^k$, $\sigma \in (Int \ \mathfrak{T}_0)^{k+1}$, $\sigma = \left[\sigma_0 \vdots \sigma_1 \vdots ... \vdots \sigma_k\right]^T$,
and $\mathbf{Y}_0^{\nu-1} \in \mathfrak{R}^{\nu N}$ be arbitrarily chosen (in view of Definition 313). We select the control vector $\mathbf{U}(t)$ with the Laplace transform $\mathbf{U}(s)$ to satisfy:

$$\mathbf{U}(s) = \Gamma_{IIOU}(s) \bullet$$

$$\bullet \left[\begin{array}{c} \mathbf{Y}(s) - G_{IIOPD}(s)\mathbf{D}(s) - G_{IIOD_0}(s)\mathbf{D}_0^{\mu-1} - \\ -G_{IIOR_0}(s)\mathbf{R}_0^{\alpha-1} - G_{IIOU_0}(s)\mathbf{U}_0^{\mu-1} - \\ -G_{IIOY_0}(s)\mathbf{Y}_0^{\nu-1} - \mathbf{Z}(s) + \mathcal{L}\left\{\sum_{j=0}^{j=k\leq\mu-1}\left|\mathbf{e}^{(j)}(t)\right|\right\} \end{array} \right] \qquad (10.98)$$

for the Laplace transform $\mathbf{Z}(s)$ of $\mathbf{Z}(t)$ defined by

$$\mathbf{Z}(t) = \left\{ \begin{array}{c} \sum_{j=0}^{j=k\leq\mu-1}\left(\left|\mathbf{e}^{(j)}(0)\right|\left(1 - \frac{t}{\sigma_m}\right), \ t \in [0, \sigma_m]\right), \\ \mathbf{0}_N, \ \forall (t \geq \sigma_m) \in \mathfrak{T}_0 \end{array} \right\} \in \mathfrak{C}(\mathfrak{T}_0),$$

$$0 < \sigma_m = \min(\sigma_1, \sigma_2, ..., \sigma_k) \leq \sigma_M, \ \mathbf{Z}(s) = \mathcal{L}\{\mathbf{Z}(t)\}. \qquad (10.99)$$

For the control given in Equation (10.98) the plant complex domain mathematical model (6.35) becomes, in view of $G_{IIOU}(s)\Gamma_{IIOU}(s) \equiv I_N$,

$$\mathbf{Y}(s) = G_{IIOPD}(s)\,\mathbf{D}(s) + \mathbf{Y}(s) - G_{IIOPD}(s)\mathbf{D}(s) -$$

$$-G_{IIOD_0}(s)\,\mathbf{D}_0^{\mu-1} - G_{IIOR_0}(s)\,\mathbf{R}_0^{\alpha-1} - G_{IIOU_0}(s)\,\mathbf{U}_0^{\mu-1} -$$

$$-G_{IIOY_0}(s)\,\mathbf{Y}_0^{\nu-1} - \mathbf{Z}(s) + \mathcal{L}\left\{ \sum_{j=0}^{j=\leq\mu-1} \left| \mathbf{e}^{(j)}(t) \right| \right\} + G_{IIOD_0}(s)\,\mathbf{D}_0^{\mu-1} +$$

$$+G_{IIOR_0}(s)\,\mathbf{R}_0^{\alpha-1} + G_{IIOU_0}(s)\,\mathbf{U}_0^{\mu-1} + G_{IIOY_0}(s)\,\mathbf{Y}_0^{\nu-1} \Longrightarrow$$

$$\mathbf{0}_N = . - \mathbf{Z}(s) + \mathcal{L}\left\{ \sum_{j=0}^{j=\leq\mu-1} \left| \mathbf{e}^{(j)}(t) \right| \right\},$$

or in the *time* domain:

$$\sum_{j=0}^{j=\leq\mu-1} \left| \mathbf{e}^{(j)}(t) \right| = \mathbf{Z}(t) = \mathbf{0}_N, \ \forall (t \geq \sigma_m) \in \mathfrak{T}_0,$$

i.e.,

$$\mathbf{Y}^{(j)}(t) = \mathbf{Y}_d^{(j)}(t), \ \forall (t \geq \sigma_j) \in \mathfrak{T}_0, \ \forall j = 0, 1, ..., k \leq \mu - 1,$$

which proves the *k-th* order global elementwise trackability of the *IIO* plant (6.1), (6.2) on $\mathfrak{D}^\eta \times \mathfrak{Y}_d^\nu$ (Definition 313). ∎

10.5.4 Conditions for imperfect natural trackability

Theorem 249 *Conditions for global complete natural trackability of the IIO plant (6.1), (6.2) on $\mathfrak{D}^k \times \mathfrak{Y}_d^k$*
  In order for the IIO plant (6.1), (6.2) to be global complete natural trackable on $\mathfrak{D}^k \times \mathfrak{Y}_d^k$ it is necessary and sufficient that the plant transfer function matrix $G_{IIOU}(s)$ relative to control has the full row rank N so that the dimension r of the control vector $\mathbf{U}$ is not less than the dimension N of the output vector $\mathbf{Y}$, i.e., that all conditions (10.89) hold.

Proof. *Necessity.* Let the *IIO* plant (6.1), (6.2) be global complete natural trackable on $\mathfrak{D}^k \times \mathfrak{Y}_d^k$. Definition 312 is applicable. It and Definition 172 confirm that the plant is global complete trackable on $\mathfrak{D}^k \times \mathfrak{Y}_d^k$ so that the conditions (10.89) are necessary due to Theorem 247.

Sufficiency. Let the conditions (10.89) be valid. They ensure the full rank N to $G_{IIOU}(s)$. The IIO plant (6.1), (6.2) is global complete trackable on $\mathfrak{D}^k \times \mathfrak{Y}_d^k$ (Theorem 247). Let the Laplace transform of the control vector be defined by Equations (10.94)-(10.95). By repeating from Equation (10.95) the sufficiency part of the proof of Theorem 244 we prove that the so defined natural tracking control implies

$$\mathbf{e}(t) = \mathbf{Y}(t) - \mathbf{Y}_d(t) = \mathbf{0}_N, \ \forall (t \geq \sigma) \in \mathfrak{T}_0.$$

This proves the global complete natural trackability of the IIO plant (6.1), (6.2) on $\mathfrak{D}^k \times \mathfrak{Y}_d^k$ in view of Definition 312. ∎

Theorem 250 *Conditions for the k-th order global elementwise natural trackability of the IIO plant (6.1), (6.2) on $\mathfrak{D}^k \times \mathfrak{Y}_d^k$*

In order for the IIO plant (6.1), (6.2) to be the k-th order global elementwise natural trackable on $\mathfrak{D}^k \times \mathfrak{Y}_d^k$, $0 \leq k \leq \mu - 1$, it is necessary and sufficient that the plant transfer function matrix $G_{IIOU}(s)$ relative to control has the full row rank N so that the dimension r of the control vector $\mathbf{U}$ is not less than the dimension N of the output vector $\mathbf{Y}$, i.e., that all conditions (10.89) hold.

Proof. *Necessity.* Let the IIO plant (6.1), (6.2) be the k-th order global elementwise natural trackable on $\mathfrak{D}^k \times \mathfrak{Y}_d^k$ for $0 \leq k \leq \mu - 1$. Definition 314 (Section 9.5) is fulfilled. The plant and control satisfy also Definition 172. It and Theorem 247 prove the necessity of the conditions (10.89).

Sufficiency. Let the conditions (10.89) be satisfied. Hence, $rank G_{IIOU}(s) = N$ (Theorem 194 and Theorem 241). Let $[\mathbf{D}(.), \mathbf{Y}_d(.)] \in \mathfrak{D}^k \times \mathfrak{Y}_d^k$, $\sigma \in (Int \ \mathfrak{T}_0)^{k+1}$, $\sigma = \begin{bmatrix} \sigma_0 \vdots \sigma_1 \vdots ... \vdots \sigma_k \end{bmatrix}^T$, $0 \leq k \leq \mu - 1$, and $\mathbf{Y}_0^{\nu-1} \in \mathfrak{R}^{\nu N}$ be arbitrarily chosen (in view of Definition 314). We select the control vector $\mathbf{U}(t)$ to satisfy the following complex domain equation:

$$G_{IIOU}(s) \mathbf{U}(s) =$$

$$= e^{-\sigma_M s} G_{IIOU}(s) \mathbf{U}(s) + \mathcal{L} \left\{ \sum_{j=0}^{j=k\leq\mu-1} \left| \mathbf{e}^{(j)}(t) \right| \right\} - \mathbf{Z}(s) \qquad (10.100)$$

for

$$0 < \sigma_M = \max(\sigma_0, \sigma_1, ... , \sigma_k) \lll 0^+, \ 0 \leq k \leq \mu - 1,$$

$$i.e., \ \sigma_M \longrightarrow 0^+, \qquad (10.101)$$

and for the Laplace transform $\mathbf{Z}(s)$ of $\mathbf{Z}(t)$ that is determined by

$$\mathbf{Z}(t) = \left\{ \begin{array}{l} \sum_{j=0}^{j=k\leq\mu-1} \left(\left|\mathbf{e}^{(j)}(0)\right| \left(1 - \frac{t}{\sigma_m}\right), \ t \in [0, \sigma_m]\right), \\ \mathbf{0}_N, \ \forall(t \geq \sigma_m) \in \mathfrak{T}_0 \end{array} \right\} \in \mathfrak{C}(\mathfrak{T}_0),$$

$$0 < \sigma_m = \min(\sigma_0, \sigma_1,... , \sigma_k) \leq \sigma_M, \ \mathbf{Z}(s) = \mathcal{L}\{\mathbf{Z}(t)\}. \qquad (10.102)$$

For such control the plant complex domain mathematical model (6.35) becomes

$$\mathbf{Y}(s) = G_{IIOD}(s)\,\mathbf{D}(s) + G_{IIOU}(s)\left[e^{-\sigma_M s}\mathbf{U}(s)\right] +$$

$$+\mathcal{L}\left\{ \sum_{j=0}^{j=k\leq\mu-1} \left|\mathbf{e}^{(j)}(t)\right| \right\} - \mathbf{Z}(s) + G_{IIOD_0}(s)\,\mathbf{D}_0^{\mu-1} +$$

$$+G_{IIOR_0}(s)\,\mathbf{R}_0^{\alpha-1} + G_{IIOU_0}(s)\,\mathbf{U}_0^{\mu-1} + G_{IIOY_0}(s)\,\mathbf{Y}_0^{\nu-1}. \qquad (10.103)$$

Let us subtract Equation (10.103) from Equation (6.35):

$$\mathbf{0}_N = .G_{IIOU}(s)\left(1 - e^{-\sigma_M s}\right)\mathbf{U}(s) + \mathbf{Z}(s) - \mathcal{L}\left\{ \sum_{j=0}^{j=k\leq\mu-1} \left|\mathbf{e}^{(j)}(t)\right| \right\}.$$

In the limit as $\sigma_M \longrightarrow 0^+$ for σ_M defined by (10.101) the preceding equation reduces to:

$$\mathbf{0}_N = \mathbf{Z}(s) - \mathcal{L}\left\{ \sum_{j=0}^{j=k\leq\mu-1} \left|\mathbf{e}^{(j)}(t)\right| \right\},$$

or in the *time* domain due to Equation (10.102):

$$\sum_{j=0}^{j=k\leq\mu-1} \left|\mathbf{e}^{(j)}(t)\right| = \mathbf{Z}(t) = \mathbf{0}_N, \ \forall(t \geq \sigma_m) \in \mathfrak{T}_0,$$

i.e.,

$$\mathbf{Y}^{(j)}(t) = \mathbf{Y}_d^{(j)}(t), \ \forall(t \geq \sigma_j) \in \mathfrak{T}_0, \ \forall j = 0, 1, ..., k \leq \mu - 1.$$

which proves the $k-th$ order global elementwise natural trackability of the *IIO* plant (6.1), (6.2) on $\mathfrak{D}^k \times \mathfrak{Y}_d^k$ (Definition 314). ∎

Summary 251 *Trackability conditions, IIO plant properties and external actions*

The trackability properties of the IIO plant (6.1), (6.2):

- Are independent of the external (control and disturbance) actions on the plant,

- Depend on the plant parameters, i.e., on the rank of the corresponding plant transfer function matrix relative to control, or equivalently, depend on the rank of the polynomial matrix $L_{IIO}(s)$.

Besides, the following statements hold:

- The trackability conditions are purely algebraic and simple for the implementation,

- The identity of the trackability conditions for all trackability properties discovers that the conditions do not depend on the selected trackability type. The conditions reflect that they depend only on the plant control characteristic that is formulated in the conditions (10.37) of all above theorems, i.e.,

$$rank G_{IIOU}(s) = full \ rank G_{IOU}(s) = full \ rank L_{IIO}(s) = N \leq r.$$

- The trackability conditions do not depend on the stability property of the plant or on the treated trackability property or on disturbance,

- The trackability conditions are different from the controllability conditions,

- The characteristics of the plant part that transmits the control action on the plant processing part, i.e., on the plant process, are those plant characteristics that determine whether the plant is trackable or not; such plant characteristics are mathematically expressed by the plant transfer function matrix relative to control, or equivalently by the polynomial matrix $L_{IIO}(s)$,

- The trackability conditions guarantee the output controllability due to (10.89) and Theorem 195 (for more details see the book [171]),

- The common trackability conditions (10.37) agree with the Fundamental control principle (Axiom 104).

The designer of the IIO plant (6.1), (6.2) should ensure its trackability.

Part IV

TRACKING CONTROL

Chapter 11

Linear tracking control (LiTC)

11.1 Common systems descriptions

Part I contains the complex domain descriptions of the plants and controllers. They can be set into the next forms that lead to their common general form and to the general form of the control system description. All are given in the complex domain.

11.1.1 Plants descriptions

Equations (2.24), (2.31)–(2.35), (Section 2.1), yield both the complex domain description $\mathbf{Y}(s)$ of the IO plant (2.15), (Section 2.1), output response $\mathbf{Y}(t)$:

$$\mathbf{Y}(s) = G_{IOD}(s)\,\mathbf{D}(s) + G_{IOU}(s)\,\mathbf{U}(s) + G_{IOD_0}(s)\,\mathbf{D}_0^{\eta-1} +$$
$$+ G_{IOU_0}(s)\,\mathbf{U}_0^{\mu-1} + G_{IOY_0}(s)\,\mathbf{Y}_0^{\nu-1}, \quad \mathbf{X}_0 = \mathbf{Y}_0^{\nu-1}. \qquad (11.1)$$

The complex domain description of the ISO plant (3.1), (3.2) output vector $\mathbf{Y}(t)$ results from Equations (3.12), (3.13), (3.23)–(3.30), all from Section 3.1:

$$\mathbf{Y}(s) = G_{ISOD}(s)\,\mathbf{D}(s) + G_{ISOU}(s)\,\mathbf{U}(s) + G_{ISOX_0}(s)\,\mathbf{X}_0. \qquad (11.2)$$

Equations (4.23), (4.35)–(4.44) determine the Laplace transform $\mathbf{Y}(s)$ of the output vector $\mathbf{Y}(t)$ of the $EISO$ plant (4.1), (4.2), all from Section 4.1:

$$\mathbf{Y}(s) = G_{EISOD}(s)\,\mathbf{D}(s) + G_{EISOU}(s)\,\mathbf{U}(s) + G_{EISOD_0}(s)\,\mathbf{D}_0^{\mu-1} +$$
$$+ G_{EISOU_0}(s)\,\mathbf{U}_0^{\mu-1} + G_{EISOX_0}(s)\,\mathbf{X}_0. \qquad (11.3)$$

Equations (5.5), (5.17)–(5.26) give the following complex domain description $\mathbf{Y}(s)$ of the output vector $\mathbf{Y}(t)$ of the $HISO$ plant (5.1), (5.2) (all from Section 5.1):

$$\mathbf{Y}(s) = G_{HISOD}(s)\,\mathbf{D}(s) + G_{HISOU}(s)\,\mathbf{U}(s) + G_{HISOD_0}(s)\,\mathbf{D}_0^{\mu-1}+$$
$$+G_{HISOU_0}(s)\,\mathbf{U}_0^{\mu-1} + G_{HISOX_0}(s)\,\mathbf{X}_{P0},$$
$$G_{HISOX_0}(s) = G_{HISOR_0}(s)\,,\ \mathbf{X}_0 = \mathbf{R}_0^{\alpha-1}. \qquad (11.4)$$

In order to get the complex domain description $\mathbf{Y}(s)$ of the IIO plant (6.1), (6.2) output vector $\mathbf{Y}(t)$ we connect Equations (6.9), (6.19), (6.22)–(6.34) (all from Section 6.1):

$$\mathbf{Y}(s) = G_{IIOD}(s)\,\mathbf{D}(s) + G_{IIOU}(s)\,\mathbf{U}(s) + G_{IIOD_0}(s)\,\mathbf{D}_0^{\mu-1}+$$
$$+G_{IIOU_0}(s)\,\mathbf{U}_0^{\mu-1} + G_{IIOR_0}(s)\,\mathbf{R}_0^{\alpha-1} + G_{IIOY_0}(s)\,\mathbf{Y}_0^{\nu-1},$$
$$G_{IIOP_0}(s) = G_{IIOR_0}(s)\,,\ \mathbf{X}_0 = \mathbf{R}_0^{\alpha-1}. \qquad (11.5)$$

The preceding Equations (11.1)-(11.5) imply the following **common general form of the Laplace transform $\mathbf{Y}(s)$ of the linear plant output vector $\mathbf{Y}(t)$**:

$$\mathbf{Y}(s) = G_{PD}(s)\,\mathbf{D}(s) + G_{PU}(s)\,\mathbf{U}(s) + G_{PD_0}(s)\,\mathbf{D}_0^{\mu-1}+$$
$$+G_{PU_0}(s)\,\mathbf{U}_0^{\mu-1} + G_{PX_0}(s)\,\mathbf{X}_0 + G_{PY_0}(s)\,\mathbf{Y}_0^{\nu-1}. \qquad (11.6)$$

11.1.2   Controllers descriptions

Equations (2.62), (2.66)–(2.68) determine the complex domain description of the IO closed loop controller (2.59) output $\mathbf{U}(t)$, (Section 2.3):

$$\mathbf{U}(s) = G_{IOCE}(s)\,\mathbf{E}(s) + G_{IOCE_0}(s)\,\mathbf{e}_0^{\mu-1} + G_{IOU_0}(s)\,\mathbf{U}^{\nu-1}. \qquad (11.7)$$

The Laplace transform $\mathbf{U}(s)$ of the output vector $\mathbf{U}(t)$ of the ISO controller (3.40), (3.43) comes out from Equations (3.46), (3.57), (3.58), (Section 3.3):

$$\mathbf{U}(s) = G_{ISOCE}(s)\,\mathbf{E}(s) + G_{ISOCX_{C0}}(s)\,\mathbf{X}_{C0}. \qquad (11.8)$$

Due to Equations (4.82), (4.88)–(4.90) the complex domain form $\mathbf{U}(s)$ of the output vector $\mathbf{U}(t)$ of the $EISO$ controller (4.69), (4.70), (4.3), reads:

$$\mathbf{U}(s) = G_{EISOCE}(s)\,\mathbf{E}(s) + G_{EISOCE_0}(s)\,\mathbf{e}_0^{\mu-1} + G_{EISOCX_{C0}}(s)\,\mathbf{X}_{C0}. \qquad (11.9)$$

The Laplace transform $\mathbf{U}(s)$ of the output vector $\mathbf{U}(t)$ of the *HISO* controller (5.38), (5.39) has the following form in view of Equations (5.43), (5.53)–(5.55) (Section 5.3):

$$\mathbf{U}(s) = G_{HISOCE}(s)\mathbf{E}(s) + G_{HISOCE_0}(s)\mathbf{e}_0^{\mu-1} + G_{HISOCX_{C0}}(s)\mathbf{X}_{C0},$$
$$G_{HISOCX_{C0}}(s) = G_{HISOCR_{C0}}(s), \quad \mathbf{X}_{C0} = \mathbf{R}_{C0}^{\alpha-1}. \tag{11.10}$$

Equations (6.54), (6.60), (6.64)–(6.68) determine the Laplace transform of the output vector of the *IIO* controller 6.51), (6.52) (Section 6.3):

$$\mathbf{U}(s) = G_{IIOCE}(s)\mathbf{E}(s) + G_{IIOCE_0}(s)\mathbf{e}_0^{\mu-1} + G_{IIOCU_0}(s)\mathbf{U}_0^{\nu-1} +$$
$$+G_{IIOCX_{C0}}(s)\mathbf{X}_{C0}, \quad G_{IIOCX_{C0}}(s) = G_{IIOCR_{C0}}(s), \quad \mathbf{X}_{C0} = \mathbf{R}_{C0\mp}^{\alpha-1}. \tag{11.11}$$

Equations (11.7)–(11.11) lead jointly to **the common general complex domain description of the control vector** being the output vector of the controller:

$$\mathbf{U}(s) = G_{CE}(s)\mathbf{E}(s) + G_{CE_0}(s)\mathbf{e}_0^{\mu-1} + G_{CU_0}(s)\mathbf{U}_0^{\nu-1} + G_{CX_{C0}}(s)\mathbf{X}_{C0}. \tag{11.12}$$

11.1.3   Control systems descriptions

The Laplace transform of the output vector of the linear feedback control system results from the common general plant output Equation (11.6):

$$\mathbf{Y}(s) = G_{PD}(s)\mathbf{D}(s) + G_{PU}(s)\mathbf{U}(s) + G_{PD_0}(s)\mathbf{D}_0^{\mu-1} +$$
$$+G_{PU_0}(s)\mathbf{U}_0^{\mu-1} + G_{PX_0}(s)\mathbf{X}_0 + G_{PY_0}(s)\mathbf{Y}_0^{\nu-1}, \tag{11.13}$$

and from the common general controller output Equation (11.12) that permits to eliminate $\mathbf{U}(s)$ from Equation (11.13):

$$\mathbf{Y}(s) = G_{PD}(s)\mathbf{D}(s) + G_{PU}(s)G_{CE}(s)\mathbf{E}(s) + G_{PU}(s)G_{CE_0}(s)\mathbf{e}_0^{\mu-1} +$$
$$+G_{PU}(s)G_{CU_0}(s)\mathbf{U}_0^{\nu-1} + G_{PU}(s)G_{CX_{C0}}(s)\mathbf{X}_{C0} +$$
$$+G_{PD_0}(s)\mathbf{D}_0^{\mu-1} + G_{PU_0}(s)\mathbf{U}_0^{\mu-1} + G_{PX_0}(s)\mathbf{X}_0 + G_{PY_0}(s)\mathbf{Y}_0^{\nu-1}.$$

We replace $\mathbf{E}(s)$ by $\mathbf{Y}_d(s) - \mathbf{Y}(s)$ and after grouping all terms containing $\mathbf{Y}(s)$ it follows:

$$[I_N + G_{PU}(s)G_{CE}(s)]\mathbf{Y}(s) = G_{PD}(s)\mathbf{D}(s) + G_{PU}(s)G_{CE}(s)\mathbf{Y}_d(s) +$$
$$+G_{PD_0}(s)\mathbf{D}_0^{\mu-1} + G_{PU}(s)G_{CE_0}(s)\mathbf{e}_0^{\mu-1} +$$
$$+[G_{PU}(s)G_{CU_0}(s) + G_{PU_0}(s)]\mathbf{U}_0^{\nu-1} +$$
$$+G_{PU}(s)G_{CX_{C0}}(s)\mathbf{X}_{C0} + G_{PX_0}(s)\mathbf{X}_{P0} + G_{PY_0}(s)\mathbf{Y}_0^{\nu-1},$$

or **the general common complex domain description of the control
system output vector:**

$$\mathbf{Y}(s) = [I_N + G_{PU}(s) G_{CE}(s)]^{-1} \bullet$$

$$\bullet \left\{ \begin{array}{c} G_{PD}(s)\mathbf{D}(s) + G_{PU}(s) G_{CE}(s)\mathbf{Y}_d(s) + \\ + G_{PD_0}(s)\mathbf{D}_0^{\mu-1} + G_{PU}(s) G_{CEo}(s)\mathbf{e}_0^{\mu-1} + \\ + [G_{PU}(s) G_{CU_0}(s) + G_{PU_0}(s)]\mathbf{U}_0^{\nu-1} + \\ + G_{PU}(s) G_{CX_{C0}}(s)\mathbf{X}_{C0} + G_{PX_0}(s)\mathbf{X}_{P0} + \\ + G_{PY_0}(s)\mathbf{Y}_0^{\nu-1} \end{array} \right\}, \qquad (11.14)$$

or

$$\mathbf{Y}(s) = F_{CS}(s)\mathbf{V}_{CS}(s), \ \ \mathbf{V}_{CS}(s) = \left[\begin{array}{c} \mathbf{I}_{CS}(s) \\ \mathbf{C}_{0CS} \end{array} \right], \qquad (11.15)$$

where

$$F_{CS}(s) = [I_N + G_{PU}(s) G_{CE}(s)]^{-1} \bullet$$

$$\bullet \left[\begin{array}{c} G_{PD}(s) \vdots G_{PU}(s) G_{CE}(s) \vdots G_{PD_0}(s) \vdots G_{PU}(s) G_{CEo}(s) \vdots \\ \vdots G_{PU}(s) G_{CU_0}(s) + G_{PU_0}(s) \vdots G_{PU}(s) G_{CX_{C0}}(s) \vdots \\ \vdots G_{PX_0}(s) \vdots G_{PY_0}(s) \end{array} \right], \qquad (11.16)$$

$$\mathbf{I}_{CS}(s) = \left[\mathbf{D}^T(s) \vdots \mathbf{Y}_d^T(s) \right]^T \qquad (11.17)$$

$$\mathbf{C}_{0CS} = \left[\left(\mathbf{D}_0^{\mu-1} \right)^T : \left(\mathbf{e}_0^{\mu-1} \right)^T : \left(\mathbf{U}_0^{\nu-1} \right)^T \vdots \mathbf{X}_{C0}^T \vdots \mathbf{X}_{P0}^T \vdots \mathbf{Y}_0^{\nu-1} \right]^T. \qquad (11.18)$$

For the tracking analysis it is useful to know the Laplace transform $\mathbf{E}(s) = \mathbf{Y}_d(s) - \mathbf{Y}(s)$ of the output error $\mathbf{e}(t) = \mathbf{Y}_d(t) - \mathbf{Y}(t)$. These equations and
Equation (11.14) furnish:

$$[I_N + G_{PU}(s) G_{CE}(s)]\mathbf{E}(s) = \mathbf{Y}_d(s) - G_{PD}(s)\mathbf{D}(s) - G_{PD_0}(s)\mathbf{D}_0^{\mu-1} -$$
$$- G_{PU}(s) G_{CEo}(s)\mathbf{e}_0^{\mu-1} - [G_{PU}(s) G_{CU_0}(s) - G_{PU_0}(s)]\mathbf{U}_0^{\nu-1} -$$
$$- G_{PU}(s) G_{CX_{C0}}(s)\mathbf{X}_{C0} - G_{PX_0}(s)\mathbf{X}_{P0} - G_{PY_0}(s)\mathbf{Y}_0^{\nu-1}.$$

The final result reads:

$$\mathbf{E}\left(s\right) = \left[I_N + G_{PU}\left(s\right)G_{CE}\left(s\right)\right]^{-1} \bullet$$

$$\bullet \left\{ \begin{array}{c} \mathbf{Y}_d(s) - G_{PD}\left(s\right)\mathbf{D}(s) - G_{PD_0}\left(s\right)\mathbf{D}_0^{\mu-1} - \\ -G_{PU}\left(s\right)G_{CE_o}\left(s\right)\mathbf{e}_0^{\mu-1} - \\ -\left[G_{PU}\left(s\right)G_{CU_0}\left(s\right) - G_{PU_0}\left(s\right)\right]\mathbf{U}_0^{\nu-1} - \\ -G_{PU}\left(s\right)G_{CX_{C0}}\left(s\right)\mathbf{X}_{C0} - G_{PX_0}\left(s\right)\mathbf{X}_{P0} - \\ -G_{PY_0}\left(s\right)\mathbf{Y}_0^{\nu-1} \end{array} \right\}, \tag{11.19}$$

or

$$\mathbf{E}\left(s\right) = F_{CSE}(s)\mathbf{V}_{CSE}(s), \quad \mathbf{V}_{CSE}(s) = \left[\begin{array}{c} \mathbf{I}_{CSE}(s) \\ \mathbf{C}_{0CSE} \end{array} \right] \tag{11.20}$$

where

$$F_{CSE}(s) = \left[I_N + G_{PU}\left(s\right)G_{CE}\left(s\right)\right]^{-1} \bullet$$

$$\bullet \left[\begin{array}{c} I_N \vdots - G_{PD}\left(s\right) \vdots - G_{PD_0}\left(s\right) \vdots - G_{PU}\left(s\right)G_{CE_o}\left(s\right) \vdots \\ \vdots \quad \left[G_{PU}\left(s\right)G_{CU_0}\left(s\right) - G_{PU_0}\left(s\right)\right] \vdots \\ \vdots - G_{PU}\left(s\right)G_{CX_{C0}}\left(s\right) \vdots - G_{PX_0}\left(s\right) \vdots - G_{PY_0}\left(s\right) \end{array} \right], \tag{11.21}$$

$$\mathbf{I}_{CSE}(s) = \left[\mathbf{Y}_d^T\left(s\right)\vdots\mathbf{D}^T(s)\right]^T, \tag{11.22}$$

$$\mathbf{C}_{0CSE} = \left[\left(\mathbf{D}_0^{\mu-1}\right)^T \vdots \left(\mathbf{e}_0^{\mu-1}\right)^T \vdots \left(\mathbf{U}_0^{\nu-1}\right)^T \vdots \mathbf{X}_{C0}^T \vdots \mathbf{X}_{P0}^T \vdots \mathbf{Y}_0^{\nu-1}\right]^T. \tag{11.23}$$

Equations (11.14) and (11.19) show how the initial conditions influence the output response and the output error.

Let $p_p\left(s\right)$ and $p_c\left(s\right)$ be the characteristic polynomial of the plant transfer function matrix $G_{PU}\left(s\right)$ and of the controller transfer function matrix $G_{CE}\left(s\right)$, respectively. Let $L_{P(.)}\left(s\right)$ and $L_{c(.)}\left(s\right)$ be the numerator polynomial matrices of the plant transfer function matrix $G_{P(.)}\left(s\right)$ and of the controller transfer function matrix $G_{C(.)}\left(s\right)$, respectively,

$$G_{P(.)}\left(s\right) = \frac{L_{P(.)}\left(s\right)}{p_p\left(s\right)} = p_p^{-1}\left(s\right)L_{P(.)}\left(s\right), \tag{11.24}$$

$$G_{C(.)}\left(s\right) = \frac{L_{C(.)}\left(s\right)}{p_c\left(s\right)} = p_{c(.)}^{-1}\left(s\right)L_{C(.)}\left(s\right). \tag{11.25}$$

The polynomial $p_{cs}(s)$,

$$p_{cs}(s) = \det \left[p_c(s) p_p(s) I_N + L_{PU}(s) L_{CE}(s) \right] \qquad (11.26)$$

is the characteristic polynomial of the closed-loop, feedback, control system. These equations permit us to set Equation (11.14) and Equation (11.19) in **the following general common forms**:

$$\mathbf{Y}(s) = p_{cs}(s)^{-1} adj \left[p_c(s) p_p(s) I_N + L_{PU}(s) L_{CE}(s) \right] \bullet$$

$$\bullet \left\{ \begin{array}{c} p_c(s) L_{PD}(s) \mathbf{D}(s) + L_{PU}(s) L_{CE}(s) \mathbf{Y}_d(s) + \\ + p_c(s) L_{PD_0}(s) \mathbf{D}_0^{\mu-1} + L_{PU}(s) L_{CEo}(s) \mathbf{e}_0^{\mu-1} + \\ + \left[L_{PU}(s) L_{CU_0}(s) + p_c L_{PU_0}(s) \right] \mathbf{U}_0^{\nu-1} + \\ + L_{PU}(s) L_{CX_{C0}}(s) \mathbf{X}_{C0} + p_c L_{PX_0}(s) \mathbf{X}_0 + \\ + p_c L_{PY_0}(s) \mathbf{Y}_0^{\nu-1} \end{array} \right\}, \qquad (11.27)$$

and

$$\mathbf{E}(s) = p_{cs}(s)^{-1} adj \left[p_c(s) p_p(s) I_N + L_{PU}(s) L_{CE}(s) \right] \bullet$$

$$\bullet \left\{ \begin{array}{c} p_c(s) p_p(s) \mathbf{Y}_d(s) - p_c(s) L_{PD}(s) \mathbf{D}(s) - \\ - p_c(s) L_{PD_0}(s) \mathbf{D}_0^{\mu-1} - L_{PU}(s) L_{CEo}(s) \mathbf{e}_0^{\mu-1} - \\ - \left[L_{PU}(s) L_{CU_0}(s) + p_c(s) L_{PU_0}(s) \right] \mathbf{U}_0^{\nu-1} - \\ - L_{PU}(s) L_{CX_{C0}}(s) \mathbf{X}_{C0} - p_c(s) L_{PX_0}(s) \mathbf{X}_0 - \\ - p_c(s) L_{PY_0}(s) \mathbf{Y}_0^{\nu-1} \end{array} \right\} \qquad (11.28)$$

The existing control theory is valid only for all zero initial conditions. In that case Equations (11.14) and (11.19) reduce to the following well known and fairly simple formulae, respectively:

$$\mathbf{Y}(s) = \left[I_N + G_{PU}(s) G_{CE}(s) \right]^{-1} \left[G_{PD}(s) \mathbf{D}(s) + G_{PU}(s) G_{CE}(s) \mathbf{Y}_d(s) \right], \qquad (11.29)$$

$$\mathbf{E}(s) = \left[I_N + G_{PU}(s) G_{CE}(s) \right]^{-1} \left[\mathbf{Y}_d(s) - G_{PD}(s) \mathbf{D}(s) \right]. \qquad (11.30)$$

Equations (11.24), (11.25), (11.29), (11.30) lead to:

$$\mathbf{Y}(s) = p_{cs}^{-1}(s) adj \left[p_c(s) p_p(s) I_N + L_p(s) L_c(s) \right] \bullet$$
$$\bullet \left[p_c(s) L_{PD}(s) \mathbf{D}(s) + L_{PU}(s) L_{CE}(s) \mathbf{Y}_d(s) \right], \qquad (11.31)$$

and

$$\mathbf{E}(s) = p_p^{-1}(s) p_{cs}^{-1}(s) adj \left[p_c(s) p_p(s) I_N + L_p(s) L_c(s) \right] \bullet$$
$$\bullet \left[p_p(s) \mathbf{Y}_d(s) - L_{PD}(s) \mathbf{D}(s) \right]. \qquad (11.32)$$

Comparing Equations (11.14), (11.19), (11.27), (11.19) valid for arbitrary initial conditions with Equations (11.29), (11.30), (11.31) and (11.32) valid for zero initial conditions, we may conclude that the initial conditions influence crucially the system behavior. Their influence is inherent for both stability and tracking.

Note 252 *Fundamental theoretical gap and its elimination*

For the Lyapunov stability study the system transfer function matrix $G_{CS0}(s)$,

$$G_{CS0}(s) = [I_N + G_{PU}(s) G_{CE}(s)]^{-1} \bullet$$
$$\bullet \left[G_{PX_0}(s) \vdots G_{PU}(s) G_{CX_{C0}}(s) \vdots G_{PY_0}(s) \right],$$

relative to the the extended initial state vector $\mathbf{S}_{CS0}$,

$$\mathbf{S}_{CS0} = \begin{bmatrix} \mathbf{X}_{P0} \\ \mathbf{X}_{C0} \\ \mathbf{Y}_0^{\nu-1} \end{bmatrix}$$

of the control system is, in general, the adequate transfer function matrix. This is due to the definitions of the Lyapunov stability properties that are related to the unperturbed system state behavior, i.e., in the system free regime, so that:

$$\mathbf{E}(s) = \mathbf{0}_N, \ \mathbf{Y}_d(s) = \mathbf{0}_N, \ \mathbf{D}_0^{\mu-1} = \mathbf{0}_{d\mu}, \ \mathbf{e}_0^{\mu-1} = \mathbf{0}_{d\mu}, \ \mathbf{U}_0^{\nu-1} = \mathbf{0}_{r\mu},$$

and

$$\mathbf{Y}(s) = -\mathbf{E}(s) = [I_N + G_{PU}(s) G_{CE}(s)]^{-1} \bullet$$
$$\bullet \left[G_{PX_0}(s) \mathbf{X}_{P0} + G_{PU}(s) G_{CX_{C0}}(s) \mathbf{X}_{C0} + G_{PY_0}(s) \mathbf{Y}_0^{\nu-1} \right]. \quad (11.33)$$

*This has not been recognized so far in the system and control theory, which has been a fundamental theoretical gap. Many discussions have been about the influence of the pole-zero cancellation on the Lyapunov (sometimes referred to as **internal**) stability. For more detailed analysis and explanation see [170, Part III: Stability study, pp. 237-356].*

Any study of the linear systems aimed to be complete and exact should rely on the formulae valid for arbitrary both initial conditions and disturbances.

11.1.4   Generating theorem

Generating theorem links the *time* domain and the complex domain. It was stated and proved in [159, Theorem 347, pp. 300-305] and later on in [170, Theorem 342, pp. 291-297].

Let us remind ourselves that a complex valued matrix function $F(.)$: $\mathfrak{C} \to \mathfrak{C}^{mxn}$ is real rational matrix function if and only if it becomes a real valued matrix for every real value of the complex variable s, i.e., for $s = \sigma \in \mathfrak{R}$, and every entry of $F(s)$ is a quotient of two polynomials in s.

Let $F(s)$ have μ different poles denoted by s_k^*, $k = 1, 2, ..., \mu$. The multiplicity of the pole s_k^* is designated by ν_k. Its real and imaginary part are denoted as $\operatorname{Re} s_k^*$ and $\operatorname{Im} s_k^*$, respectively.

Theorem 253 *Generating theorem [159, Theorem 347, pp. 300-305], [170, Theorem 342, pp. 291-297]*

Let $F(.) : \mathfrak{R} \to \mathfrak{R}^{mxn}$, $F(t) = [F_{ij}(t)]$, have the Laplace transform $F(.)$: $\mathfrak{C} \to \mathfrak{C}^{mxn}$, $F(s) = [F_{ij}(s)]$, that is real rational matrix function. In order for the norm $\|F(t)\|$ of the original $F(t)$:

- *a)   To be **bounded**, i.e.:*

$$\exists \alpha \in \mathfrak{R}^+ \implies \|F(t)\| < \alpha, \ \forall t \in \mathfrak{T}_0,$$

it is necessary and sufficient that:

1. The real parts of all poles of $F(s)$ are nonpositive,

$$\operatorname{Re} s_i^* \leq 0, \ \forall i = 1, 2, ..., \ \mu;$$

2. All imaginary poles of $F(s)$ are simple (i.e., with the multiplicity ν_i equal to one),

$$\operatorname{Re} s_i^* = 0, \nu_i = 1;$$

3. $F(s)$ is the zero matrix in the infinity, i.e., it is strictly proper,

$$F(\infty) = O_{mn} \in \mathfrak{R}^{mxn};$$

- *b)   and in order for $\|F(t)\|$ to **vanish asymptotically**, i.e., in order for the following condition to hold:*

$$\lim[\|F(t)\| : \ t \longrightarrow \infty] = 0,$$

it is necessary and sufficient that:

1. The real parts of all poles of $F(s)$ are negative,

$$\operatorname{Re} s_i^* < 0, \ \forall i = 1, 2, ..., \ \mu;$$

and

2. $F(s)$ is the zero matrix in the infinity, i.e., it is strictly proper,

$$F(\infty) = O_{pn} \in \mathfrak{R}^{p \times n}.$$

Generating theorem is the key stone of the Lyapunov stability [159, Section 13.4, pp. 299-323], [170, Section 13.4, pp. 291-315], of the Bounded Input stability [159, Section 14.3, pp. 339-362], [170, Section 14.3, pp. 330-356], and of tracking control synthesis [188, Chapter 10, pp. 209-221] of the linear systems.

11.1.5 Tracking conditions and control

At first we present the general tracking theorem in the complex domain. It permits nonvanishing and increasing disturbances to act on the plant, i.e. that the real parts of the poles of their Laplace transforms are nonnegative.

Theorem 254 *Criterion for the linear control system tracking*
For the linear feedback control system determined by Equation (11.14), equivalently by Equation (11.27), to exhibit tracking over $[\mathfrak{D} \times \mathfrak{Y}_d] \cap \mathfrak{L}$ it is necessary and sufficient that the real parts of all poles of the Laplace transform $\mathbf{E}(s)$ determined by Equation (11.19), equivalently by Equation (11.28), of the output error vector $\mathbf{e}$ are negative for every $[\mathbf{D}(.), \mathbf{Y}_d(.)] \in [\mathfrak{D} \times \mathfrak{Y}_d] \cap \mathfrak{L}$.
Then, and only then, the linear feedback control system exhibits
  - *Global stablewise tracking over $[\mathfrak{D} \times \mathfrak{Y}_d] \cap \mathfrak{L}$, and*
  - *Global exponential tracking over $[\mathfrak{D} \times \mathfrak{Y}_d] \cap \mathfrak{L}$.*
The zero equilibrium state of the feedback control system is then globally exponentially stable.

Proof. Linearity of the *IO* linear feedback control system (11.14), Definitions 121, 125, 129, continuity and boundedness of every $[\mathbf{D}(.), \mathbf{Y}_d(.)] \in [\mathfrak{D} \times \mathfrak{Y}_d] \cap \mathfrak{L}$ by the definition, (1.22) (Section 1.5), (1.23), (1.24), (1.26), (1.27) (Section 1.5), and the Generating theorem 253 imply the statement of the theorem on tracking. The stability statement results from Definitions 121, 125, 129 and the well-known global asymptotic and exponential stability definitions (see also Note 135, Section 8.3). ∎

Comment 255 *This theorem emphasizes the need to use the full transfer function matrix $F_{CSE}(s)$ (11.21) of the feedback control system with respect to the output error* **e**.

The method used so far to apply only the feedback control system transfer function matrix $G_{CSE}(s)$ (11.34),

$$G_{CSE}(s). = [I_N + G_{PU}(s) G_{CE}(s)]^{-1} \left[I_N \vdots G_{PD}(s) \right] \qquad (11.34)$$

is invalid in the general case that permits arbitrary initial conditions.

The use of only $G_{CSE}(s)$ (11.34) is insufficient for the analysis or synthesis of the control system from the tracking point of view, while the use of $F_{CSE}(s)$ (11.21) is both necessary and sufficient.

The poles of $F_{CSE}(s)$ (11.21), of $\mathbf{D}(s)$ and of $\mathbf{Y}_d(s)$ constitute the set of all poles of $\mathbf{E}(s)$ in view of Equation (11.19), equivalently Equation (11.28). The only information about the disturbance vector function $\mathbf{D}(.)$ is that it belongs to $\mathfrak{D}$ and that the signs of the real parts of all poles of its Laplace transform $\mathbf{D}(s)$ are known. If some poles of $\mathbf{D}(s)$ have nonnegative real parts, then they should be known as well as their multiplicity.

If all disturbances are vanishing, i.e., the real parts of all poles of their Laplace transforms are negative, then Theorem 254 reduces to the following in view of Definition 20 (Section 1.5):

Theorem 256 *For the linear feedback control system determined by Equation (11.14), equivalently by Equation (11.27), which is subjected only to the vanishing disturbances, to exhibit tracking over $[\mathfrak{D}_- \mathsf{x} \mathfrak{Y}_{d-}] \cap \mathfrak{L}$, (Subsection 1.5), it is necessary and sufficient that the real parts of all poles of the system full transfer function matrix $F_{CSE}(s)$ (11.21) are negative.*

Then, and only then, the linear feedback control system exhibits

- *Global stablewise tracking over $[\mathfrak{D}_- \mathsf{x} \mathfrak{Y}_{d-}] \cap \mathfrak{L}$, and*
- *Global exponential tracking over $[\mathfrak{D}_- \mathsf{x} \mathfrak{Y}_{d-}] \cap \mathfrak{L}$.*

The zero equilibrium state of the feedback control system is then globally exponentially stable.

The exponential tracking is the best tracking quality that can be achieved with this approach. It is asymptotic infinite *time* tracking. The controller is linear and *time*-invariant. Its full transfer function matrix should be such that Theorem 254 in general, or Theorem 256 if all disturbances are vanishing in the course of *time*, is satisfied. The control action is smooth and robust relative to the disturbance action.

The control synthesis should ensure that the real parts of all poles:

- Of the Laplace transform $\mathbf{E}(s)$ determined by Equation (11.19), equivalently by Equation (11.28), of the output error vector $\mathbf{e}$ are negative for every $[\mathbf{D}(.), \mathbf{Y}_d(.)] \in [\mathfrak{D} \times \mathfrak{Y}_d] \cap \mathcal{L}$ if the disturbances are not vanishing,

- Of the system full transfer function matrix $F_{CSE}(s)$ (11.21) are negative if the disturbances are vanishing.

This reduces the control synthesis to the application of any known method for the linear system stabilization..

The criterion demands the test of the real parts of all poles of the full transfer function matrix $F_{CSE}(s)$ (11.21) of the control system relative to the output error vector $\mathbf{e}(t)$ to be negative. It usually means that only the denominator polynomial of $F_{CSE}(s)$ (11.21) is to be known.

Chapter 12

Lyapunov Tracking Control (LyTC)

12.1 General form of the linear systems

This section extends the theory established in [188, Chapter 11, pp. 223-248] to the general linear systems that incorporate all five classes of the systems studied herein.

12.1.1 Introduction

The Lyapunov tracking theory [175], [188] has been established and developed by extending and generalizing the Lyapunov stability theory [230]. Both theories demand the mathematical model of a (physical, or mathematical) system to be in a generalized, extended, state form. What follows shows that such form is the *EISO* system form defined in Equations (4.1), (4.2) (Section 4.1) in terms of the total values of all variables, i.e.,

$$\frac{d\mathbf{X}(t)}{dt} = A\mathbf{X}(t) + P^{(\mu)}\mathbf{I}^\mu(t), \ \forall t \in \mathfrak{T}_0, \ \mu \geq 0, \tag{12.1}$$

$$\mathbf{Y}(t) = C\mathbf{X}(t) + Q\mathbf{I}(t), \ \forall t \in \mathfrak{T}_0. \tag{12.2}$$

We allow now $\mu = 0$ in order for Equations (12.1) and (12.2) to incorporate the *ISO* mathematical model (3.1), (3.2).

In terms of the deviations **i** (2.54), **x** (3.37), and **y** (2.56), the *EISO* system form is determined in Equations (4.67), (4.68) (Section 4.2), i.e.,

$$\frac{d\mathbf{x}(t)}{dt} = A\mathbf{x}(t) + P^{(\mu)}\mathbf{i}^\mu(t), \; \forall t \in \mathfrak{T}_0, \; \mu \geq 0, \tag{12.3}$$

$$\mathbf{y}(t) = C\mathbf{x}(t) + Q\mathbf{i}(t), \; \forall t \in \mathfrak{T}_0. \tag{12.4}$$

12.1.2 *EISO* form of the *IO* systems

The *EISO* form (12.1), (12.2) of the *IO* system (2.15) is defined by Equations (4.6)–(4.11) (Section 4.1).

12.1.3 *EISO* form of the *ISO* systems

The *ISO* system (3.1), (3.2) represents a special case of the *EISO* form (12.1), (12.2) in which $\mu = 0$.

12.1.4 *EISO* form of the *EISO* systems

The *EISO* system (4.1), (4.2) is by the definition in the *EISO* form (12.1), (12.2).

12.1.5 *EISO* form of the *HISO* systems

The state vector $\mathbf{X} = \mathbf{S}_{HISO}$ of the *HISO* system (5.1), (5.2), (Section 5.1), is determined in Equation (5.3), i.e.,

$$\mathbf{X} = \mathbf{S}_{HISO} = \mathbf{R}^{\alpha-1} = \left[\mathbf{R}^T \; \vdots \; \mathbf{R}^{(1)^T} \; \vdots \; ... \; \vdots \; \mathbf{R}^{(\alpha-1)^T} \right]^T \in \mathfrak{R}^n, \; n = \alpha\rho. \tag{12.5}$$

Equation (5.1) and Equation (12.5) lead to:

$$\mathbf{X} = \left[\mathbf{X}_1^T \; \vdots \; \mathbf{X}_2^T \; \vdots \; ... \; \vdots \; \mathbf{X}_\alpha^T \right]^T = \left[\mathbf{R}^T \; \vdots \; \mathbf{R}^{(1)^T} \; \vdots \; ... \; \vdots \; \mathbf{R}^{(\alpha-1)^T} \right]^T,$$

$$\mathbf{X}_k = \mathbf{R}^{(k-1)}, \; \forall k = 1, 2, ..., \alpha. \tag{12.6}$$

$$\mathbf{X}_1^{(1)} = \mathbf{X}_2,$$
$$\mathbf{X}_2^{(1)} = \mathbf{X}_3,$$
$$...$$
$$\mathbf{X}_{\alpha-1}^{(1)} = \mathbf{X}_\alpha,$$
$$\mathbf{X}_\alpha^{(1)}(t) = - \sum_{k=0}^{k=\alpha-1} A_\alpha^{-1} A_k \mathbf{X}_{k+1}(t) + \sum_{k=0}^{k=\mu} A_\alpha^{-1} H_k \mathbf{I}^{(k)}(t).$$

These equations, Equations (12.6) and Equation (5.2) result in the *EISO* system (12.1), (12.2) in which:

$$\alpha > 1 \Longrightarrow A =$$

$$\begin{bmatrix} O_\rho & I_\rho & ... & O_\rho & O_\rho \\ O_\rho & O_\rho & ... & O_\rho & O_\rho \\ O_\rho & O_\rho & ... & I_\rho & O_\rho \\ ... & ... & ... & ... & ... \\ O_\rho & O_\rho & ... & O_\rho & I_\rho \\ -A_\alpha^{-1} A_0 & -A_\alpha^{-1} A_1 & ... & -A_\alpha^{-1} A_{\alpha-2} & -A_\alpha^{-1} A_{\alpha-1} \end{bmatrix} \in \mathfrak{R}^{\alpha\rho \times \alpha\rho},$$

$$\alpha = 1 \Longrightarrow A = -A_1^{-1} A_0 \in \mathfrak{R}^{\rho \times \rho}, \tag{12.7}$$

$$P^{(\mu)} = \left\{ \begin{bmatrix} O_{(\alpha-1)N,(\mu+1)M} \\ A_\alpha^{-1} H^{(\mu)} \end{bmatrix} \in \mathfrak{R}^{n \times (\mu+1)M}, \ \alpha > 1, \atop A_1^{-1} H^{(\mu)} \in \mathfrak{R}^{\rho \times (\mu+1)M}, \ \alpha = 1, \right\}, \tag{12.8}$$

or, equivalently,

$$P^{(\mu)} = \left\{ \begin{bmatrix} O_{(\alpha-1)\rho,\rho} \\ I_\rho \end{bmatrix} A_\alpha H^{(\mu)}, \ \alpha > 1, \atop I_\rho A_1^{-1} H^{(\mu)}, \ \alpha = 1, \right\}, \tag{12.9}$$

$$P_{inv} = \left\{ \begin{bmatrix} O_{(\alpha-1)\rho,\rho} \\ I_\rho \end{bmatrix} \in \mathfrak{R}^{n \times \rho}, \ \alpha > 1, \atop I_\rho \in \mathfrak{R}^{\rho \times \rho}, \ \alpha = 1, \right\}, \tag{12.10}$$

i.e.,

$$P^{(\mu)} = P_{inv} A_\alpha H^{(\mu)} \in \mathfrak{R}^{n \times (\mu+1)M} \tag{12.11}$$

$$C = \begin{bmatrix} R_{y0} \vdots R_{y1} \vdots R_{y2} \vdots ... \vdots R_{y,\alpha-1} \end{bmatrix} \in \mathfrak{R}^{N \times n}, \ Q = Q \in \mathfrak{R}^{N \times M}. \tag{12.12}$$

The *EISO* form (12.1), (12.2) of the *HISO* system (5.1), (5.2) is determined by Equations (12.5)–(12.12).

12.1.6   *EISO* form of the *IIO* systems

Equation (6.6) determines the state vector $\mathbf{X} = \mathbf{S}_{IIO}$ of the *IIO* system (6.1), (6.2), Section 6.1:

$$\mathbf{X} = \mathbf{S}_{IIO} = \begin{bmatrix} \mathbf{R}^{\alpha-1} \\ \mathbf{Y}^{\nu-1} \end{bmatrix} = \begin{bmatrix} \mathbf{R}^T \vdots \mathbf{R}^{(1)^T} \vdots \ldots \vdots \mathbf{R}^{(\alpha-1)^T} \\ \mathbf{Y}^T \vdots \mathbf{Y}^{(1)^T} \vdots \ldots \vdots \mathbf{Y}^{(\nu-1)^T} \end{bmatrix}^T \in \mathfrak{R}^n,$$

$$n = \alpha\rho + \nu N. \tag{12.13}$$

More explicitly,

$$\mathbf{X} = \begin{bmatrix} \mathbf{X}_1^T \vdots \mathbf{X}_2^T \vdots \ldots \vdots \mathbf{X}_\alpha^T \vdots \mathbf{X}_{\alpha+1}^T \vdots \mathbf{X}_{\alpha+2}^T \vdots \cdots \vdots \mathbf{X}_{\alpha+\nu}^T \end{bmatrix}^T \in \mathfrak{R}^n. \tag{12.14}$$

Equation (6.1) and Equation (6.2) lead to the following sets of substate equations due to Equation (12.14):

$$\mathbf{X}_1^{(1)} = \mathbf{X}_2,$$
$$\mathbf{X}_2^{(1)} = \mathbf{X}_3,$$
$$\ldots$$
$$\mathbf{X}_{\alpha-1}^{(1)} = \mathbf{X}_\alpha,$$

$$\mathbf{X}_\alpha^{(1)}(t) = -\sum_{k=0}^{k=\alpha-1} A_\alpha^{-1} A_k \mathbf{X}_{k+1}(t) + \sum_{k=0}^{k=\mu} A_\alpha^{-1} H_k \mathbf{I}^{(k)}(t), \tag{12.15}$$

$$\mathbf{X}_{\alpha+1}^{(1)} = \mathbf{X}_{\alpha+2},$$
$$\mathbf{X}_{\alpha+2}^{(1)} = \mathbf{X}_{\alpha+3},$$
$$\ldots$$
$$\mathbf{X}_{\alpha+\nu-1}^{(1)} = \mathbf{X}_{\alpha+\nu},$$

$$\mathbf{X}_{\alpha+\nu}^{(1)}(t) = \left\{ \begin{array}{c} \sum_{k=0}^{k=\alpha-1} E_\nu^{-1} R_{yk} \mathbf{X}_{k+1}(t) - \sum_{k=0}^{k=\nu-1} E_\nu^{-1} E_k \mathbf{X}_{\alpha+k+1}(t) + \\ + \sum_{k=0}^{k=\mu} E_\nu^{-1} Q_k \mathbf{I}^{(k)}(t) \end{array} \right\},$$
$$\tag{12.16}$$

$$\mathbf{Y} = \mathbf{X}_{\alpha+1} = \underbrace{\left[\underbrace{O_{N,\rho} \vdots O_{N,\rho} \vdots \cdots \vdots O_{N,\rho}}_{\alpha-times} \vdots I_N \vdots \underbrace{O_N \vdots O_N \vdots \cdots \vdots O_N}_{(\nu-1)-times} \right]}_{C} \mathbf{X} =$$

$$= C\mathbf{X}. \tag{12.17}$$

Equations (12.13)–(12.17) determine the *EISO* form (12.1), (12.2) of the *IIO* system (6.1), (6.2), in which:

$$\alpha > 1 \Longrightarrow A_{11} =$$

$$\begin{bmatrix} O_\rho & I_\rho & \ldots & O_\rho & O_\rho \\ O_\rho & O_\rho & \ldots & O_\rho & O_\rho \\ O_\rho & O_\rho & \ldots & I_\rho & O_\rho \\ \ldots & \ldots & \ldots & \ldots & \ldots \\ O_\rho & O_\rho & \ldots & O_\rho & I_\rho \\ -A_\alpha^{-1}A_0 & -A_\alpha^{-1}A_1 & \ldots & -A_\alpha^{-1}A_{\alpha-2} & -A_\alpha^{-1}A_{\alpha-1} \end{bmatrix} \in \mathfrak{R}^{\alpha\rho\times\alpha\rho},$$

$$\alpha = 1 \Longrightarrow A_{11} = -A_1^{-1}A_0 \in \mathfrak{R}^{\rho\times\rho},$$

(12.18)

$$A_{12} = O_{\alpha\rho,\nu N}, \tag{12.19}$$

$$P_1^{(\mu)} = \left\{ \begin{bmatrix} O_{(\alpha-1)N,(\mu+1)M} \\ A_\alpha^{-1}H^{(\mu)} \end{bmatrix} \in \mathfrak{R}^{n\times(\mu+1)M}, \ \alpha > 1, \\ A_1^{-1}H^{(\mu)} \in \mathfrak{R}^{\rho\times(\mu+1)M}, \ \alpha = 1, \right\}, \tag{12.20}$$

$$C_1 = O_{N,\alpha\rho} \in \mathfrak{R}^{N\times\alpha\rho}, \tag{12.21}$$

$$\nu > 1 \Longrightarrow A_{21} =$$

$$\begin{bmatrix} O_{N,\rho} & O_{N,\rho} & \ldots & O_{N,\rho} & O_{N,\rho} \\ O_{N,\rho} & O_{N,\rho} & \ldots & O_{N,\rho} & O_{N,\rho} \\ O_{N,\rho} & O_{N,\rho} & \ldots & O_{N,\rho} & O_{N,\rho} \\ \ldots & \ldots & \ldots & \ldots & \ldots \\ O_{N,\rho} & O_{N,\rho} & \ldots & O_{N,\rho} & O_{N,\rho} \\ -E_\nu^{-1}R_{y0} & -E_\nu^{-1}R_{y1} & \ldots & -E_\nu^{-1}R_{y,\alpha-2} & -E_\nu^{-1}R_{y,\alpha-1} \end{bmatrix} \in \mathfrak{R}^{\nu N\times\alpha\rho},$$

$$\alpha = 1 \Longrightarrow A_{21} = -E_1^{-1}E_0 \in \mathfrak{R}^{N\times\rho},$$

(12.22)

$$\nu > 1 \Longrightarrow E_{22} =$$

$$\begin{bmatrix} O_N & I_N & \ldots & O_N & O_N \\ O_N & O_N & \ldots & O_N & O_N \\ O_N & O_N & \ldots & I_N & O_N \\ \ldots & \ldots & \ldots & \ldots & \ldots \\ O_N & O_N & \ldots & O_N & I_N \\ -E_\nu^{-1}E_0 & -E_\nu^{-1}E_1 & \ldots & -E_\nu^{-1}E_{\nu-2} & -E_\nu^{-1}E_{\nu-1} \end{bmatrix} \in \mathfrak{R}^{\nu N\times\nu N},$$

$$\alpha = 1 \Longrightarrow E_{22} = -E_1^{-1}E_0 \in \mathfrak{R}^{N\times N},$$

(12.23)

$$P_2^{(\mu)} = \left\{ \begin{bmatrix} O_{(\nu-1)N,(\mu+1)M} \\ E_\nu^{-1}H^{(\mu)} \end{bmatrix} \in \mathfrak{R}^{n\times(\mu+1)M}, \ \nu > 1, \\ E_1^{-1}H^{(\mu)} \in \mathfrak{R}^{N\times(\mu+1)M}, \ \nu = 1, \right\}, \tag{12.24}$$

$$C_2 = \left[I_N \vdots \underbrace{O_N \vdots O_N \vdots \cdots \vdots O_N}_{(\nu-1)-times} \right] \in \mathfrak{R}^{N \times \nu N}, \qquad (12.25)$$

$$A = blockdiag \left[\begin{array}{cc} A_{11} & O_{\alpha\rho,\nu N} \\ A_{21} & E_{22} \end{array} \right], \qquad (12.26)$$

$$P^{(\mu)} = \left[\begin{array}{c} P_1^{(\mu)} \\ P_2^{(\mu)} \end{array} \right], \qquad (12.27)$$

$$C = \left[C_1 \vdots C_2 \right] \in \mathfrak{R}^{N \times (\alpha\rho + \nu N)} = \mathfrak{R}^{N \times n}, \qquad (12.28)$$

$$Q = O_{N,M}. \qquad (12.29)$$

12.2 Lyapunov tracking theory basis

This section continues the presentation of Chapter 13 in [170, pp. 239-315].

12.2.1 Lyapunov matrix theorem

It is well known that the *EISO* system defined in Equations (12.1) and (12.2) in terms of the total values of all variables and from it deduced its description (12.3), (12.4) in terms of deviations have the same stability properties. By the definition, the Lyapunov stability properties concern the behavior of the system described in terms of the deviations in the free regime, i.e., for $\mathbf{i}(t) \equiv \mathbf{0}_M$. We recall that the *stable*, equivalently *Hurwitz*, matrix A means that the real parts of all its eigenvalues are negative, equivalently that the zero equilibrium state $\mathbf{x}_e \equiv \mathbf{0}_n$ of the system (12.3), (12.4) is globally asymptotically, hence, globally exponentially, stable.

The basic result for the linear systems, and very useful for the nonlinear systems, is the following famous fundamental Lyapunov matrix theorem:

Theorem 257 *The Lyapunov matrix theorem for the EISO system (12.1), (12.2), equivalently for the EISO system (12.3), (12.4)*

In order for the zero equilibrium vector $\mathbf{x}_e = \mathbf{0}_n$ of the EISO system (12.3), (12.4) to be asymptotically stable, equivalently for the matrix A (12.3) to be stable matrix, it is necessary and sufficient that for any positive definite symmetric matrix G, $G = G^T \in \mathfrak{R}^{n \times n}$, the matrix solution H of the Lyapunov matrix equation

$$A^T H + HA = -G \qquad (12.30)$$

is also positive definite symmetric matrix and the unique solution to (12.30).

The Kronecker matrix product of two matrices $M = [m_{ij}] \in \mathfrak{R}^{\mu \times s}$ and $U = [u_{ij}] \in \mathfrak{R}^{\nu \times \sigma}$ is denoted by $\otimes$ and defined by

$$
M \otimes U = \begin{bmatrix}
m_{11}U & m_{12}U & \cdots & m_{1s}U \\
m_{21}U & m_{21}U & \cdots & m_{2s}U \\
- & - & \cdots & - \\
m_{\mu 1}U & m_{\mu 2}U & \cdots & m_{\mu s}U
\end{bmatrix} \in \mathfrak{R}^{\mu \nu \times s \sigma}. \tag{12.31}
$$

If the positive definite matrix $G = G^T$ is given then we can effectively solve the Lyapunov matrix equation (12.30) for the matrix H by applying the Kronecker matrix product of the matrix A and the identity matrix I of the dimension $n \times n$:

$$
A^T \otimes I + I \otimes A^T. \tag{12.32}
$$

We define the vectors $\mathbf{h}$ and $\mathbf{g}$ induced by the matrices $H = H^T = [h_{ij}] \in \mathfrak{R}^{n \times n}$ and $G = G^T = [g_{ij}] \in \mathfrak{R}^{n \times n}$, respectively,

$$
\mathbf{h} = [h_{11} \ h_{12} \cdots h_{1p} \ h_{21} \ h_{22} \cdots h_{2n} \ \cdots \ h_{n1} \ \cdots \ h_{n2} \cdots h_{nn}]^T \in \mathfrak{R}^{nn}, \tag{12.33}
$$

$$
\mathbf{g} = [g_{11} \ g_{12} \cdots g_{1n} \ g_{21} \ g_{22} \cdots g_{2n} \ \cdots \ g_{n1} \ \cdots \ g_{n2} \cdots g_{nn}]^T \in \mathfrak{R}^{nn}. \tag{12.34}
$$

These vectors enable us

- To determine in the straightforward procedure the Lyapunov function $v(.)$ of the *EISO* system (12.1), (12.2), equivalently of the *EISO* system (12.3), (12.4),

and

- To set the Lyapunov matrix Equation (12.30) in the vector form (12.35) by using the Kronecker matrix product between the matrix A and the identity matrix I,

$$
\left(A^T \otimes I + I \otimes A^T \right) \mathbf{h} = -\mathbf{g}. \tag{12.35}
$$

This equation corrects the typographical error in [170, Equation (13.75), p. 285].

If the eigenvalues $\lambda_i(A)$ of the matrix A obey

$$
\lambda_j(A) + \lambda_k(A) \neq 0, \quad \forall j, k = 1, 2, ..., n, \tag{12.36}
$$

then Equation (12.35) is solvable in $\mathbf{h}$ and the solution reads:

$$
\mathbf{h} = -\left(A^T \otimes I_n + I_n \otimes A^T \right)^{-1} \mathbf{g}. \tag{12.37}
$$

This equation corrects the typographical error in [170, Equation (13.77), p. 286] and determines the vector $\mathbf{h}$, which, together with (12.33), (12.34), defines completely the matrix H, $H = H^T \in \mathfrak{R}^{n \times n}$, and its quadratic form $v(\mathbf{X}) = \mathbf{X}^T H \mathbf{X}$ being the Lyapunov function of the $EISO$ system (12.1), (12.2), equivalently of the $EISO$ system (12.3), (12.4).

The condition (12.36) means that:

- The matrix A does not have an eigenvalue with the zero real part (i.e., on the imaginary axis of the complex plane),

- The complex eigenvalues $\lambda_j(A)$ and $\lambda_k(A)$ of the matrix A with positive real parts satisfying

$$Re\lambda_j(A) = -Re\lambda_k(A), \ \ Re\lambda_j(A)Re\lambda_k(A) < 0,$$

must have imaginary parts obeying

$$Im\lambda_j(A) \neq -Im\lambda_k(A),$$

and vice versa, and

- The real eigenvalues of the matrix A with the equal absolute values may not have the opposite signs.

If the matrix A is stable matrix, then it satisfies all these conditions, i.e., then Equation (12.35) is solvable in $\mathbf{h}$. Equation (12.37) determines the solution $\mathbf{h}$ that induces directly the matrix H via (12.33).

The quadratic form $v(\mathbf{X})$,

$$v(\mathbf{X}) = \mathbf{X}^T H \mathbf{X}, \ H = H^T, \tag{12.38}$$

with positive definite symmetric matrix $H = H^T$ is an adequate form for the stability analysis and for the tracking study of the linear systems via the Lyapunov method (for another useful form of the Lyapunov function $v(.)$ see [188, Chapter 11, pp. 223-248]). If H is the matrix solution of the Lyapunov matrix Equation (12.30) for an arbitrary positive definite symmetric matrix $G = G^T$ then the matrix A is stable matrix and the $EISO$ system (12.3), (12.4) is stable, i.e., its zero equilibrium state is globally exponentially stable.

12.2.2   Arbitrary scalar Lyapunov function

The special case of the $EISO$ system (12.1) and (12.2), equivalently of (12.3), (12.4), specified by Equations (4.3), (Section 4.1), is the $EISO$ plant (12.1) and (12.2), (Section 4.1), i.e.,

$$\frac{d\mathbf{X}(t)}{dt} = A\mathbf{X}(t) + D^{(\mu)}\mathbf{D}^\mu(t) + \mathbf{B}^{(\mu)}\mathbf{U}^\mu(t), \ \forall t \in \mathfrak{T}_0, \tag{12.39}$$

$$\mathbf{Y}(t) = C\mathbf{X}(t) + V\mathbf{D}(t) + U\mathbf{U}(t), \ \forall t \in \mathfrak{T}_0. \tag{12.40}$$

It appears reasonable to use a tentative Lyapunov function $v(.)$ dependent only on the output error vector $\mathbf{e} = \mathbf{Y}_d - \mathbf{Y}$, $v(.) : \mathfrak{R}^N \longrightarrow \mathfrak{R}_+$. Let the function $v(.)$ be continuously differentiable, $v(\mathbf{e}) \in \mathfrak{C}^1(\mathfrak{R}^N)$. Its total time derivative $v^{(1)}[\mathbf{e}(t)]$ along motions of the *EISO* plant (12.39), (12.40) is expressed via its gradient $gradv(\mathbf{e})$,

$$gradv(\mathbf{e}) = \left[\begin{array}{cccc} \frac{\partial v(\mathbf{e})}{\partial e_1} & \frac{\partial v(\mathbf{e})}{\partial e_2} & \cdots & \frac{\partial v(\mathbf{e})}{\partial e_{N-1}} & \frac{\partial v(\mathbf{e})}{\partial e_N} \end{array} \right]^T,$$

so that

$$v^{(1)}[\mathbf{e}(t)] = [gradv(\mathbf{e})]^T \frac{d\mathbf{Y}_d(t)}{dt} - [gradv(\mathbf{e})]^T \frac{d\mathbf{Y}(t)}{dt}. \tag{12.41}$$

The derivative $\mathbf{Y}^{(1)}(t)$ follows from (12.40),

$$\mathbf{Y}^{(1)}(t) = C\mathbf{X}^{(1)}(t) + V\mathbf{D}^{(1)}(t) + U\mathbf{U}^{(1)}(t) =$$
$$= C\mathbf{X}^{(1)}(t) + V^{(\mu)}\mathbf{D}^\mu(t) + U^{(\mu)}\mathbf{U}^\mu(t),$$

$$where \ V^{(\mu)} = \left[O_{N,d} \vdots V \vdots O_{N,d} \vdots \cdots \vdots O_{N,d} \right] \in \mathfrak{R}^{N \times (\mu+1)d},$$

$$and \ U^{(\mu)} = \left[O_{N,r} \vdots U \vdots O_{N,r} \vdots \cdots \vdots O_{N,r} \right] \in \mathfrak{R}^{N \times (\mu+1)r}, \tag{12.42}$$

and from (12.39),

$$\mathbf{Y}^{(1)}(t) = C\left[A\mathbf{X}(t) + D^{(\mu)}\mathbf{D}^\mu(t) + B^{(\mu)}\mathbf{U}^\mu(t) \right] + V^{(\mu)}\mathbf{D}^\mu(t) + U^{(\mu)}\mathbf{U}^\mu(t)$$

i.e.,

$$\mathbf{Y}^{(1)}(t) = CA\mathbf{X}(t) + \left(CB^{(\mu)} + U^{(\mu)} \right)\mathbf{U}^\mu(t) + \left(CD^{(\mu)} + V^{(\mu)} \right)\mathbf{D}^\mu(t). \tag{12.43}$$

This transforms (12.41) into

$$v^{(1)}[\mathbf{e}(t)] = [gradv(\mathbf{e})]^T \left[\frac{d\mathbf{Y}_d}{dt} - CA\mathbf{X}(t) - \left(CB^{(\mu)} + U^{(\mu)} \right)\mathbf{U}^\mu(t) \right] -$$
$$- [gradv(\mathbf{e})]^T \left(CD^{(\mu)} + V^{(\mu)} \right)\mathbf{D}^\mu(t). \tag{12.44}$$

Assumption 258 *The state vector* $\mathbf{X}$ *of the EISO plant (12.39), (12.40) is measurable.*

Assumption 259 *i) The dimension r of the control vector* **U** *is not less than the dimension N of the output vector* **Y**,

$$r \geq N. \tag{12.45}$$

ii) Either

$$rank U^{(\mu)} = N, \ if \ U^{(\mu)} \neq O_{N,(\mu+1)r} \ and \ CB^{(\mu)} = O_{N,(\mu+1)r}, \tag{12.46}$$

or

$$rank CB^{(\mu)} = N \ if \ CB^{(\mu)} \neq O_{N,(\mu+1)r} \ and \ U^{(\mu)} = O_{N,(\mu+1)r}, \tag{12.47}$$

or

$$rank \left(CB^{(\mu)} + U^{(\mu)} \right) = N \ if \ CB^{(\mu)} \neq O_{N,(\mu+1)r} \ and \ U^{(\mu)} \neq O_{N,(\mu+1)r}. \tag{12.48}$$

The condition i) is reasonable. In fact it is necessary in the case when every output variable should be controlled independently of other output variables (Fundamental control principle 104). Under these conditions the control **U** is determined as follows.

Let Assumption 259 be satisfied. It ensures that the control **U**(t) is the solution of the following subsidiary linear differential equation

$$\left(CB^{(\mu)} + U^{(\mu)} \right) \mathbf{U}^{\mu}(t) = \mathbf{W}(t), \tag{12.49}$$

which exists due to Assumption 259. The problem of the synthesis of the control vector **U**(t) is reduced to the synthesis of the vector **W**(t).

Equation (12.49) transforms Equation (12.44) into:

$$v^{(1)} \left[\mathbf{e}\left(t \right) \right] = \left[grad v(\mathbf{e}) \right]^{T} \left[\frac{d\mathbf{Y}_{d}}{dt} - CA\mathbf{X}(t) - \mathbf{W}(t) \right] -$$
$$- \left[grad v(\mathbf{e}) \right]^{T} \left(CD^{(\mu)} + V^{(\mu)} \right) \mathbf{D}^{\mu}(t), \ \forall \, [t, \mathbf{D}(.), \mathbf{Y}_{d}(.)] \in \mathfrak{T}_{0} \times \mathfrak{D}^{\mu} \times \mathfrak{Y}_{d}^{\mu}. \tag{12.50}$$

Let

$$\mathbf{D}_{M} \geq \sup \left\{ \left| \left(CD^{(\mu)} + V^{(\mu)} \right) \mathbf{D}^{\mu}(t) \right| : [\mathbf{D}(.), t] \in \mathfrak{D}^{\mu} \times \mathfrak{T}_{0} \right\}. \tag{12.51}$$

The vector majorization (12.51) of

$$\left| \left(CD^{(\mu)} + V^{(\mu)} \right) \mathbf{D}^{\mu}(t) \right|$$

over $[\mathbf{D}(.), t] \in \mathfrak{D}^{\mu} \times \mathfrak{T}_0$ ensures the full control robustness relative to the disturbance vector $\mathbf{D}(t)$ and eliminates the need to measure the disturbance instantaneous value. We define $\mathbf{W}(t)$ by

$$\mathbf{W}(t) = \frac{d\mathbf{Y}_d}{dt} - CA\mathbf{X}(t)+$$

$$+ \|gradv(\mathbf{e})\|^{-2} [gradv(\mathbf{e})] \left\{ \begin{array}{l} |gradv(\mathbf{e})|^T \mathbf{D}_M + 2\beta\zeta v(\mathbf{e})+ \\ +\mu k signv(\mathbf{e}_0) + 2\eta k v^{1/2}(\mathbf{e}) \end{array} \right\}$$

$$\beta, k \in \mathfrak{R}^+, \zeta, \mu, \eta \in \{0, 1\}, \ \zeta + \mu + \eta = 1, \tag{12.52}$$

which demands the measurability of the plant state vector $\mathbf{X}(t)$. If $\mathbf{X}(t)$ is measurable then this algorithm ensures very robust good quality tracking as shown in what follows. When $\mathbf{e}(t) \longrightarrow \mathbf{0}_N$ then $\mathbf{W}(t) \longrightarrow \infty \mathbf{1}_N$ that is the drawback of the scalar Lyapunov function approach.

Equations (12.49) and (12.52) determine the control $\mathbf{U}(t)$.

Equations (12.50), (12.51) and (12.52) lead to

$$v^{(1)}[\mathbf{e}(t)] \le - \left\{ 2\beta\zeta v[\mathbf{e}(t)] + \mu k signv(\mathbf{e}_0) + 2\eta k v^{1/2}[\mathbf{e}(t)] \right\},$$

$$\forall [t, \mathbf{D}(.), \mathbf{Y}_d(.)] \in \mathfrak{T}_0 \times \mathfrak{D}^{\mu} \times \mathfrak{Y}_d^{\mu}, \tag{12.53}$$

where the parameters are free for the selection. Their choice determines the type and the quality of the tracking.

12.2.3   Quadratic form as a Lyapunov function

The selection of the Lyapunov function $v(.)$ is to be a quadratic form,

$$v(\mathbf{e}) = \mathbf{e}^T H \mathbf{e}, \ H = H^T > O_N, \ gradv(\mathbf{e}) = 2H\mathbf{e}. \tag{12.54}$$

Equation (12.52) takes the following form

$$\mathbf{W}(t) = \frac{d\mathbf{Y}_d}{dt} - CA\mathbf{X}(t)+$$

$$+ \|H\mathbf{e}\|^{-2} [H\mathbf{e}] \left\{ \begin{array}{l} 2|H\mathbf{e}|^T \mathbf{D}_M + 4\beta\zeta \mathbf{e}^T H\mathbf{e}+ \\ +\mu k sign(\mathbf{e}_0^T H\mathbf{e}_0) + 2\eta k (\mathbf{e}^T H\mathbf{e})^{1/2} \end{array} \right\}. \tag{12.55}$$

Inequality (12.53) becomes the following due to (12.54):

$$v^{(1)}[\mathbf{e}(t)] \le - \left\{ 2\beta\zeta \mathbf{e}^T(t) H\mathbf{e}(t) + \mu k sign(\mathbf{e}_0^T H\mathbf{e}_0) + 2\eta k \left[\mathbf{e}^T(t) H\mathbf{e}(t) \right]^{1/2} \right\}$$

$$\forall [\mathbf{D}(.), t] \in \mathfrak{D}^{\mu} \times \mathfrak{Y}_d^{\mu} \times \mathfrak{T}_0, \ \mathbf{e}^T(t) H\mathbf{e}(t) = v[\mathbf{e}(t)]. \tag{12.56}$$

Case 260 *Control synthesis for the global exponential tracking*
   If

$$\zeta = 1 \Longrightarrow \mu = \eta = 0,$$

then (12.56) reduces to

$$v^{(1)}\left[\mathbf{e}\left(t\right)\right] \le -2\beta v\left[\mathbf{e}\left(t\right)\right], \ \forall\left(t, \mathbf{D}(.), \mathbf{e}_0\right) \in \mathfrak{T}_0 \mathsf{x} \mathfrak{D}^{\mu} \mathsf{x} \mathfrak{R}^{N}.$$

The solution reads

$$\left\|\mathbf{e}\left(t\right)\right\| \le \alpha e^{-\beta t}\left\|\mathbf{e}_0\right\|, \forall\left(t, \mathbf{D}(.), \mathbf{e}_0\right) \in \mathfrak{T}_0 \mathsf{x} \mathfrak{D}^{\mu} \mathsf{x} \mathfrak{R}^{N},$$

$$\alpha = \sqrt{\lambda_M(H)\lambda_m^{-1}(H)}. \tag{12.57}$$

$\lambda_M(H)$ *and* $\lambda_m(H)$ *are the maximal and minimal eigenvalue of the matrix* H, *respectively. They are positive real numbers because the matrix* H *is positive definite and symmetric. Inequality (12.57) proves the robust global exponential tracking on* $\mathfrak{D}^1 \times \mathfrak{Y}_d^1$.

Case 261 *Control synthesis for the global stablewise tracking with the finite reachability time*
   If

$$\mu = 1 \Longrightarrow \zeta = \eta = 0,$$

then (12.56) becomes the following due to (12.54):

$$v^{(1)}\left[\mathbf{e}\left(t\right)\right] \le -ksignv(\mathbf{e}_0), \ \forall\left(t, \mathbf{D}(.), \mathbf{e}_0\right) \in \mathfrak{T}_0 \mathsf{x} \mathfrak{D}^{\mu} \mathsf{x} \mathfrak{R}^{N}, \ v(\mathbf{e}_0) = \mathbf{e}_0^T H \mathbf{e}_0.$$

The solution reads

$$\left\|\mathbf{e}\left(t\right)\right\| \le \left\{ \begin{array}{ll} \lambda_m^{-1/2}(H)\sqrt{\mathbf{e}_0^T H \mathbf{e}_0 - kt}, & t \le k^{-1}\mathbf{e}_0^T H \mathbf{e}_0, \\ 0, & t \ge k^{-1}\mathbf{e}_0^T H \mathbf{e}_0 \end{array} \right\},$$

$$\forall\left(t, \mathbf{D}(.), \mathbf{e}_0\right) \in \mathfrak{T}_0 \mathsf{x} \mathfrak{D}^{1} \mathsf{x} \mathfrak{R}^{N}.$$

This proves the robust stablewise tracking on $\mathfrak{D}^1 \times \mathfrak{Y}_d^1$ *with the finite scalar reachability time* τ_R

$$\tau_R = k^{-1}\mathbf{e}_0 H \mathbf{e}_0 = \tau_R\left(\mathbf{e}_0; k\right),$$

which depends on $\mathbf{e}_0$. *If the reachability time* τ_R *is given, then we determine the gain* k *from*

$$k = \tau_R^{-1}\mathbf{e}_0 H \mathbf{e}_0 = k\left(\mathbf{e}_0; \tau_R\right).$$

These equations express the relationship among the initial error vector $\mathbf{e}_0$, *the reachability time* τ_R *and the gain* k.

Case 262 *Control synthesis for the global stablewise tracking with the finite reachability time*
 If

$$\eta = 1 \Longrightarrow \zeta = \mu = 0,$$

then (12.56) takes the following form due to (12.54):

$$v^{(1)}\left[\mathbf{e}\left(t\right)\right] \leq -2kv^{1/2}\left[\mathbf{e}\left(t\right)\right], \ \forall \left[t, \mathbf{D}(.), \mathbf{Y}_d(.)\right] \in \mathfrak{T}_0 \times \mathfrak{D}^\mu \times \mathfrak{Y}_d^\mu.$$

We find the solution in the form

$$\left\|\mathbf{e}\left(t\right)\right\| \leq \left\{ \begin{array}{ll} \lambda_m^{-1/2}(H)\left[\sqrt{\mathbf{e}_0 H \mathbf{e}_0} - kt\right]^2, & t \in [0, \tau_R], \\ 0, & t \geq \tau_R \end{array} \right\},$$

$$\tau_R = k^{-1}\sqrt{\mathbf{e}_0 H \mathbf{e}_0}, \ \forall \left[t, \mathbf{D}(.), \mathbf{Y}_d(.)\right] \in \mathfrak{T}_0 \times \mathfrak{D}^\mu \times \mathfrak{Y}_d^\mu.$$

This proves the stablewise tracking on $\mathfrak{D}^1$ *with the finite reachability time* τ_R

$$\tau_R = k^{-1}\sqrt{\mathbf{e}_0 H \mathbf{e}_0} = \tau_R\left(\mathbf{e}_0; k\right),$$

which depends on $\mathbf{e}_0$*. If* τ_R *is given, then we calculate the gain* k *from*

$$k = \tau_R^{-1}\sqrt{\mathbf{e}_0 H \mathbf{e}_0} = k\left(\mathbf{e}_0; \tau_R\right).$$

The smaller the reachability time τ_R*, the bigger the gain* k *for the given initial output error vector* $\mathbf{e}_0$*, and vice versa.*

12.2.4 Introduction to *VLF* concept

The concept of vector Lyapunov functions (VLF) is due to R. Bellman [23] *in the linear systems setting and V. M. Matrosov* [235] *in the general nonlinear systems framework. Matrosov developed further the VLF concept by generalizing it in the framework of the large-scale nonlinear systems* [236], [237]. It is the basic mathematical tool for studying stability properties of complex (interconnected and large-scale) dynamical systems [146], [236], [241], [287]. The *VLF* is the mathematical tool to effectively construct a scalar Lyapunov function for the complex dynamical systems and to reduce their stability test to simple algebraic conditions.

 The application of a scalar Lyapunov function for control synthesis faces the complex mathematical problem of how to separate the control from the Lyapunov function gradient and how to adjust it to the tracking task. In order to cope with this drawback of the scalar Lyapunov function approach it was suggested in [160], [174], [185], [189], [188, Chapter 11.1, pp. 223-227, Chapter 11.2, pp. 239-242, Chapter 11.3, pp. 247, 248] to use the *VLF* concept in the exact vector form without any need for the scalar Lyapunov function application to the whole system in order to ensure its tracking. We present it in its simplified form adequate to the need of the tracking control synthesis in the framework of the linear systems.

12.2.5 Definitions of *VLF*s

All vector and matrix equalities, inequalities and powers hold elementwise (Section 1.3). We accept the generalization of Lyapunov's concept of definite functions proposed and developed in [188, Chapter 11.1, pp. 223-227, Chapter 11.2, pp. 239-242, Chapter 11.3, pp. 247, 248].

Definition 263 *Definition of vector definite functions*

A vector function $\mathbf{v}(.) : R^N \rightarrow R^N$, $\mathbf{v}(\mathbf{e}) = [v_1(\mathbf{e}) \quad v_2(\mathbf{e}) \quad ... \quad v_N(\mathbf{e})]^T$, $v_i(.) : \mathfrak{R}^N \longrightarrow \mathfrak{R}$, $\forall i = 1, 2, ..., N$, *is*

a)   **Positive (negative) definite** *if and only if there is a neighborhood* $\mathfrak{S}$ *of* $\mathbf{e} = \mathbf{0}_N$, $\mathfrak{S} \subseteq \mathfrak{R}^N$, *such that (i) through (iii) hold:*

(i) $\mathbf{v}(.)$ *is defined and continuous on* $\mathfrak{S}$: $\mathbf{v}(\mathbf{e}) \in \mathfrak{C}(\mathfrak{S})$,

(ii) $\mathbf{v}(\mathbf{e}) \geq \mathbf{0}_N$, $(\mathbf{v}(\mathbf{e}) \leq \mathbf{0}_N)$, $\forall \mathbf{e} \in \mathfrak{S}$,

(iii) $v_i(\mathbf{e}) = 0$ *for* $\mathbf{e} \in \mathfrak{S}$ *if and only if* $e_i = 0$, $\forall i = 1, 2, ..., N$.

b)   **Global positive (negative) definite** *if and only if (i) through (iii) hold for* $\mathfrak{S} = \mathfrak{R}^N$.

c)   **Elementwise positive (negative) definite** *if and only if it is positive (negative) definite and*

(iv) $v_i(.) : \mathfrak{R} \longrightarrow \mathfrak{R} \Longrightarrow v_i(\mathbf{e}) \equiv v_i(e_i)$, $\forall i = 1, 2, ..., N$.

d)   **Global elementwise positive (negative) definite** *if and only if (i) through (iv) hold for* $\mathfrak{S} = \mathfrak{R}^N$.

e)   **Radially strictly increasing on** $\mathfrak{S}$ *if and only if*

(v) $\mathbf{v}(\lambda_1 \mathbf{e}) < \mathbf{v}(\lambda_2 \mathbf{e})$, $0 < \lambda_1 < \lambda_2$, $\forall (\mathbf{e} \neq \mathbf{0}_N) \in \mathfrak{S}$.

f)   **Radially unbounded** *if and only if the corresponding above property is global and*

(vi) $\mathbf{v}(\lambda \mathbf{e}) \longrightarrow \infty \mathbf{1}_N$ *as* $\lambda \longrightarrow \infty$, $\forall (\mathbf{e} \neq \mathbf{0}_N) \in \mathfrak{R}^N$.

The conditions (i) and (ii) do not imply positive definiteness on $\mathfrak{S}$ of any entry $v_i(.) : \mathfrak{R}^N \longrightarrow \mathfrak{R}_+$ of $\mathbf{v}(.) : \mathfrak{R}^N \longrightarrow \mathfrak{R}_+^N$. However, they imply the positive semi-definiteness on $\mathfrak{S}$ of every entry $v_i(.)$ of $\mathbf{v}(.)$ because $v_i(.)$ is defined and continuous on $\mathfrak{R}^N$ and nonnegative on $\mathfrak{S}$. The conditions (i) through (iii) imply both $\mathbf{v}(\mathbf{e}) = \mathbf{0}_N$ for $\mathbf{e} \in \mathfrak{S}$ if and only if $\mathbf{e} = \mathbf{0}_N$ and positive definiteness of $\mathbf{v}(.)$.

The conditions (i) through (iv) imply positive definiteness on $\mathfrak{S}_i$, $\mathfrak{S}_i \subseteq \mathfrak{R}^1$, of the entry $v_i(.)$ of $\mathbf{v}(.)$ because $v_i(.)$ is defined and continuous on $\mathfrak{R}^1$, and positive out of the origin on $\mathfrak{S}_i$, $\forall i = 1, 2, ..., N$. Then $\mathfrak{S}$ is the Cartesian product of all $\mathfrak{S}_i$: $\mathfrak{S} = \mathfrak{S}_1 \times \mathfrak{S}_2 \times ... \times \mathfrak{S}_N$.

Definition 263 is compatible with Lyapunov's original definition of the scalar definite functions [230], as well as with the concept of the matrix definite functions introduced in [124].

The condition (iii) under a) can be relaxed if we accept the use of a scalar overall positive definite function $v : \mathfrak{R}^N \longrightarrow \mathfrak{R}^N$, $v(\mathbf{e}) \in \mathfrak{C}(\mathfrak{R}^N)$,

$$\sum_{i=0}^{i=N} v_i(\mathbf{e}) \geq 0, \ \forall \mathbf{e} \in \mathfrak{R}^N, \ \sum_{i=0}^{i=N} v_i(\mathbf{e}) = 0 \Longleftrightarrow \mathbf{e} = \mathbf{0}_N.$$

In this case the functions $v_1(.)$, $v_2(.)$, ..., $v_N(.)$ can be each, but need not be each, (global) (radially unbounded) positive definite functions. However, their sum must be, respectively, (global) (radially unbounded) positive definite function, which permits that some of them are only positive semidefinite functions permitting their dependence only on a subvector of the vector $\mathbf{e}$.

We allow only continuity without continuous differentiability of the Lyapunov functions. In order to determine their (right-hand) derivative we use their Dini derivatives (on Dini derivatives see the references [175, Appendix B, pp. 383-387], [195, Definition 3.5, p. 54], [238], [321]).

Definition 264 *Definition of vector Lyapunov functions [188, Definition 315, pp. 224, 225]*

A vector function $\mathbf{v}(.) : R^N \to R^N$ is

*a)   **An error vector Lyapunov function of a given dynamical system** if and only if both (i) and (ii) hold:*

(i) $\mathbf{v}(.)$ is positive definite,

(ii) There is a neighborhood $\mathfrak{B}$ of $\mathbf{e} = \mathbf{0}_N$, $\mathfrak{B} \subseteq \mathfrak{R}^N$, such that the following is valid,

$$D^+\mathbf{v}(\mathbf{e}) \leq 0, \ \forall \left(\mathbf{e}, \mathbf{e}^{(1)} \right) \in \mathfrak{B} \times \mathfrak{B}. \tag{12.58}$$

If and only if additionally there is a positive definite vector function $\boldsymbol{\Psi}(.) : \mathfrak{R}^{2N} \to \mathfrak{R}^N$ such that

$$D^+\mathbf{v}(\mathbf{e}) \leq -\boldsymbol{\Psi}(\mathbf{e}^1), \ \forall \mathbf{e}^1 \in \mathfrak{B} \times \mathfrak{B}, \tag{12.59}$$

*then the function $v(.)$ is a **strict error vector Lyapunov function of the system**.*

*b)   **An elementwise error vector Lyapunov function of the
system** if and only if both (1) and (2) hold:*

(1) $\mathbf{v}(.)$ is elementwise positive definite,

*(2) There is a neighborhood $\mathfrak{B}$ of $\mathbf{e} = \mathbf{0}_N$, $\mathfrak{B} \subseteq \mathfrak{R}^N$, such that
(12.58) is valid.*

*If and only if, additionally, there is an elementwise positive definite vector
function $\mathbf{\Psi}(.) : R^N \to R^N$ such that (12.59) holds then the function $\mathbf{v}(.)$ is
a **strict elementwise error vector Lyapunov function of the system**.*

This definition is compatible with the concept of vector Lyapunov functions by R. Bellman [23] and V. M. Matrosov [235]–[237], as well as with the
concept of matrix Lyapunov functions introduced in [124].

Note 265 *The vector function $\mathbf{v}(.)$: $\mathfrak{R}^N \longrightarrow \mathfrak{R}^N$ induces $D^+\mathbf{v}(.)$:
$\mathfrak{R}^{2N} \longrightarrow \mathfrak{R}^N$. This means that $\mathbf{v}(.)$ depends on $\mathbf{e}$, while $D^+\mathbf{v}(.)$ is a function
of $\mathbf{e}^1 = \left[\mathbf{e}^T \ \ \mathbf{e}^{(1)^T}\right]^T$.*

12.2.6   **VLF** generalization of the classical stability theorems

The Greek letter ϕ denotes the empty set. Let $\mathbf{c} \in \mathfrak{R}^{+^N}$. The set $\mathfrak{V}_\mathbf{c}$, $\mathfrak{V}_\mathbf{c} \subseteq
\mathfrak{R}^N$, is the largest open connected neighborhood of $\mathbf{e} = \mathbf{0}_N$ such that a
vector function $\mathbf{v}(.)$ and the vector $\mathbf{c}$, which determines the set set $\mathfrak{V}_\mathbf{c}$, obey
elementwise

$$\mathbf{v}(\mathbf{e}) < \mathbf{c}, \forall \mathbf{e} \in \mathfrak{V}_\mathbf{c}. \tag{12.60}$$

$Cl\mathfrak{V}_\mathbf{c}$ is the closure of the set $\mathfrak{V}_\mathbf{c}$, and $\partial\mathfrak{V}_\mathbf{c}$ is its boundary if the boundary
exists. $\mathfrak{N}_{\mathbf{a}_i}$ is the $\mathbf{a}_i$-neighborhood of $\mathbf{e} = \mathbf{0}_N$ defined by

$$\mathfrak{N}_{\mathbf{a}_i} = \left\{\mathbf{e} : \mathbf{e} \in \mathfrak{R}^N, \ |\mathbf{e}| < \mathbf{a}_i\right\}, \ \mathbf{a}_i \in \mathfrak{R}^{+^N}. \tag{12.61}$$

Condition 266 *The sets $\mathfrak{V}_{\mathbf{c}_i}$, $\mathbf{c}_i \in \mathfrak{R}^{+^N}$, $i = 1, 2$, satisfy a) through c):*

*a) $Cl\mathfrak{V}_{\mathbf{c}_1} \subset Cl\mathfrak{V}_{\mathbf{c}_2}$, $\partial\mathfrak{V}_{\mathbf{c}_1} \cap \partial\mathfrak{V}_{\mathbf{c}_2} = \phi$, $\forall \mathbf{c}_i \in \mathfrak{R}^{+^N}$, $i = 1, 2$, $\mathbf{0}_N <
\mathbf{c}_1 < \mathbf{c}_2$,*

b) $\mathbf{c}_i \to \infty\mathbf{1}_N \Longrightarrow \mathfrak{V}_{\mathbf{c}_i} \to \mathfrak{R}^N$, $i = 1, 2$,

c) $\forall \mathbf{c}_i \in \mathfrak{R}^{+^N}$, $\exists \mathbf{a}_i \in \mathfrak{R}^{+^N} \Longrightarrow \mathfrak{V}_{\mathbf{c}_i} \subseteq \mathfrak{N}_{\mathbf{a}_i}$, $i = 1, 2$.

Note 267 *If the vector positive definite function $\mathbf{v}(.)$ is radially strictly increasing on $\mathfrak{S}$, then the sets $\mathfrak{V}_{\mathbf{c}_i}$ associated with $\mathbf{v}(.)$ satisfy a) of Condition
266 on $\mathfrak{S}$.*

Theorem 268 *[188, Theorem 319, p. 226] Let Condition 266 hold. In order for* $\mathbf{e} = \mathbf{0}_N$ *of the system to be, respectively, {elementwise} asymptotically stable it is sufficient that there is a strict (elementwise) vector Lyapunov function* $\mathbf{v}(.)$ *of the system.*

If, additionally, $\mathfrak{B} = \mathfrak{S} = \mathfrak{R}^N$, $\mathbf{v}(.)$ *is also global strict {elementwise} vector Lyapunov function and radially unbounded, then* $\mathbf{e} = \mathbf{0}_N$ *is globally (elementwise) asymptotically stable.*

12.2.7 VLF forms

Example 269 *Let the vector function* $\mathbf{v}(.) : \mathfrak{R}^{(k+1)N} \longrightarrow \mathfrak{R}^{(k+1)N}$ *depend on the extended error vector* $\mathbf{e}^k$ *in the following form:*

$$\mathbf{v}(\mathbf{e}^k) = \frac{1}{2} E^k \mathbf{e}^k, \ E^k = blocdiag \left\{ E^{(0)} \ \ E^{(1)} \ ... \ E^{(k)} \right\}, \ k \in \{0, 1, 2, .., \nu - 1\},$$

$$E^{(0)} = E, \ E^{(i)} = diag \left\{ e_1^{(i)} \ \ e_2^{(i)} \ ... \ e_N^{(i)} \right\} \in \mathfrak{R}^{N \times N}, \ i \in \{0, 1, 2, .., k\},$$

$$E^k \in \mathfrak{R}^{(k+1)N \times (k+1)N}.$$

This vector function dependent on $\mathbf{e}^k$, $\mathbf{v}(\mathbf{e}^k)$, *is an example of a global strict elementwise positive definite error vector function [160], [189] as a VLF candidate.*

Example 270 *Other possible forms of VLF follow:*

$$\mathbf{v}(\mathbf{e}^k) = \frac{1}{2} E^k H \mathbf{e}^k \in \mathfrak{R}^{(k+1)N}, \ H = H^T > O_{(k+1)N}, \ H \in \mathfrak{R}^{(k+1)N \times (k+1)N},$$

where $H > O_{(k+1)N}$ *denotes that* H, $H \in \mathfrak{R}^{(k+1)N \times (k+1)N}$, *is positive definite, or*

$$\mathbf{v}(\mathbf{e}^k) = \left| \mathbf{e}^k \right| \in \mathfrak{R}_+^{(k+1)N},$$

or

$$\mathbf{v}(\mathbf{e}^k) = V(\mathbf{e}^k) H \mathbf{V}(\mathbf{e}^k) \in \mathfrak{R}_+^{(k+1)N},$$

$$V(\mathbf{e}^k) = diag \left\{ v_1 \left(\mathbf{e}^k \right) \ \ v_2 \left(\mathbf{e}^k \right) \ ... \ v_N \left(\mathbf{e}^k \right) \right\},$$

$$\mathbf{V}(\mathbf{e}^k) = \left[v_1 \left(\mathbf{e}^k \right) \ \ v_2 \left(\mathbf{e}^k \right) \ ... \ v_N \left(\mathbf{e}^k \right) \right]^T \in \mathfrak{C} \left(\mathfrak{R}^{(k+1)N} \right),$$

or simply

$$\mathbf{v}(\mathbf{e}^k) = \mathbf{V}(\mathbf{e}^k) \in \mathfrak{R}_+^{(k+1)N}.$$

$\mathbf{V}(.) : \mathfrak{R}^{(k+1)N} \longrightarrow \mathfrak{R}^{(k+1)N}$ *is positive definite vector function on* $\mathfrak{R}^{(k+1)N}$.

12.2.8   Choice of a vector Lyapunov function

The usage of the vector Lyapunov function

$$\mathbf{v(e)} = \frac{1}{2}E\mathbf{e} = \frac{1}{2}\left[e_1^2 : e_2^2 : \dots : e_N^2\right]^T$$

leads to

$$\mathbf{v}^{(1)}(\mathbf{e}) = E\mathbf{e}^{(1)} = E\left[\mathbf{Y}_d^{(1)}(t) - \mathbf{Y}^{(1)}(t)\right] =$$

$$= E\left[\begin{array}{c} \mathbf{Y}_d^{(1)}(t) - CA\mathbf{X}(t) - \left(CB^{(\mu)} + U^{(\mu)}\right)\mathbf{U}^\mu(t) - \\ - \left(CD^{(\mu)} + V^{(\mu)}\right)\mathbf{D}^\mu(t). \end{array}\right]$$

due to (12.43). Let

$$\left(CB^{(\mu)} + U^{(\mu)}\right)\mathbf{U}^\mu(t) = \mathbf{Y}_d^{(1)}(t) - CA\mathbf{X}(t) + \mathbf{W}(t)$$

so that

$$\mathbf{v}^{(1)}(\mathbf{e}) = -E\left[\mathbf{W}(t) + \left(CD^{(\mu)} + V^{(\mu)}\right)\mathbf{D}^\mu(t)\right].$$

Let $\mathbf{\Psi}(.) : \mathfrak{T}_0 \longrightarrow \mathfrak{R}^{+N}$ obey

$$\mathbf{\Psi}(t) > \left|\left(CD^{(\mu)} + V^{(\mu)}\right)\mathbf{D}^\mu(t)\right|, \ \forall (t, \mathbf{D}(.)) \in \mathfrak{T}_0 \times \mathfrak{D}^\mu,$$

and

$$\mathbf{W}(t) = S(\mathbf{e})\mathbf{\Psi}(t) + \beta\zeta\mathbf{e} + \mu K signe, \ \beta, k \in \mathfrak{R}^+, \ \zeta, \mu \in \{0,1\}, \ \zeta + \mu = 1.$$

Notice that $\mathbf{e} \longrightarrow \mathbf{0_N}$ does not imply $\mathbf{W}(t) \longrightarrow \infty \mathbf{1_N}$. The VLF approach is free of the scalar Lyapunov function drawback. The control is the solution to the following differential equation:

$$\left(CB^{(\mu)} + U^{(\mu)}\right)\mathbf{U}^\mu(t) = \mathbf{Y}_d^{(1)}(t) - CA\mathbf{X}(t) +$$

$$+ S(\mathbf{e})\mathbf{\Psi}(t) + \beta\zeta\mathbf{e} + \mu K signe,$$

and the derivative $\mathbf{v}^{(1)}(\mathbf{e})$ satisfies

$$\mathbf{v}^{(1)}(\mathbf{e}) \le -E(\beta\zeta\mathbf{e} + \mu K signe) = -.2\beta\zeta\mathbf{v(e)} - \mu K|\mathbf{e}| \qquad (12.62)$$

Case 271 *Control synthesis for the global elementwise exponential tracking*

This is the case in which

$$\zeta = 1 \Longrightarrow \mu = 0,$$

and the solution to (12.62) is found in the form

$$|\mathbf{e}(t; \mathbf{e}_0)| \leq \exp(-\beta t) |\mathbf{e}_0|, \ \forall (t, \mathbf{e}_0) \in \mathfrak{T}_0 \times \mathfrak{R}^N.$$

This proves the global elementwise exponential tracking.

Case 272 *Control synthesis for the global elementwise stablewise tracking with finite vector reachability time* τ_R^N

For

$$\mu = 1 \Longrightarrow \zeta = 0,$$

the solution to (12.62) is determined as

$$|\mathbf{e}(t; \mathbf{e}_0)| \begin{cases} \leq (|\mathbf{e}_0| - tK\mathbf{1}_N), \ \forall t\mathbf{1}_N \in [\mathbf{0}_N, \ \mathbf{t}_R^N] \\ = \mathbf{0}_N, \ \forall t\mathbf{1}_N \in [\tau_R^N, \ \infty\mathbf{1}_N[\end{cases} \right\}, \ \forall \mathbf{e}_0 \in \mathfrak{R}^N,$$

$$\mathbf{t}_R^N = K^{-1} |\mathbf{e}_0|.$$

The tracking is global elementwise stablewise with the finite vector reachability time $\mathbf{t}_R^N$. *The output error convergence is in the linear form with the constant speed* K *to the zero error vector. There are not oscillations, overshoot and undershoot in the output error vector.*

The equation

$$\mathbf{t}_R^N = K^{-1} |\mathbf{e}_0|$$

shows the elementwise trade off among τ_R^N, K *and* $|\mathbf{e}_0|$.

Comment 273 *Lyapunov tracking control of the linear plant is nonlinear.*

Chapter 13

Natural Tracking Control (NTC)

This chapter folllows, refines, extends and generalizes [188, Chapter 12, pp. 249-292].

13.1 High quality tracking criteria

13.1.1 *Time* vectors and *time* sets

The use of the *time vector* $\mathbf{t}^{(k+1)N}$ (8.44) and of *the vector reachability time* $\mathbf{t}_R^N$ (8.47) (Section 8.6), simplifies the treatment of the vector calculus and formulae related to the elementwise tracking with the finite vector reachability time $\mathbf{t}_R^{(k+1)N}$ (8.50) (Section 8.6).

We introduced *the scalar reachability time* $t_{Ri(j)}$, $t_{Ri(j)} \in \mathfrak{T}_0 \cup \{\infty\}$, of the j-th derivative $e_i^{(j)}$ of the i-th entry e_i of the error vector $\mathbf{e}$, They induce the *time sets* $\mathfrak{T}_{Ri(j)}$ and $\mathfrak{T}_{Ri(j)\infty}$, (8.48), (Section 8.6). They are entries of the reachability *time* $\mathbf{t}_{R(j)}^N$ (8.49) of the j−th output vector derivative and they induce *the vector reachability time* $t_R^{(k+1)N}$ (8.51), $t_R^{(k+1)N} \in \mathfrak{T}_0^{(k+1)N} \cup \{\infty\}^{(k+1)N}$, and the *time* set products $\mathfrak{T}_R^{(k+1)N}$, $\mathfrak{T}_{R\infty}^{(k+1)N}$ (8.55) and $\mathfrak{T}_{RF}^{(k+1)N}$ (8.56), all defined in Section 8.6.

13.1.2 Subsidiary reference output

Most often the plant initial real extended output vector $\mathbf{Y}_0^k$ is different from its initial desired extended output vector $\mathbf{Y}_{d0}^k$, i.e., the plant initial extended output error vector $\mathbf{e}_0^k$ is nonzero: $\mathbf{e}_0^k \neq \mathbf{0}_{(k+1)N}$. The perfect tracking on $\mathfrak{T}_0^{(k+1)N}$ is impossible for such initial conditions. The ideal tracking control strategy is unrealizable on $\mathfrak{T}_0^{(k+1)N}$ under such initial conditions. The wise theoretical and engineering strategy is to recognize the reality, that is that $\mathbf{Y}_0^k \neq \mathbf{Y}_{d0}^k$, and to incorporate information about it in our demand on tracking, hence on control. We can request the following if an initial output vector $\mathbf{Y}_0^k$ is nondesired, i.e., if $\mathbf{Y}_0^k \neq \mathbf{Y}_{do}^k$: the control should force the plant real behavior $\mathbf{Y}\left(\mathbf{t}^{(k+1)N}\right)$ to be satisfactory on $\mathfrak{T}_0^{(k+1)N}$ despite it begins from $\mathbf{Y}_0^k \neq \mathbf{Y}_{d0}^k$, i.e., to be sufficiently close to the desired plant output $\mathbf{Y}_d\left(\mathbf{t}^{(k+1)N}\right)$ by converging to it until the finite vector reachability moment $\mathbf{t}_R^{(k+1)N}$, at the moment $\mathbf{t}_R^{(k+1)N}$ the real output $\mathbf{Y}\left(\mathbf{t}^{(k+1)N}\right)$ is to become equal to $\mathbf{Y}_d\left(\mathbf{t}^{(k+1)N}\right)$ and they should rest equal since that moment on. Besides, we can specify the form of the transient process from the initial moment $\mathbf{t}^{(k+1)N} = \mathbf{0}_{(k+1)N}$ until the moment $\mathbf{t}_R^{(k+1)N}$. The satisfactory transient behavior on the time set $\mathfrak{T}_R^{(k+1)N}$ can be defined, for example, by a vector function $\mathbf{f}\left(.; \mathbf{f}_0^k\right)$. The application of the appropriately defined vector function $\mathbf{f}\left(.; \mathbf{f}_0^k\right)$ enables us to cope with the problem of the uncertainty of the initial errors.

Definition 274 *Let* $\mathbf{f}\left(t; \mathbf{f}_0^k\right)$ *obey*

$$\mathbf{f}\left(.\right) : \mathfrak{T}_0^N \longrightarrow \mathfrak{R}^N, \; \mathbf{f}\left(\mathbf{t}^N; \mathbf{f}_0^k\right) \in \mathfrak{C}^i(\mathfrak{T}_0^N), \; i \in \{0, 1, 2, .., k, ...\}, \tag{13.1}$$

$$\mathbf{t}^{(k+1)N} = \mathbf{0}_{(k+1)N} \Longrightarrow \mathbf{f}_0^k = \mathbf{f}^k\left(\mathbf{0}_{(k+1)N}; \mathbf{f}_0^k\right) = -\mathbf{e}^k\left(\mathbf{0}_{(k+1)N}; \mathbf{e}_0^k\right) = -\mathbf{e}_0^k, \tag{13.2}$$

$$\mathbf{f}_0^k = \mathbf{0}_{(k+1)N} \Longrightarrow \mathbf{f}^k\left(\mathbf{t}^{(k+1)N}; \mathbf{0}_{(k+1)N}\right) \equiv \mathbf{0}_{(k+1)N}, \tag{13.3}$$

$$\mathbf{f}^k\left(\mathbf{t}^{(k+1)N}; \mathbf{f}_0^k\right) = \mathbf{0}_{(k+1)N}, \; \left\{ \begin{array}{c} \forall \mathbf{t}^{(k+1)N} = \mathfrak{T}_{RF}^{(k+1)N} \\ if \; \mathfrak{T}_{RF}^{(k+1)N} \subset \mathfrak{T}_0^{(k+1)N} \end{array} \right\}, \tag{13.4}$$

$$\mathbf{f}^k\left(\mathbf{t}^{(k+1)N}; \mathbf{f}_0^k\right) \longrightarrow \mathbf{0}_{(k+1)N} \; \left\{ \begin{array}{c} as \; \mathbf{t}^{(k+1)N} \longrightarrow \infty \mathbf{1}_{(k+1)N} \\ if \; \mathfrak{T}_{RF}^{(k+1)N} = \{\infty\}^{(k+1)N} \end{array} \right\}. \tag{13.5}$$

Figure 13.1 [188, Fig 12.1, p. 254] explains the symbol of the switch used in the block diagram of the **f**-function generator shown in Figure 13.2 [188, Fig. 12.2, p. 255].

Note 275 *An example of the function* $\mathbf{f}\left(.\right)$ *(13.1)–(13.5) is given in Appendix C.1.*

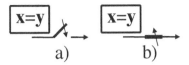

Figure 13.1: (a) Switch closes if and only if $x = y$. (b) Switch opens if and only if $x \neq y$.

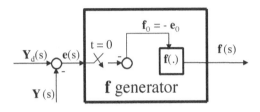

Figure 13.2: Block diagram of the **f**-generator.

Definition 274 of the subsidiary vector function $\mathbf{f}(.)$ enables us to introduce also the subsidiary reference output vector $\mathbf{Y}_R$ that will assume the role of the desired output vector $\mathbf{Y}_d$ during the transient process. Its first crucial characteristic should be $\mathbf{Y}_R^k(\mathbf{0}_{(k+1)N}) = \mathbf{Y}^k(\mathbf{0}_{(k+1)N})$. Its second characteristic is to satisfy $\mathbf{Y}_R^k(\mathbf{t}^{(k+1)N}) = \mathbf{Y}_d^k\left(\mathbf{t}^{(k+1)N}\right)$ at the moment $\mathbf{t}_R^{(k+1)N}$ and on.

The control should force the plant *to exhibit the perfect tracking of* $\mathbf{Y}_R(t)$ on $\mathfrak{T}_0$ for any initial conditions $\mathbf{Y}_0^k$ and any disturbances so that after some finite *time* $\mathbf{Y}_R(t)$ becomes and rests for ever equal to $\mathbf{Y}_d(t)$.

Definition 276 *Let* ***the reference output vector variable*** $\mathbf{Y}_R$ *be such that,*

$$\mathbf{Y}_R^k(\mathbf{t}^{(k+1)N}) = \mathbf{Y}_d^k\left(\mathbf{t}^{(k+1)N}\right) +$$

$$+ \left\{ \begin{array}{ll} \mathbf{0}_{(k+1)N}, & \left(\begin{array}{c} \forall \mathbf{t}^{(k+1)N} \in \mathfrak{T}_0^{(k+1)N} \\ if\ \mathbf{e}_0^k = \mathbf{0}_{(k+1)N}, \end{array} \right) \\ \mathbf{f}^k\left(\mathbf{t}^{(k+1)N}\right), & \left(\begin{array}{c} \forall \mathbf{t}^{(k+1)N} \in \mathfrak{T}_0^{(k+1)N} \\ if\ \mathbf{e}_0^k \neq \mathbf{0}_{(k+1)N}, \end{array} \right) \end{array} \right\} \implies \quad (13.6)$$

$$\mathbf{Y}_R^k(\mathbf{0}_{(k+1)N}) = \mathbf{Y}^k(\mathbf{0}_{(k+1)N}) = \mathbf{Y}_d^k\left(\mathbf{0}_{(k+1)N}\right) + \mathbf{f}_0^k = \mathbf{Y}_d^k\left(\mathbf{0}_{(k+1)N}\right) - \mathbf{e}_0^k,$$
$$(13.7)$$

*and let **the induced subsidiary error vector** ϵ be defined by:*

$$\epsilon = [\epsilon_1 \quad \epsilon_2 \quad ... \quad \epsilon_N]^T \in \mathfrak{R}^N, \tag{13.8}$$

$$\epsilon^k(\mathbf{t}^{(k+1)N}) = \mathbf{Y}_R^k(\mathbf{t}^{(k+1)N}) - \mathbf{Y}^k(\mathbf{t}^{(k+1)N}) \implies$$
$$\epsilon^k(\mathbf{t}^{(k+1)N}) = \mathbf{e}^k(\mathbf{t}^{(k+1)N}) +$$

$$+ \left\{ \begin{array}{ll} \mathbf{0}_{(k+1)N}, & \left(\begin{array}{c} \forall \mathbf{t}^{(k+1)N} \in \mathfrak{T}_0^{(k+1)N} \\ if \ \ \mathbf{e}_0^k = \mathbf{0}_{(k+1)N}, \end{array} \right) \\ \mathbf{f}^k(\mathbf{t}^{(k+1)N}; \mathbf{f}_0^k), & \left(\begin{array}{c} \forall \mathbf{t}^{(k+1)N} \in \mathfrak{T}_0^{(k+1)N} \\ if \ \ \mathbf{e}_0^k \neq \mathbf{0}_{(k+1)N}, \end{array} \right) \end{array} \right\}, \tag{13.9}$$

Figure 13.1.2 [188, Fig. 12.3, p. 256] presents the block diagram of the generator of the subsidiary error vector ϵ.

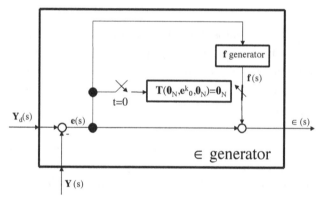

Block diagram of the ϵ-generator.

From (13.2), (13.6) and (13.7) follows:

$$\mathbf{Y}_R^k(\mathbf{0}_{(k+1)N}) = \mathbf{Y}_d^k(\mathbf{0}_{(k+1)N}) - \left\{ \begin{array}{ll} \mathbf{0}_{(k+1)N} & if \ \mathbf{e}_0^k = \mathbf{0}_{(k+1)N}, \\ \mathbf{e}^k(0) & if \ \mathbf{e}_0^k \neq \mathbf{0}_{(k+1)N}, \end{array} \right\}, \tag{13.10}$$

i.e.,

$$\mathbf{Y}_R^k(\mathbf{0}_{(k+1)N}) = \left\{ \begin{array}{ll} \mathbf{Y}_d^k(\mathbf{0}_{(k+1)N}) & iff \ \mathbf{e}_0^k = \mathbf{0}_{(k+1)N} \\ \mathbf{Y}^k(\mathbf{0}_{(k+1)N}) & iff \ \mathbf{e}_0^k \neq \mathbf{0}_{(k+1)N} \end{array} \right\}, \tag{13.11}$$

and

$$\epsilon_0^k = \epsilon_0^k\left(\mathbf{e}_0^k\right) = \mathbf{e}_0^k + \mathbf{f}^k(0; \mathbf{f}_0^k).$$

It follows that

$$\epsilon_0^k \left(\mathbf{e}_0^k \right) = \left\{ \begin{array}{ll} \mathbf{e}_0^k & if \ \ \mathbf{e}_0^k = \mathbf{0}_{(k+1)N}, \\ \mathbf{0}_{(k+1)N} & if \ \ \mathbf{e}_0^k \neq \mathbf{0}_{(k+1)N} \neq \mathbf{0}_N, \end{array} \right\} \Longrightarrow$$

$$\epsilon_0^k \left(\mathbf{e}_0^k \right) = \mathbf{0}_{(k+1)N}, \ \forall \mathbf{e}_0^k \in \mathfrak{R}^{(k+1)N}, \ k \in \{0, 1, 2, ...\}. \qquad (13.12)$$

This shows that the initial value $\epsilon(\mathbf{0}_N; \mathbf{e}_0^k) = \epsilon_0(\mathbf{e}_0^k)$ of the subsidiary error vector $\epsilon \left(\mathbf{t}^N; \mathbf{e}_0^k \right)$ and the initial values $\epsilon_0^{(1)} \left(\mathbf{e}_0^k \right), \epsilon_0^{(2)} \left(\mathbf{e}_0^k \right), ..., \epsilon_0^{(k)} \left(\mathbf{e}_0^k \right)$ of the derivatives $\epsilon^{(1)}(\mathbf{t}^N), \epsilon^{(2)}(\mathbf{t}^N), ..., \epsilon^{(k)}(\mathbf{t}^N)$ of $\epsilon \left(\mathbf{t}^N \right)$ are all equal to the zero vector $\mathbf{0}_N$ for every real initial error vector $\mathbf{e}_0 \in \mathfrak{R}^N$ and for every initial vector value $\mathbf{e}_0^{(1)} \in \mathfrak{R}^N$, $\mathbf{e}_0^{(2)} \in \mathfrak{R}^N$, ..., $\mathbf{e}_0^{(k)} \in \mathfrak{R}^N$ of the derivatives $\mathbf{e}^{(1)}(t)$, $\mathbf{e}^{(2)}(t)$, ..., $\mathbf{e}^{(k)}(t)$ of $\mathbf{e}(t)$.

Comment 277 *The reference output variable* $\mathbf{Y}_R$ *will replace the desired output variable* $\mathbf{Y}_d$. *The subsidiary error vector* ϵ *will then replace the real error vector* $\mathbf{e}$. *Equations (13.6)–(13.12) establish relations among them. This establishes the basis for control design relative to the real plant situation rather then to its desired, but unreal, situation. It is the wise engineering control strategy.*

The zero time evolution of the extended subsidiary error vector from the initial moment on, $\epsilon^k(\mathbf{t}^{(k+1)N}) \equiv \mathbf{0}_{(k+1)N}$, *ensures the perfect tracking of the subsidiary reference output* $\mathbf{Y}_R^k(\mathbf{t}^{(k+1)N})$ *on* $\mathfrak{T}_0^{(k+1)N}$ *due to*

$$\epsilon^k(\mathbf{t}^{(k+1)N}) = \mathbf{Y}_R^k(\mathbf{t}^{(k+1)N}) - \mathbf{Y}^k(\mathbf{t}^{(k+1)N}) \equiv \mathbf{0}_{(k+1)N} \Longleftrightarrow$$

$$\epsilon^k(\mathbf{t}_0^{(k+1)N}) = \mathbf{0}_{(k+1)N} \Longrightarrow \mathbf{Y}^k(\mathbf{t}^{(k+1)N}) \equiv \mathbf{Y}_R^k(\mathbf{t}^{(k+1)N}).$$

This further means that the plant will exhibit the perfect tracking of the desired output $\mathbf{Y}_d^k(\mathbf{t}^{(k+1)N})$ *at latest at and after the tracking reachability time vector* $\mathbf{t}_R^{(k+1)N}$, *i.e., on* $\mathfrak{T}_{RF}^{(k+1)N}$ *due to Equations (13.4) and (13.6):*

$$\mathbf{Y}^k(\mathbf{t}^{(k+1)N}; \mathbf{Y}_0^k) \equiv \mathbf{Y}_R^k(\mathbf{t}^{(k+1)N}; \mathbf{Y}_0^k) = \mathbf{Y}_d^k(\mathbf{t}^{(k+1)N}; \mathbf{Y}_{d0}^k),$$

$$\forall \mathbf{t}^{(k+1)N} \in \mathfrak{T}_{RF}^{(k+1)N}, \ \forall \left(\mathbf{Y}_0^k, \mathbf{Y}_{d0}^k \right) \in \mathfrak{R}^{(k+1)N} \times \mathfrak{R}^{(k+1)N}.$$

13.1.3 Tracking quality criterion

We associate the matrix Ξ with the subsidiary vector ϵ by

$$\Xi = diag \left\{ \epsilon_1 \ \epsilon_2 \ ... \ \epsilon_N \right\}. \qquad (13.13)$$

Tracking quality criteria can be expressed in various forms [188, Section 12.1, pp. 249-258]. For the NTC synthesis it appears suitable to define a demanded quality of tracking in terms of the properties of the solution to a differential equation in the subsidiary output error ϵ (13.8), (13.9) and its derivatives and/or its integral so that the following holds:

$$
\mathbf{T}\left(\mathbf{t}^N, \epsilon\left(\mathbf{t}^N\right), \epsilon^{(1)}\left(\mathbf{t}^N\right), ..., \epsilon^{(k)}\left(\mathbf{t}^N\right), \int_{\mathbf{t}_0^N=\mathbf{0}_N}^{\mathbf{t}^N} \Xi\left(\mathbf{t}^N\right) dt^N\right) =
$$

$$
= \mathbf{T}\left(\mathbf{t}^N, \epsilon^k(\mathbf{t}^{(k+1)N}), \int_{\mathbf{t}_0^N=\mathbf{0}_{(k+1)N}}^{\mathbf{t}^N} \Xi\left(\mathbf{t}^N\right) dt^N\right) = \mathbf{0}_N, \ \forall \mathbf{t}^{(k+1)N} \in \mathfrak{T}_0^{(k+1)N}.
$$

$$(13.14)$$

The control aim is to force the plant behavior to satisfy the requested tracking quality (13.14). The main task of the control synthesis is to determine such natural tracking control (NTC).

Let us specify the crucial properties of *the vector tracking operator $T(.)$* (13.14). They determine the class of *the tracking algorithms* that fulfill the requested high quality of the tracking properties. They will be used as the basis for the NTC synthesis.

Property 278 *If (13.14) holds, then the operator $T(.)$ guarantees that the solution $\epsilon\left(t; \epsilon_0^k\right)$ of (13.14) is continuous in time on $\mathfrak{T}_0$,*

$$
\mathbf{T}\left(\mathbf{t}^N, \epsilon^k\left(\mathbf{t}^{(k+1)N}\right), \int_{\mathbf{t}_0^N=\mathbf{0}_N}^{\mathbf{t}^N} \Xi\left(\mathbf{t}^N\right) dt^N\right) = \mathbf{0}_N, \ \forall \mathbf{t}^{(k+1)N} \in \mathfrak{T}_0^{(k+1)N}
$$

$$
\Longrightarrow \epsilon\left(\mathbf{t}^N; \epsilon_0^k\right) \in \mathfrak{C}(\mathfrak{T}_0^N), \ \forall \epsilon_0^k \in \mathfrak{R}^{(k+1)N}. \tag{13.15}
$$

Property 279 *The operator $T(.)$ has the property to vanish at the origin at every moment,*

$$
\epsilon^k = \mathbf{0}_{(k+1)N} \Longrightarrow \mathbf{T}\left(\mathbf{t}^N, \mathbf{0}_{(k+1)N}, \int_{\mathbf{t}_0^N=\mathbf{0}_N}^{\mathbf{t}^N} O_N dt^N\right) = \mathbf{0}_N, \ \forall \mathbf{t}^N \in \mathfrak{T}_0^N. \tag{13.16}
$$

Property 280 *The solution of (13.14) for all zero initial conditions is iden-*

tically equal to the zero vector,

$$\epsilon_0^k = \mathbf{0}_{(k+1)N} \; and$$

$$\mathbf{T}\left(\mathbf{t}^N, \epsilon^k\left(\mathbf{t}^{(k+1)N}; \mathbf{0}_{(k+1)N}\right), \int_{\mathbf{t}_0^N = \mathbf{0}_N}^{\mathbf{t}^N} \Xi\left(\mathbf{t}^N\right) dt^N\right) = \mathbf{0}_N, \; \forall \mathbf{t}^N \in \mathfrak{T}_0^N \Longrightarrow$$

$$\epsilon^k\left(\mathbf{t}^{(k+1)N}; \mathbf{0}_{(k+1)N}\right) = \mathbf{0}_{(k+1)N}, \; \forall \mathbf{t}^{(k+1)N} \in \mathfrak{T}_0^{(k+1)N}. \qquad (13.17)$$

We will present several characteristic simple forms of the tracking algorithm $T(.)$. They satisfy (13.15) through (13.17), i.e., they obey Properties 278-280. They satisfy also Definition 274 and Definition 276.

Example 281 *The zero order linear elementwise tracking algorithm*
The simplest tracking algorithm is the zero order linear:

$$\mathbf{T}(t, \epsilon) = K_0 \epsilon(t) = \mathbf{0}_N, \; \forall t \in \mathfrak{T}_0,$$
$$K_0 = diag\{k_{01} \; k_{02} \; ... k_{0N}\} > O_N, \qquad (13.18)$$

which yields

$$\epsilon(t; \mathbf{0}_N) \equiv \mathbf{e}(t; \mathbf{e}_0) + \mathbf{f}(t; \mathbf{f}_0) \equiv \mathbf{0}_N \Longrightarrow$$
$$\mathbf{e}(t; \mathbf{e}_0) \equiv -\mathbf{f}(t; -\mathbf{e}_0), \; \forall \mathbf{e}_0 \in \mathfrak{R}^N. \qquad (13.19)$$

The definition of the vector function $\mathbf{f}(.; \mathbf{f}_0)$ determines the finite vector reachability time.

Example 282 *The first order linear elementwise exponential tracking algorithm*
The following tracking algorithm

$$\mathbf{T}\left(t, \epsilon^1\right) = T_1 \epsilon^{(1)}(t) + K_0 \epsilon(t) = \mathbf{0}_N, \; \forall t \in \mathfrak{T}_0,$$
$$T_1 = diag\{t_1 \; t_2 \; ... t_N\} > O_N, \qquad (13.20)$$

determines the global exponential tracking, which is illustrated by the solution in the form of the exponential function:

$$\epsilon(t; \epsilon_0) = e^{-tK_0 T_1^{-1}} \epsilon_0, \; \forall (\epsilon_0, t) \in \mathfrak{R}^N \times \mathfrak{T}_0,$$
$$e^{-tK_0 T_1^{-1}} = \exp\left(-tK_0 T_1^{-1}\right) = diag\left\{e^{-tk_{01}t_1^{-1}} \; e^{-tk_{02}t_2^{-1}} \; ... \; e^{-tk_{0N}t_N^{-1}}\right\}.$$

This is the stablewise tracking algorithm. For $\epsilon_0 = \mathbf{0}_N$:

$$\epsilon(t; \mathbf{0}_N) \equiv \mathbf{e}(t; \mathbf{e}_0) + \mathbf{f}(t; \mathbf{f}_0) \equiv \mathbf{0}_N \Longrightarrow$$
$$\mathbf{e}(t; \mathbf{e}_0) \equiv -\mathbf{f}(t; -\mathbf{e}_0), \ \forall \mathbf{e}_0 \in \mathfrak{R}^N. \tag{13.21}$$

The reachability time is finite. It is determined when the vector function $\mathbf{f}(.)$ is fully defined, Equation (13.4). The convergence to the zero error vector is elementwise and determined by the function $\mathbf{f}(.)$.

Let

$$E_k \in \mathfrak{R}^{N \times N}, \ k = 0, 1, ..., \eta, \ E^{(\eta)} = [E_0 \quad E_1 \ \dots \ E_\eta] \in \mathfrak{R}^{N \times (\eta+1)N}.$$

Example 283 *The higher order linear elementwise exponential tracking algorithm*

We define the higher order linear elementwise exponential tracking algorithm by

$$\mathbf{T}(t, \epsilon^\eta) = \sum_{k=0}^{k=\eta \leq \nu - 1} E_k \epsilon^{(k)}(t) = E^{(\eta)} \epsilon^\eta(t) = \mathbf{0}_N,$$
$$\epsilon^{\eta-1}(t) = \mathbf{0}_{\eta N}, \ \forall t \in \mathfrak{T}_0 \ if \ \epsilon_0^{\eta-1} = \mathbf{0}_{\eta N}, , \tag{13.22}$$

with the matrices $H_k \in \mathfrak{R}^{N \times N}$ such that the real parts of the roots of its characteristic polynomial $f(s)$,

$$f(s) = \det\left(\sum_{k=0}^{k=\eta \leq \nu} E_k s^k\right) = \det\left(E^{(\eta)} S_N^{(\eta)}(s)\right),$$

are negative.

Being the linear differential equation with the constant coefficients the above differential equation has the unique exponential solution for every initial condition $\epsilon^{\eta-1}(0) \in \mathfrak{R}^{\eta N}$. This is the stablewise tracking algorithm. Since $\epsilon^{\eta-1}(0; \mathbf{e}_0^{\eta-1}) = \mathbf{0}_{\eta N}$ for every $\mathbf{e}_0^{\eta-1} \in \mathfrak{R}^{\eta N}$ then Equations (13.21) hold.

The reachability time is determined in the full definition of the subsidiary vector function $\mathbf{f}(.)$.

Example 284 *The sharp elementwise stablewise tracking with the finite vector reachability time t_R^N*

The algorithm for the elementwise stablewise tracking with the finite vector reachability time t_R^N,

$$\mathbf{t}_R^N = \mathbf{t}_{R[0]}^N = T_1 K_0^{-1} |\epsilon_0|,$$

$$|\epsilon_0| = [|\epsilon_{10}| \quad |\epsilon_{20}| \quad ... \quad |\epsilon_{N0}|]^T \in \mathfrak{R}^N,$$

reads

$$\mathbf{T}\left(\mathbf{t}^N, \epsilon^1\right) = T_1 \epsilon^{(1)}(\mathbf{t}^N) + K_0 signe(\mathbf{0}_N) = \mathbf{0}_N, \ \forall \mathbf{t}^N \in \mathfrak{T}_0^N, \qquad (13.23)$$

where is defined in (1.15), Section 1.3. The solution $\epsilon(\mathbf{t}^N; \epsilon_0)$,

$$\epsilon(\mathbf{t}^N; \epsilon_0) = \left\{ \begin{array}{ll} \epsilon_0 - T_1^{-1} K_0 S\left(\epsilon_0\right) \mathbf{t}^N, & \mathbf{t}^N \in \mathfrak{T}_R^N, \\ \mathbf{0}_N, & \mathbf{t}^N \in \mathfrak{T}_{RF}^N \end{array} \right\} \Longrightarrow$$

$$\mathbf{t}_R^N = T_1 K_0^{-1} |\epsilon_0|,$$

$$S\left(\epsilon_0\right) = diag\left\{signe_{10} \quad signe_{20} \quad ... \quad signe_{N0}\right\},$$

to (13.23) determines the output error behavior that approaches sharply the zero error vector in the linear form (along a straight line) with the nonzero constant velocity $T_1^{-1} K_0 S\left(\epsilon_0\right) \mathbf{1}_N$ *if* $\epsilon_0 \neq \mathbf{0}_N$. *Then the convergence to the zero error vector is elementwise, strictly monotonous, continuous and*

$$|\epsilon(t; \epsilon_0)| \leq |\epsilon_0|, \ \forall t \in \mathfrak{T}_0 \Longrightarrow$$

$$\|\epsilon(t; \epsilon_0)\| \leq \|\epsilon_0\|, \ \forall t \in \mathfrak{T}_0 \Longrightarrow$$

$$\forall \varepsilon \in \mathfrak{R}^+, \ \exists \delta \in \mathfrak{R}^+, \ \delta = \delta\left(\varepsilon\right) = \varepsilon \Longrightarrow$$

$$\|\epsilon_0\| < \varepsilon \Longrightarrow \|\epsilon(t; \epsilon_0)\| \leq \varepsilon, \ \epsilon(t; \mathbf{0}_N) = \mathbf{0}_N, \ \forall t \in \mathfrak{T}_0.$$

The tracking is stablewise.

The bigger K_0, the smaller t_R^N for the fixed T_1 and ϵ_0, and vice versa. The smaller T_1, the smaller t_R^N for the fixed K_0 and ϵ_0, and vice versa. The bigger $|\epsilon_0|$, the bigger t_R^N for the fixed T_1 and K_0, and vice versa. These relationships hold elementwise.

Example 285 **The first power smooth elementwise stablewise tracking with the finite vector reachability time** t_R^N

If the control acting on the plant ensures

$$\mathbf{T}\left(\mathbf{t}^N, \epsilon^1\right) = T_1 \epsilon^{(1)}(\mathbf{t}^N) + 2K_0 \left|\Xi\left(\mathbf{t}^N\right)\right|^{1/2} signe_0 = \mathbf{0}_N,$$

$$\forall \mathbf{t}^N \in \mathfrak{T}_0^N, \qquad (13.24)$$

where

$$\left|\Xi\left(\mathbf{t}^N\right)\right|^{1/2} = diag\left\{\left|\epsilon_1\left(t\right)\right|^{1/2} \vdots \left|\epsilon_2\left(t\right)\right|^{1/2} \vdots ... \vdots \left|\epsilon_N\left(t\right)\right|^{1/2}\right\},$$

then the plant exhibits the elementwise stablewise tracking with the finite vector reachability time t_R^N,

$$\mathbf{t}_R^N = T_1 K_0^{-1} \left|\Xi_0\right|^{1/2} \mathbf{1}_N,$$

which is determined by the output error behavior

$$\epsilon(\mathbf{t}^N; \epsilon_0) = \left\{\begin{array}{ll} \left[\left|\Xi_0\right|^{1/2} - T_1^{-1} K_0 T\right]^2 sign\epsilon_0, \forall \mathbf{t}^N \in \mathfrak{T}_R^N, \\ \mathbf{0}_N, \hspace{3.5cm} \forall \mathbf{t}^N \in \mathfrak{T}_{RF}^N \end{array}\right\},$$

$$T = diag\{t \quad t \quad ... t\} \in \mathfrak{T}_0^{N \times N}, \quad \mathbf{t}_R^N = T_1 K_0^{-1} \left|\epsilon_0\right|^{1/2},$$

$$\left|\epsilon_0\right|^{1/2} = \left[\left|\epsilon_1\left(0\right)\right|^{1/2} \vdots \left|\epsilon_2\left(0\right)\right|^{1/2} \vdots ... \vdots \left|\epsilon_N\left(0\right)\right|^{1/2}\right]^T, \tag{13.25}$$

which implies

$$sign\epsilon(\mathbf{t}^N) = sign\epsilon_0, \quad \forall \mathbf{t}^N \in [\mathbf{0}_N, \mathbf{t}_R^N[.$$

The output error vector approaches smoothly elementwise the zero output vector in the finite vector reachability time t_R^N. *The convergence is strictly monotonous and continuous. It is also without any oscillation, overshoot or undershoot. Then the solution (13.25) obeys the following:*

$$\|\epsilon(t; \epsilon_0)\| \le \|\epsilon_0\|, \quad \forall t \in \mathfrak{T}_0 \Longrightarrow$$
$$\forall \varepsilon \in \mathfrak{R}^+, \quad \exists \delta \in \mathfrak{R}^+, \quad \delta = \delta\left(\varepsilon\right) = \varepsilon \Longrightarrow$$
$$\|\epsilon_0\| < \varepsilon \Longrightarrow \|\epsilon(t; \epsilon_0)\| \le \varepsilon, \quad \epsilon(t; \mathbf{0}_N) = \mathbf{0}_N, \quad \forall t \in \mathfrak{T}_0.$$

Therefore, such tracking is stablewise.

The bigger K_0, *the smaller* t_R^N *for fixed* T_1 *and* ϵ_0, *and vice versa. The smaller* T_1, *the smaller* t_R^N *for fixed* K_0 *and* ϵ_0, *and vice versa. The bigger* $\left|\epsilon_0\right|$, *the bigger* t_R^N *for fixed* T_1 *and* K_0, *and vice versa. These claims are in the elementwise sense.*

Example 286 *Higher power smooth elementwise stablewise track-ing with the finite vector reachability time* t_R^N

Let the tracking algorithm be

$$\mathbf{T}\left(\mathbf{t}^N, \epsilon^1\right) = T_1 \epsilon^{(1)}(\mathbf{t}^N) + K_0 \left|\Xi\left(\mathbf{t}^N\right)\right|^{I-K^{-1}} sign \epsilon_0 = \mathbf{0}_N,$$
$$\forall \mathbf{t}^N \in \mathfrak{T}_0^N,$$
$$K = diag\{k_1\ k_2\ ...k_N\},\ k_i \in \{2,\ 3,\ ...\},\ \forall i = 1, 2,\ ...\ ,\ N,$$
$$\left|\Xi\left(\mathbf{t}^N\right)\right|^{I-K^{-1}} = diag\left\{\left|\epsilon_1(t)\right|^{1-k_1^{-1}}\quad \left|\epsilon_2(t)\right|^{1-k_2^{-1}}\quad ...\quad \left|\epsilon_N(t)\right|^{1-k_N^{-1}}\right\}.$$
$$(13.26)$$

The solution $\epsilon(\mathbf{t}^N; \epsilon_0)$ *to* $T\left(\mathbf{t}^N, \epsilon^1\right) = \mathbf{0}_N, \forall \mathbf{t}^N \in \mathfrak{T}_0^N,$ *reads*

$$\epsilon(\mathbf{t}^N; \epsilon_0) =$$
$$= \frac{1}{2}S(\epsilon_0)\left\{I_N + S\left[\left|\epsilon_0\right|^{K^{-1}} - T_1^{-1}K_0 \mathbf{t}^N\right]\right\}\left[\left|\epsilon_0\right|^{K^{-1}} - T_1^{-1}K_0 \mathbf{t}^N\right]^K =$$
$$= \left\{\begin{array}{ll} \left[\left|E_0\right|^{K^{-1}} - T_1^{-1}K_0 T\right]^K sign \epsilon_0, & \mathbf{t}^N \in \mathfrak{T}_R^N, \\ \mathbf{0}_N, & \mathbf{t}^N \in \mathfrak{T}_{RF}^N \end{array}\right\} \Longrightarrow$$
$$\mathbf{t}_R^N = T_1 K_0^{-1} \left|\epsilon_0\right|^{K^{-1}}, \qquad (13.27)$$

where

$$\left[\left|\epsilon_0\right|^{K^{-1}} - T_1^{-1}K_0 \mathbf{t}^N\right]^K = \begin{bmatrix} \left[\left|\epsilon_{10}\right|^{k_1^{-1}} - t\tau_1^{-1}k_{01}\right]^{k_1} \\ \left[\left|\epsilon_{20}\right|^{k_2^{-1}} - t\tau_2^{-1}k_{02}\right]^{k_2} \\ \cdots\cdots \\ \left[\left|\epsilon_{N0}\right|^{k_N^{-1}} - t\tau_N^{-1}k_{0N}\right]^{k_N} \end{bmatrix} \in \mathfrak{R}^N. \quad (13.28)$$

This expresses the elementwise nonlinear convergence to the zero error vector $\epsilon = \mathbf{0}_N$. *Then the convergence is strictly monotonous and continuous, without any oscillation, overshoot or undershoot. The errors enter the zero values smoothly. Besides, (13.27) and (13.28) imply*

$$\|\epsilon(t; \epsilon_0)\| \le \|\epsilon_0\|,\ \forall t \in \mathfrak{T}_0,$$
$$\forall \varepsilon \in \mathfrak{R}^+,\ \exists \delta \in \mathfrak{R}^+,\ \delta = \delta(\varepsilon) = \varepsilon \Longrightarrow$$
$$\|\epsilon_0\| < \varepsilon \Longrightarrow \|\epsilon(t; \epsilon_0)\| \le \varepsilon,,\ \epsilon(t; \mathbf{0}_N) = \mathbf{0}_N,\ \forall t \in \mathfrak{T}_0.$$

The tracking is stablewise.

The bigger K_0, *the smaller* t_R^N *for fixed* T_1 *and* ϵ_0, *and vice versa. The smaller* T_1, *the smaller* t_R^N *for fixed* K_0 *and* ϵ_0, *and vice versa. The bigger* $|\epsilon_0|$, *the bigger* t_R^N *for fixed* T_1 *and* K_0, *and vice versa.*

Example 287 *Sharp absolute error vector value tracking element-wise and stablewise with the finite vector reachability time t_R^N*
   Let us define

$$\sigma\left(\epsilon_i^{(k)}, \epsilon_i^{(k+1)}\right) =$$

$$= \left\{ \begin{array}{l} -1, \ \epsilon_i^{(k)} < 0, \forall \epsilon_i^{(k+1)} \in \mathfrak{R}; \ or \ \epsilon_i^{(k)} = 0 \ and \ \epsilon_i^{(k+1)} < 0, \\ \quad 0, \ \epsilon_i^{(k)} = 0 \ and \ \epsilon_i^{(k+1)} = 0, \\ \ 1, \ \epsilon_i^{(k)} > 0, \forall \epsilon_i^{(k+1)} \in \mathfrak{R}; \ or \ \epsilon_i^{(k)} = 0 \ and \ \epsilon_i^{(k+1)} > 0 \end{array} \right\}, \qquad (13.29)$$

$$\Sigma\left(\epsilon^{(k)}, \epsilon^{(k+1)}\right) =$$

$$= diag\left\{ \sigma\left(\epsilon_1^{(k)}, \epsilon_1^{(k+1)}\right) \ \vdots \ \sigma\left(\epsilon_2^{(k)}, \epsilon_2^{(k+1)}\right) \ \vdots \ ... \ \vdots \ \sigma\left(\epsilon_N^{(k)}, \epsilon_N^{(k+1)}\right) \right\},$$

$$\forall k = 1, 2, ... \qquad (13.30)$$

   The solution to the following tracking algorithm (in which we use (13.29) and (13.30)):

$$\mathbf{T}\left(\mathbf{t}^N, \epsilon, \epsilon^{(1)}\right) = T_1 \Sigma\left(\epsilon, \epsilon^{(1)}\right) \epsilon^{(1)} + K_0 sign\,|\epsilon_0| = \mathbf{0}_N,$$

$$\forall \mathbf{t}^N \in \mathfrak{T}_0^N, \qquad (13.31)$$

reads:

$$\left|\epsilon(\mathbf{t}^N; \epsilon_0)\right| = \left\{ \begin{array}{ll} |\epsilon_0| - T_1^{-1} K_0 T sign\,|\epsilon_0|, & \mathbf{t}^N \in [\mathbf{0}_N, T_1 K_0^{-1} |\epsilon_0|], \\ \mathbf{0}_N, & \mathbf{t}^N \in [T_1 K_0^{-1} |\epsilon_0|, \ \infty \mathbf{1}_N[. \end{array} \right\},$$

$$\mathbf{t}_R^N = T_1 K_0^{-1} |\epsilon_0|, \qquad (13.32)$$

which permits

$$|\epsilon(t; \epsilon_0)| \le |\epsilon_0|, \ \forall t \in \mathfrak{T}_0, \Longrightarrow$$

$$\|\epsilon(t; \epsilon_0)\| \le \|\epsilon_0\|, \ \forall t \in \mathfrak{T}_0 \Longrightarrow$$

$$\forall \varepsilon \in \mathfrak{R}^+, \ \exists \delta \in \mathfrak{R}^+, \ \delta = \delta\left(\varepsilon\right) = \varepsilon \Longrightarrow$$

$$\|\epsilon_0\| < \varepsilon \Longrightarrow \|\epsilon(t; \epsilon_0)\| \le \varepsilon, \ \epsilon(t; \mathbf{0}_N) = \mathbf{0}_N, \ \forall t \in \mathfrak{T}_0.$$

The tracking is stablewise and elementwise with the finite vector reachability time $t_R^N = T_1 K_0^{-1} |\epsilon_0|$. It is strictly monotonous and continuous without oscillation, overshoot and undershoot.

Example 288 *The exponential absolute error vector value tracking elementwise and stablewise with the finite vector reachability time* t_R^{2N}

The tracking algorithm is in terms of the elementwise absolute value of the subsidiary error vector,

$$\mathbf{T}\left(\epsilon, \epsilon^{(1)}\right) = T_1 D^+ |\epsilon| + K\left(|\epsilon| + K_0 sign\, |\epsilon_0|\right) =$$

$$= T_1 \Sigma\left(\epsilon, \epsilon^{(1)}\right) \epsilon^{(1)} + K\left(|\epsilon| + K_0 sign\, |\epsilon_0|\right) = \mathbf{0}_N,$$

$$\forall \mathbf{t}^N \in \mathfrak{T}_0^N. \tag{13.33}$$

The solution of the differential equation written in the matrix diagonal form

$$D^+ \left[|\Xi| + K_0 S\left(|\epsilon_0|\right)\right] = -T_1^{-1} K \left[|\Xi| + K_0 S\left(|\epsilon_0|\right)\right] \Longrightarrow$$

$$\left[|\Xi| + K_0 S\left(|\epsilon_0|\right)\right]^{-1} D^+ \left[|\Xi| + K_0 S\left(|\epsilon_0|\right)\right] = -T_1^{-1} K \Longrightarrow$$

$$D^+ \left\{\ln\left[|\Xi| + K_0 S\left(|\epsilon_0|\right)\right]\right\} = -T_1^{-1} K,$$

reads in the matrix form

$$\ln\left\{\left[|\Xi| + K_0 S\left(|\epsilon_0|\right)\right]\left[|\Xi_0| + K_0 S\left(|\epsilon_0|\right)\right]^{-1}\right\} = -T_1^{-1} KT,\ \forall T \in \mathfrak{T}_0^{N \times N}.$$

The final form of the solution is

$$|\Xi(t; \Xi_0)| =$$

$$= \left\{ \begin{array}{c} e^{-T_1^{-1}KT}\left[|\Xi_0| + K_0 S\left(|\epsilon_0|\right)\right] - K_0 S\left(|\epsilon_0|\right),\ \forall T \in [O_N, T_R], \\ \forall T \in [T_R,\ \infty I_N[, \\ O_N, \\ where\ 0\infty = 0, \end{array} \right\},$$

$$\tag{13.34}$$

$$T_R = \left\{ \begin{array}{c} T_1 K^{-1} \ln\left\{K_0^{-1} S^{-1}\left(|\epsilon_0|\right)\left[|\Xi_0| + K_0 S\left(|\epsilon_0|\right)\right]\right\},\ \epsilon_0 \neq \mathbf{0}_N, \\ O_N, \qquad\qquad\qquad\qquad\qquad\qquad \epsilon_0 = \mathbf{0}_N \end{array} \right\},$$

$$\tag{13.35}$$

$$\mathbf{t}_R^N = \begin{bmatrix} t_{R1} & t_{R2} & \ldots & t_{RN} \end{bmatrix}^T \Longleftrightarrow T_R = diag\left\{t_{R1}\ t_{R2}\ \ldots\ t_{RN}\right\}. \tag{13.36}$$

We can set the solution (13.34) in the equivalent vector form

$$\left|\epsilon(\mathbf{t}^N; \epsilon_0)\right| = \left\{ \begin{array}{c} e^{-T_1^{-1}KT}\left[|\epsilon_0| + K_0 sign\left(|\epsilon_0|\right)\right] - K_0 sign\left(|\epsilon_0|\right), \\ \forall \mathbf{t}^N \in \left[\mathbf{0}_N, \mathbf{t}_R^N\right],\ i.e.,\ \forall T \in [O_N,\ T_R], \\ \mathbf{0}_N,\ \forall \mathbf{t}^N \in [\mathbf{t}_R^N,\ \infty \mathbf{1}_N[,\ i.e.,\ \forall T \in [T_R,\ \infty I_N]. \end{array} \right\}.$$

The solution is continuous and monotonous without oscillation, overshoot and undershoot, and obeys

$$|\epsilon(t; \epsilon_0)| \leq |\epsilon_0|, \ \forall t \in \mathfrak{T}_0 \Longrightarrow$$
$$\forall \varepsilon \in \mathfrak{R}^+, \ \exists \delta \in \mathfrak{R}^+, \ \delta = \delta(\varepsilon) = \varepsilon \Longrightarrow$$
$$\|\epsilon_0\| < \varepsilon \Longrightarrow \|\epsilon(t; \epsilon_0)\| \leq \varepsilon, \ \epsilon(t; \mathbf{0}_N) = \mathbf{0}_N, \ \forall t \in \mathfrak{T}_0.$$

The tracking is stablewise. It converges with the exponential rate to the zero error vector and reaches elementwise the origin in finite vector reachability time t_R^N (13.35), (13.36).

We apply the preceding tracking algorithms to the *NTC* synthesis.

13.2   NTC concept and definition

How does the nature, i.e., the brain as the extravagant product of the nature, create control of any organ? Who knows? Can the brain explain its own work, functioning, process? These very exciting questions rest without answer. However, we can reply to the question *which information the brain uses to create the control*. It uses information about the error e of the real organ behavior $Y(.)$ relative to its desired behavior $Y_d(.)$. But, this is not the sufficient information for the brain to create the appropriate control. For example, in order to control the position of a hand, of a finger, of a leg, the brain uses information about the difference between their desired and real positions, which is information about their position errors. This is the classical process to use *the plant (organ) output information necessary* to create the negative feedback control. But it is not sufficient information. The brain simultaneously uses information about the forces of the muscles acting on the organs, on the hand, on the finger, on the leg. The muscle force is *a control variable*. The brain, as the central part of *the controller created by the nature*, uses information about *the (just realized) control itself.* It is simultaneously an input and output variable of *the natural controller* (of the brain). This is an inherent characteristic of the control created by the brain, i.e., by the nature.

The brain, in general the nature, does not have any information, any knowledge, about a mathematical model of the controlled organ. This is another essential characteristic of the control created by the brain, i.e., by the nature.

Definition 289 *Natural Control (NC) [188, Definition 349, p. 249]*

A control **U** *is* **Natural Control (NC)** *if, and only if:*

1.   it obeys the Time Continuity and Uniqueness Principle (TCUP, Principle 10, Chapter 1),

2.   its synthesis and effective implementation use information about both the output error vector **e** *(and possibly its derivatives and/or its integral) and the control action* **U** *itself,*

3.   its synthesis and effective implementation do not use information either about the plant mathematical model or about the mathematical description of the plant internal dynamics, i.e., about the plant state, or about the real instantaneous values of disturbances,

$$\mathbf{U} = \mathbf{U}(\mathbf{e}, \mathbf{U}), \ \mathbf{U}(t) \in \mathfrak{C}(\mathfrak{T}_0). \tag{13.37}$$

The controller should possess an internal local feedback from its output to its input in order to generate *Natural Control*. A mathematical rather than a physical consideration determines clearly and precisely the sign, the character and the strength of such local feedback. We refer to [118]–[121], [148]–[154], [160], [169]–[173], [188, Definition 350, p. 250], [190]–[194], [249]–[257] for the following definition.

Definition 290 *Natural Tracking Control (NTC)*

Natural Control is **Natural Tracking Control (NTC)** *if and only if it ensures a (demanded) type of tracking determined by a tracking algorithm described by an operator* **T** *(.),*

$$\mathbf{U} = \mathbf{U}(\mathbf{e}, \mathbf{U}; \mathbf{T}), \mathbf{U}(t) \in \mathfrak{C}(\mathfrak{T}_0). \tag{13.38}$$

We will show and further broaden the fundamentals of the *NTC* theory. The papers demonstrated the usefulness of the mathematical possibility to replace the internal plant dynamics together with the external disturbance action by the control applied to compensate completely their influence on the plant behavior. The mathematics showed that such control demands *the unit positive local feedback without delay in the controller*. It is well known that the unit positive feedback without delay is forbidden in the control theory because such isolated feedback system is totally unstable and will explode immediately in reality. The feedback *NT* controller is in the closed loop of the overall control system. Its local unit positive feedback operates in the full harmony with the global negative feedback of the overall control system. This control principle is the basis of the life of every human cell and of the whole human organism. Such control is *self adaptive control and fully robust*.

13.3 *NTC* origin and development

The root of the *Natural Tracking Control* (*NTC*) concept is in the papers [135], [143, Comment 3, Corollary 1 and Comment 4, p. 335, the second passage of 9. Conclusion, pp. 335, 336], [144, Section 7, Theorem 8, pp. 325, 326 and 8. Conclusion, p. 326] and [145, Note 11, p. S-38]. It was called *Self-Adaptive Control* [144, Section 7, Theorem 8, pp. 325, 326 and 8. Conclusion, p. 326].

Z. B. Ribar[then 1] and the author (of this book) had simulated effectively on an analog computer the *NTC* of a second order linear plant in the Laboratory of Automatic Control, Faculty of Mechanical Engineering, Belgrade University, Serbia (Spring 1988). The papers [148]–[156], [190]–[194], [247]–[257] introduced the names and the *Natural Trackability* concept and the *Natural Tracking Control* (*NTC*) concept. In the papers [148]–[156], [190]–[194], [247]–[257].

William Pratt Mounfield, Jr.[2] was the first to work out all the examples by solving the difficult problem of digital simulations of the plant behavior controlled by *time*-continuous *NTC* that incorporates the local unit positive feedback.

Other developments of the *NTC* and of its various applications to control of continuous-*time* technical plants can be found in the Ph. D./D. Sci. dissertations by A. Kökösy[3] [214] and D. V. Lazitch[4] [224], in the papers by N. Nedić[5a] (Neditch) and D. Pršić[5b] (Prshitch) [260]–[262], [273], Z. B. Ribar[now6] et al. [279], [281], and in the M. Sci. thesis by M. R. Jovanović[then was 7a, now 7b] (Yovanovitch) [324].

The author further developed [175], [188] the *Natural Trackability* concept and the *Natural Tracking Control* (*NTC*) concept.

---

[1] Assistant at the Faculty of Mechanical Engineering, University of Belgrade, Belgrade, Serbia.

[2] President, M&M Technologies, Inc.

[3] The Leading Researcher and Assistant Professor, Institut Supérieur de l'Électronique et du Numérique (ISEN), Lille, France.

[4] Professor at the Faculty of Mechanical Engineering, University of Belgrade, Belgrade, Serbia.

[5a] Professor at the Faculty of Mechanical and Civil Engineering, University of Kraguyevats, Kralyevo, Serbia.

[5b] Assistant Professor at the Faculty of Mechanical and Civil Engineering, University of Kraguyevats, Kralyevo, Serbia.

[6] Professor at the Faculty of Mechanical Engineering, University of Belgrade, Belgrade, Serbia.

[7a] Postgraduate student at the Faculty of Mechanical Engineering, University of Belgrade, Belgrade, Serbia.

[7b] Professor at the University of Southern California, Los Angeles, CA.

13.4 NTC of linear systems

13.4.1 General consideration

The mathematical model of the plant (12.39), (12.40), (Section 12.2), reads:

$$\frac{d\mathbf{X}(t)}{dt} = A\mathbf{X}(t) + D^{(\mu)}\mathbf{D}^\mu(t) + \mathbf{B}^{(\mu)}\mathbf{U}^\mu(t), \ \forall t \in \mathfrak{T}_0, \ \mathbf{X} \in R^n, \quad (13.39)$$

$$\mathbf{Y}(t) = C\mathbf{X}(t) + V\mathbf{D}(t) + U\mathbf{U}(t), \ \forall t \in \mathfrak{T}_0, \ \mathbf{Y} \in R^N. \quad (13.40)$$

Equations (4.23), (4.36)–(4.57) (Section 4.1), determine the transfer function matrix $G_U(s)$ of the plant (13.39), (13.40) relative to the control $\mathbf{U}$,

$$G_U(s) = C(sI - A)^{-1}\mathbf{B}^{(\mu)}S_r^{(\mu)}(s) + U = \quad (13.41)$$

$$. = \left[C(sI - A)^{-1}\mathbf{B}^{(\mu)} + U^{(\mu)}\right]S_r^{(\mu)}(s) = p^{-1}(s)L_U(s), \quad (13.42)$$

and the transfer function matrix $G_{U\mu}(s)$ (4.50), Subsection 4.1.2, of the plant (13.39), (13.40) relative to the extended control vector $\mathbf{U}^\mu$,

$$G_{U\mu}(s) = C(sI - A)^{-1}\mathbf{B}^{(\mu)} + U^{(\mu)} = p^{-1}(s)N_U(s), \quad (13.43)$$

We accept the validity of Property 278 through Property 280, (Section 13.1).

Theorem 291 *General NTC synthesis*

Let Equations (13.1)–(13.5), (13.6)–(13.9) (Section 13.1) be valid.

In order for the trackable plant (13.39), (13.40) to be controlled by the natural tracking control $\mathbf{U}$ and to exhibit tracking on $\mathfrak{D}^1 \times \mathfrak{Y}_d^1$ determined by the tracking algorithm $\mathbf{T}(.)$, (13.14) (Section 13.1), it is necessary and sufficient that

1) $N \leq r$,

2)   There exists $s^* \in \mathfrak{R}$ for which the plant transfer function matrix

$G_U(s)$ has the full rank N, i.e.,

$$\exists s^* \in \mathfrak{R} \Longrightarrow$$

$$rank G_U(s^*) = rank \left[C(s^*I - A)^{-1} \mathbf{B}^{(\mu)} S_r^{(\mu)}(s^*) + U \right] =$$

$$= rank \left\{ \left[C(s^*I - A)^{-1} \mathbf{B}^{(\mu)} + U^{(\mu)} \right] S_r^{(\mu)}(s^*) \right\} =$$

$$= rank L_U(s^*) = N. \tag{13.44}$$

3)　　　The control obeys the following equation in the complex domain:

$$\mathbf{U}(s) = \left[e^{-\varepsilon s} \mathbf{U}(s) \right] + \Gamma(s) \mathcal{L} \left\{ \mathbf{T} \left(\mathbf{t}^N, \epsilon^k(\mathbf{t}^{(k+1)N}), \int_{\mathbf{t}_0^N}^{\mathbf{t}^N} \Xi(\mathbf{t}^N) d\mathbf{t}^N \right) \right\}$$

$$0 < \varepsilon <<< 1, \ \varepsilon \longrightarrow 0^+,$$

$$\Gamma(s) = G_U^T(s) \left[G_U(s) G_U^T(s) \right]^{-1}, \tag{13.45}$$

where $\mathcal{L} \left\{ \mathbf{T} \left(\mathbf{t}^N, \epsilon^k(\mathbf{t}^{(k+1)N}), \int_{\mathbf{t}_0^N = \mathbf{0}_{(k+1)N}}^{\mathbf{t}^N} \Xi(\mathbf{t}^N) d\mathbf{t}^N \right) \right\}$ is the Laplace transform of $\mathbf{T} \left(\mathbf{t}^N, \epsilon^k(\mathbf{t}^{(k+1)N}), \int_{\mathbf{t}_0^N = \mathbf{0}_{(k+1)N}}^{\mathbf{t}^N} \Xi(\mathbf{t}^N) d\mathbf{t}^N \right)$.

4) If the matrix A is nonsingular then for the trackable plant (13.39), (13.40) to be controlled by the natural tracking control $\mathbf{U}$ and to exhibit tracking on $\mathfrak{D}^1 \times \mathfrak{Y}_d^1$ determined by the tracking algorithm $\mathbf{T}(.)$, (13.14) (Section 13.1), it is necessary and sufficient that the condition 1)–3) hold, or equivalently that the condition 1) is valid together with the following:

$$\det A \neq 0 \Longrightarrow$$

$$rank G_{U\mu}(s) = N = rank \left(U^{(\mu)} - C A^{-1} \mathbf{B}^{(\mu)} \right). \tag{13.46}$$

$$G_{U\mu}(s) \mathcal{L} \left\{ \mathbf{U}^{(\mu)}(t) \right\} = G_{U\mu}(s) \mathcal{L} \left\{ \mathbf{U}^{(\mu)}(t^-) \right\} + \mathbf{T}(s, \epsilon(s)),$$

$$\mathbf{T}(s, \epsilon(s)) = \mathcal{L} \left\{ \mathbf{T} \left(\mathbf{t}^N, \epsilon^k(\mathbf{t}^{(k+1)N}), \int_{\mathbf{t}_0^N}^{\mathbf{t}^N} \Xi(\mathbf{t}^N) d\mathbf{t}^N \right) \right\}, \tag{13.47}$$

equivalently in the time domain:

$$\Gamma_\mu \mathbf{U}^\mu(t) = \Gamma_\mu \mathbf{U}^\mu(t^-) + \mathbf{T} \left(\mathbf{t}^N, \epsilon^k(\mathbf{t}^{(k+1)N}), \int_{\mathbf{t}_0^N = \mathbf{0}_{(k+1)N}}^{\mathbf{t}^N} \Xi(\mathbf{t}^N) d\mathbf{t}^N \right), \tag{13.48}$$

$$\Gamma_\mu = \left(U^{(\mu)} - C A^{-1} \mathbf{B}^{(\mu)} \right), \tag{13.49}$$

Proof. *Necessity.* Let the trackable plant described by Equations (13.39), (13.40) be controlled by the natural tracking control $\mathbf{U}$ and let it exhibit tracking on $\mathfrak{D}^1 \times \mathfrak{Y}_d^1$ determined by (13.14). The plant trackability implies (Theorem 227, Section 10.3.4) the necessity of the condition 1) and that

$$rank G_U(s) = N \ on \ \mathbb{C} \tag{13.50}$$

$$\Longrightarrow \det\left[G_U(s)G_U^T(s)\right] \neq 0 \ on \ \mathbb{C}. \tag{13.51}$$

The full rank N of $G_U(s)$, Equations (13.50), and the statement under 1) of Theorem 194 (Subsection 10.1.1) imply the necessity of Equations (13.44).

Equations (4.23), (4.35), (4.36), (4.42)–(4.44) (Section 4.1) determine $\mathcal{L}\left\{\mathbf{Y}(t)\right\} = \mathbf{Y}(s)$ in the form of Equation (10.51) (Section 10.3.1):

$$\mathcal{L}\left\{\mathbf{Y}(t)\right\} = G_{EISOPD}(s)\mathcal{L}\left\{\mathbf{D}(t)\right\} + G_U(s)\mathcal{L}\left\{\mathbf{U}(t)\right\} +$$

$$+ G_{EISOPD_0}(s)\mathbf{D}_0^{\mu-1} + G_{EISOPU_0}(s)\mathbf{U}_0^{\mu-1} + G_{EISOPX_0}(s)\mathbf{X}_0. \tag{13.52}$$

The plant natural tracking control $\mathbf{U}$ ensures the tracking on $\mathfrak{D}^1 \times \mathfrak{Y}_d^1$ determined by Equation (13.14) that may be subtracted from Equation (13.52):

$$\mathcal{L}\left\{\mathbf{Y}(t)\right\} = G_{EISOPD}(s)\mathcal{L}\left\{\mathbf{D}(t)\right\} + G_U(s)\mathcal{L}\left\{\mathbf{U}(t)\right\} +$$

$$+ G_{EISOPD_0}(s)\mathbf{D}_0^{\mu-1} + G_{EISOPU_0}(s)\mathbf{U}_0^{\mu-1} + G_{EISOPX_0}(s)\mathbf{X}_0 -$$

$$- \mathcal{L}\left\{\mathbf{T}\left(t, \epsilon(t), \epsilon^{(1)}(t), ..., \int_{t_0=0}^{t}\epsilon(t)dt\right)\right\} \tag{13.53}$$

For the moment t^-,

$$t^- = t - \varepsilon, \ 0 < \varepsilon <<< 1, \varepsilon \longrightarrow 0^+, \forall t \in \mathfrak{T}_0, \tag{13.54}$$

Equation (13.52) reads

$$\mathcal{L}\left\{\mathbf{Y}\left(t^-\right)\right\} = G_{EISOPD}(s)\mathcal{L}\left\{\mathbf{D}(t^-)\right\} + G_U(s)\mathcal{L}\left\{\mathbf{U}(t^-)\right\} +$$

$$+ G_{EISOPD_0}(s)\mathbf{D}_0^{\mu-1} + G_{EISOPU_0}(s)\mathbf{U}_0^{\mu-1} + G_{EISOPX_0}(s)\mathbf{X}_0. \tag{13.55}$$

Since the control $\mathbf{U}(t)$ is natural control, it is continuous in $t \in \mathfrak{T}_0$. This, the system linearity and $[\mathbf{D}(.), \mathbf{Y}_d(.)] \in \mathfrak{D}^1 \times \mathfrak{Y}_d^1$ imply (Principle 10, Section 1.2) continuity of all system variables in $t \in \mathfrak{T}_0$,

$$\mathbf{D}(t) = \mathbf{D}(t^-), \ \mathbf{X}(t) = \mathbf{X}(t^-), \ \mathbf{Y}(t) = \mathbf{Y}(t^-), \ \mathbf{X}^{(1)}(t^-) = \mathbf{X}^{(1)}(t),$$

$$\mathcal{L}\left\{\mathbf{D}(t)\right\} = \mathcal{L}\left\{\mathbf{D}(t^-)\right\}, \ \mathcal{L}\left\{\mathbf{X}(t)\right\} = \mathcal{L}\left\{\mathbf{X}(t^-)\right\}, \ \mathcal{L}\left\{\mathbf{Y}(t)\right\} = \mathcal{L}\left\{\mathbf{Y}(t^-)\right\}. \tag{13.56}$$

These equations set Equation (13.55) into

$$\mathcal{L}\left\{\mathbf{Y}(t)\right\} = G_{EISOPD}(s)\,\mathcal{L}\left\{\mathbf{D}(t\}\right) + G_U(s)\,\mathcal{L}\left\{\mathbf{U}(t^-)\right\} +$$
$$+ G_{EISOPD_0}(s)\,\mathbf{D}_0^{\mu-1} + G_{EISOPU_0}(s)\,\mathbf{U}_0^{\mu-1} + G_{EISOPX_0}(s)\,\mathbf{X}_0.$$

Subtracting this equation from Equation (13.53) the result is:

$$\mathbf{0}_N = G_U(s)\left[\mathcal{L}\left\{\mathbf{U}(t)\right\} - \mathcal{L}\left\{\mathbf{U}(t^-)\right\}\right] -$$
$$-\mathcal{L}\left\{\mathbf{T}\left(t, \epsilon(t), \epsilon^{(1)}(t), ..., \int_{t_0=0}^{t} \epsilon(t)dt\right)\right\},$$

i.e.,

$$G_U(s)\,\mathbf{U}(s) =$$
$$= G_U(s)\left[e^{-\varepsilon s}\mathbf{U}(s)\right] + \mathcal{L}\left\{\mathbf{T}\left(t, \epsilon(t), \epsilon^{(1)}(t), ..., \int_{t_0=0}^{t} \epsilon(t)dt\right)\right\},$$

which together with (13.54) and Equation (13.50) proves Equation 13.45 and the necessity of the condition 3).

Let the matrix A be nonsingular. We recall Equations (4.47)–(4.57), Section 10.3.1, to get:

$$\mathcal{L}\left\{\mathbf{Y}(t)\right\} = G_{D\mu}(s)\,\mathcal{L}\left\{\mathbf{D}^{\mu}(t)\right\} + G_{U\mu}(s)\,\mathcal{L}\left\{\mathbf{U}^{\mu}(t)\right\}. \tag{13.57}$$

The plant natural tracking control $\mathbf{U}$ ensures the tracking on $\mathfrak{D}^1 \times \mathfrak{Y}_d^1$ determined by Equation (13.14). This guarantees solvability of the algebraic Equation (13.57) for $\mathcal{L}\left\{\mathbf{U}^{\mu}(t)\right\}$, which implies $rank G_{U\mu}(s) = N$ and proves the first equation in (13.46). The Laplace transform of

$$\mathbf{T}\left(t, \epsilon(t), \epsilon^{(1)}(t), ..., \int_{t_0=0}^{t} \epsilon(t)dt\right) = \mathbf{0}_N$$

may be subtracted from Equation (13.57):

$$\mathcal{L}\left\{\mathbf{Y}(t)\right\} = G_{D\mu}(s)\,\mathcal{L}\left\{\mathbf{D}^{\mu}(t)\right\} + G_{U\mu}(s)\,\mathcal{L}\left\{\mathbf{U}^{\mu}(t)\right\} -$$
$$-\mathcal{L}\left\{\mathbf{T}\left(t, \epsilon(t), \epsilon^{(1)}(t), ..., \int_{t_0=0}^{t} \epsilon(t)dt\right)\right\} \tag{13.58}$$

For t^- (13.54), Equation (13.57) becomes

$$\mathcal{L}\left\{\mathbf{Y}(t^-)\right\} = G_{D\mu}(s)\,\mathcal{L}\left\{\mathbf{D}^{\mu}(t^-)\right\} + G_{U\mu}(s)\,\mathcal{L}\left\{\mathbf{U}^{\mu}(t^-)\right\},$$

which takes the following form due to Equations (13.56):

$$\mathcal{L}\left\{\mathbf{Y}(t)\right\} = G_{D\mu}(s)\,\mathcal{L}\left\{\mathbf{D}^{\mu}(t)\right\} + G_{U\mu}(s)\,\mathcal{L}\left\{\mathbf{U}^{\mu}(t^{-})\right\}.$$

The difference between Equation (13.58) and this equation is:

$$\mathbf{0}_{N} = G_{U\mu}(s)\,\mathcal{L}\left\{\mathbf{U}^{\mu}(t)\right\} - G_{U\mu}(s)\,\mathcal{L}\left\{\mathbf{U}^{\mu}(t^{-})\right\} -$$

$$-\mathcal{L}\left\{\mathbf{T}\left(t, \epsilon(t), \epsilon^{(1)}(t), ..., \int_{t_0=0}^{t}\epsilon(t)dt\right)\right\},$$

or :

$$G_{U\mu}(s)\,\mathcal{L}\left\{\mathbf{U}^{\mu}(t)\right\} = G_{U\mu}(s)\,\mathcal{L}\left\{\mathbf{U}^{\mu}(t^{-})\right\} +$$

$$+\mathcal{L}\left\{\mathbf{T}\left(t, \epsilon(t), \epsilon^{(1)}(t), ..., \int_{t_0=0}^{t}\epsilon(t)dt\right)\right\}.$$

This proves Equation (13.47).

Let us verify in the *time* domain the case $det A \neq 0$. Equations (13.39), (13.40), (4.45), (4.46) permit:

$$\mathbf{X}(t) = A_{P}^{-1}\left(\frac{d\mathbf{X}(t)}{dt} - D^{(\mu)}\mathbf{D}^{\mu}(t) - \mathbf{B}^{(\mu)}\mathbf{U}^{\mu}(t)\right),$$

$$\mathbf{Y}(t) = C\mathbf{X}(t) + V\mathbf{D}(t) + U\mathbf{U}(t) =$$

$$= CA_{P}^{-1}\left(\frac{d\mathbf{X}(t)}{dt} - D^{(\mu)}\mathbf{D}^{\mu}(t) - \mathbf{B}^{(\mu)}\mathbf{U}^{\mu}(t)\right) + V^{(\mu)}\mathbf{D}^{\mu}(t) +$$

$$+U^{(\mu)}\mathbf{U}^{\mu}(t) \Longrightarrow$$

$$\begin{aligned}\mathbf{Y}(t) \;=\;& CA_{P}^{-1}\mathbf{X}^{(1)}(t) + \left(V^{(\mu)} - CA_{P}^{-1}D^{(\mu)}\right)\mathbf{D}^{\mu}(t) + \\ & + \left(U^{(\mu)} - CA_{P}^{-1}\mathbf{B}^{(\mu)}\right)\mathbf{U}^{\mu}(t). \end{aligned} \qquad (13.59)$$

The plant exhibits tracking on $\mathfrak{D}^{1}\mathbf{x}\mathfrak{Y}_{d}^{1}$ determined by (13.14), which permits to put Equation (13.59) into

$$\mathbf{Y}(t) = CA_{P}^{-1}\mathbf{X}^{(1)}(t) + \left(V^{(\mu)} - CA_{P}^{-1}D^{(\mu)}\right)\mathbf{D}^{\mu}(t) +$$

$$+ \left(U^{(\mu)} - CA_{P}^{-1}\mathbf{B}^{(\mu)}\right)\mathbf{U}^{\mu}(t) - \mathbf{T}\left(t, \epsilon(t), \epsilon^{(1)}(t), .., \int_{t_0=0}^{t}\epsilon(t)dt\right), \quad (13.60)$$

and guarantees the existence of $\mathbf{U}(t)$ obeying differential Equation (13.60) in $\mathbf{U}(t)$, which implies

$$rank\left(U^{(\mu)} - CA_{P}^{-1}\mathbf{B}^{(\mu)}\right) = N. \qquad (13.61)$$

This proves the second equation in (13.46) and completes the proof of the necessity of Equations (13.46).

At the moment t^-, Equation (13.54), Equations (13.39), (13.40), (4.45), (4.46) lead to:

$$
\begin{aligned}
\mathbf{Y}(t^-) &= CA_P^{-1}\mathbf{X}^{(1)}(t^-) + \left(V^{(\mu)} - CA_P^{-1}D^{(\mu)}\right)\mathbf{D}^\mu(t^-) + \\
&\quad + \left(U^{(\mu)} - CA_P^{-1}\mathbf{B}^{(\mu)}\right)\mathbf{U}^\mu(t^-).
\end{aligned}
\tag{13.62}
$$

We apply Equations (13.56) to $\mathbf{Y}(t^-) = \mathbf{Y}(t)$, $\mathbf{X}^{(1)}(t^-) = \mathbf{X}^{(1)}(t)$, $\mathbf{D}(t^-) = \mathbf{D}(t)$ in Equation (13.62) that becomes:

$$
\begin{aligned}
\mathbf{Y}(t) &= CA_P^{-1}\mathbf{X}^{(1)}(t) + \left(V^{(\mu)} - CA_P^{-1}D^{(\mu)}\right)\mathbf{D}^\mu(t) + \\
&\quad + \left(U^{(\mu)} - CA_P^{-1}\mathbf{B}^{(\mu)}\right)\mathbf{U}^\mu(t^-).
\end{aligned}
$$

After subtracting this from Equation (13.60) the obtained result reads:

$$
\begin{aligned}
\mathbf{0}_N &= \left(U^{(\mu)} - CA_P^{-1}\mathbf{B}^{(\mu)}\right)\mathbf{U}^\mu(t) - \mathbf{T}\left(t, \epsilon(t), \epsilon^{(1)}(t), ..., \int_{t_0=0}^t \epsilon(t)dt\right) - \\
&\quad - \left(U^{(\mu)} - CA_P^{-1}\mathbf{B}^{(\mu)}\right)\mathbf{U}^\mu(t^-),
\end{aligned}
$$

or

$$
\Gamma_\mu \mathbf{U}^\mu(t) = \Gamma_\mu \mathbf{U}^\mu(t^-) + \mathbf{T}\left(\mathbf{t}^N, \epsilon^k(\mathbf{t}^{(k+1)N}), \int_{t_0^N=\mathbf{0}_{(k+1)N}}^{t^N} \Xi\left(\mathbf{t}^N\right)dt^N\right),
$$

$$
\Gamma_\mu = (U^{(\mu)} - CA_P^{-1}\mathbf{B}^{(\mu)}).
$$

These equations are Equations (13.48), (13.49). The proof of the necessity of the conditions 1)–4) is complete.

Sufficiency. Let all the conditions 1)–3) hold. Equation (13.52) enables us to set Equation (13.45) in the following form:

$$
\begin{aligned}
&\mathbf{Y}(s) - G_{EISOPD}(s)\mathbf{D}(s) - G_{EISOPD_0}(s)\mathbf{D}_0^{\mu-1} - G_{EISOPU_0}(s)\mathbf{U}_0^{\mu-1} \\
&\quad - G_{EISOPX_0}(s)\mathbf{X}_0 = e^{-\varepsilon s}\mathbf{Y}(s) - e^{-\varepsilon s}G_{EISOPD}(s)\mathbf{D}(s) - \\
&\quad - e^{-\varepsilon s}G_{EISOPD_0}(s)\mathbf{D}_0^{\mu-1} - e^{-\varepsilon s}G_{EISOPX_0}(s)\mathbf{X}_0 + \mathbf{T}(s, \epsilon(s))
\end{aligned}
$$

so that in the limit as $\varepsilon \longrightarrow 0^+$:

$$
\mathbf{0}_N = \mathbf{T}(s, \epsilon(s)).
$$

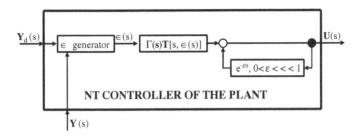

Figure 13.3: Block diagram of the natural tracking controller of the plant.

The inverse Laplace transform of this equation is

$$\mathbf{T}\left(\mathbf{t}^N, \epsilon^k(\mathbf{t}^{(k+1)N}), \int_{\mathbf{t}_0^N = \mathbf{0}_{(k+1)N}}^{\mathbf{t}^N} \Xi\left(\mathbf{t}^N\right) d\mathbf{t}^N\right) = \mathbf{0}_N, \ \forall \mathbf{t}^{(k+1)N} \in \mathfrak{T}_0^{(k+1)N}.$$

This proves that the plant exhibits tracking on $\mathfrak{D}^1 \mathbf{x} \mathfrak{Y}_d^1$ determined by (13.14).

If $det\ A \neq 0$ then by repeating the preceding procedure to Equation (13.47) the result is

$$\mathbf{T}\left(\mathbf{t}^N, \epsilon^k(\mathbf{t}^{(k+1)N}), \int_{\mathbf{t}_0^N = \mathbf{0}_{(k+1)N}}^{\mathbf{t}^N} \Xi\left(\mathbf{t}^N\right) d\mathbf{t}^N\right) = \mathbf{0}_N, \ \forall \mathbf{t}^{(k+1)N} \in \mathfrak{T}_0^{(k+1)N}.$$

Notice that $rankG_{U\mu}(0) = rank\left(-CA_P^{-1}B^{(\mu)} + U^{(\mu)}\right) = N$, Equations (13.46), shows that $s = s^*$ in Equations (13.44) is zero, $s = s^* = 0$. The proof of sufficiency of the condition under 4) follows directly from 3) for $s = s^* = 0$. The proof is complete. ∎

The full block diagram of the *Natural Tracking (NT) Controller* of the plant (13.39), (13.40) is in Figure 13.3 [188, Fig. 12.4, p. 271, Fig. 12.5, p. 286].

Claim 292 *Full rank of $G_U(s)$ and full rank of $G_{U\mu}(s)$*

If the rank of $G_U(s)$ is its full rank N then the rank of $G_{U\mu}(s)$ is also its full rank N.

Proof. $G_U(s) = G_{U\mu}(s) S_r^{(\mu)}(s)$ (4.53), $rankS_r^{(\mu)}(s) \equiv r \geq N$ and the rule on the rank of the matrix product,

$$N = full\ rankG_{U\mu}(s) S_r^{(\mu)}(s) \leq \min\left[rankG_{U\mu}(s), rankS_r^{(\mu)}(s)\right] =$$

$$= \min\left[rankG_{U\mu}(s), r\right] = \min\left[\min\left(N, r\right), r\right] = N$$

imply $rankG_{U\mu}(s) = N$. ∎

Comment 293 *If the plant is stable then the simplest choice of $s^* \in \mathfrak{R}$ is $s^* = 0$ to test whether*

$$rankG_U(0) = rank\left(-CA^{-1}B_0 + U\right) =$$

$$= rank\left(-CA^{-1}B^{(\mu)} + U^{(\mu)}\right)S_r^{(\mu)}(0) =$$

$$= rank\left(-CA^{-1}B^{(\mu)} + U^{(\mu)}\right) = rankG_{\mu U}(0) = N?$$

13.4.2   Control synthesis for specific tracking qualities

Assumption 294 *Equations (13.1)–(13.5), (13.6)–(13.12) (Section 13.1) are valid.*

We accept the validity of Assumption 294 throughout this subsection:

The proofs of all following theorems result directly from Theorem 291 and the corresponding *NTC* law (Section 13.1).

In the sequel

$$0 < \varepsilon <<< 1, \ \varepsilon \longrightarrow 0^+.$$

Theorem 295 *NTC for the zero order linear elementwise tracking Example 281 (Section 13.1)*

In order for the trackable plant (13.39), (13.40) to be controlled by the natural tracking control **U** *and to exhibit tracking on $\mathfrak{D}^1 \times \mathfrak{Y}_d^1$ determined by the zero order linear elementwise tracking law (13.18) it is necessary and sufficient that*

1) $N \leq r$,

2)

$$rankG_U(s) = rank\left[C(sI - A)^{-1}B^{(\mu)}S_r^{(\mu)}(s) + U\right] = N \qquad (13.63)$$

in general, and

$$rankG_{U\mu}(s) = rank\left[C(sI - A)^{-1}B^{(\mu)} + U^{(\mu)}\right] =$$

$$= rank\left(U^{(\mu)} - CA^{-1}B^{(\mu)}\right) = rank\left(U - CA^{-1}B_0\right) = N \qquad (13.64)$$

in the case the matrix A is nonsingular,

3)   The control obeys in general

$$\mathbf{U}(s) = e^{-\varepsilon s}\mathbf{U}(s) + \Gamma(s)\mathbf{T}[\epsilon(s)], \qquad (13.65)$$

$$\mathbf{T}[\epsilon(s)] = K_0\epsilon(s), \qquad (13.66)$$

or, if $\det A \neq 0$,

$$\Gamma_\mu \mathbf{U}^\mu(t) = \Gamma_\mu \mathbf{U}^\mu(t^-) + K_0 \epsilon(t). \tag{13.67}$$

Theorem 296 *NTC for the first order linear elementwise exponential tracking Example 282 (Section 13.1)*

In order for the trackable plant (13.39), (13.40) to be controlled by the natural tracking control $\mathbf{U}$ and to exhibit tracking on $\mathfrak{D}^1 \times \mathfrak{Y}_d^1$ determined by NTC law (13.20) it is necessary and sufficient that

    1) $N \leq r$,

    2)   Equation (13.63) holds in general, or Equation (13.64) is valid in the case the matrix A is nonsingular,

    3)   The control obeys

$$\mathbf{U}(s) = e^{-\varepsilon s}\mathbf{U}(s) + \Gamma(s)\mathcal{L}\left\{T_1\epsilon(t) + K_0\epsilon(t)\right\},$$

in general, or, if $\det A \neq 0$,

$$\Gamma_\mu \mathbf{U}^\mu(t) = \Gamma_\mu \mathbf{U}^\mu(t^-) + [T_1\epsilon(t) + K_0\epsilon(t)], \tag{13.68}$$

Theorem 297 *NTC for the higher order linear elementwise exponential tracking Example 283 (Section 13.1)*

In order for the trackable plant (13.39), (13.40) to be controlled by the natural tracking control $\mathbf{U}$ and to exhibit tracking on $\mathfrak{D}^\eta \times \mathfrak{Y}_d^\eta$ determined by the tracking algorithm (13.22) it is necessary and sufficient that

    1) $N \leq r$,

    2)   Equation (13.63) holds in general, or Equation (13.64) is valid in the case the matrix A is nonsingular, and

    3)   The control obeys

$$\mathbf{U}(s) = e^{-\varepsilon s}\mathbf{U}(s) + \Gamma(s)E^{(\eta)}S_N^{(\eta)}(s), \tag{13.69}$$

in general, or, if $\det A \neq 0$,

$$\Gamma_\mu \mathbf{U}^\mu(t) = \Gamma_\mu \mathbf{U}^\mu(t^-) + E^{(\eta)}\epsilon^\eta(t). \tag{13.70}$$

Theorem 298 *NTC for the sharp elementwise stablewise tracking with the finite vector reachability time $\mathbf{t}_R^N$, Example 284 (Section 13.1)*

In order for the trackable plant (13.39), (13.40) to be controlled by the natural tracking control $\mathbf{U}$ and to exhibit tracking on $\mathfrak{D}^1 \times \mathfrak{Y}_d^1$ determined by the tracking algorithm (13.23) it is necessary and sufficient that

    1) $N \leq r$,

2)   Equation (13.63) holds in general, or Equation (13.64) is valid in the case the matrix A is nonsingular, and

3)   The control obeys

$$\mathbf{U}(s) = e^{-\varepsilon s}\mathbf{U}(s) + \Gamma(s)\,\mathbf{T}\,(s,\epsilon(\mathbf{s}))\,,\ 0 < \varepsilon <<< 1,\ \varepsilon \longrightarrow 0^{+}, \qquad (13.71)$$

$$\mathbf{T}\,(s,\epsilon(\mathbf{s})) = \mathfrak{L}\left\{T_1\epsilon^{(1)}(t) + K_0 sign\epsilon(0)\right\}, \qquad (13.72)$$

in general, or, if $det\,A \neq 0$,

$$\Gamma_{\mu}\mathbf{U}^{\mu}(t) = \Gamma_{\mu}\mathbf{U}^{\mu}(t^{-}) + T_1\epsilon^{(1)}(t) + K_0 sign\epsilon(0). \qquad (13.73)$$

Theorem 299 *The NTC synthesis for the first power smooth elementwise stablewise tracking with the finite vector reachability time* $\mathbf{t}_R^N$ *Example 285 (Section 13.1)*

In order for the trackable plant (13.39), (13.40) to be controlled by the natural tracking control $\mathbf{U}$ *and to exhibit tracking on* $\mathfrak{D}^1 \times \mathfrak{Y}_d^1$ *determined by Equation (13.24) it is necessary and sufficient that*

1) $N \leq r$,

2)   Equation (13.63) holds in general, or Equation (13.64) is valid in the case the matrix A is nonsingular, and

3)   The control obeys

$$\mathbf{U}(s) = e^{-\varepsilon s}\mathbf{U}(s) + \Gamma(s)\,\mathbf{T}\,(s,\epsilon(\mathbf{s}))\,, \qquad (13.74)$$

$$\mathbf{T}\,(s,\epsilon(\mathbf{s})) = \mathfrak{L}\left\{T_1\epsilon(t) + 2K_0\left|\Xi(t)\right|^{1/2}sign\epsilon_0\right\}, \qquad (13.75)$$

in general, or, if $det\,A \neq 0$,

$$\Gamma_{\mu}\mathbf{U}^{\mu}(t) = \Gamma_{\mu}\mathbf{U}^{\mu}(t^{-}) + T_1\epsilon(t) + 2K_0\left|\Xi(t)\right|^{1/2}sign\epsilon_0. \qquad (13.76)$$

Theorem 300 *The NTC synthesis for the higher power smooth elementwise stablewise tracking with the finite vector reachability time* $\mathbf{t}_R^N$, *Example 286 (Section 13.1)*

In order for the trackable plant (13.39), (13.40) to be controlled by the natural tracking control $\mathbf{U}$ *and to exhibit tracking on* $\mathfrak{D}^1 \times \mathfrak{Y}_d^1$ *determined by the tracking algorithm (13.26) it is necessary and sufficient that*

1) $N \leq r$,

2)   Equation (13.63) holds in general, or Equation (13.64) is valid in the case the matrix A is nonsingular, and

3)   The control obeys

$$\mathbf{U}(s) = e^{-\varepsilon s}\mathbf{U}(s) + \Gamma(s)\,\mathbf{T}\,(s,\epsilon(\mathbf{s}))\,, \qquad (13.77)$$

$$\mathbf{T}\,(s,\epsilon(\mathbf{s})) = \mathfrak{L}\left\{T_1\epsilon(t) + K_0 sign\epsilon(t)\right\}, \qquad (13.78)$$

in general, or, if $\det A \neq 0$,

$$\Gamma_\mu \mathbf{U}^\mu(t) = \Gamma_\mu \mathbf{U}^\mu(t^-) + T_1 \epsilon(t) + K_0 |E(t)|^{I-K^{-1}} sign \epsilon_0. \tag{13.79}$$

Theorem 301 *The NTC synthesis for the sharp absolute error vector value tracking elementwise and stablewise with the finite vector reachability time* $\mathbf{t}_R^N$, *Example 287 (Section 13.1)*

In order for the trackable plant (13.39), (13.40) to be controlled by the natural tracking control $\mathbf{U}$ and to exhibit tracking on $\mathfrak{D}^1 \times \mathfrak{Y}_d^1$ determined by the tracking algorithm (13.31) it is necessary and sufficient that

1) $N \leq r$,

2)   Equation (13.63) holds in general, or Equation (13.64) is valid in the case the matrix A is nonsingular, and

3)   The control obeys

$$\mathbf{U}(s) = e^{-\varepsilon s} \mathbf{U}(s) + \Gamma(s) \mathbf{T}(s, \epsilon(\mathbf{s})), \tag{13.80}$$

$$\mathbf{T}(s, \epsilon(\mathbf{s})) = \mathfrak{L} \left\{ T_1 \Sigma \left[\epsilon(t), \epsilon^{(1)}(t) \right] \epsilon^{(1)}(t) + K_0 sign |\epsilon_0| \right\}, \tag{13.81}$$

in general, or, if $\det A \neq 0$,

$$\Gamma_\mu \mathbf{U}^\mu(t) = \Gamma_\mu \mathbf{U}^\mu(t^-) + T_1 \Sigma \left(\epsilon, \epsilon^{(1)} \right) \epsilon^{(1)} + K_0 sign |\epsilon_0|. \tag{13.82}$$

Theorem 302 *NTC synthesis for the exponential absolute error vector value tracking elementwise and stablewise with the finite vector reachability time* $\mathbf{t}_R^N$, *Example 288 (Section 13.1)*

In order for the trackable plant (13.39), (13.40) to be controlled by the natural tracking control $\mathbf{U}$ and to exhibit tracking on $\mathfrak{D}^1 \times \mathfrak{Y}_d^1$ determined by the NTC algorithm (13.33) it is necessary and sufficient that

1) $N \leq r$,

2)   Equation (13.63) holds in general, or Equation (13.64) is valid in the case the matrix A is nonsingular, and

3)   The control obeys

$$\mathbf{U}(s) = e^{-\varepsilon s} \mathbf{U}(s) + \Gamma(s) \mathbf{T}(s, \epsilon(\mathbf{s})), \tag{13.83}$$

$$\mathbf{T}(s, \epsilon(\mathbf{s})) = \mathfrak{L} \left\{ T_1 \Sigma \left(\epsilon, \epsilon^{(1)} \right) \epsilon^{(1)} + K (|\epsilon| + K_0 sign |\epsilon_0|) \right\}, \tag{13.84}$$

in general, or, if $\det A \neq 0$,

$$\Gamma_\mu \mathbf{U}^\mu(t) = \Gamma_\mu \mathbf{U}^\mu(t^-) + T_1 \Sigma \left(\epsilon, \epsilon^{(1)} \right) \epsilon^{(1)} + K (|\epsilon| + K_0 sign |\epsilon_0|) = \mathbf{0}_N. \tag{13.85}$$

Comment 303 *On the tracking algorithm order*

If we wished to apply a higher order linear or nonlinear NTC algorithms then we should cope with the problem of the measurement of the higher output error vector derivatives that are often unmeasurable. The simple nonlinear NTC algorithms are only of the first order. They fulfill the high tracking qualities including the finite reachability time together with the full robustness with respect to the variations of both the plant state and the external disturbances.

The future research can explore needs for the second order or higher order tracking algorithms.

Comment 304 *NTC algorithms, disturbances and the plant state*

The NTC algorithms do not need the measurement either of values of the plant state variables or of disturbances. They are expressed in terms of the plant transfer function matrix relative to the control, which is the control-output system characteristics.

Part V

APPENDIX

Appendix A

Notation

The meaning of the notation is explained in the text at its first use.

A.1 Abbreviations

$\mathcal{C}$ controller

  CS control system

  Cl closure

  $HISO$ system   Higher order Input-State-Output system defined by (5.1), (5.2)

  iff if and only if

  I Input

  II Input-Internal (dynamics)

  IIO Input-Internal dynamics-Output

  IIO system   the Input-Internal and Output dynamics system defined by (6.1), (6.2)

  In the interior

  IO Input-Output

  IO system   the Input-Output system defined by (2.1)

  IS Input-State

  ISO Input-State-Output

  ISO system   the Input-State-Output system defined by (3.1), (3.2)

  $LiTC$ Linear Tracking Control

  $LyTC$ Lyapunov Tracking Control

  $MIMO$ Multiple-Input-Multiple-Output

  NTC Natural Tracking Control

  $\mathcal{O}$ object

$\mathcal{P}$ *plant*

PCUP   Physical Continuity and Uniqueness Principle

EISO system   Extended Input-State-Output system (4.1), (4.2)

SISO   Single-Input Single-Output

System   Continuous-time time-invariant linear dynamical system

TCUP   Time Continuity and Uniqueness Principle.

A.2   Indexes

A.2.1   SUBSCRIPTS

d the subscript d denotes "desired"

   e *equilibrium,*

   i the subscript i denotes "the i-th"

   j the subscript j denotes "the j-th"

  nd *nondegenerate*

  P for *plant*

  R stands for *Rosenbrock*

  rd *reduced*

  rnd *row nondegenerate*

 $zero$ the subscript $zero$ denotes "the zero value"

  0   the subscript 0 (zero) associated with a variable (.) denotes its initial value $(.)_0$; however, if $(.) \subset \mathfrak{T}$ then the subscript 0 (zero) associated with (.) denotes the *time* set $\mathfrak{T}_0$, $(.)_0 = \mathfrak{T}_0$

A.2.2   SUPERSCRIPT

$i \in \{0, 1, ..., \eta, \mu\}$ is the highest derivative of the disturbance vector acting on the plant in general

  $k \in \{0, 1, ..., \mu\}$ is the highest derivative of the control vector acting on the plant in general

  $l \in \{0, 1, ..., m\}$ is the highest order of the tracking in general

  $m \in \{1, \alpha, \nu, \alpha + \nu\}$ is the plant order in general

  0   is the highest derivative of both the disturbance vector and control vector acting on the *ISO* plant

  1   is the order of the *ISO* and *EISO* plant

  α is the order of the *HISO* and *IIO* plant

  η is the highest derivative of the disturbance vector acting on the *IO* plant

μ is the highest derivative of the control vector acting on the *HISO*, *IO*, *IIO* and *EISO* plant, as well as he highest derivative of the disturbance vector acting on the *IIO* and on the *EISO* plant

ν is the order of the *IO* plant

A.3   Letters

Lower-case block or italic letters are used for scalars. Lower-case bold block letters denote vectors. Upper-case block letters denote matrices, or points. Upper-case Fraktur letters designate sets or spaces.

The notation ";$t_{(.)0}$" will be omitted as an argument of a variable if and only if a choice of the initial moment $t_{(.)0}$ does not have any influence on the value of the variable.

BLACKBOARD BOLD LETTERS

$\mathbb{C}$ *set of complex numbers s*

$\mathbb{C}^k$ *k-dimensional complex vector space*

A.3.1   CALLIGRAPHIC LETTERS

$\mathcal{C}$ *controller*, or *controllability matrix* $\mathcal{C}$,

$$\mathcal{C} = \left[B \vdots AB \vdots A^2 B \cdots \vdots A^{n-1} B \right]$$

$\mathcal{CS}$ *control system*

$\mathcal{D}_{(.)}$ *tracking domain related to the tracking property symbolized by (.)*

$\mathcal{I}$ *integral output space*, $\mathcal{I} = \mathfrak{T} \times \mathfrak{R}^N$

$\mathcal{L}^{\mp}\{\mathbf{i}(.)\}$ *Left (-), right (+), respectively, the Laplace transform of a function* $\mathbf{i}(.)$

$$\mathcal{L}^{\mp}\{\mathbf{i}(t)\} = \mathbf{I}^{\mp}(s) = \int_{0^{\mp}}^{\infty} \mathbf{i}(t)e^{-st}dt = \lim \left[\int_{\mp\zeta}^{\infty} \mathbf{i}(t)e^{-st}dt : \zeta \longrightarrow 0^+ \right]$$

$\mathcal{P}$ *plant*

$\mathcal{S}$ *state*

$\mathcal{S}(.)$ *motion*

A.3.2   FRAKTUR LETTERS

Capital Fraktur letters are used for spaces or for sets.

$\mathfrak{A} \subseteq \mathfrak{R}^n$ *a nonempty subset of* $\mathfrak{R}^n$

$\mathfrak{B} \subseteq \mathfrak{R}^n$ *a nonempty subset of* $\mathfrak{R}^n$

$\mathfrak{B}_\xi(\mathbf{z})$ *an open hyperball with the radius* ξ *centered at the point* $\mathbf{z}$ *in* the corresponding space

$$\mathfrak{B}_\xi(\mathbf{z}) = \{\mathbf{w} : \ \|\mathbf{w} - \mathbf{z}\| < \xi\}$$

$\mathfrak{B}_\xi$ *an open hyperball with the radius* ξ *centered at the origin* of the corresponding space

$$\mathfrak{B}_\xi = \mathfrak{B}_\xi(\mathbf{0})$$

$\mathfrak{C}$ *the family of all defined and continuous functions on* $\mathfrak{T}_0$

$\mathfrak{C}^k(\mathfrak{S})$ *the family of all functions defined, continuous and k-times continuously differentiable on the set* $\mathfrak{S} \subseteq \mathfrak{T} \cup \mathfrak{R}^i$, $\mathfrak{C}^k(\mathfrak{R}^i) = \mathfrak{C}^{ki}$,

$\mathfrak{C}^k(\mathfrak{T}_0)$ *the family of all functions defined, continuous and k-times continuously differentiable on* $\mathfrak{T}_0$

$\mathfrak{C}^0(\mathfrak{S})$ *the family of all functions defined and continuous on the set* $\mathfrak{S}$, $\mathfrak{C}^0(\mathfrak{R}^i) = \mathfrak{C}^{0,i} = \mathfrak{C}(\mathfrak{R}^i)$

$\mathfrak{C}^{k-}(\mathfrak{R}^i)$ *the family of all functions defined everywhere and k-times continuously differentiable on* $\mathfrak{R}^i \backslash \{\mathbf{0}_i\}$, *which have defined and continuous derivatives at the origin* $\mathbf{0}_i$ *of* $\mathfrak{R}^i$ up to the order $(k-1)$, which are defined and continuous at *at the origin* $\mathbf{0}_i$ and have defined the left and the right $k - th$ order derivative *at the origin* $\mathbf{0}_i$

$\mathfrak{D}^k$ is a given, or to be determined, *family of all bounded k-times continuously differentiable on* $\mathfrak{T}_0$ *permitted disturbance vector total functions* $\mathbf{D}(.)$, or *deviation functions* $\mathbf{d}(.)$, $\mathfrak{D}^k \subset \mathfrak{C}^k$, the Laplace transforms of which are strictly proper real rational complex functions

$$\mathfrak{D}^k = \left\{ \mathbf{D}(.) : \mathbf{D}^{(k)}(t) \in \mathfrak{C}, \ \exists \zeta \in \mathfrak{R}^+ \Longrightarrow \left\| \mathbf{D}^k(t) \right\| < \zeta, \ \forall t \in \mathfrak{T}_0 \right\}$$

or

$$\mathfrak{D}^k = \left\{ \mathbf{d}(.) : \mathbf{d}^{(k)}(t) \in \mathfrak{C}, \ \exists \xi \in \mathfrak{R}^+ \Longrightarrow \left\| \mathbf{d}^k(t) \right\| < \xi, \ \forall t \in \mathfrak{T}_0 \right\}$$

$\mathfrak{D}^k_-$ is a subfamily of $\mathfrak{D}^k$, $\mathfrak{D}^k_- \subset \mathfrak{D}^k$, such that the real part of every pole of the Laplace transform $\mathbf{D}(s)$ of every $\mathbf{D}(.) \in \mathfrak{D}^k_-$ is negative, $\mathfrak{D}_- = \mathfrak{D}^0_-$

$\mathfrak{D}^0 = \mathfrak{D}$ is the *family of all bounded continuous permitted disturbance vector total functions* $\mathbf{D}(.)$ *or deviation functions* $\mathbf{d}(.)$, $\mathfrak{D} \subset \mathfrak{C}$, the Laplace transforms of which are strictly proper real rational complex functions

$\mathfrak{I}^k$ is a given, or to be determined, family of all bounded and k-times continuously differentiable permitted input vector functions $\mathbf{I}(.)$

$$\mathfrak{I}^k \subset \mathfrak{C}^k \cap \mathfrak{L}$$

$\mathfrak{I}^0 = \mathfrak{I}$ is the family of all bounded continuous permitted input vector functions $\mathbf{I}(.)$

$$\mathfrak{I} \subset \mathfrak{C} \cap \mathfrak{L}.$$

$\mathfrak{I}^k_-$ is a subfamily of $\mathfrak{D}^k$, $\mathfrak{I}^k_- \subset \mathfrak{I}^k$, such that the real part of every pole of the Laplace transform $\mathbf{I}(s)$ of every $\mathbf{I}(.) \in \mathfrak{I}^k_-$ is negative, $\mathfrak{I}_- = \mathfrak{I}^0_-$

$\mathfrak{L}$ *the family of all strictly proper real rational complex functions, the original of which are bounded time-dependent functions*

$$\mathfrak{L} = \left\{ \mathbf{I}(.) : \begin{pmatrix} \exists \gamma(\mathbf{I}) \in \mathfrak{R}^+ \implies \|\mathbf{I}(t)\| < \gamma(\mathbf{I}), \; \forall t \in \mathfrak{T}_0, \\ \mathcal{L}^{\mp}\{\mathbf{I}(t)\} = \mathbf{I}^{\mp}(s) = \begin{bmatrix} I_1^{\mp}(s) & I_2^{\mp}(s) & \dots & I_M^{\mp}(s) \end{bmatrix}^T, \\ I_k^{\mp}(s) = \dfrac{\displaystyle\sum_{j=0}^{j=\zeta_k} a_{kj} s^j}{\displaystyle\sum_{j=0}^{j=\psi_k} b_{kj} s^j}, 0 \le \zeta_k < \psi_k, \; \forall k = 1, 2, ..., M, \end{pmatrix} \right\}$$

$\mathfrak{R}$ *the set of all real numbers*

$\mathfrak{R}^+$ *the set of all positive real numbers*

$\mathfrak{R}_+$ *the set of all nonnegative real numbers*

$\mathfrak{R}^{\nu N}$ *the extended output space of the IO system*, which is simultaneously the space of its internal dynamics - its *internal dynamics space*

$\mathfrak{R}^n$ *an n-dimensional real vector space, the state space of the ISO system*

$\mathfrak{R}^{N\nu} \backslash \mathfrak{B}_\varepsilon$ *the set of all vectors $\mathbf{y}^{\nu-1}$ in $\mathfrak{R}^{N\nu}$ out of $\mathfrak{B}_\varepsilon$,*

$$\mathfrak{R}^{N\nu} \backslash \mathfrak{B}_\varepsilon = \left\{ \mathbf{y}^{\nu-1} : \mathbf{y}^{\nu-1} \in \mathfrak{R}^{N\nu}, \; \mathbf{y}^{\nu-1} \notin \mathfrak{B}_\varepsilon \right\}$$

$\mathfrak{T}$ *the accepted reference time set, the arbitrary element of which is an arbitrary moment t and the *time* unit of which is second s, $1_t = s$, $t \langle s \rangle$*

$$\mathfrak{T} = \{ t : t[T] \langle s \rangle, \; numt \in \mathfrak{R}, \; dt > 0 \}, \; \inf \mathfrak{T} = -\infty, \; \sup \mathfrak{T} = \infty$$

$\mathfrak{T}_{0\mp}$ *the subset of $\mathfrak{T}$, which has the minimal element $min\mathfrak{T}_{0\mp}$ that is the initial instant $t_{0\mp}$, $numt_{0\mp} = 0^{\mp}$*

$$\mathfrak{T}_{0\mp} = \left\{ t : t \in \mathfrak{T}, \; t \ge t_{0\mp}, \; numt_{0\mp} = 0^{\mp} \right\}, \mathfrak{T}_{0\mp} \subset \mathfrak{T},$$
$$min\mathfrak{T}_{0\mp} = t_{0\mp} \in \mathfrak{T}, \; \sup \mathfrak{T}_{0\mp} = \infty$$

$\mathfrak{T}_{0F}$ the time set over which the system is to work,

$$\mathfrak{T}_{0F} = \left\{ t:\ t \in \mathfrak{T}, \left\langle \begin{array}{l} 0 \leq t \leq t_F \ \ iff \ t_F < \infty, \\ 0 \leq t < t_F \ \ iff \ t_F = \infty \end{array} \right\rangle \right\} \subseteq \mathfrak{T}_0$$

$\mathfrak{T}_R$ the time set over which the real output can deviate from the desired output; i.e., the reachability time set

$$\mathfrak{T}_R = \{ t \in \mathfrak{T}_0 : t_0 \leq t \leq t_R,\ t_R > t_0 \} \subset \mathfrak{T}_{0F}$$

$\mathfrak{T}_{RF}$ the time set over which the real output should be equal to the desired output; the post reachability time set

$$\mathfrak{T}_{RF} = \left\{ t \in Cl\mathfrak{T}_0 : \left\langle \begin{array}{l} t_R \leq t \leq t_F < \infty, \\ t_R \leq t < t_F = \infty \end{array} \right\rangle \right\} \subset \mathfrak{T}_{0F},$$

$$\mathfrak{T}_R \cup \mathfrak{T}_{RF} = \mathfrak{T}_{0F},$$

$$\mathfrak{T}_{R\infty} = \{ t:\ t_R \leq t < \infty \} = [t_R, \infty[$$

$\mathfrak{T}_0^i$ the Cartesian set product of the time set $\mathfrak{T}_0$ multiplied by itself i-times

$$\mathfrak{T}_0^i = \underbrace{\mathfrak{T}_0 \times \mathfrak{T}_0 \times ... \times \mathfrak{T}_0}_{i-times},$$

$$Cl\ \mathfrak{T}_0^i = \underbrace{Cl\ \mathfrak{T}_0 \times Cl\ \mathfrak{T}_0 \times ... \times\ Cl\ \mathfrak{T}_0}_{i-times}, \ In\ \mathfrak{T}_0^i = \underbrace{In\ \mathfrak{T}_0 \times In\ \mathfrak{T}_0 \times ... \times In\ \mathfrak{T}_0}_{i-times}$$

$\mathfrak{T}^{(k+1)N}$ the time set product the elements of which are time vectors
$\mathbf{t}^{(k+1)N}$

$$\mathfrak{T}^{(k+1)N} = \left\{ \mathbf{t}^{(k+1)N} :\ -\infty\mathbf{1}_{(k+1)N} < \mathbf{t}^{(k+1)N} < \infty\mathbf{1}_{(k+1)N} \right\}$$

$\mathfrak{T}_R^{(k+1)N}$ the time set product the elements of which are time vectors
$\mathbf{t}^{(k+1)N}$ not greater elementwise than the vector reachability time $\mathbf{t}_R^{(k+1)N}$

$$\mathfrak{T}_R^{(k+1)N} = \left\{ \mathbf{t}^{(k+1)N} :\ \mathbf{t}_0^{(k+1)N} = \mathbf{0}_{(k+1)N} \leq \mathbf{t}^{(k+1)N} \leq \mathbf{t}_R^{(k+1)N} < \infty\mathbf{1}_{(k+1)N} \right\}$$

$$\mathfrak{T}_{R\infty}^{(k+1)N} = \left\{ \mathbf{t}^{(k+1)N} :\ \mathbf{t}_R^{(k+1)N} \leq \mathbf{t}^{(k+1)N} < \infty\mathbf{1}_{(k+1)N} \right\}$$

$\mathfrak{T}_{RF}^{(k+1)N}$ the time set product the elements of which are time vectors
$\mathbf{t}^{(k+1)N}$ not less elementwise than the vector reachability time $\mathbf{t}_R^{(k+1)N}$

$$\mathfrak{T}_{RF}^{(k+1)N} = \left\{ \mathbf{t}^{(k+1)N} : \left\langle \begin{array}{l} \mathbf{t}_R^{(k+1)N} \leq \mathbf{t}^{(k+1)N} \leq \mathbf{t}_F^{(k+1)N} < \infty\mathbf{1}_{(k+1)N}, \ or \\ \mathbf{t}_R^{(k+1)N} \leq \mathbf{t}^{(k+1)N} < \mathbf{t}_F^{(k+1)N} = \infty\mathbf{1}_{(k+1)N} \end{array} \right\rangle \right\}$$

$\mathfrak{Y}_d^k$ a given, or to be determined, *family of all bounded k-times continuously differentiable realizable desired total output vector functions* $\mathbf{Y}_d(.)$, $\mathfrak{Y}_d^k \subset \mathfrak{C}^{kN}$, the Laplace transforms of which are strictly proper real rational complex functions

$$\mathfrak{Y}_d^k = \left\{ \mathbf{Y}_d(.) : \mathbf{Y}_d(t) \in \mathfrak{C}^{kN}, \; \exists \kappa \in \mathfrak{R}^+ \implies \left\| \mathbf{Y}_d^k(t) \right\| < \kappa, \; \forall t \in \mathfrak{T}_0 \right\}$$

$\mathfrak{Y}_{d-}^k$ is a subfamily of $\mathfrak{Y}_d^k$, $\mathfrak{Y}_{d-}^k \subset \mathfrak{Y}_d^k$, such that the real part of every pole of the Laplace transform $\mathbf{Y}_d(s)$ of every $\mathbf{Y}_d(.) \in \mathfrak{Y}_{d-}^k$ is negative, $\mathfrak{Y}_{d-} = \mathfrak{Y}_{d-}^0$

$\mathfrak{Y}_{d0}^k$ *the set of the desired output initial conditions* $\mathbf{Y}_{d0}^k = \mathbf{Y}_d^k(t_0)$ *of* $\mathbf{Y}_d^k(t)$ *of every* $\mathbf{Y}_d(.) \in \mathfrak{Y}_d^k$

$$\mathfrak{Y}_{d0}^k = \left\{ \mathbf{Y}_{d0}^k : \mathbf{Y}_{d0}^k = \mathbf{Y}_d^k(t_0), \; \mathbf{Y}_d(.) \in \mathfrak{Y}_d^k \right\} \tag{A.1}$$

$\mathfrak{Y}_d = \mathfrak{Y}_d^0$ is the *family of all bounded continuous realizable desired total output vector functions* $\mathbf{Y}_d(.)$, $\mathfrak{Y}_d = \mathfrak{Y}_d^0 \subset \mathfrak{C}^{0d}$, the Laplace transforms of which are strictly proper real rational complex functions

A.3.3   GREEK LETTERS

α *a nonnegative integer*

    β *a nonnegative integer*

    $\gamma = \max\{\beta, \mu\}$

    δ_{ij} *the Kronecker delta,* $\delta_{ij} = 1$ *for* $i = j$, *and* $\delta_{ij} = 0$ *for* $i \neq j$

    $\delta^j \in \mathfrak{R}^{+i}$ *elementwise positive vector* δ^j *of the dimension* j

    $\partial \mathfrak{D}$ *the boundary of a set* $\mathfrak{D}$

    ϵ *the subsidiary error variable,* $\epsilon = Y_R - Y$

    ϵ *the subsidiary error vector,* $\epsilon = \mathbf{Y}_R - \mathbf{Y}$

    η *a natural number*

    θ *a nonnegative integer*

    $\lambda_m(H)$ *the minimal eigenvalue of the symmetric matrix* H

    $\lambda_M(H)$ *the maximal eigenvalue of the symmetric matrix* H

    μ *a nonnegative integer*

    ν *a nonnegative integer*

    τ *a subsidiary notation for* *time t*

    ϕ *the empty set*

    $\Xi = diag\{\epsilon_1 \;\; \epsilon_2 \;\; ... \;\; \epsilon_N\}$

$\Phi_{IO}(t,t_0) \in \mathfrak{R}^{N \times n}$ the *IO* system *output fundamental matrix* (Equation (2.27), Subsection 2.1.2)

$$\Phi_{IO}(t,t_0) = \mathcal{L}^{-1}\left\{\left(A^{(\nu)}S_N^{(\nu)}(s)\right)^{-1}\right\},$$

$$\Phi_{IO}(s) = \left(A^{(\nu)}S_N^{(\nu)}(s)\right)^{-1}$$

$\Phi_{ISO}(t,t_0) \equiv \Phi(t,t_0) \in \mathfrak{R}^{n \times n}$ *the fundamental matrix (3.4) of the ISO system* (3.1), (3.2)

$$\Phi(t,t_0) = e^{At}\left(e^{At_0}\right)^{-1} = e^{A(t-t_0)} \in \mathfrak{R}^{n \times n}$$

$$\Phi(s) = \mathcal{L}\{\Phi(t,t_0)\} = \mathcal{L}\left\{e^{A(t-t_0)}\right\} = (sI - A)^{-1}$$

has the following well-known properties, Equations (3.5)–(3.7) (all in Section 3.1):

$$det\Phi(t,t_0) \neq 0, \ \forall t \in \mathfrak{T}_0$$

$$\Phi(t,t_0)\Phi(t_0,t) = e^{A(t-t_0)}e^{A(t_0-t)} \equiv e^{A0} = I_n$$

$$\Phi^{(1)}(t,t_0) = A\Phi(t,t_0) = \Phi(t,t_0)A$$

ρ *a natural number*

A.3.4   ROMAN LETTERS

$A \in \mathfrak{R}^{n \times n}$ *the matrix describing the internal dynamics of the ISO system*
   $A_k \in \mathfrak{R}^{N \times N}$ *the matrix associated with the $k-th$ derivative* $\mathbf{Y}^{(k)}$ *of the output vector* $\mathbf{Y}$ *of the IO system*
   $A^{(\nu)} \in \mathfrak{R}^{N \times (\nu+1)N}$ *the extended matrix describing the IO system internal dynamics,* $A^{(\nu)} = \left[A_0 \ \vdots \ A_1 \ \vdots \ ... \ \vdots \ A_\nu\right]$

   $B_k \in \mathfrak{R}^{N \times M}$ *the matrix associated with the $k-th$ derivative* $\mathbf{I}^{(k)}$ *of the input vector* $\mathbf{I}$ *of the IO system*
   $B^{(\mu)} \in \mathfrak{R}^{N \times (\mu+1)r}$ *the extended matrix describing the transmission of the influence of the control vector* $\mathbf{U}(t)$ *on the system dynamics,* $B^{(\mu)} = \left[B_0 \ \vdots \ B_1 \ \vdots \ ... \ \vdots \ B_\mu\right]$

   $\mathbf{B}^{(\mu)} \in \mathfrak{R}^{n \times (\mu+1)r}$ *the extended matrix describing the transmission of the influence of the control vector* $\mathbf{U}(t)$ *on the system dynamics,* $\mathbf{B}^{(\mu)} = \left[\mathbf{B}_0 \ \vdots \ \mathbf{B}_1 \ \vdots \ ... \ \vdots \ \mathbf{B}_\mu\right]$

$C \in \mathfrak{R}^{N \times n}$ *the matrix of the ISO system, which describes the transmission of the state vector action on the system output vector* $\mathbf{Y}$

$\mathbf{C}_0$ is *the vector of all initial conditions* acting on the system

$$\mathbf{C}_0 = \begin{bmatrix} I_0 \\ I_0^{(1)} \\ \dots \\ I_0^{(\mu-1)} \\ X_0 \\ Y_0 \\ Y_0^{(1)} \\ \dots \\ Y_0^{(\nu-1)} \end{bmatrix} \in \mathfrak{R}^{\mu M + n + \nu N}$$

d *a natural number*

$\mathbf{d} \in \mathfrak{R}^d$ *the disturbance deviation vector,* (2.53), (Section 2.2)

$$\mathbf{d} = \mathbf{D} - \mathbf{D}_N$$

$\mathbf{D} \in \mathfrak{R}^d$ *the total disturbance vector*

$\mathbf{D}_N \in \mathfrak{R}^d$ *the nominal disturbance vector*

$D \in R^{N \times d}$ *the ISO system matrix describing the transmission of the influence of* $\mathbf{I}(t)$ *on the system output*

$\mathbf{e}$ *the output error vector* $\mathbf{e} \in R^N$

$$\mathbf{e} = \mathbf{Y_d} - \mathbf{Y} = -\mathbf{y}, \mathbf{e} = \begin{bmatrix} e_1 & e_2 & \dots & e_N \end{bmatrix}^T$$

$\mathbf{E}(s)$ the Laplace transform of the output error vector $\mathbf{e}(t)$

$F(.): \mathfrak{T}_0 \longrightarrow \mathfrak{R}^{N \times N}$ *a matrix function associated with* $\mathbf{f}(.)$

$$\mathbf{f} = \begin{bmatrix} f_1 & f_2 & \dots & f_N \end{bmatrix}^T \implies F = diag\{f_1 \quad f_2 \quad \dots \quad f_N\}$$

$F(s)$ is *the full (complete) transfer function matrix of a time-invariant continuous-time linear dynamical system*

$F_{IIO}(s) \in \mathfrak{C}^{N \times [(\gamma+1)M + \delta\rho + \nu N]}$,   is *the full transfer function matrix of the IIO system (Equation (6.19) in Subsection 6.1.2)*

$F_{IIOIS}(s) \in \mathfrak{C}^{\rho \times [(\beta+1)M + \alpha\rho]}$ is *the full (complete) IS transfer function matrix of the IIO system (Equation (6.10) in Subsection 6.1.2)*

$F_{IO}(s)$ is *the full transfer function matrix of the IO system (Equation (2.25) in Subsection 2.1.2)*

$F_{ISO}(s) \in \mathfrak{C}^{N \times (M+n)}$ is *the full ISO transfer function matrix of the ISO system (Equation (3.20) in Subsection 3.1.2)*

$F_{ISOIS}(s) \in \mathfrak{C}^{n \times (M+n)}$ the full (complete) IS transfer function matrix of the ISO system (Equation (3.14) in Subsection 3.1.2)

$G = G^T \in R^{p \times p}$ the symmetric matrix of the quadratic form $v(\mathbf{w}) = \mathbf{w}^T G \mathbf{w}$

$G(s)$ is the transfer function matrix of a time-invariant continuous-time linear dynamical system

$G_{IIOISU}(s) \in \mathfrak{C}^{\rho \times M}$ the IS transfer function matrix of the IIO system (Equation (6.13) in Subsection 6.1.2)

$G_{IIOISU_0}(s) \in \mathfrak{C}^{\rho \times \mu M}$ the IS transfer function matrix relative to $\mathbf{U}_0^{\mu-1}$ of the IIO system (Equation (6.16) in Subsection 6.1.2)

$G_{IIOISR_0}(s) \in \mathfrak{C}^{\rho \times \alpha \rho}$ the II transfer function matrix relative to $\mathbf{R}_0^{\alpha-1}$ of the IIO system (Equation(6.15) in Subsection 6.1.2)

$G_{IIOD}(s) \in \mathfrak{C}^{N \times d}$ the transfer function matrix relative to $\mathbf{D}$ of the IIO system (Equation (6.22) in Subsection 6.1.2)

$G_{IIOD_0}(s) \in \mathfrak{C}^{N \times d\mu}$ the transfer function matrix relative to $\mathbf{D}_0^{\mu-1}$ of the IIO system (Equation (6.30) in Subsection 6.1.2)

$G_{IIOR_0}(s) \in \mathfrak{C}^{N \times \alpha \rho}$ the transfer function matrix relative to $\mathbf{R}_0^{\alpha-1}$ of the IIO system (Equation (6.31) in Subsection 6.1.2)

$G_{IIOU}(s) \in \mathfrak{C}^{N \times M}$ is the transfer function matrix of the IIO system (Equation (6.23) in Subsection 6.1.2)

$G_{IIOU_0}(s) \in \mathfrak{C}^{N \times (\gamma+1)M}$ the transfer function matrix relative to $\mathbf{U}_0^{\mu-1}$ of the IIO system (Equation (6.32) in Subsection 6.1.2)

$G_{IIOY_0}(s) \in \mathfrak{C}^{N \times \nu N}$ the transfer function matrix relative to $\mathbf{Y}_0^{\nu-1}$ of the IIO system (Equation (6.33) in Subsection 6.1.2)

$G_{IOU}(s)$ is the transfer function matrix of the IO system (Equation (2.30) in Subsection 2.1.2)

$G_{IOU_0}(s) \in \mathfrak{C}^{N \times \mu M}$ the transfer function matrix relative to $\mathbf{U}_0^{\mu-1}$ of the IO system (Equation (2.33) in Subsection 2.1.2)

$G_{IOY_0}(s) \in \mathfrak{C}^{N \times \nu N}$ the transfer function matrix relative to $\mathbf{Y}_0^{\nu-1}$ of the IO system (Equation (2.34) in Subsection 2.1.2)

$G_{ISOD}(s) \in \mathfrak{C}^{N \times d}$ is the IO transfer function matrix of the ISO system (Equation (3.23) in Subsection 3.1.2)

$G_{ISOISU}(s) \in \mathfrak{C}^{n \times M}$ the IS transfer function matrix of the ISO system (Equation (3.18) in Subsection 3.1.2)

$G_{ISOX_oIS}(s) \in \mathfrak{C}^{N \times n}$ is the IS transfer function matrix relative to $\mathbf{X}_0$ of the ISO system (Equation 3.19 in Subsection 3.1.2)

$G_{ISOU}(s) \in \mathfrak{C}^{N \times M}$ is the IO transfer function matrix of the ISO system (Equation (3.24) in Subsection 3.1.2)

$H \in R^{N \times r}$ a matrix

$H = H^T \in R^{p \times p}$ *the symmetric matrix of the quadratic form* $v(\mathbf{w}) =$ $\mathbf{w}^T H \mathbf{w}$

i *an arbitrary natural number, or the imaginary unit* $\sqrt{-1}$, *or the input deviation variable*

$\mathbf{i} \in \mathfrak{R}^M$ *the input deviation vector,* $\mathbf{i} = [i_1 \quad i_2 \ ... \ i_M]^T$, (2.54), *(Section 2.2)*

$$\mathbf{i} = \mathbf{I} - \mathbf{I}_N$$

$\mathbf{i}^\mu(t) \in \mathfrak{R}^{(\mu+1)M}$ *the extended input vector at a moment* t, $\mathbf{i}^\mu(t) =$ $\left[\mathbf{i}(t) \vdots \mathbf{i}^{(1)}(t) \vdots ... \vdots \mathbf{i}^{(\mu)}(t) \right]^T$

$\mathbf{i}_{0\mp}^{\mu-1} \in \mathfrak{R}^{\mu M}$ *the initial extended input vector at the initial moment* $t_0 = 0$, $\mathbf{i}_{0\mp}^{\mu-1} = \mathbf{i}^{\mu-1}(0^\mp) = \left[\mathbf{i}_{0(\mp)} \vdots \mathbf{i}_{0(\mp)}^{(1)} \vdots ... \vdots \mathbf{i}_{0(\mp)}^{(\mu-1)} \right]^T \in \mathfrak{R}^{\mu M}$

I *the identity matrix of the n-th order,* $I = diag\{1 \quad 1 \ ... \ 1\} \in \mathfrak{R}^{n \times n}$, *or the total input variable,* $I_n = I$

I_N *the identity matrix of the N-th order,* $I_N = diag\{1 \quad 1 \ ... \ 1\} \in$ $\mathfrak{R}^{N \times N}$

$\mathbf{I} \in \mathfrak{R}^M$ *the total input vector,* $\mathbf{I} = [I_1 \quad I_2 \ ... \ I_M]^T$

I_i *is the i-th order identity matrix*

$\mathbf{I}_N \in \mathfrak{R}^M$ *the nominal input vector,* $\mathbf{I}_N = [I_{N1} \quad I_{N2} \ ... \ I_{NM}]^T$,

$Int\mathfrak{S}$ *the interior of the set* $\mathfrak{S}$

$Int\mathfrak{T}_0$ *the interior of the set* $\mathfrak{T}_0$

$$Int\mathfrak{T}_0 = \{t : \ t \in \mathfrak{T}_0, \ t > 0\}$$

$\text{Im} \, s$ the imaginary part of $s = \sigma + j\omega$, $\text{Im} \, s = j\omega$

j *an arbitrary natural number, or* $j = \sqrt{-1}$ is the imaginary unit

$J \in \mathfrak{R}^{n \times M}$ *a matrix*

k *an arbitrary natural number*

$M(.)$ *a complex valued matrix function* of any type

$M(s)$ *a complex valued matrix* of any type

m *a nonnegative integer*

n *a natural number*

N *a natural number,* if N is the dimension of the output vector and if n is the dimension of the state vector then $N \leq n$

O *is the origin of* $\mathfrak{R}^n$; *or the zero matrix of the appropriate order*

p *a natural number*

$P \in R^{n \times N}$ *a matrix*

$P_k \in \mathfrak{R}^{p \times M}$ *a matrix*

$P^{(\beta)} \in \mathfrak{R}^{\rho \times M(\beta+1)}$ *an extended matrix describing the transmission of the influence of* $\mathbf{i}^{\beta}(t)$ *on the internal dynamics of the IIO system,* $P^{(\alpha)} = [P_0 \vdots P_1 \vdots ... P_\beta]$

    q *a natural number*

    $Q \in R^{N \times N}$ *a matrix*

    $Q_k \in \mathfrak{R}^{\rho \times \rho}$ *a matrix*

    $Q^{(\alpha)} \in \mathfrak{R}^{\rho \times \rho(\alpha+1)}$ *the extended matrix describing the internal dynamics of the IIO system,* $Q^{(\alpha)} = \left[Q_0 \vdots Q_1 \vdots ... Q_\alpha \right]$

    $\mathbf{r} \in \mathfrak{R}^{\rho}$ *a subsidiary deviation vector, which is the internal dynamics deviation vector, (5.35), (Section 5.2)*

$$\mathbf{r} = \mathbf{R} - \mathbf{R}_N$$

    $\mathbf{R} \in \mathfrak{R}^{\rho}$ *a subsidiary total vector, which is the internal dynamics total vector of the HISO system and of the IIO system*

    $\mathbf{R}_N \in \mathfrak{R}^{\rho}$ *a subsidiary nominal vector, which is the internal dynamics nominal vector of the HISO system and of the IIO system*

    $R_k \in \mathfrak{R}^{N \times \rho}$ *a matrix*

    $R_y^{(\alpha-1)} \in \mathfrak{R}^{N \times \alpha \rho}$ *the extended matrix describing the action of the extended internal dynamics vector* $\mathbf{R}^{\alpha-1}$ *on the output dynamics of the IIO system,* $R_y^{(\alpha-1)} = \left[R_{y0} \vdots R_{y1} \vdots ... R_{y,(\alpha-1)} \right]$

    S *a state variable*

    $\mathbf{S}$ *a state vector*

$$\mathbf{S} = \left[S_1 \vdots S_2 \vdots S_3 \vdots ... \vdots S_K \right]^T \in \mathfrak{R}^K$$

    $\mathrm{Re}\, s$ *the real part of* $s = \sigma + j\omega$, $\mathrm{Re}\, s = \sigma$

    s *the basic time unit: second, or a complex variable or a complex number* $s = \sigma + j\omega$

    $sign(.) : \mathfrak{R} \rightarrow \{-1, 0, 1\}$ *the scalar signum function,*

$$sign(x) = |x|^{-1} x \ if \ x \neq 0, \ and \ sign(0) = 0$$

    $S_i^{(k)}(.) : \mathfrak{C} \longrightarrow \mathfrak{C}^{i(k+1) \times i}$ *the matrix function of* s *defined by (1.9) in Section 1.3:*

$$S_i^{(k)}(s) = \left[s^0 I_i \vdots s^1 I_i \vdots s^2 I_i \vdots ... \vdots s^k I_i \right]^T \in \mathfrak{C}^{i(k+1) \times i},$$

$$(k, i) \in \{(\mu, M), (\nu, N)\}$$

t *time (temporal variable)*, or *an arbitrary time value (an arbitrary moment, an arbitrary instant)*; and formally mathematically t denotes for short also the numerical *time* value *numt* if it does not create a confusion

$$t[\mathrm{T}]\,\langle s\rangle\,,\ numt \in \Re,\ dt > 0\,,\ or\ equivalently:\ t \in \mathfrak{T}$$

It has been the common attitude to use the notation t of *time* and of its arbitrary temporal value also for its numerical value *numt*, e.g. $t = 0$ is used in the sense *numt* $= 0$. We do the same throughout the book if there is not any confusion because we can replace t everywhere by $t1_t^{-1}$, $\left(t1_t^{-1}\right) \in \Re$, that we denote again by t, *numt* $= num\left(t1_t^{-1}\right)$

t_0 a conventionally accepted *initial value of time (initial instant, initial moment)*, $t_0 \in \mathfrak{T}$, *numt*$_0 = 0$, i.e. simply $t_0 = 0$ in the sense *numt*$_0 = 0$

$t_{\inf}$ *the first instant*, which has not happened, $t_{\inf} = -\infty$

$t_{\sup}$ *the last instant*, which will not occur, $t_{\sup} = \infty$

$t_{ZeroTotal}$ *the total zero value of time*, which has not existed and will not happen

t_{zero} a conventionally accepted *relative zero value of time*

$\mathbf{t}^{(k+1)N}$ the $(k+1)N$-time vector $\mathbf{t}^{(k+1)N}$

$$\mathbf{t}^{(k+1)N} = t\mathbf{1}_{(k+1)N} = [t\ t...t]^T \in \mathfrak{T}_0^{(k+1)N} \cup \{\infty\}^{(k+1)N}\,,\ \ k \in \{0, 1, 2, ..., \alpha\}\,,$$
$$\mathbf{t} = \mathbf{t}^N = t\mathbf{1}_N = [t\ t...t]^T \in \mathfrak{T}_0^N$$
$$\mathbf{t}_0 = \mathbf{t}_0^N = t_0\mathbf{1}_N = 0\mathbf{1}_N = [0\ 0...0]^T = \mathbf{0}_N \in In\ \mathfrak{T}^N$$

$\mathbf{t}_{R(j)}^N$ *time vector related to the j-th derivative $\mathbf{Y}^{(j)}$ of the output vector $\mathbf{Y} \in \Re^N$*

$$\mathbf{t}_{R(j)}^N = \begin{bmatrix} t_{R1,(j)} \\ t_{R2,(j)} \\ ... \\ t_{RN,(j)} \end{bmatrix} \in In\ \mathfrak{T}_{0F}^N,\ j \in \{0, 1, 2, ..\}$$

$\mathbf{t}_R^N$ *vector reachability time related to the output vector $\mathbf{Y} \in \Re^N$*

$$\mathbf{t}_R^N = \mathbf{t}_{R(0)}^N = \begin{bmatrix} t_{R1} \\ t_{R2} \\ ... \\ t_{RN} \end{bmatrix} = \begin{bmatrix} t_{R1,(0)} \\ t_{R2,(0)} \\ ... \\ t_{RN,(0)} \end{bmatrix} \in In\ \mathfrak{T}_{0F}^N$$

$t_R^{(k+1)N}$ *the finite vector reachability time (FVRT) related to the extended output vector* $\mathbf{Y}^k \in \mathfrak{R}^{(k+1)N}$

$$\mathbf{t}_R^{(k+1)N} = \begin{bmatrix} \mathbf{t}_R^N \\ \mathbf{t}_{R(1)}^N \\ \mathbf{t}_{R(2)}^N \\ ... \\ \mathbf{t}_{R(k)}^N \end{bmatrix} = \begin{bmatrix} \mathbf{t}_{R(0)}^N \\ \mathbf{t}_{R(1)}^N \\ \mathbf{t}_{R(2)}^N \\ ... \\ \mathbf{t}_{R(k)}^N \end{bmatrix} \in In \ \mathfrak{T}_{0F}^{(k+1)N}$$

$$\mathbf{t}_R^{(k+1)N} = [t_{R1,(0)} \ ... \ t_{RN,(0)} \ \ t_{R1,(1)} ... \ t_{RN,(1)} \ ... \ t_{R1,(k)} \ ... \ t_{RN,(k)}]^T,$$
$$\forall k = 0, \ 1, ..., \alpha - 1$$

T   *the temporal dimension,* "*the time* dimension", *which is the physical dimension of time*

$T \in \mathfrak{R}^+$ *the period of a periodic behavior*

$T_k \in \mathfrak{R}^{N \times M}$ *a matrix*

$T^{(\mu)} \in \mathfrak{R}^{N \times M(\mu+1)}$ *the extended matrix describing the action of the extended input vector* $\mathbf{i}^\mu$ *on the output dynamics of the IIO system,* $T^{(\mu)} = \begin{bmatrix} T_0 \vdots T_1 \vdots ... T_\mu \end{bmatrix}$

$\mathbf{U}_{[t_0,t_1]}^\mu$ extended control on the time interval $[t_0, t_1]$

$v(.) : \mathfrak{R}^p \to \mathfrak{R}$ *a quadratic form,* $v(\mathbf{w}) = \mathbf{w}^T \mathbf{W} \mathbf{w}$

$\mathbf{V}(s)$ is *the Laplace transform of all actions on the system; it is composed of the Laplace transform* $\mathbf{I}(s)$ *of the input vector* $\mathbf{I}(t)$ *and of all (input and output) initial conditions*

$$\mathbf{V}(s) = \begin{bmatrix} \mathbf{I}(s) \\ \mathbf{C}_0 \end{bmatrix}$$

$\mathbf{w} \in \mathfrak{R}^p$ *a subsidiary real valued vector*

$$\mathbf{w} = [w_1 \ \ w_2 \ ... \ wp]^T \in \left\{ [\mathbf{r}^{\alpha-1} \ \ \mathbf{y}^{\nu-1}]^T, \ \mathbf{x}, \ \mathbf{y}^{\nu-1} \right\},$$
$$p \in \{\rho, \ n, \ N\}$$

$W = W^T \in R^{p \times p}$ *the symmetric matrix of the quadratic form* $v(\mathbf{w})$
$v(\mathbf{w}) = \mathbf{w}^T W \mathbf{w}$, $W \in \{G = G^T, H = H^T\}$

$x \in \mathfrak{R}$ *a real valued scalar state deviation variable*

$\mathbf{x} \in \mathfrak{R}^n$ *the state vector deviation of the ISO system,* (3.37), *(Section 3.2)*

$$\mathbf{x} = [x_1 \ \ x_2 \ ... \ x_n]^T, \ \mathbf{x} = \mathbf{X} - \mathbf{X}_N = \mathbf{X} - \mathbf{X}_d$$

$\mathbf{X} \in \mathfrak{R}^n$ the total state vector of the ISO system, $\mathbf{X} = [X_1\ X_2\ ..\ X_n]^T$

$\mathbf{X}_N \in \mathfrak{R}^n$ the total nominal state vector of the ISO system, $\mathbf{X}_N = [X_{N1}\ X_{N2}\ ...\ X_{Nn}]^T$

$y \in \mathfrak{R}$ a real valued scalar output deviation variable

$\mathbf{y} \in \mathfrak{R}^N$ a real valued vector output deviation variable - the output deviation vector of both the plant and of its control system, $\mathbf{y} = [y_1\ y_2\ ...\ y_N]^T$, (2.56), (Section 2.2)

$$\mathbf{y} = \mathbf{Y} - \mathbf{Y}_d = -\varepsilon$$

$\mathbf{Y} \in \mathfrak{R}^N$ a real total valued vector output - the total output vector of both the plant and of its control system, $\mathbf{Y} = [Y_1\ Y_2\ ...\ Y_N]^T$

$\mathbf{Y}_d \in \mathfrak{R}^N$ a desired (a nominal) total valued vector output - the desired total output vector of both the plant and of its control system $\mathbf{Y}_d = [Y_{d1}\ Y_{d2}\ ...\ Y_{dN}]^T$

$\mathbf{y}_{0\mp}^{\nu-1} \in \mathfrak{R}^{\nu N}$ the initial extended output vector at the initial moment $t_0 = 0$, $\mathbf{y}_{0\mp}^{\nu-1} = \mathbf{y}^{\nu-1}(0^\mp) = \left[\mathbf{y}_{0(\mp)}^T \vdots \mathbf{y}_{0(\mp)}^{(1)^T} \vdots ... \vdots \mathbf{y}_{0(\mp)}^{(\nu-1)^T} \right]^T$, $\mathbf{y}_{0\mp}^0 = \mathbf{y}^0(0^\mp) = \mathbf{y}_{0\mp} = \mathbf{y}(0^\mp)$

$Z_k^{(\varsigma-1)}(.) : \mathfrak{C} \to \mathfrak{C}^{(\varsigma+1)k \times k}$ the matrix function of s defined by (1.11) in the Section 1.3:

$$Z_k^{(\varsigma-1)}(s) = \begin{bmatrix} O_k & O_k & O_k & ... & O_k \\ s^0 I_k & O_k & O_k & ... & O_k \\ ... & ... & ... & ... & ... \\ s^{\varsigma-1} I_k & s^{\varsigma-2} I_k & s^{\varsigma-3} I_k & ... & s^0 I_k \end{bmatrix}, \varsigma \geq 1,$$

$$Z_k^{(\varsigma-1)}(s) \in \mathfrak{C}^{(\varsigma+1)k \times k}, \ (\varsigma, k) \in \{(\mu, M),\ (\nu, N)\}$$

See Note 43 (Subsection 2.1.2) on $Z_k^{(\varsigma-1)}(.)$ for $\zeta \leq 0$

A.4   Name

Stable (stability) matrix: a square matrix is *stable (stability) matrix* if, and only if, the real parts of all its eigenvalues are negative.

A.5   Symbols, vectors, sets, and matrices

$(.)$ an arbitrary variable, or an index

$|(.)| : \mathfrak{R} \to \mathfrak{R}_+$ the absolute value (module) of a (complex valued) scalar variable $(.)$

$\|.\| : \mathfrak{R}^n \to \mathfrak{R}_+$ *an accepted norm on $\mathfrak{R}^n$, which is the Euclidean norm on $\mathfrak{R}^n$ iff not stated otherwise:*

$$\|\mathbf{x}\| = \|\mathbf{x}\|_2 = \sqrt{\mathbf{x}^T\mathbf{x}} = \sqrt{\sum_{i=1}^{i=n} x_i^2}$$

$\|.\|_1 : \mathfrak{R}^n \to \mathfrak{R}_+$ *is the taxicab norm or Manhattan norm:*

$$\|\mathbf{x}\|_1 = \sum_{i=1}^{i=n} |x_i|$$

$\sim$ *equivalent*

$\langle 1.. \rangle$ shows *the units 1... of a physical variable*

$[\alpha, \beta] \subset \mathfrak{R}$ *a compact interval*, $[\alpha, \beta] = \{x : x \in \mathfrak{R}, \alpha \leq x \leq \beta\}$

$[\alpha, \beta [\subseteq \mathfrak{R}$ *a left closed, right open interval*, $[\alpha, \beta[= \{x : x \in \mathfrak{R}, \alpha \leq x < \beta\}$

$] \alpha, \beta] \subseteq \mathfrak{R}$ *a left open, right closed interval*, $[\alpha, \beta[= \{x : x \in \mathfrak{R}, \alpha < x \leq \beta\}$

$] \alpha, \beta [\subseteq \mathfrak{R}$ *an open interval*, $]\alpha, \beta[= \{x : x \in \mathfrak{R}, \alpha < x < \beta\}$, $(\sigma, \infty[\in \{]\sigma, \infty[, [\sigma, \infty[\}$

$(\alpha, \beta) \subseteq \mathfrak{R}$ *a general interval*, $(\alpha, \beta) \in \{[\alpha, \beta], [\alpha, \beta[,]\alpha, \beta],]\alpha, \beta[\}$

$\mathfrak{A} \backslash \mathfrak{B}$ is the set difference between the set $\mathfrak{A}$ and the set $\mathfrak{B}$

$$\mathfrak{A} \backslash \mathfrak{B} = \{x : x \in \mathfrak{A}, v \notin \mathfrak{B}\}$$

$\lambda_i(A)$ *the eigenvalue $\lambda_i(A)$ of the matrix A*

$[A..]$ shows *the physical dimension A... of a physical variable*

$\left[A_1 \vdots A_2 \vdots ... \vdots A_\nu \right]$ *a structured matrix* composed of the submatrices $A_1, A_2, ..., A_\nu$

$\mathbf{0}_k = [0 \ 0 \ ...0]^T \in \mathfrak{R}^k,$ *the elementwise zero vector*, $\mathbf{0}_n = \mathbf{0}$

$\mathbf{1}_k = [1 \ 1...1]^T \in \mathfrak{R}^k,$ *the elementwise unity vector*, $\mathbf{1}_n = \mathbf{1}$

$\mathbf{w} = \varepsilon$ *the elementwise vector equality*

$$\varepsilon = [\varepsilon_1 \ \varepsilon_2 \ ... \ \varepsilon_N]^T \in \mathfrak{R}^N, \ \mathbf{w} = [w_1 \ w_2 \ ... \ w_N]^T,$$
$$\mathbf{w} = \varepsilon \iff w_i = \varepsilon_i, \forall i = 1, 2, ..., N$$

$\mathbf{w} \neq \varepsilon$ *the elementwise vector inequality*

$$\mathbf{w} \neq \varepsilon \iff w_i \neq \varepsilon_i, \ \forall i = 1, 2, ..., N$$

$\forall$ for every

$adj\,A$ the adjoint matrix of the nonsingular square matrix A, $det\,A \neq$ $0 \Longrightarrow A\,adj\,A = (det\,A)\,I$

$det\,A$ the determinant of the matrix A, $det\,A = |A|$

A^{-1} the inverse matrix of the nonsingular square matrix A, $det\,A \neq$ $0 \Longrightarrow A^{-1} = adj\,A/det\,A$

$blockdiag\,\{\cdot \quad \cdot \quad \cdot \;\ldots\; \cdot\}$ block diagonal matrix, the entries of which are matrices, e.g.,

$$blockdiag\,\{E_k \;\; B_k\} = \begin{bmatrix} E_k & O_{N,r} \\ O_{N,d} & B_k \end{bmatrix}, \; k = 0,1,..,\nu$$

$Cl\,\mathfrak{S}$ the closure of a set $\mathfrak{S}$

$d(\mathbf{v},\mathfrak{S})$ the scalar distance of a vector $\mathbf{v}$ from a set $\mathfrak{S}$

$$d(\mathbf{v},\mathfrak{S}) = \inf[\|\mathbf{v} - \mathbf{w}\| : \mathbf{w} \in \mathfrak{S}]$$

$DenF(s)$ the denominator matrix polynomial of the real rational matrix $F(s) = [DenF(s)]^{-1}\,NumF(s)$, or $NumF(s)\,[DenF(s)]^{-1}$

$\partial\mathfrak{S}$ the boundary of a set $\mathfrak{S}$

$gradv\,(\mathbf{y}^{\nu-1})$ the gradient of $v\,(\mathbf{y}^{\nu-1})$

$$gradv\,(\mathbf{y}^{\nu-1}) = \left[\frac{\partial v\,(\mathbf{y}^{\nu-1})}{\partial y_1} .. \frac{\partial v\,(\mathbf{y}^{\nu-1})}{\partial y_N} .... \frac{\partial v\,(\mathbf{y}^{\nu-1})}{\partial y_1^{(\nu-1)}} .. \frac{\partial v\,(\mathbf{y}^{\nu-1})}{\partial y_N^{(\nu-1)}}\right]^T$$

$In\,\mathfrak{T}_{0F}$ the interior of $\mathfrak{T}_{0F}$

$$In\,\mathfrak{T}_{0F} = \{t: \; t \in Cl\mathfrak{T}, \; 0 < t < t_F\} \tag{A.2}$$

$Im\lambda_i(A)$ the imaginary part of the eigenvalue $\lambda_i(A)$ of the matrix A

$\min(\delta,\Delta)$ denotes the smaller between δ and Δ

$$\min(\delta,\Delta) = \left\{\begin{array}{l} \delta, \; \delta \leq \Delta, \\ \Delta, \; \Delta \leq \delta \end{array}\right\}$$

$NumF(s)$ the numerator matrix polynomial of the real rational matrix $F(s) = [DenF(s)]^{-1}\,NumF(s)$, or $NumF(s)\,[DenF(s)]^{-1}$

$Re\lambda_i(A)$ the real part of the eigenvalue $\lambda_i(A)$ of the matrix A

$\exists$ there exist(s)

$\exists!$ there exists exactly one

$\in$ belong(s) to, are (is) members (a member) of, respectively

$\subset$ *a proper subset of* (it can not be equal to)

$\subseteq$ *a subset of* (it can be equal to)

$\simeq$ *equivalent*

$\sqrt{-1}$ *the imaginary unit* denoted by i, $i = \sqrt{-1}$

inf   *infimum*

max   *maximum*

min   *minimum*

numx   *the numerical value of* x, if $x = 50V$ then *numx* = 50

phdim $x(.)$ *the physical dimension of a variable* $x(.)$

$$x(.) = t \implies phdim\ x(.) = phdim\ t = \mathrm{T},\ but\ dim\ t = 1$$

sup   *supremum*

$\|.\|$ can be any *norm* if not otherwise specified

A.6   Units

$1_{(.)}$ *the unit of a physical variable* (.)

1_t *the time unit of the reference time axis* T, $1_t = s$

Appendix B

Equivalent definitions

B.1 Equivalent tracking definitions

B.1.1 Equivalent imperfect tracking definitions

$\mathfrak{N}_\varepsilon \left[\mathbf{0}_{(k+1)N}; \mathbf{D}; \mathbf{U}; \mathbf{Y}_d^k \right]$ is a time-invariant connected ε-neighborhood of the extended zero error vector $\mathbf{e}^k = \mathbf{0}_{(k+1)N}$, which is the open hyperball $\mathfrak{B}_\varepsilon$ centered at the origin $\mathbf{0}_{(k+1)N}$ with the radius ε,

and

$\mathfrak{N} \left(\varepsilon; \mathbf{0}_{(k+1)N}; \mathbf{D}; \mathbf{U}; \mathbf{Y}_d^k \right)$ is a *time*-invariant connected neighborhood of the extended zero error vector $\mathbf{e}^k = \mathbf{0}_{(k+1)N}$, which is determined by $\varepsilon \in \mathfrak{R}^+$, $\mathbf{D}(.) \in \mathfrak{D}^j$, $\mathbf{U}(.) \in \mathfrak{U}_d^l$, and $\mathbf{Y}_d(.) \in \mathfrak{Y}_d^k$, the outer radius [175, Definition 32, p. 15] of which cannot be greater than ε, and obeys the following:

$$\forall \mathfrak{N}_\varepsilon \left[\mathbf{0}_{(k+1)N}; \mathbf{D}; \mathbf{U}; \mathbf{Y}_d^k \right] \subseteq \mathfrak{R}^{(k+1)N} \Longrightarrow$$

$$\exists \mathfrak{N} \left(\varepsilon; \mathbf{0}_{(k+1)N}; \mathbf{D}; \mathbf{U}; \mathbf{Y}_d^k \right) \subseteq \mathfrak{R}^{(k+1)N}, \ 0 < \varepsilon_1 < \varepsilon_2 \Longrightarrow$$

$$\mathfrak{N} \left(\varepsilon_1; \mathbf{0}_{(k+1)N}; \mathbf{D}; \mathbf{U}; \mathbf{Y}_d^k \right) \subseteq \mathfrak{N} \left(\varepsilon_2; \mathbf{0}_{(k+1)N}; \mathbf{Y}_{d0}^k; \mathbf{D}; \mathbf{U}; \mathbf{Y}_d^k \right),$$

$$\varepsilon \longrightarrow 0^+ \Longrightarrow \mathfrak{N} \left(\varepsilon; \mathbf{0}_{(k+1)N}; \mathbf{Y}_{d0}^k; \mathbf{D}; \mathbf{U}; \mathbf{Y}_d^k \right) \longrightarrow \left\{ \mathbf{0}_{(k+1)N} \right\},$$

$$\mathfrak{N} \left(\varepsilon; \mathbf{0}_{(k+1)N}; \mathbf{D}; \mathbf{U}; \mathbf{Y}_d^k \right) \subseteq \mathfrak{N}_\varepsilon \left[\mathbf{0}_{(k+1)N}; \mathbf{D}; \mathbf{U}; \mathbf{Y}_d^k \right], \qquad \text{(B.1)}$$

The neighborhood $\mathfrak{N} \left(\varepsilon; \mathbf{0}_{(k+1)N}; \mathbf{D}; \mathbf{U}; \mathbf{Y}_d^k \right)$ can be accepted to be the δ-hyperball $\mathfrak{B}_{\delta(\varepsilon)}$ centered at the $\mathbf{e}^k = \mathbf{0}_{(k+1)N}$ with the radius δ for $0 < \delta =$

$\delta\left(\varepsilon\right)\leq\varepsilon,\ \forall\varepsilon\in\mathfrak{R}^{+}$. In general

$$\mathfrak{B}_{\delta(\varepsilon)}\left(\mathbf{0}_{(k+1)N};\mathbf{D};\mathbf{U};\mathbf{Y}_{d}^{k}\right)\subseteq\mathfrak{N}\left(\varepsilon;\mathbf{0}_{(k+1)N};\mathbf{D};\mathbf{U};\mathbf{Y}_{d}^{k}\right)\subseteq$$

$$\subseteq\mathfrak{B}_{\varepsilon}\left(\mathbf{0}_{(k+1)N};\mathbf{D};\mathbf{U};\mathbf{Y}_{d}^{k}\right).$$

Definition 305 *Tracking of the extended desired output behavior* $\mathbf{Y}_{d}^{k}(t)$ *on the set product* $\mathfrak{D}^{j}\times\mathfrak{Y}_{d}^{k}$ *of the system controlled by a control* $\mathbf{U}(.)\in\mathfrak{U}^{l}$

a) *The system exhibits* **the asymptotic output tracking of** $\mathbf{Y}_{d}^{k}(t)$, $\mathbf{Y}_{d}(.)\in\mathfrak{Y}_{d}^{k}$, *on* $\mathfrak{D}^{j}\times\mathfrak{Y}_{d}^{k}$, *for short* **the tracking of** $\mathbf{Y}_{d}^{k}(t)$ *on* $\mathfrak{D}^{j}\times\mathfrak{Y}_{d}^{k}$, *if and only if for every* $[\mathbf{D}(.),\mathbf{Y}_{d}(.)]\in\mathfrak{D}^{j}\times\mathfrak{Y}_{d}^{k}$ *there exists a connected neighborhood* $\mathfrak{N}\left(\mathbf{0}_{(k+1)N};\mathbf{D};\mathbf{U};\mathbf{Y}_{d}^{k}\right)\subseteq\mathfrak{R}^{(k+1)N}$ *of the plant extended zero output error vector* $\mathbf{e}^{k}=\mathbf{0}_{(k+1)N}$ *and for every* $\varsigma>0$ *there exists a non-negative real number* τ, $\tau=\tau\left(\varsigma,\mathbf{0}_{(k+1)N};\mathbf{D};\mathbf{U};\mathbf{Y}_{d}^{k}\right)\in\mathfrak{R}_{+}$, *such that* $\mathbf{e}_{0}^{k}$ *from* $\mathfrak{N}\left(\mathbf{0}_{(k+1)N};\mathbf{D};\mathbf{U};\mathbf{Y}_{d}^{k}\right)$ *guarantees that the extended output error vector* $\mathbf{e}^{k}(t;\mathbf{e}_{0}^{k};\mathbf{D};\mathbf{U})$ *belongs to the* ς*-neighborhood* $\mathfrak{N}_{\varsigma}\left(t,\mathbf{0}_{(k+1)N};\mathbf{D};\mathbf{U};\mathbf{Y}_{d}^{k}\right)$ *of* $\mathbf{e}^{k}(t;\mathbf{e}_{0}^{k})$ *for all time*

$$t\in]\tau\left(\varsigma,\mathbf{0}_{(k+1)N};\mathbf{D};\mathbf{U};\mathbf{Y}_{d}^{k}\right),\infty[,$$

i.e.,

$$\forall\varsigma>0,\ \forall[\mathbf{D}(.),\mathbf{Y}_{d}(.)]\in\mathfrak{D}^{j}\times\mathfrak{Y}_{d}^{k},$$

$$\exists\mathfrak{N}\left(\mathbf{0}_{(k+1)N};\mathbf{D};\mathbf{U};\mathbf{Y}_{d}^{k}\right)\subseteq\mathfrak{R}^{(k+1)N},\ \mathbf{e}_{0}^{k}\in\mathfrak{N}\left(\mathbf{0}_{(k+1)N};\mathbf{D};\mathbf{U};\mathbf{Y}_{d}^{k}\right)\implies$$

$$\left\{\begin{array}{c}\mathbf{e}^{k}(t;\mathbf{e}_{0}^{k};\mathbf{D};\mathbf{U})\in\mathfrak{N}_{\varsigma}\left(t,\mathbf{0}_{(k+1)N};\mathbf{D};\mathbf{U};\mathbf{Y}_{d}^{k}\right),\\\forall t\in]\tau\left(\varsigma,\mathbf{0}_{(k+1)N};\mathbf{D};\mathbf{U};\mathbf{Y}_{d}^{k}\right),\ \infty[\end{array}\right\}.\qquad\text{(B.2)}$$

The zero $(k=0)$ *order tracking is simply called* **the tracking**.

b) *The largest connected neighborhood* $\mathfrak{N}\left(\mathbf{0}_{(k+1)N};\mathbf{D};\mathbf{U};\mathbf{Y}_{d}^{k}\right)$ *of* $\mathbf{e}^{k}=\mathbf{0}_{(k+1)N}$ *that obeys (B.2), is* **the** k**-th-order tracking domain**

$$\mathcal{D}_{T}^{k}\left(\mathbf{0}_{(k+1)N};\mathbf{D};\mathbf{U};\mathbf{Y}_{d}^{k}\right)$$

of $\mathbf{e}^{k}=\mathbf{0}_{(k+1)N}$ *for every* $[\mathbf{D}(.),\mathbf{Y}_{d}(.)]\in\mathfrak{D}^{j}\times\mathfrak{Y}_{d}^{k}$, *i.e., on* $\mathfrak{D}^{j}\times\mathfrak{Y}_{d}^{k}$.

c) *The tracking of* $\mathbf{Y}_{d}^{k}(t)$ *on* $\mathfrak{D}^{j}\times\mathfrak{Y}_{d}^{k}$ *is* **global** *(in the whole) if and only if* $\mathcal{D}_{T}^{k}\left(\mathbf{Y}_{d0}^{k};\mathbf{D};\mathbf{U};\mathbf{Y}_{d}^{k}\right)=\mathfrak{R}^{(k+1)N}$ *for every* $[\mathbf{D}(.),\mathbf{Y}_{d}(.)]\in\mathfrak{D}^{j}\times\mathfrak{Y}_{d}^{k}$.

Definition 306 *Stablewise tracking of the extended desired output behavior* $\mathbf{Y}_d^k(t)$ *on* $\mathfrak{D}^j \times \mathfrak{Y}_d^k$ *of the system controlled by a control* $\mathbf{U}(.) \in \mathfrak{U}^l$

a) The system exhibits **stablewise output tracking of** $\mathbf{Y}_d^k(t)$ *on* $\mathfrak{D}^j \times \mathfrak{Y}_d^k$, *for short* **the stablewise tracking of** $\mathbf{Y}_d^k(t)$ *on* $\mathfrak{D}^j \times \mathfrak{Y}_d^k$, *if and only if it exhibits tracking of* $\mathbf{Y}_d^k(t)$ *on* $\mathfrak{D}^j \times \mathfrak{Y}_d^k$, *and for every connected neighborhood* $\mathfrak{N}_\varepsilon\left(\mathbf{0}_{(k+1)N}\right) = \mathfrak{B}_\varepsilon \subseteq \mathfrak{R}^{(k+1)N}$, *of* $\mathbf{0}_{(k+1)N}$, *there is a connected neighborhood* $\mathfrak{N}\left(\varepsilon; \mathbf{0}_{(k+1)N}; \mathbf{D}; \mathbf{U}; \mathbf{Y}_d^k\right)$, *(B.1), of the plant extended zero output error vector* $\mathbf{e}_0^k = \mathbf{0}_{(k+1)N}$ *such that for an initial* $\mathbf{e}_0^k \in \mathfrak{N}\left(\varepsilon; \mathbf{0}_{(k+1)N}; \mathbf{D}; \mathbf{U}; \mathbf{Y}_d^k\right)$ *the instantaneous* $\mathbf{Y}^k(t)$ *stays in* $\mathfrak{N}_\varepsilon\left(\mathbf{0}_{(k+1)N}\right) = \mathfrak{B}_\varepsilon$ *for all* $t \in \mathfrak{T}_0$; *i.e.*,

$$\forall \mathfrak{N}_\varepsilon\left(\mathbf{0}_{(k+1)N}\right) = \mathfrak{B}_\varepsilon \subseteq \mathfrak{R}^{(k+1)N},$$

$$\forall \left[\mathbf{D}(.), \mathbf{Y}_d(.)\right] \in \mathfrak{D}^j \times \mathfrak{Y}_d^k, \ \exists \mathfrak{N}\left(\varepsilon; \mathbf{0}_{(k+1)N}; \mathbf{D}; \mathbf{U}; \mathbf{Y}_d^k\right) \subseteq \mathfrak{R}^{(k+1)N},$$

$$\mathfrak{N}\left(\varepsilon; \mathbf{0}_{(k+1)N}; \mathbf{D}; \mathbf{U}; \mathbf{Y}_d^k\right) \subseteq \mathcal{D}_T^k\left(\mathbf{0}_{(k+1)N}; \mathbf{D}; \mathbf{U}; \mathbf{Y}_d^k\right) \cap \mathfrak{B}_\varepsilon,$$

$$\mathbf{e}_0^k \in \mathfrak{N}\left(\varepsilon; \mathbf{0}_{(k+1)N}; \mathbf{D}; \mathbf{U}; \mathbf{Y}_d^k\right) \implies$$

$$\mathbf{e}_0^k \in \mathfrak{N}\left(\varepsilon; \mathbf{0}_{(k+1)N}; \mathbf{D}; \mathbf{U}; \mathbf{Y}_d^k\right) \implies \mathbf{e}^k(t; \mathbf{e}_0^k; \mathbf{D}; \mathbf{U}) \in \mathfrak{B}_\varepsilon, \ \forall t \in \mathfrak{T}_0. \quad \text{(B.3)}$$

b) The largest connected neighborhood $\mathfrak{N}_L\left(\varepsilon; \mathbf{0}_{(k+1)N}; \mathbf{D}; \mathbf{U}; \mathbf{Y}_d^k\right)$ *of the extended zero output error vector* $\mathbf{e}^k = \mathbf{0}_{(k+1)N}$, *(B.3), is* **the** ε-**tracking domain** *denoted by* $\mathcal{D}_{ST}^k\left(\varepsilon; \mathbf{0}_{(k+1)N}; \mathbf{D}; \mathbf{U}; \mathbf{Y}_d^k\right)$ *of the stablewise tracking of* $\mathbf{e}^k = \mathbf{0}_{(k+1)N}$ *on* $\mathfrak{D}^j \times \mathfrak{Y}_d^k$.

The domain $\mathcal{D}_{ST}^k\left(\mathbf{0}_{(k+1)N}; \mathbf{D}; \mathbf{U}; \mathbf{Y}_d^k\right)$ *of the stablewise tracking of* $\mathbf{e}^k = \mathbf{0}_{(k+1)N}$ *on* $\mathfrak{D}^j \times \mathfrak{Y}_d^k$ *is the union of all* $\mathcal{D}_{ST}^k\left(\varepsilon; \mathbf{0}_{(k+1)N}; \mathbf{D}; \mathbf{U}; \mathbf{Y}_d^k\right)$ *over* $\varepsilon \in \mathfrak{R}^+$,

$$\mathcal{D}_{ST}^k\left(\mathbf{0}_{(k+1)N}; \mathbf{D}; \mathbf{U}; \mathbf{Y}_d^k\right) = \cup \left[\mathcal{D}_{ST}^k\left(\varepsilon; \mathbf{0}_{(k+1)N}; \mathbf{D}; \mathbf{U}; \mathbf{Y}_d^k\right) : \varepsilon \in \mathfrak{R}^+\right].$$
(B.4)

Let $[0, \varepsilon_M)$ *be the maximal interval over which* $\mathcal{D}_{ST}^k\left(\varepsilon; \mathbf{Y}_{d0}^k; \mathbf{D}; \mathbf{U}; \mathbf{Y}_d^k\right)$ *is continuous in* $\varepsilon \in \mathfrak{R}_+$,

$$\mathcal{D}_{ST}^k\left(\varepsilon; \mathbf{Y}_{d0}^k; \mathbf{D}; \mathbf{U}; \mathbf{Y}_d^k\right) \in \mathfrak{C}\left([0, \varepsilon_M)\right), \ \left[\mathbf{D}(.), \mathbf{Y}_d(.)\right] \in \mathfrak{D}^j \times \mathfrak{Y}_d^k.$$

The **strict domain** $\mathcal{D}_{SST}^k\left(\mathbf{Y}_{d0}^k; \mathbf{D}; \mathbf{U}; \mathbf{Y}_d^k\right)$ *of the stablewise k-th-order tracking of* $\mathbf{Y}_d(t)$ *on* $\mathfrak{D}^j \times \mathfrak{Y}_d^k$ *is the union of all stablewise tracking domains*

$$\mathcal{D}_{ST}^k\left(\varepsilon; \mathbf{Y}_{d0}^k; \mathbf{D}; \mathbf{U}; \mathbf{Y}_d^k\right):$$

over $\varepsilon \in [0, \varepsilon_M)$,

$$\mathcal{D}_{SST}^k \left(\mathbf{Y}_{d0}^k; \mathbf{D}; \mathbf{U}; \mathbf{Y}_d^k \right) = \cup \left\{ \mathcal{D}_{ST}^k \left(\varepsilon; \mathbf{Y}_{d0}^k; \mathbf{D}; \mathbf{U}; \mathbf{Y}_d^k \right) : \varepsilon \in [0, \varepsilon_M) \right\},$$
$$[\mathbf{D}(.), \mathbf{Y}_d(.)] \in \mathfrak{D}^j \times \mathfrak{Y}_d^k. \tag{B.5}$$

*c) The stablewise tracking of $\mathbf{Y}_d^k(t)$ on $\mathfrak{D}^j \times \mathfrak{Y}_d^k$ is **global (in the whole)** if and only if it is both the global tracking of $\mathbf{Y}_d^k(t)$ on $\mathfrak{D}^j \times \mathfrak{Y}_d^k$ and the stablewise tracking of $\mathbf{Y}_d^k(t)$ on $\mathfrak{D}^j \times \mathfrak{Y}_d^k$ with $\mathcal{D}_{ST}^k \left(\mathbf{Y}_{d0}^k; \mathbf{D}; \mathbf{U}; \mathbf{Y}_d^k \right) = \mathfrak{R}^{(k+1)N}$ for every $[\mathbf{D}(.), \mathbf{Y}_d(.)] \in \mathfrak{D}^j \times \mathfrak{Y}_d^k$.*

B.1.2   Equivalent exponential tracking definition

Definition 307 ***Exponential output tracking of* $\mathbf{Y}_d^k(t)$ *on* $\mathfrak{D}^j \times \mathfrak{Y}_d^k$ *of the system controlled by a control* $\mathbf{U}(.) \in \mathfrak{U}^l$**

*a) The system exhibits **exponential asymptotic output tracking of** $\mathbf{Y}_d^k(t)$, $\mathbf{Y}_d^k(.) \in \mathfrak{Y}_d^k$, on $\mathfrak{D}^j \times \mathfrak{Y}_d^k$, for short the **exponential tracking of** $\mathbf{Y}_d^k(t)$ **on** $\mathfrak{D}^j \times \mathfrak{Y}_d^k$ if and only if for every $[\mathbf{D}(.), \mathbf{Y}_d(.)] \in \mathfrak{D}^j \times \mathfrak{Y}_d^k$ there exist positive real numbers $a \geq 1$ and $b > 0$, and a connected neighborhood $\mathfrak{N} \left(\mathbf{0}_{(k+1)N}; a, b; \mathbf{D}; \mathbf{U}; \mathbf{Y}_d^k \right)$ of $\mathbf{e}^k = \mathbf{0}_{(k+1)N}$, $a = a(\mathbf{D}, \mathbf{U}, \mathbf{Y}_d^k)$ and $b = b(\mathbf{D}, \mathbf{U}, \mathbf{Y}_d^k)$, such that $\mathbf{e}_0^k \in \mathfrak{N} \left(\mathbf{0}_{(k+1)N}; a, b; \mathbf{D}; \mathbf{U}; \mathbf{Y}_d^k \right)$ guarantees that $\mathbf{e}^k(t)$ approaches exponentially $\mathbf{e}^k = \mathbf{0}_{(k+1)N}$ all the time; i.e.,*

$$\forall [\mathbf{D}(.), \mathbf{Y}_d(.)] \in \mathfrak{D}^j \times \mathfrak{Y}_d^k, \ \exists a \in [1, \infty[, \ \exists b \in \mathfrak{R}^+,$$
$$a = a(\mathbf{D}, \mathbf{U}, \mathbf{Y}_d^k), \quad b = b(\mathbf{D}, \mathbf{U}, \mathbf{Y}_d^k),$$
$$\exists \mathfrak{N} \left(\mathbf{0}_{(k+1)N}; a, b; \mathbf{D}; \mathbf{U}; \mathbf{Y}_d^k \right),$$
$$\mathbf{e}_0^k \in \mathfrak{N} \left(\mathbf{0}_{(k+1)N}; a, b; \mathbf{D}; \mathbf{U}; \mathbf{Y}_d^k \right) \implies$$
$$\left\| \mathbf{e}^k(t; \mathbf{e}_0^k; \mathbf{D}; \mathbf{U}) \right\| \leq a \left\| \mathbf{e}_0^k \right\| exp\,(-bt), \quad \forall t \in \mathfrak{T}_0. \tag{B.6}$$

*b) The largest connected neighborhood $\mathfrak{N} \left(\mathbf{Y}_{d0}^k; a, b; \mathbf{D}; \mathbf{U}; \mathbf{Y}_d^k \right)$ of $\mathbf{Y}_{d0}^k$ is the **domain** $\mathcal{D}^k \left(\mathbf{Y}_{d0}^k; a, b; \mathbf{D}; \mathbf{U}; \mathbf{Y}_d^k \right)$ **of the exponential tracking of** $\mathbf{Y}_d^k(t)$ **on** $\mathfrak{D}^j \times \mathfrak{Y}_d^k$ relative to a and b. When a and b are fixed then they can be omitted,*

$$\mathcal{D}^k \left(\mathbf{Y}_{d0}^k; a, b; \mathbf{D}; \mathbf{U}; \mathbf{Y}_d^k \right) = \mathcal{D}^k \left(\mathbf{Y}_{d0}^k; \mathbf{D}; \mathbf{U}; \mathbf{Y}_d^k \right).$$

*c) The exponential tracking of $\mathbf{Y}_d^k(t)$ on $\mathfrak{D}^j \times \mathfrak{Y}_d^k$ is **global (in the whole)** if and only if the domain $\mathcal{D}^k \left(\mathbf{Y}_{d0}^k; \mathbf{D}; \mathbf{U}; \mathbf{Y}_d^k \right) = \mathfrak{R}^{(k+1)N}$ for every $[\mathbf{D}(.), \mathbf{Y}_d(.)] \in \mathfrak{D}^j \times \mathfrak{Y}_d^k$.*

B.2 Equivalent trackability definitions

B.2.1 Equivalent definitions of perfect trackability

We present the following due to [175, Chapter 6, pp. 65-68], [188, Definition 235, p. 171]:

Definition 308 *Definition of the l-th order perfect trackability and elementwise perfect trackability of $\mathbf{Y}_d(.)$ for the given $\mathbf{D}(.)$*

a) The desired output vector function $\mathbf{Y}_d(.)$ of the m-th order plant is the l-th order perfect trackable under the action of the given $\mathbf{D}(.)$ if and only if there exists a control vector function $\mathbf{U}(.)$ such that the plant real output extended error vector $\mathbf{e}^l(t)$ is always equal to the extended zero output error vector $\mathbf{e} = \mathbf{0}_{(l+1)N}$, as soon as $\mathbf{e}^{m-1}(0) = \mathbf{0}_{mN}$,

$$\text{given } \mathbf{D}(.), \ \exists \mathbf{U}(.) \text{ and } \mathbf{e}_0^{m-1} = \mathbf{0}_{mN} \Longrightarrow \mathbf{e}^l(t) = \mathbf{0}_{(l+1)N}, \ \forall t \in \mathfrak{T}_0. \quad (B.7)$$

The zero order ($l = 0$) perfect trackability is called simply **perfect trackability**.

b) **The perfect trackability is elementwise** *if and only if the control vector $\mathbf{U}$ can act simultaneously on every entry Y_i, $\forall i = 1, 2, \ldots, N$, of $\mathbf{Y}$ mutually independently at every $t \in \mathfrak{T}_0$.*

Definition 309 *The l-th order perfect trackability and elementwise perfect trackability of the plant on $\mathfrak{D}^i \times \mathfrak{Y}_d^k$*

a) The m-th order dynamical plant is the l-th order perfect trackable on $\mathfrak{D}^i \times \mathfrak{Y}_d^k$ if and only if for every $[\mathbf{D}(.), \mathbf{Y}_d(.)] \in \mathfrak{D}^i \times \mathfrak{Y}_d^k$ there exists a control vector function $\mathbf{U}(.)$ such that the plant real output extended error vector $\mathbf{e}^l(t)$ is always equal to the extended zero output error vector $\mathbf{e}^l = \mathbf{0}_{(l+1)N}$, as soon as $\mathbf{e}^{m-1}(0) = \mathbf{0}_{mN}$:

$$\mathbf{e}^{m-1}(0) = \mathbf{0}_{mN}, \ \forall [\mathbf{D}(.), \mathbf{Y}_d(.)] \in \mathfrak{D}^i \times \mathfrak{Y}_d^k,$$
$$\exists \mathbf{U}(.) \Longrightarrow \mathbf{e}^l(t) = \mathbf{0}_{(l+1)N}, \ \forall t \in \mathfrak{T}_0, \quad (B.8)$$

The zero order ($l = 0$) perfect trackability on $\mathfrak{D}^i \times \mathfrak{Y}_d^k$ is called simply **perfect trackability on $\mathfrak{D}^i \times \mathfrak{Y}_d^k$**.

b) **The perfect trackability is elementwise** *if and only if the control vector $\mathbf{U}$ can act on every entry Y_i, $\forall i = 1, 2, \ldots, N$, of $\mathbf{Y}$ mutually independently.*

Definition 310 *The l-th order perfect natural trackability and elementwise perfect natural trackability on* $\mathfrak{D}^i \times \mathfrak{Y}_d^k$

a) The m-th order dynamical plant is **the l-th order perfect natural trackable on** $\mathfrak{D}^i \times \mathfrak{Y}_d^k$ *if and only if for every pair* $[\mathbf{D}(.), \mathbf{Y}_d(.)] \in \mathfrak{D}^i \times \mathfrak{Y}_d^k$ *there exists a control vector function* $\mathbf{U}(.)$ *obeying TCUP on* $\mathfrak{T}_0$, *which can be synthesized without using information about the form and the value of any* $\mathbf{D}(.) \in \mathfrak{D}^i$ *and about the plant state, such that the plant real output extended error vector* $\mathbf{e}^l(t)$ *is always equal to the extended zero output error vector* $\mathbf{e} = \mathbf{0}_{(l+1)N}$, *as soon as* $\mathbf{e}^{m-1}(0) = \mathbf{0}_{mN}$, *i.e., that (9.3) holds,*

$$\mathbf{e}_0^{m-1} = \mathbf{0}_{mN}, \ \forall [\mathbf{D}(.), \mathbf{Y}_d(.)] \in \mathfrak{D}^i \times \mathfrak{Y}_d^k \Longrightarrow$$
$$\exists \mathbf{U}(.) \in \mathfrak{C}(\mathfrak{T}_0) \Longrightarrow \mathbf{e}^l(t) = \mathbf{0}_{(l+1)N}, \ \forall t \in \mathfrak{T}_0. \tag{B.9}$$

The zero order $(l = 0)$ *perfect natural trackability on* $\mathfrak{D}^i \times \mathfrak{Y}_d^k$ *is simply called* **the perfect natural trackability on** $\mathfrak{D}^i \times \mathfrak{Y}_d^k$.

b) **The perfect natural trackability is elementwise** *if and only if the control vector* $\mathbf{U}$ *can act simultaneously on every entry* Y_i, $\forall i = 1, 2, ..., N$, *of* $\mathbf{Y}$ *mutually independently.*

B.2.2 Equivalent definition of imperfect trackability

Definition 311 *The l-th order trackability on* $\mathfrak{D}^i \times \mathfrak{Y}_d^k$

a) The m-th order dynamical plant is **the l-th order trackable on** $\mathfrak{D}^i \times \mathfrak{Y}_d^k$ *if and only if there is* $\Delta \in \mathfrak{R}^+$, *or* $\Delta = \infty$, *such that for every disturbance vector function* $\mathbf{D}(.) \in \mathfrak{D}^i$, *for every plant output desired response* $\mathbf{Y}_d(.) \in \mathfrak{Y}_d^k$, *and for every instant* $\sigma \in Int\ \mathfrak{T}_0$, *there is a control vector function* $\mathbf{U}(.)$ *such that for every plant extended initial output vector* $\mathbf{e}_0^{m-1}$ *in the* Δ *neighborhood of the extended zero initial output error vector* $\mathbf{e}_0^{m-1} = 0_{mN}$, *the extended real output error vector* $\mathbf{e}^l(t)$ *becomes equal to the zero extended output error vector* $\mathbf{0}_{(l+1)N}$ *at latest at the moment* σ, *after which they rest equal forever, i.e.,*

$$\exists \Delta \in]0, \ \infty], \ \forall [\mathbf{D}(.), \mathbf{Y}_d(.)] \in \mathfrak{D}^i \times \mathfrak{Y}_d^k, \ \forall \sigma \in Int\ \mathfrak{T}_0,$$
$$\exists \mathbf{U}(.), \ \mathbf{U}(t) = \mathbf{U}(t; \sigma; \mathbf{D}; \mathbf{Y}_d) \Longrightarrow$$
$$\left\| \mathbf{e}_0^{m-1} \right\| < \Delta \Longrightarrow \mathbf{e}^l(t) = \mathbf{0}_{(l+1)}, \ \forall (t \geq \sigma) \in \mathfrak{T}_0. \tag{B.10}$$

Such control is **the l-th order tracking control on** $\mathfrak{D}^i \times \mathfrak{Y}_d^k$, *for short,* **the l-th order tracking control.**

The zero, $(l = 0)$, *order trackability on* $\mathfrak{D}^i \times \mathfrak{Y}_d^k$ *is simply called* **trackability on** $\mathfrak{D}^i \times \mathfrak{Y}_d^k$.

*The zero, (l = 0), order tracking control on $\mathfrak{D}^i \times \mathfrak{Y}_d^k$ is simply called **the tracking control on $\mathfrak{D}^i \times \mathfrak{Y}_d^k$**, for short, **the tracking control**.*

*b) The l-th order trackability on $\mathfrak{D}^i \times \mathfrak{Y}_d^k$ is **global (in the whole)** if and only if $\Delta = \infty$.*

*c) The l-th order trackability on $\mathfrak{D}^i \times \mathfrak{Y}_d^k$ is **uniform** over $\mathfrak{D}^i \times \mathfrak{Y}_d^k$ if and only if $\mathbf{U}(.)$ depends on $\mathfrak{D}^i \times \mathfrak{Y}_d^k$ but not on an individual pair $[\mathbf{D}(.), \mathbf{Y}_d(.)]$ from $\mathfrak{D}^i \times \mathfrak{Y}_d^k$, $\mathbf{U}(t) = \mathbf{U}(t; \sigma; \mathfrak{D}^k; \mathfrak{Y}_d^k)$.*

*d) If and only if, additionally to a), the output variables are **mutually functionally interrelated** by P functional constraints, $P < N$, then the l-th order trackability on $\mathfrak{D}^i \times \mathfrak{Y}_d^k$ of $\mathbf{Y}_d(.)$ is **incomplete with $f = N - P$ degrees of freedom**.*

*If and only if, additionally to a), all output variables can be controlled simultaneously mutually independently $(P = 0)$, then the l-th order trackability on $\mathfrak{D}^i \times \mathfrak{Y}_d^k$ of $\mathbf{Y}_d(.)$ is **with complete, i.e., with N, degrees of freedom**, for short, it is **complete**.*

B.2.3 Equivalent definition of natural trackability

Definition 312 *The l-th order natural trackability on $\mathfrak{D}^i \times \mathfrak{Y}_d^k$ [175, Deinition 171, pp. 83-85], [188, Definition 259, p. 180]*

*a) The m-th order dynamical plant is **the l-th order natural trackable on $\mathfrak{D}^i \times \mathfrak{Y}_d^k$** if and only if there is $\Delta \in \mathfrak{R}^+$, or $\Delta = \infty$, such that for every disturbance vector function $\mathbf{D}(.) \in \mathfrak{D}^i$, for every plant output desired response $\mathbf{Y}_d(.) \in \mathfrak{Y}_d^k$, and for every instant $\sigma \in Int\ \mathfrak{T}_0$, there is a control vector function $\mathbf{U}(.)$ obeying $TCUP$ on $\mathfrak{T}_0$, which can be synthesized without using information about the form and the value of $\mathbf{D}(.) \in \mathfrak{D}^i$ and about the plant state, such that for every plant extended initial output error vector $\mathbf{e}_0^{m-1}$ in the Δ neighborhood of the extended zero output error vector $\mathbf{e}^{m-1} = \mathbf{0}_{mN}$, the extended real output error vector $\mathbf{e}^l(t)$ becomes equal to the extended zero output error vector $\mathbf{e}^l = \mathbf{0}_{(l+1)N}$ at latest at the moment σ, after which they rest equal forever, i.e.,*

$$\exists \Delta \in]0,\ \infty], \ \forall [\mathbf{D}(.), \mathbf{Y}_d(.)] \in \mathfrak{D}^i \times \mathfrak{Y}_d^k, \ \forall \sigma \in Int\ \mathfrak{T}_0, \ \sigma \longrightarrow 0^+,$$
$$\exists \mathbf{U}(.), \ \mathbf{U}(t) = \mathbf{U}(t; \sigma; \mathbf{Y}_d) \in \mathfrak{C}(\mathfrak{T}_0) \Longrightarrow$$
$$\left\| \mathbf{e}_0^{m-1} \right\| < \Delta \Longrightarrow \mathbf{e}^l(t) = \mathbf{0}_{(l+1)N}, \ \forall (t \geq \sigma) \in \mathfrak{T}_0. \tag{B.11}$$

*Such control is **the l-th order natural tracking control on $\mathfrak{D}^i \times \mathfrak{Y}_d^k$**, for short **the l-th order natural tracking control**.*

*The zero, (l = 0), order natural trackability on $\mathfrak{D}^i \times \mathfrak{Y}_d^k$ is called **natural trackability on $\mathfrak{D}^i \times \mathfrak{Y}_d^k$**.*

*The zero, (l = 0), order natural tracking control on $\mathfrak{D}^i \times \mathfrak{Y}_d^k$ is called for short **natural tracking control on** $\mathfrak{D}^i \times \mathfrak{Y}_d^k$, or shorter **natural tracking control (NTC)**.*

*b) The l-th order natural trackability on $\mathfrak{D}^i \times \mathfrak{Y}_d^k$ is **global (in the whole)** if and only if $\Delta = \infty$.*

*c) The l-th order natural trackability on $\mathfrak{D}^i \times \mathfrak{Y}_d^k$ is **uniform** over $\mathfrak{Y}_d^k$ if and only if control $\mathbf{U}(.)$ depends on $\mathfrak{Y}_d^k$ but not on an individual $\mathbf{Y}_d(.)$ from $\mathfrak{Y}_d^k$, $\mathbf{U}(t) = \mathbf{U}(t; \sigma; \mathfrak{Y}_d^k)$.*

*d) If and only if, additionally to a), the output variables are **mutually functionally interrelated** by P functional constraints then the l-th-order natural trackability on $\mathfrak{D}^i \times \mathfrak{Y}_d^k$ of $\mathbf{Y}_d(.)$ is **incomplete with** $f = N - P$ **degrees of freedom,**.*

*If and only if, additionally to a), all output variables can be controlled simultaneously mutually independently, $(P = 0)$, then the l-th-order natural trackability on $\mathfrak{D}^i \times \mathfrak{Y}_d^k$ of $\mathbf{Y}_d(.)$ is **complete**.*

*The term **incomplete** means **incomplete (f<N) degrees of freedom**.*

*The term **complete** means **complete**, i.e., **full, (f=N), degrees of freedom**.*

B.2.4   Equivalent definition of elementwise trackability

Definition 313 *The l-th order **elementwise trackability on** $\mathfrak{D}^i \times \mathfrak{Y}_d^k$ [175, Definition 175, pp. 86, 87], [188, Definition 263, pp 181, 182]*

*a) The m-th order dynamical plant is **the l-th order elementwise trackable on** $\mathfrak{D}^i \times \mathfrak{Y}_d^k$ if and only if there is $\boldsymbol{\Delta}^{mN} \in \mathfrak{R}^{+mN}$, or $\boldsymbol{\Delta}^{mN} = \infty \mathbf{1}_{mN}$, such that for every disturbance vector function $\mathbf{D}(.) \in \mathfrak{D}^i$, for every plant output desired response $\mathbf{Y}_d(.) \in \mathfrak{Y}_d^k$, and for every vector instant $\sigma \in (Int\ \mathfrak{T}_0)^{(l+1)N}$, there is a control vector function $\mathbf{U}(.)$ such that for every plant extended initial output error vector $\mathbf{e}_0^{m-1}$ in the $\boldsymbol{\Delta}^{mN}$ elementwise neighborhood of the plant extended zero error vector $\mathbf{e}^{m-1} = \mathbf{0}_{mN}$, the plant real extended output output error vector $\mathbf{e}^l(t)$ becomes elementwise equal to the plant extended zero error vector $\mathbf{e}^l = \mathbf{0}_{(l+1)N}$ at latest at the vector moment σ, after which they rest equal forever, i.e.,*

$$\exists \boldsymbol{\Delta}^{mN} \in]\mathbf{0}_{mN},\ \infty \mathbf{1}_{mN}],\ \forall [\mathbf{D}(.), \mathbf{Y}_d(.)] \in \mathfrak{D}^i \times \mathfrak{Y}_d^k,$$

$$\forall \sigma \in (Int\ \mathfrak{T}_0)^{(l+1)N},\ \exists \mathbf{U}(.),$$

$$\mathbf{U}(t) = \mathbf{U}(t; \sigma; \mathbf{D}; \mathbf{Y}) \in \mathfrak{C}(\mathfrak{T}_0) \implies \left| \mathbf{e}_0^{m-1} \right| < \boldsymbol{\Delta}^{mN} \implies$$

$$\mathbf{e}^l(\mathbf{t}^{(l+1)N}) = \mathbf{0}_{(l+1)N},\ \forall \left(\mathbf{t}^{(l+1)N} \geq \sigma \right) \in \mathfrak{T}_0^{(l+1)N}. \tag{B.12}$$

*Such control is **the l-th order elementwise tracking control on** $\mathfrak{D}^i \times \mathfrak{Y}_d^k$.*

*The zero, $(l = 0)$, order elementwise trackability on $\mathfrak{D}^i \times \mathfrak{Y}_d^k$ is called **elementwise trackability on** $\mathfrak{D}^i \times \mathfrak{Y}_d^k$.*

*The zero, $(l = 0)$, order elementwise tracking control on $\mathfrak{D}^i \times \mathfrak{Y}_d^k$ is called for short **elementwise tracking control on** $\mathfrak{D}^i \times \mathfrak{Y}_d^k$, or shorter, **elementwise tracking control.***

*b) The l-th order elementwise trackability on $\mathfrak{D}^i \times \mathfrak{Y}_d^k$ is **global (in the whole)** if and only if $\mathbf{\Delta}^{mN} = \infty \mathbf{1}_{mN}$.*

*The l-th order elementwise trackability on $\mathfrak{D}^i \times \mathfrak{Y}_d^k$ is **uniform** over $\mathfrak{D}^i \times \mathfrak{Y}_d^k$ if and only if $\mathbf{U}(.)$ depends on $\mathfrak{D}^i \times \mathfrak{Y}_d^k$ but not on an individual pair $[\mathbf{D}(.), \mathbf{Y}_d(.)]$ from $\mathfrak{D}^i \times \mathfrak{Y}_d^k$, $\mathbf{U}(t) = \mathbf{U}(t; \sigma; \mathfrak{D}^i; \mathfrak{Y}_d^k)$.*

B.2.5 Elementwise natural trackability

The natural trackability concept can also satisfy the demand for different reachability *times* to be associated with different output variables.

Definition 314 *The l-th order elementwise natural trackability on $\mathfrak{D}^i \times \mathfrak{Y}_d^k$*

*a) The m-th order dynamical plant is **the l-th order elementwise natural trackable on** $\mathfrak{D}^i \times \mathfrak{Y}_d^k$ if and only if there is $\mathbf{\Delta}^{mN} \in \mathfrak{R}^{+mN}$, or $\mathbf{\Delta}^{mN} = \infty \mathbf{1}_{mN}$, such that for every disturbance vector function $\mathbf{D}(.) \in \mathfrak{D}^i$, for every plant desired output response $\mathbf{Y}_d(.) \in \mathfrak{Y}_d^k$, and for every vector instant $\sigma \in (Int\ \mathfrak{T}_0)^{(l+1)N}$, there is control vector function $\mathbf{U}(.)$ obeying $TCUP$ on $\mathfrak{T}_0$, which can be synthesized without using information about the form and value of $\mathbf{D}(.) \in \mathfrak{D}^k$ and about the plant state, such that for every plant initial output vector $\mathbf{Y}_0^{m-1}$ in the $\mathbf{\Delta}^{mN}$- elementwise neighborhood of the plant initial desired output vector $\mathbf{Y}_{d0}^{m-1}$, the plant extended real output response $\mathbf{Y}^l(t)$ becomes elementwise equal to $\mathbf{Y}_d^l(t)$ at latest at the vector moment σ, after which they rest equal forever, i.e.,*

$$\exists \mathbf{\Delta}^{mN} \in]\mathbf{0}_{mN},\ \infty \mathbf{1}_{mN}],\ \forall [\mathbf{D}(.), \mathbf{Y}_d(.)] \in \mathfrak{D}^i \times \mathfrak{Y}_d^k,$$

$$\forall \sigma \in (Int\ \mathfrak{T}_0)^{(l+1)N},\ \exists \mathbf{U}(.),$$

$$\mathbf{U}(t) = \mathbf{U}(t; \sigma; \mathbf{Y}_d) \in \mathfrak{C}(\mathfrak{T}_0)\ \ and\ \ \left| \mathbf{Y}_0^{m-1} - \mathbf{Y}_{d0}^{m-1} \right| < \mathbf{\Delta}^{mN} \Longrightarrow$$

$$\mathbf{Y}^l(\mathbf{t}^{(l+1)N}) = \mathbf{Y}_d^l(\mathbf{t}^{(l+1)N})\ \forall \left(\mathbf{t}^{(l+1)N} \geq \sigma \right) \in \mathfrak{T}_0^{(l+1)N}. \tag{B.13}$$

*Such control is **the l-th order elementwise natural tracking control on** $\mathfrak{D}^i \times \mathfrak{Y}_d^k$, for short, **the l-th order elementwise natural tracking control**.*

The zero, $(l = 0)$, order elementwise natural trackability on $\mathfrak{D}^i \times \mathfrak{Y}_d^k$ is called **elementwise natural trackability on $\mathfrak{D}^i \times \mathfrak{Y}_d^k$**. The zero, $(l = 0)$, order elementwise natural tracking control on $\mathfrak{D}^i \times \mathfrak{Y}_d^k$ is called **elementwise natural tracking control on $\mathfrak{D}^i \times \mathfrak{Y}_d^k$**, for short, **elementwise natural tracking control**.

b) The l-th order elementwise natural trackability on $\mathfrak{D}^i \times \mathfrak{Y}_d^k$ is **global** (**in the whole**) if and only if $\mathbf{\Delta}^{mN} = \infty \mathbf{1}_{mN}$.

c) The l-th order elementwise natural trackability on $\mathfrak{D}^i \times \mathfrak{Y}_d^k$ is **uniform** over $\mathfrak{Y}_d^k$ if and only if $\mathbf{U}(.)$ depends on $\mathfrak{Y}_d^k$ but not on an individual $\mathbf{Y}_d(.)$ from $\mathfrak{Y}_d^k$, $\mathbf{U}(t) = \mathbf{U}(t; \sigma; \mathfrak{Y}_d^k)$.

Appendix C

Example

C.1 Example of f(.)-function

An example of the function $\mathbf{f}(.)$ (13.1)–(13.5), (Section 13.1), Figure C.1 given in [183, pp. 141-146], reads:

$$\mathbf{f(t)} = \left\{ \begin{array}{l} \mathbf{k} + A\sin\left(\Omega\mathbf{t} + \varphi\right),\ \mathbf{t} \in [\mathbf{t}_0,\ \mathbf{t}_R], \\ \mathbf{0}_N,\ \mathbf{t} \in [\mathbf{t}_R,\ \infty\mathbf{1}_N[\end{array} \right\}, \tag{C.1}$$

$$\mathbf{f(t_0)} = \mathbf{k} + A_{(..)}\sin\left(\Omega\mathbf{t}_0 + \varphi\right) = -\varepsilon_0,$$
$$\mathbf{f}^{(1)}(\mathbf{t_0}) = A\Omega\cos\left(\Omega\mathbf{t}_0 + \varphi\right) = -\varepsilon_0^{(1)}, \tag{C.2}$$

$$\mathbf{f(t_R)} = \mathbf{k} + A\sin\left(\Omega\mathbf{t}_R + \varphi\right) = \mathbf{0}_N,$$
$$\mathbf{f}^{(1)}(\mathbf{t_R}) = A\Omega\cos\left(\Omega\mathbf{t}_R + \varphi\right) = \mathbf{0}_N, \tag{C.3}$$

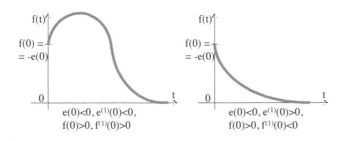

Figure C.1: The diagram of the possible form of the subsidiary function $f(.)$ if $\mathbf{e}_0 \neq \mathbf{0}_N$, Equations (13.1)-(13.5) (the scalar case).

$$\varphi = \left[\varphi_1 \vdots \varphi_2 \vdots \cdots \vdots \varphi_N \right]^T,$$

$$\omega = \left[\omega_1 \vdots \omega_2 \vdots \cdots \vdots \omega_N \right]^T \implies \Omega = diag \left\{ \omega_1 \vdots \omega_2 \vdots \cdots \vdots \omega_N \right\},$$

$$cos\left(\Omega t_R + \varphi\right) = \left[\begin{array}{l} cos\left(\omega_1 t_{R1} + \varphi_1\right) \vdots cos\left(\omega_2 t_{R2} + \varphi_2\right) \vdots \cdots \\ \qquad\qquad \cdots \vdots cos\left(\omega_N t_{RN} + \varphi_N\right) \end{array} \right]^T \quad (C.4)$$

Appendix D

Proofs

D.1 Proof of Theorem 67

Proof. Let Equation (4.4) hold and let $\mathbf{X}$ be defined by (4.5) so that

$$\nu > 1 \Longrightarrow$$
$$\mathbf{X}_1(t) = \mathbf{Y}(t),$$
$$\mathbf{X}_2(t) = \mathbf{Y}^{(1)}(t) = \mathbf{X}_1^{(1)}(t)$$
$$\mathbf{X}_3(t) = \mathbf{Y}^{(2)}(t) = \mathbf{X}_2^{(1)}(t)$$

$$\cdots \cdots$$

$$\mathbf{X}_{\nu-1}(t) = \mathbf{Y}^{(\nu-2)}(t) = \mathbf{X}_{\nu-2}^{(1)}(t)$$
$$\mathbf{X}_\nu(t) = \mathbf{Y}^{(\nu-1)}(t) = \mathbf{X}_{\nu-1}^{(1)}(t)$$
$$\nu = 1 \Longrightarrow$$
$$\mathbf{X}(t) = \mathbf{X}_1(t) = \mathbf{Y}(t). \qquad (D.1)$$

We can solve the preceding equations for the derivatives:

$$\nu > 1 \Longrightarrow$$
$$\mathbf{X}_1^{(1)}(t) = \mathbf{X}_2(t),$$
$$\mathbf{X}_2^{(1)}(t) = \mathbf{X}_3(t),$$
$$\cdot \quad \cdot \quad \cdot \quad \cdot$$
$$\mathbf{X}_{\nu-2}^{(1)}(t) = \mathbf{X}_{\nu-1}(t)$$
$$\mathbf{X}_{\nu-1}^{(1)}(t) = \mathbf{X}_\nu(t)$$
$$\nu = 1 \Longrightarrow$$
$$\mathbf{X}_1^{(1)}(t) = \mathbf{Y}^{(1)}(t). \tag{D.2}$$

The equations (D.1) and (D.2) transform IO Equation (2.1) into

$$\nu > 1 \Longrightarrow A_\nu \mathbf{X}_\nu^{(1)}(t) + \sum_{k=0}^{k=\nu-1} A_k \mathbf{X}_{k+1}(t) = H^{(\mu)} \mathbf{I}^\mu(t),$$

$$\nu = 1 \Longrightarrow A_1 \mathbf{X}_1^{(1)}(t) + A_0 \mathbf{X}_1(t) = H^{(\mu)} \mathbf{I}^{\mu:}(t) = H^{(1)} \mathbf{I}^{1:}(t),$$

which implies the following due to $det A_\nu \neq 0$ determined in (2.1):

$$\nu > 1 \Longrightarrow$$
$$\mathbf{X}_\nu^{(1)}(t) = -A_\nu^{-1} \left[\sum_{k=0}^{k=\nu-1} A_k \mathbf{X}_{k+1}(t) - H^{(\mu)} \mathbf{I}^{\mu:}(t) \right],$$

$$\nu = 1 \Longrightarrow A_1 \mathbf{X}_1^{(1)}(t) + A_0 \mathbf{X}_1(t) = H^{(\mu)} \mathbf{I}^{\mu:}(t),$$

i.e.,

$$\nu > 1 \Longrightarrow$$
$$\mathbf{X}_\nu^{(1)}(t) = \left\{ \begin{array}{l} -A_\nu^{-1} A_0 \mathbf{X}_1(t) - A_\nu^{-1} A_1 \mathbf{X}_2(t) - \ \ldots \\ \ldots - A_\nu^{-1} A_{\nu-1} \mathbf{X}_\nu(t) + A_\nu^{-1} H^{(\beta)} \mathbf{I}^\mu(t) \end{array} \right\},$$

$$\nu = 1 \Longrightarrow$$
$$\mathbf{X}_1^{(1)}(t) = -A_1^{-1} A_0 \mathbf{X}_1(t) + A_1^{-1} H^{(\mu)} \mathbf{I}^{\mu:}(t). \tag{D.3}$$

This and (D.2) yield

$$\nu > 1 \Longrightarrow$$

$$\mathbf{X}_1^{(1)}(t) = \mathbf{X}_2(t),$$

$$\mathbf{X}_2^{(1)}(t) = \mathbf{X}_3(t),$$

$$\ldots$$

$$\mathbf{X}_{\nu-2}^{(1)}(t) = \mathbf{X}_{\nu-1}(t)$$

$$\mathbf{X}_{\nu-1}^{(1)}(t) = \mathbf{X}_\nu(t)$$

$$\mathbf{X}_\nu^{(1)}(t) = \left\{ \begin{array}{c} -A_\nu^{-1}A_0\mathbf{X}_1(t) - A_\nu^{-1}A_1\mathbf{X}_2(t) - \ldots \\ \ldots - A_\nu^{-1}A_{\nu-1}\mathbf{X}_\nu(t) + A_\nu^{-1}H^{(\mu)}\mathbf{I}^{\mu:}(t) \end{array} \right\},$$

$$\nu = 1 \Longrightarrow$$

$$\mathbf{X}_1^{(1)}(t) = -A_1^{-1}A_0\mathbf{X}_1(t) + A_1^{-1}H^{(\mu)}\mathbf{I}^{\mu:}(t), \tag{D.4}$$

or,

$$\nu > 1 \Longrightarrow \begin{bmatrix} \mathbf{X}_1^{(1)}(t) \\ \mathbf{X}_2^{(1)}(t) \\ \ldots \\ \mathbf{X}_{\nu-2}^{(1)}(t) \\ \mathbf{X}_{\nu-1}^{(1)}(t) \\ \mathbf{X}_\nu^{(1)}(t) \end{bmatrix} =$$

$$= \underbrace{\begin{bmatrix} block\ diag\ \{I_{(\nu-1)N} \quad -A_\nu^{-1}\} \bullet \\ \bullet \begin{bmatrix} O_N & I_N & O_N & O_N & \ldots & O_N \\ O_N & O_N & I_N & O_N & \ldots & O_N \\ O_N & O_N & O_N & I_N & \ldots & O_N \\ \ldots & \ldots & \ldots & \ldots & \ldots & \ldots \\ O_N & O_N & O_N & O_N & \ldots & I_N \\ A_0 & A_1 & A_2 & A_3 & \ldots & A_{\nu-1} \end{bmatrix} \end{bmatrix}}_{A} \underbrace{\begin{bmatrix} \mathbf{X}_1(t) \\ \mathbf{X}_2(t) \\ \ldots \\ \mathbf{X}_{\nu-2}(t) \\ \mathbf{X}_{\nu-1}(t) \\ \mathbf{X}_\nu(t) \end{bmatrix}}_{\mathbf{X}} +$$

$$+ \underbrace{\begin{bmatrix} O_{N,(\mu+1)M} \\ O_{N,(\mu+1)M} \\ O_{N,(\mu+1)M} \\ \ldots \\ O_{N,(\mu+1)M} \\ A_\nu^{-1}H^{(\mu)} \end{bmatrix}}_{P} \mathbf{I}^\mu(t),$$

$$\nu = 1 \Longrightarrow \mathbf{X}_1^{(1)}(t) = -A_1^{-1}A_0\mathbf{X}_1(t) + A_1^{-1}H^{(\mu)}\mathbf{I}^{\mu\vdots}(t),$$

which imply (4.6)–(4.11). ■

D.2   Proof of Theorem 72

Proof. The matrix A of the *EISO* system (4.1), (4.2), (4.4)–(4.11) is determined by (4.6) .

    I)   a) Let $\nu = 1$ and $0 \leq \mu \leq 1$ due to(2.1). Then $N = n$ and the matrix $\left[(sI_n - A) \vdots P_{inv} \right]$ has the following form due to (4.9):

$$rank \left[(sI_n - A) \vdots P_{inv} \right] = rank \left[sI_n + A_1^{-1}A_0 \vdots I_n \right] =$$

$$= rankI_n = n \Longrightarrow \forall s = s_i \left(A \right) \in \mathbb{C}, \forall i = 1, 2, ..., n; \ i.e.,$$
$$\forall s \in \mathbb{C}, \ \forall A_k \in \mathfrak{R}^{n \times n}, \ k = 0, 1.$$

This proves the statement under I-a).

    b) Let $\nu = 1$ and $0 \leq \mu \leq 1$ due to(2.1). Let (4.59) be valid. For the statement under I-b) we use the matrix $\left[(sI_n - A) \vdots P^{(\mu)} \right]$ that has the following form due to (4.8):

$$\left[(sI_n - A) \vdots P^{(\mu)} \right] = \left[sI_n + A_1^{-1}A_0 \vdots A_1^{-1}H^{(\mu)} \right] \Longrightarrow$$

$$rank \left[(sI_n - A) \vdots P^{(\mu)} \right] = rank \left[sI_n + A_1^{-1}A_0 \vdots A_1^{-1}H^{(\mu)} \right].$$

Necessity. Let $A_0 = O_n$ *and* $s = 0$ in $\left[(sI_n - A) \vdots P^{(\mu)} \right]$:

$$\left[(0I_n - O_n) \vdots P^{(\mu)} \right] = \left[O_n \vdots A_1^{-1}H^{(\mu)} \right] \Longrightarrow$$

Let the matrix

$$\left[(sI_n - A) \vdots P^{(\mu)} \right] = \left[O_n \vdots A_1^{-1}H^{(\mu)} \right]$$

have the full rank n :

$$n = rank \left[(sI_n - A) \vdots P^{(\mu)} \right] = rank \left[O_n \vdots A_1^{-1}H^{(\mu)} \right] =$$

$$= rank A_1^{-1} H^{(\mu)} = rank H^{(\mu)} \text{ due to } det A_1^{-1} \neq 0.$$

The rank of $H^{(\mu)}$ equals n. Equation (4.59) holds, which proves its necessity.

Sufficiency. Let $rank H^{(\mu)} = N = n$ due to (4.59). This and $det A_1^{-1} \neq 0$ yield

$$N = n = rank H^{(\mu)} = rank A_1^{-1} H^{(\mu)} =$$

$$= rank \left[sI_n + A_1^{-1} A_0 \vdots A_1^{-1} H^{(\mu)} \right] = rank \left[(sI_n - A) \vdots P^{(\mu)} \right],$$

$$0 \leq \mu \leq 1, \ \forall s \in \mathbb{C}, \ \forall A_k \in \mathfrak{R}^{n \times n}, \ k = 0, 1,$$

This proves that for the rank of the matrix $\left[(sI_n - A) \vdots P^{(\mu)} \right]$ to be full, i.e., to be equal to n, it is sufficient that the rank of the matrix $H^{(1)}$ is full, i.e., equal to n. Hence,

$$rank H^{(1)} = n \implies rank \left[(sI_n - A) \vdots P^{(\mu)} \right] \equiv n.$$

This proves the statement under I-b).

II) Let $\nu > 1$ and $\mu \geq 0$, $\mu < \infty$.

a) The matrix $\left[(sI_n - A) \vdots P_{inv} \right]$ has the following form due to (4.9):

$$\nu > 1, \ \mu \geq 0 \implies \left[(sI_n - A) \vdots P_{inv} \right] =$$

$$= \left[sI_n - A \vdots \begin{array}{c} O_{(\nu-1)N,N} \\ I_N \end{array} \right] =$$

$$= \begin{bmatrix} sI_N & -I_N & O_N & \cdots & O_N & O_N & O_N \\ O_N & sI_N & -I_N & \cdots & O_N & O_N & O_N \\ O_N & O_N & sI_N & \cdots & O_N & O_N & O_N \\ \cdots & \cdots & \cdots & \cdots & \cdots & \cdots & \cdots \\ O_N & O_N & O_N & \cdots & sI_N & -I_N & O_N \\ A_\nu^{-1} A_0 & A_\nu^{-1} A_1 & A_\nu^{-1} A_2 & \cdots & A_\nu^{-1} A_{\nu-2} & sI_N + A_\nu^{-1} A_{\nu-1} & I_N \end{bmatrix}.$$

This implies

$$rank \left[(sI_n - A) \vdots P_{inv} \right] =$$

$$rank \left[sI_n - A \vdots \begin{bmatrix} O_{(\nu-1)N,N} \\ I_N \end{bmatrix} \right] =$$

$$= rank \begin{bmatrix} sI_N & -I_N & ... & O_N & O_N & O_N \\ O_N & sI_N & ... & O_N & O_N & O_N \\ O_N & O_N & ... & O_N & O_N & O_N \\ ... & ... & ... & ... & ... & ... \\ O_N & O_N & ... & sI_N & -I_N & O_N \\ A_\nu^{-1}A_0 & A_\nu^{-1}A_1 & ... & A_\nu^{-1}A_{\nu-2} & sI_N + A_\nu^{-1}A_{\nu-1} & I_N \end{bmatrix} =$$

$$= rank \begin{bmatrix} -I_N & O_N & ... & O_N & O_N & O_N \\ sI_N & -I_N & ... & O_N & O_N & O_N \\ O_N & sI_N & ... & O_N & O_N & O_N \\ ... & ... & ... & ... & ... & ... \\ O_N & O_N & ... & sI_N & -I_N & O_N \\ A_\nu^{-1}A_1 & A_\nu^{-1}A_2 & ... & A_\nu^{-1}A_{\nu-2} & -sI_N + A_\nu^{-1}A_{\nu-1} & I_N \end{bmatrix} =$$

$$= \nu N = n, \ \forall (s, A) \in \mathbb{C} \times \mathfrak{R}^{n\times n}.$$

This proves the invariance of the matrix $\left[(sI_n - A) \vdots P_{inv} \right]$ relative to every $(s, A) \in \mathbb{C} \times \mathfrak{R}^{n\times n}$. The first statement under II) is true.

b) Let $\nu > 1$ and $0 \le \mu < \infty$.

Necessity. Let $A_k = O_N, \ \forall k = 0, 1, ..., \nu - 1$ and $s = 0$ in $\left[(sI_n - A) \vdots P^{(\mu)} \right]$ and let $rank \left[(sI_n - A) \vdots P^{(\mu)} \right] = n$:

$$\left[(sI_n - A) \vdots P^{(\mu)} \right] = \begin{bmatrix} O_N & -I_N & ... & O_N & O_N & O_N \\ O_N & O_N & ... & O_N & O_N & O_N \\ O_N & O_N & ... & O_N & O_N & O_N \\ ... & ... & ... & ... & ... & ... \\ O_N & O_N & ... & O_N & -I_N & O_N \\ O_N & O_N & ... & O_N & O_N & A_\nu^{-1}H^{(\mu)} \end{bmatrix} \Longrightarrow$$

$$n = \nu N = rank \left[(sI_n - A) \vdots P^{(\mu)} \right] =$$

$$= rank \begin{bmatrix} O_N & -I_N & ... & O_N & O_N & O_N \\ O_N & O_N & ... & O_N & O_N & O_N \\ O_N & O_N & ... & O_N & O_N & O_N \\ ... & ... & ... & ... & ... & ... \\ O_N & O_N & ... & O_N & -I_N & O_N \\ O_N & O_N & ... & O_N & O_N & A_\nu^{-1}H^{(\mu)} \end{bmatrix} =$$

$$= rank \begin{bmatrix} -I_N & \cdots & O_N & O_N & O_N \\ O_N & \cdots & O_N & O_N & O_N \\ O_N & \cdots & O_N & O_N & O_N \\ \cdots & \cdots & \cdots & \cdots & \cdots \\ O_N & \cdots & O_N & -I_N & O_N \\ O_N & \cdots & O_N & O_N & A_\nu^{-1} H^{(\mu)} \end{bmatrix} =$$

$$= (\nu - 1) N + rank A_\nu^{-1} H^{(\mu)} \Longrightarrow$$

$$N = rank A_\nu^{-1} H^{(\mu)} = rank H^{(\mu)} \text{ due to } det A_\nu^{-1}.$$

This proves the validity of the condition (4.62), i.e., its necessity.
Sufficiency. Let the condition (4.62) hold. The matrix

$$\left[(sI_n - A) \vdots P^{(\mu)} \right] = \left[sI_n - A \vdots \begin{matrix} O_{(\nu-1)N,(\mu+1)M} \\ A_\nu^{-1} H^{(\mu)} \end{matrix} \right]$$

has the following form in view of (4.7)–(4.9):

$$\left[sI_n - A \vdots \begin{matrix} O_{(\nu-1)N,(\mu+1)M} \\ A_\nu^{-1} H^{(\mu)} \end{matrix} \right] =$$

$$= \begin{bmatrix} sI_N & -I_N & O_N & \cdots & O_N & O_N & O_{N,(\mu+1)M} \\ O_N & sI_N & -I_N & \cdots & O_N & O_N & O_{N,(\mu+1)M} \\ O_N & O_N & sI_N & \cdots & O_N & O_N & O_{N,(\mu+1)M} \\ \cdots & \cdots & \cdots & \cdots & \cdots & \cdots & \cdots \\ O_N & O_N & O_N & \cdots & sI_N & -I_N & O_{N,(\mu+1)M} \\ A_\nu^{-1} A_0 & A_\nu^{-1} A_1 & A_\nu^{-1} A_2 & \cdots & A_{\nu-2} & sI_N + A_\nu^{-1} A_{\nu-1} & A_\nu^{-1} H^{(\mu)} \end{bmatrix}.$$

Having in mind that for the matrix

$$\left[(sI_n - A) \vdots \begin{matrix} O_{(\nu-1)N,(\mu+1)M} \\ A_\nu^{-1} H^{(\mu)} \end{matrix} \right]$$

to have the full rank n for every eigenvalue $s_i(A)$ of the matrix $s_i(A)$ it is
sufficient that its following submatrix has the full rank n:

$$\begin{bmatrix} -I_N & O_N & \cdots & O_N & O_N & O_{N,(\mu+1)M} \\ sI_N & -I_N & \cdots & O_N & O_N & O_{N,(\mu+1)M} \\ O_N & sI_N & \cdots & O_N & O_N & O_{N,(\mu+1)M} \\ \cdots & \cdots & \cdots & \cdots & \cdots & \cdots \\ O_N & O_N & \cdots & sI_N & -I_N & O_{N,(\mu+1)M} \\ A_\nu^{-1} A_1 & A_\nu^{-1} A_2 & \cdots & A_\nu^{-1} A_{\nu-2} & sI_N + A_\nu^{-1} A_{\nu-1} & A_\nu^{-1} H^{(\mu)} \end{bmatrix}$$

which is true because the matrix $H^{(\mu)}$ has the rank N due to $rank H^{(\mu)} = N$ (4.62) and implies $N = rank H^{(\mu)} = rank A_\nu^{-1} H^{(\mu)}$ due to $det A_\nu^{-1} \neq 0$:

$$rank \left[sI_n - A \ \vdots \ \begin{bmatrix} O_{(\nu-1)N,(\mu+1)M} \\ A_\nu^{-1} H^{(\mu)} \end{bmatrix} \right] =$$

$$= rank \begin{bmatrix} -I_N & O_N & \cdots & O_N & O_N & O_{N,(\mu+1)M} \\ sI_N & -I_N & \cdots & O_N & O_N & O_{N,(\mu+1)M} \\ O_N & sI_N & \cdots & O_N & O_N & O_{N,(\mu+1)M} \\ \cdots & \cdots & \cdots & \cdots & \cdots & \cdots \\ O_N & O_N & \cdots & sI_N & -I_N & O_{N,(\mu+1)M} \\ A_\nu^{-1}A_1 & A_\nu^{-1}A_2 & \cdots & A_\nu^{-1}A_{\nu-2} & sI_N + A_\nu^{-1}A_{\nu-1} & A_\nu^{-1}H^{(\mu)} \end{bmatrix} =$$

$$= rank \begin{bmatrix} -I_N & O_N & \cdots & O_N & O_N \\ sI_N & -I_N & \cdots & O_N & O_N \\ O_N & sI_N & \cdots & O_N & O_N \\ \cdots & \cdots & \cdots & \cdots & \cdots \\ O_N & O_N & \cdots & sI_N & -I_N \end{bmatrix} + rank A_\nu^{-1} H^{(\mu)} =$$

$$= (\nu - 1) N + rank A_\nu^{-1} H^{(\mu)} = (\nu - 1) N + N = \nu N = n,$$
$$\forall s_i (A) \in \mathbb{C}, \ i.e., \ \forall s \in \mathbb{C}, \ \forall A \in \mathfrak{R}^n, \tag{D.5}$$

This proves the second statement under II) and completes the proof. ∎

D.3 Proof of Theorem 126

Proof. Let $[\mathbf{D}(.), \mathbf{U}(.), \mathbf{Y}_d(.)]$ be arbitrary from $\mathfrak{D}^j \times \mathfrak{U}^l \times \mathfrak{Y}_d^k$. Let the system exhibits stablewise tracking of the desired output $\mathbf{Y}_d^k (t)$ on $\mathfrak{D}^j \times \mathfrak{Y}_d^k$. The conditions of Definition 125 hold and the condition (8.18) is valid. Let us assume that $\mathbf{Y}_d^k (t)$ is not realizable. We disprove this assumption by showing that it leads to a contradiction. Let $\mathbf{Y}_0^k = \mathbf{Y}_{d0}^k$. Hence, $\mathbf{Y}_0^k \in \mathfrak{N} \left(\varepsilon; \mathbf{Y}_{d0}^k; \mathbf{D}; \mathbf{U}; \mathbf{Y}_d^k \right)$ for every $\varepsilon \in \mathfrak{R}^+$, which implies

$$\left\| \mathbf{Y}_d^k(t; \mathbf{Y}_{d0}^k) - \mathbf{Y}^k(t; \mathbf{Y}_0^k; \mathbf{D}; \mathbf{U};) \right\| < \varepsilon, \forall t \in \mathfrak{T}_0, \forall \varepsilon \in \mathfrak{R}^+,$$

due to (8.18). If $\mathbf{Y}_d^k (t)$ were unrealizable, then there would be a moment $t \in \mathfrak{T}_0$ and a number $\xi \in \mathfrak{R}^+$ such that $\left\| \mathbf{Y}_d^k(t; \mathbf{Y}_{d0}^k) - \mathbf{Y}^k(t; \mathbf{Y}_0^k; \mathbf{D}; \mathbf{U}) \right\| \geq \xi$. This would contradict the fact that

$$\mathbf{Y}_0^k \in \mathfrak{N} \left(\varepsilon; \mathbf{Y}_{d0}^k; \mathbf{D}; \mathbf{U}; \mathbf{Y}_d^k \right) \text{ implies}$$

$$\left\| \mathbf{Y}_d^k(t; \mathbf{Y}_{d0}^k) - \mathbf{Y}^k(t; \mathbf{Y}_0^k; \mathbf{D}; \mathbf{U}) \right\| < \varepsilon, \forall t \in \mathfrak{T}_0, \forall \varepsilon \in \mathfrak{R}^+ \implies \forall \varepsilon \in [0, \xi].$$

Hence, the assumption that $\mathbf{Y}_d^k(t)$ is not realizable is invalid. Therefore, $\mathbf{Y}_d^k(t)$ is realizable. ∎

Hence the assumption that $Y_1^p(Y)$ is not realizable as the additional function $Y_2^p(Y)$ is realizable.

Appendix E

Transformations

E.1 Transformation of *IO* into *ISO* system

The state space theory of the linear dynamical and control systems has been mainly established and effective for the *ISO* systems (3.1), (3.2) (Section 3.1). In order to transform the *IO* system (2.1), i.e., (2.15) (Section 2.1.1) into the *ISO* systems (3.1), (3.2) the well-known formal mathematical transformation has been used. It has to satisfy the condition that the transformed system should not contain any derivative of the input vector despite the influence of derivatives of the input vector on the original system and the condition that the only accepted derivative is the first derivative of the state vector and only in the state equation. We will illustrate it for the *IO* system (2.1) subjected to the external action of the input vector $\mathbf{I}$ and its derivatives, i.e., subjected to the action of the extended input vector $\mathbf{I}^\mu$.

The *IO* system (2.1):

$$\sum_{k=0}^{k=\nu} A_k \mathbf{Y}^{(k)}(t) = \sum_{k=0}^{k=\mu} H_k \mathbf{I}^{(k)}(t), \ det A_\nu \neq 0, \ \forall t \in \mathfrak{T}_0, \ \nu \geq 1, \ 0 \leq \mu \leq \nu,$$

(E.1)

can be formally mathematical transformed into mathematically equivalent *ISO* system (3.1), (3.2),

$$\frac{d\mathbf{X}(t)}{dt} = A\mathbf{X}(t) + H\mathbf{I}(t), \ \forall t \in \mathfrak{T}_0, \ A \in \mathfrak{R}^{n \times n}, \ \mathbf{U} \in \mathfrak{R}^M, \ P \in \mathfrak{R}^{n \times M}, \quad \text{(E.2)}$$

$$\mathbf{Y}(t) = C\mathbf{X}(t) + Q\mathbf{I}(t), \ \forall t \in \mathfrak{T}_0, \ C \in \mathfrak{R}^{N \times n}, \ C \neq O_{N,n}, \ Q \in \mathfrak{R}^{N \times M}.$$

(E.3)

by applying the following formal mathematical transformations:

$$\mathbf{X}_1 = \mathbf{Y} - H_\nu \mathbf{I}, \tag{E.4}$$

$$\mathbf{X}_2 = \overset{\bullet}{\mathbf{X}}_1 + A_{\nu-1}\,\mathbf{Y} - H_{\nu-1}\mathbf{I}, \tag{E.5}$$

$$\mathbf{X}_3 = \overset{\bullet}{\mathbf{X}}_2 + A_{\nu-2}\,\mathbf{Y} - H_{\nu-2}\mathbf{I}, \tag{E.6}$$

$$.... \tag{E.7}$$

$$\mathbf{X}_{\nu-1} = \overset{\bullet}{\mathbf{X}}_{\nu-2} + A_2\,\mathbf{Y} - H_2\mathbf{I} \tag{E.8}$$

$$\mathbf{X}_\nu = \overset{\bullet}{\mathbf{X}}_{\nu-1} + A_1\mathbf{Y} - H_1\mathbf{I}, \tag{E.9}$$

where $H_k = O_{N,r}$ for $k = \mu+1, \mu+2, ..., \nu$ if $\mu < \nu$. The vectors $\mathbf{X}_1, \mathbf{X}_2, ...$ $\mathbf{X}_\nu \in \mathfrak{R}^N$ are the mathematical state subvectors of the vector $\mathbf{X} \in \mathfrak{R}^n$ that is *the mathematical state vector* of the *IO* system (E.1) and of the equivalent *ISO* system (E.2), (E.3),

$$\mathbf{X} = \begin{bmatrix} \mathbf{X}_1^T & \mathbf{X}_2^T & ... & \mathbf{X}_\nu^T \end{bmatrix}^T \in \mathfrak{R}^n, \; n = \nu N. \tag{E.10}$$

Comment 315 *The state subvectors* $\mathbf{X}_1, \mathbf{X}_2, ... \mathbf{X}_\nu$ *(E.4) - (E.9) and the state vector* $\mathbf{X}$ *(E.10) do not any physical sense, i.e., they are physically meaningless, if* $\mu > 0$, *equivalently if* $H_k = O_{N,r}$ *for* $k \in \{1, 2, ..., \nu\}$. *This is the consequence of their definitions to be linear combinations of the input vector, the output vector and the derivative of the preceding state subvector if it exists. Their physical nature and properties are most often inherently different.*

The transformations (E.4)–(E.9) are formal mathematical, physically useless in general. They lead to the following matrices of the *ISO* system (E.2), (E.3) mathematically formally equivalent to the *IO* system (E.1):

$$A = \begin{bmatrix} -A_{\nu-1} & I_N & ... & O_N & O_N \\ -A_{\nu-2} & O_N & ... & O_N & O_N \\ ... & ... & ... & ... & ... \\ -A_1 & O_N & ... & O_N & I_N \\ -A_0 & O_N & ... & O_N & O_N \end{bmatrix}, \tag{E.11}$$

$$H = \begin{bmatrix} H_{\nu-1} - A_{\nu-1}H_\nu \\ H_{\nu-2} - A_{\nu-2}H_\nu \\ .... \\ H_1 - A_1 H_\nu \\ H_0 - A_0 H_\nu \end{bmatrix}, \tag{E.12}$$

$$C = \begin{bmatrix} I_N & O_N & O_N & ... & O_N & O_N & O_N \end{bmatrix},$$ (E.13)

$$Q = H_\nu.$$ (E.14)

Conclusion 316 *The aim of the book and the transformations (E.4)–(E.9)*

The aim of the book to further develop and generalize the control theory with the simultaneous physical and mathematical, i.e., the full engineering, sense, excludes the use of the pure formal mathematical transformations (E.4)–(E.9) if $\mu > 0$.

E.2   *ISO* and *EISO* forms of *IIO* system

The *ISO* and *EISO* forms of the *IIO* system

The compact form of the overall mathematical model of the *IIO* system (6.1), (6.2), (Section 6.1), reads in terms of the total coordinates:

$$\begin{bmatrix} A_\alpha & O_{\rho,N} \\ O_{N,\alpha} & E_\nu \end{bmatrix} \begin{bmatrix} \mathbf{R}^{(\alpha)}(t) \\ \mathbf{Z}^{(\nu)}(t) \end{bmatrix} + \begin{bmatrix} A^{(\alpha-1)} & O_{\rho,\nu+1} \\ -R^{(\alpha-1)} & E^{(\nu-1)} \end{bmatrix} \begin{bmatrix} \mathbf{R}^{\alpha-1}(t) \\ \mathbf{Z}^{\nu-1}(t) \end{bmatrix} =$$
$$= \begin{bmatrix} H^{(\mu)} \\ Q^{(\mu)} \end{bmatrix} \mathbf{I}^\mu(t), \ \mathbf{Y}(t) = \mathbf{Z}(t),$$ (E.15)

where we use a subsidiary vector $\mathbf{Z}$,

$$\mathbf{Z}(t) = \mathbf{Y}(t) = \mathbf{S}_{\alpha+1}(t), \ \mathbf{Z}^{(k)}(t) = \mathbf{Y}^{(k)}(t) = \mathbf{S}_{\alpha+k+1}(t), \ k = 0, 1, .., \nu - 1,$$
$$\mathbf{Z}^{\nu-1}(t) = \mathbf{Y}^{\nu-1}(t) = \mathbf{S}_O(t).$$ (E.16)

In terms of the deviations the system model is given by (6.47), (6.48) (Section 6.2), which can be set in the form of (E.15), (E.16):

$$\begin{bmatrix} A_\alpha & O_{\rho,N} \\ O_{N,\alpha} & E_\nu \end{bmatrix} \begin{bmatrix} \mathbf{r}^{(\alpha)}(t) \\ \mathbf{z}^{(\nu)}(t) \end{bmatrix} + \begin{bmatrix} A^{(\alpha-1)} & O_{\rho,\nu+1} \\ -R^{(\alpha-1)} & E^{(\nu-1)} \end{bmatrix} \begin{bmatrix} \mathbf{r}^{\alpha-1}(t) \\ \mathbf{z}^{\nu-1}(t) \end{bmatrix} =$$
$$= \begin{bmatrix} H^{(\mu)} \\ Q^{(\mu)} \end{bmatrix} \mathbf{i}^\mu(t), \ \mathbf{y}(t) = \mathbf{z}(t),$$ (E.17)

$$\mathbf{z}(t) = \mathbf{y}(t) = \mathbf{s}_{\alpha+1}(t), \ \mathbf{z}^{(k)}(t) = \mathbf{y}^{(k)}(t) = \mathbf{s}_{\alpha+k+1}(t), \ k = 0, 1, .., \nu - 1,$$ (E.18)

$$\mathbf{z}^{\nu-1}(t) = \mathbf{y}^{\nu-1}(t) = \mathbf{s}_O(t).$$

We continue to use the system model (E.17), (E.18) in terms of the deviations by recalling the fact that the system models (E.15), (E.16) and (E.17), (E.18) have the same properties.

Condition 93 (Section 6.1) and (6.6) (Subsection 6.1.1) permit us to transform Equation (E.17) into

$$
\begin{bmatrix} s_\alpha^{(1)}(t) \\ s_{\alpha+\nu}^{(1)}(t) \end{bmatrix} + \begin{bmatrix} A_\alpha^{-1} A^{(\alpha-1)} & O_{\rho,\nu+1} \\ -E_\nu^{-1} R^{(\alpha-1)} & E_\nu^{-1} E^{(\nu-1)} \end{bmatrix} \begin{bmatrix} r^{\alpha-1}(t) \\ z^{\nu-1}(t) \end{bmatrix} =
$$

$$
= \begin{bmatrix} A_\alpha^{-1} H^{(\mu)} \\ E_\nu^{-1} Q^{(\mu)} \end{bmatrix} i^\mu(t),
$$

$$
y(t) = \left[\underbrace{\overbrace{O_{N,\rho} : ... : O_{N,\rho}}^{O_{N,\rho} \ repeats \ \alpha-times} : I_N :}_{O_{N,\alpha\rho}=C_I} \underbrace{\overbrace{O_N : ... : O_N}^{O_N \ repeats \ (\nu-1)-times}}_{O_{(\nu-1)N}} \right] \begin{bmatrix} r^{\alpha-1}(t) \\ z^{\nu-1}(t) \end{bmatrix}.
$$

$$(E.19)$$

In view of Equation (6.6), the following equations result:

$$s_1 = r_1$$

$$s_2 = s_1^{(1)} = r^{(1)} \implies s_1^{(1)} = s_2$$

$$s_3 = s_2^{(1)} = r^{(2)} \implies s_2^{(1)} = s_3$$

$$\cdots$$

$$s_\alpha = s_{\alpha-1}^{(1)} = r^{(\alpha-1)} \implies s_{\alpha-1}^{(1)} = s_\alpha$$

$$s_{\alpha+1} = y$$

$$s_{\alpha+2} = s_{\alpha+1}^{(1)} = y^{(1)} \implies s_{\alpha+1}^{(1)} = s_{\alpha+2},$$

$$s_{\alpha+3} = s_{\alpha+2}^{(1)} = y^{(2)} \implies s_{\alpha+2}^{(1)} = s_{\alpha+3},$$

$$\cdots$$

$$s_{\alpha+\nu} = s_{\alpha+\nu-1}^{(1)} = y^{(\nu-1)} \implies s_{\alpha+\nu-1}^{(1)} = s_{\alpha+\nu}, \qquad (E.20)$$

Equation (E.19) determines:

$$\mathbf{s}_\alpha^{(1)}(t) = -A_\alpha^{-1} A^{(\alpha-1)} \underbrace{\left[\mathbf{s}_1^T(t) \; \mathbf{s}_2^T(t) \; ... \; \mathbf{s}_\alpha^T(t)\right]^T}_{\mathbf{s}_I(t)} + A_\alpha^{-1} H^{(\mu)} \mathbf{i}^\mu(t),$$

$$\mathbf{s}_{\alpha+\nu}^{(1)}(t) = \begin{pmatrix} E_\nu^{-1} R^{(\alpha-1)} \underbrace{\left[\mathbf{s}_1^T(t) \; \mathbf{s}_2^T(t) \; ... \; \mathbf{s}_\alpha^T(t)\right]^T}_{\mathbf{s}_I(t)} - \\ -E_\nu^{-1} E^{(\nu-1)} \underbrace{\left[\mathbf{s}_{\alpha+1}^T(t) \; \mathbf{s}_{\alpha+2}^T(t) \; ... \; \mathbf{s}_{\alpha+\nu}^T(t)\right]^T}_{\mathbf{s}_O(t)} + \\ +E_\nu^{-1} Q^{(\mu)} \mathbf{i}^\mu(t) \end{pmatrix}, \qquad (\text{E.21})$$

$$\mathbf{s}_I(t) = \left[\mathbf{s}_1^T(t) \; \mathbf{s}_2^T(t) \; ... \; \mathbf{s}_\alpha^T(t)\right]^T \in \mathfrak{R}^{\alpha\rho}, \qquad (\text{E.22})$$

$$\mathbf{s}_O(t) = \left[\mathbf{s}_{\alpha+1}^T(t) \; \mathbf{s}_{\alpha+2}^T(t) \; ... \; \mathbf{s}_{\alpha+\nu}^T(t)\right]^T \in \mathfrak{R}^{\nu N}. \qquad (\text{E.23})$$

Equations (E.19)–(E.21) imply:

$$A = \begin{bmatrix} A_{11} & O_{\alpha\rho,\nu N} \\ A_{21} & A_{22} \end{bmatrix} \in \mathfrak{R}^{n \times n}, \; n = \alpha\rho + \nu N, \qquad (\text{E.24})$$

$$A_{11} = \begin{bmatrix} O_\rho & I_\rho & ... & O_\rho & O_\rho \\ O_\rho & O_\rho & ... & O_\rho & O_\rho \\ \vdots & \vdots & \vdots & \vdots & \vdots \\ O_\rho & O_\rho & ... & O_\rho & I_\rho \\ -A_\alpha^{-1} A_0 & -A_\alpha^{-1} A_1 & ... & -A_\alpha^{-1} A_{\alpha-2} & -A_\alpha^{-1} A_{\alpha-1} \end{bmatrix},$$

$$A_{11} \in \mathfrak{R}^{\alpha\rho \times \alpha\rho}, \qquad (\text{E.25})$$

$$A_{21} = \begin{bmatrix} O_{N,\rho} & O_{N,\rho} & ... & O_{N,\rho} & O_{N,\rho} \\ O_{N,\rho} & O_{N,\rho} & ... & O_{N,\rho} & O_{N,\rho} \\ \vdots & \vdots & \vdots & \vdots & \vdots \\ E_\nu^{-1} R_{y0} & E_\nu^{-1} R_{y1} & ... & E_\nu^{-1} R_{y,\alpha-2} & E_\nu^{-1} R_{y,\alpha-1} \end{bmatrix},$$

$$A_{21} \in \mathfrak{R}^{\nu N \times \alpha\rho}, \qquad (\text{E.26})$$

$$A_{22} = \begin{bmatrix} O_N & I_N & ... & O_N & O_N \\ O_N & O_N & ... & O_N & O_N \\ \vdots & \vdots & \vdots & \vdots & \vdots \\ O_N & O_N & ... & O_N & I_N \\ -E_\nu^{-1} E_0 & -E_\nu^{-1} E_1 & ... & -E_\nu^{-1} E_{\nu-2} & -E_\nu^{-1} E_{\nu-1} \end{bmatrix},$$

$$A_{22} \in \mathfrak{R}^{\nu N \times \nu N}, \qquad (\text{E.27})$$

$$C = \left[\underbrace{O_{N,\alpha\rho}}_{C_I} \vdots \underbrace{I_N \vdots O_{N,(\nu-1)N}}_{C_O} \right] = \left[C_I \vdots C_O \right] \in \mathfrak{R}^{N \times n}, \qquad (E.28)$$

$$P^{(\mu)} = W = \left[\begin{array}{c} W_1 \\ W_2 \end{array} \right] \in \mathfrak{R}^{n \times (\mu+1)M}, \qquad (E.29)$$

$$W_1 = \left[\begin{array}{ccccc} O_{\rho,M} & O_{\rho,M} & \cdots & O_{\rho,M} & O_{\rho,M} \\ O_{\rho,M} & O_{\rho,M} & \cdots & O_{\rho,M} & O_{\rho,M} \\ \vdots & \vdots & \vdots & \vdots & \vdots \\ O_{\rho,M} & O_{\rho,M} & \cdots & O_{\rho,M} & O_{\rho,M} \\ A_\alpha^{-1} H_0 & A_\alpha^{-1} H_1 & \cdots & A_\alpha^{-1} H_{\mu-1} & A_\alpha^{-1} H_\mu \end{array} \right],$$

$$W_1 \in \mathfrak{R}^{\alpha\rho \times (\mu+1)M}, \qquad (E.30)$$

$$W_2 = \left[\begin{array}{ccccc} O_{N,M} & O_{N,M} & \cdots & O_{N,M} & O_{N,M} \\ O_{N,M} & O_{N,M} & \cdots & O_{N,M} & O_{N,M} \\ \vdots & \vdots & \vdots & \vdots & \vdots \\ O_{N,M} & O_{N,M} & \cdots & O_{N,M} & O_{N,M} \\ E_\nu^{-1} Q_0 & E_\nu^{-1} Q_1 & \cdots & E_\nu^{-1} Q_{\mu-1} & E_\nu^{-1} Q_\mu \end{array} \right]$$

$$W_2 \in \mathfrak{R}^{\nu N \times (\mu+1)M}, \qquad (E.31)$$

$$\mathbf{W}(t) = \mathbf{I}^\mu(t) \in \mathfrak{R}^{(\mu+1)M}, \ \mathbf{w}(t) = \mathbf{i}^\mu(t) \in \mathfrak{R}^{(\mu+1)M}. \qquad (E.32)$$

Altogether,

$$\frac{d\mathbf{S}(t)}{dt} = A\mathbf{S}(t) + W\mathbf{W}(t) = A\mathbf{S}(t) + P^{(\mu)}\mathbf{I}^\mu(t), \qquad (E.33)$$

$$\mathbf{Y}(t) = C\mathbf{S}(t), \qquad (E.34)$$

These equations represent the *ISO* form for $\mathbf{I}(t)$ replaced by $\mathbf{W}(t)$, and *EISO* form of the *IIO* system (6.47), (6.48 for $\mathbf{I}^\mu(t)$ replaced by $\mathbf{W}(t)$. In terms of the deviations of all variables which in the free regime, i.e., for $\mathbf{w}(t) \equiv \mathbf{0}_m$, Equations (E.33), (E.34) take the following form:

$$\frac{d\mathbf{s}(t)}{dt} = A\mathbf{s}(t), \qquad (E.35)$$

$$\mathbf{y}(t) = C\mathbf{s}(t). \qquad (E.36)$$

Bibliography

[1] A. B. Açìkmeşe and M. Corles, "Robust output tracking for uncertain/nonlinear systems subject to almost constant disturbances", *Automatica*, vol. 38, pp. 1919-1926, 2002.

[2] T. Ahmed-Ali and F. Lamnabhi-Lagarrigue, "Tracking control of nonlinear systems with disturbance attenuation", *C. R. Acad. Sci.*, Paris, France, t. 325, Série I, pp. 329-338, 1997.

[3] G. Ambrosino, G. Celentano, and F. Garofalo, "Robust model tracking control for a class of nonlinear plants", *IEEE Transactions on Automatic Control*, Vol. AC-30, No. 3, pp. 275-279, 1985.

[4] B. D. O. Anderson and J. B. Moore, *Linear Optimal Control*, Englewood Cliffs, NJ: Prentice Hall, 1971.

[5] N. P. I. Aneke, H. Nijmeijer and A. G. de Jager, "Tracking control of second-order chained form systems by cascaded backstepping", *Int. J. Robust and Nonlinear Control*, Vol. 13, pp. 95-115, 2003.

[6] P. J. Antsaklis and O. R. Gonzalez, "Compensator Structure and Internal Models in Tracking and Regulation," *Proceedings of the 23rd IEEE Conference on Decision and Control*, Las Vegas, NV, pp. 1-2, December 12-14, 1984

[7] P. J. Antsaklis and A. N. Michel, *Linear Systems*, New York: The McGraw Hill Companies, Inc., 1997, Boston: Birkhaüser, 2006.

[8] P. J. Antsaklis and A. N. Michel, *A Linear Systems Primer*, Boston: Birkhaüser, 2007.

[9] M. A. Arteaga, "Tracking control of flexible robot arms with nonlinear observer," *Automatica*, Vol. 36, pp. 1329-1337, 2000.

[10] M. A. Arteaga and B. Siciliano, "On tracking control of flexible robot arms," *IEEE Trans. on Automatic Control*, Vol. 45, No. 3, pp. 520-527, March 2000.

[11] M. Athanassiades, "Bang-bang control for tracking systems", *IRE Transactions on Automatic Control*, Vol. AC-7, No. 3, pp. 77-78, 1962.

[12] M. A. Athans and P. L. Falb, *Optimal Control*, New York: McGraw–Hill, 1966, 1994, 2007.

[13] E.-W. Bai and Y.-F. Huang, "Variable gain parameter estimation algorithms for fast tracking and smooth steady state," *Automatica*, Vol. 36, pp. 1001-1008, 2000.

[14] J. A. Ball, P. Kachroo, and A. J. Krener, "H_∞ tracking control for a class of nonlinear systems," *IEEE Trans. Automatic Control*, Vol. 44, No. 6, pp. 1202-1206, June 1999.

[15] A. Balluchi and A. Bicchi, "Necessary and sufficient conditions for robust perfect tracking under variable structure control," *Int. J. Robust and Nonlinear Control*, Vol. 13, pp. 141-151, 2003.

[16] S. Barnett, *Introduction to Mathematical Control Theory*, Oxford: Clarendon Press, 1975.

[17] Y. Bar-Shalom and T. E. Fortmann, *Tracking and Data Association*, Boston, MA, USA: Academic Press, Inc., 1988.

[18] Y. Bar-Shalom and T. E. Fortmann, *Tracking and Data Association*, Boston: Academic Press, 1988.

[19] Y. Bar-Shalom and X. R. Li, *Estimation and Tracking, Principles, Techniques and Software*, Norwood, MA: Artech House, 1993.

[20] Y. Bar-Shalom and X. R. Li, *Multitarget-Multisensor Tracking, Principles and Techniques*, Storrs, CT: YBBS Publishing, 1995.

[21] Y. Bar-Shalom, X. R. Li and T. Kirubarajan, *Estimation with Applications to Tracking and Navigation*, New York: Wiley-Interscience, 2001.

[22] Y. Bar-Shalom, P. K Willet, and X. Tian, *Tracking, and Data Fusion*, Storrs, CT: YBBS Publishing, 2011.

[23] R. Bellman, "Vector Lyapunov functions", *J.S.I.A.M. Control*, Ser. A, Vol. 1, No.1, pp. 32-34, 1962.

[24] A. Benzaoiua, F. Mesquine and M. Benhayoun, *Saturated Control of Linear Systems*, Berlin: Springer, 2018.

[25] L. D. Berkovitz, *Optimal Control Theory*, New York: Springer Verlag, 2010.

[26] L. D. Berkovitz and N. G. Medhin, *Nonlinear Optimal Control*, Boca Raton, FL: Taylor & Francis-CRC, 2013.

[27] J. E. Bertram and P. E. Sarachik, "On optimal computer control", *Proc. of the First International Congress of the Federation of Automatic Control*, London: Butterworths, pp. 419-422, 1961.

[28] G. Besançon, "Global output feeddback tracking control for a class of Lagrangian systems", *Automatica*, Vol. 36, pp. 1915-1921, 2000.

[29] S. P. Bhattacharyya, "Frequency domain conditions for disturbance rejection", *IEEE Transactions on Automatic Control*, Vol. AC-25, No. 6, 1211-1213, December 1980.

[30] S. P. Bhattacharyya, "Transfer function conditions for output feedback disturbance rejection", *IEEE Transactions on Automatic Control*, Vol. AC-27, No. 4, pp. 974-977, August 1982.

[31] S. P. Bhattacharyya, A. C. del Nero Gomes and J. W. Howze, "The structure of robust disturbance rejection", *IEEE Transactions on Automatic Control*, Vol. AC-28, No. 9, 874-881, September 1983.

[32] S. P. Bhattacharyya, A. Datta and L. H. Keel, *Linear Control Theory: Structure, Robustness, and Optimization*, Boca Raton, FL: CRC Press, Taylor & Francis Group, 2009.

[33] D. Biswa, *Numerical Methods for Linear Control Systems*, London: Elsevier Inc., 2004.

[34] S. S. Blackman, *Multiple-Target Tracking with radar Applications*, Norwood, MA: Artech House, 1999.

[35] S. S. Blackman and R. Popoli, *Design and Analysis of Modern Tracking Systems*, Norwood, MA: Artech House, 1999.

[36] J. H. Blakelock, *Automatic control of aircrafts and missiles*, New York: Wiley, 1991.

[37] P. Borne, G. Dauphin-Tanguy, J.-P. Richard, F. Rotella and I. Zambettakis, *Commande et Optimisation des Processus*, Paris: Éditions TECHNIP, 1990.

[38] R. D. Braatz, "On Internal Stability and Unstable Pole-Zero Cancellations", *IEEE Control Systems Magazine*, vol. 32, No. 5, October 2012, pp. 15, 16.

[39] R. W. Brockett and M. D. Mesarović, "The reproducibility of multivariable systems", *J. Mathematics Analysis and Applications*, Vol. 1, pp. 548-563, 1965.

[40] W. L. Brogan, *Modern Control Theory*, New York: Quantum Publishers, Inc., 1974.

[41] G. S. Brown and D. P. Campbell, *Principles of Servomechanisms*, New York: Wiley, 1948.

[42] Z. M. Buchevats and Ly. T. Gruyitch, *Linear Discrete-time Systems*, Boca Raton, FL: CRC Press, 2018.

[43] F. M. Callier and C. A. Desoer, *Multivariable Feedback Systems*, New York: Springer-Verlag, 1982

[44] F. M. Callier and C. A. Desoer, *Linear System Theory*, New York: Springer-Verlag, 1991.

[45] G. E. Carlson, *Signal and Linear Systems Analysis and Matlab*, second edition, New York: Wiley, 1998.

[46] Y.-C. Chang, "Robust tracking control for nonlinear MIMO systems via fuzzy approaches", *Automatica*, Vol. 36, pp. 1535-1545, 2000.

[47] B. Chen and J. K. Tugnait, "Tracking of multiple maneuvering targets in clutter using IMM/JPDA filtering and fixed-lag smoothing", *Automatica*, Vol. 37, pp. 239-249, 2001.

[48] C.-T. Chen, *Linear System Theory and Design*, New York: Holt, Rinehart and Winston, Inc., 1984; Oxford: Oxford University Press, 2013.

[49] X. P. Cheng and R. V. Patel, "Neural network based tracking conrol of a flexible macro-micro manipulator system", *Neural Networks*, Vol. 16, pp. 271-286, 2003.

[50] H. Chestnut and R. W. Mayer, *Servomechanisms and Regulating System Design*, New York: Wiley, 1955.

[51] S.-I. Cho and I.-J. Ha, "A learning approach to tracking in mechanical systems with friction," *IEEE Trans. Automatic Control*, Vol. 45, No. 1, pp. 111-116, January 2000.

[52] D. Chwa, "Sliding-mode tracking control of nonholonomic wheeled mobile robots in polar coordinates", *IEEE Transactions on Control Systems Technology*, Vol. 12, No. 4, pp. 637-644, July 2004.

[53] M. J. Corless and A. E. Frazho, *Linear Systems and Control*, Boca Raton, FL: CRC Press, 2003.

[54] F. E. Daum, "Bounds on performance for multiple target tracking", *IEEE Transactions on Automatic Control*, Vol. 35, No. 4, pp. 443-446, 1990.

[55] F. E. Daum, "Bounds on track purity for multiple target tracking," *Proc. 28th Conference on Decision and Control*, Tampa, FL, pp. 1423-1424, December 1989.

[56] R. Davies, C. Edwards and S. K. Spurgeon, "Robust tracking with a sliding mode", in *Variable Structure and Lyapunov Control*, Ed. A. S. I. Zinober, London: Springer Verlag, pp. 51-73, 1994.

[57] J. H. Davis and R. M. Hirschorn, "Tracking control of a flexible robot link," *IEEE Trans. Automatic Control*, Vol. 33, No. 3, pp. 238-248, March 1998.

[58] E. J. Davison, "The robust decentralized control of a servomechanism problem for composite systems with input-output interconnections", *IEEE Transactions on Automatic Control*, Vol. AC-24, No. 4, pp. 325-327, 1979.

[59] E. J. Davison and B. R. Copeland, "Gain margin and time lag tolerance constraints applied to the stabilization problem and robust servomechanism problem", *IEEE Transactions on Automatic Control*, Vol. AC-30, No. 3, pp. 229-239, 1985.

[60] E. J. Davison and I. Ferguson, "The design of controllers for the multi-variable robust servomechanism problem using parameter optimization methods", *IEEE Transactions on Automatic Control*, Vol. AC-26, No. 1, pp. 93-110, 1981.

[61] E. J. Davison and P. Patel, "Application of the robust servomechanism controller to systems with periodic tracking/disturbance signals", *Int. J. Control*, Vol. 47, No. 1, pp. 111-127, 1988.

[62] E. J. Davison and B. M. Scherzinger, "Perfect control of the robust servomechanism problem", *IEEE Transactions on Automatic Control*, Vol. AC-32, No. 8, pp. 689-702, August 1987.

[63] J. J. D'Azzo and C. H. Houpis, *Linear Control System Analysis & Design*, New York: McGraw-Hill Book Company, 1988.

[64] J. J. D'Azzo, C. H. Houpis and S. N. Sheldon, *Linear Control System Analysis & Design with Matlab*, Boca Raton, FL: CRC Press, 2003.

[65] L. Debnath, *Integral Transformations and Their Applications*, Boca Raton, FL: CRC Press, 1995.

[66] C. A. Desoer, *Notes for A Second Course on Linear Systems*, New York: Van Nostrand Reinhold Company, 1970.

[67] C. A. Desoer and C.-A. Lin, "Tracking and disturbance rejection of MIMO nonlinear systems with PI controller", *IEEE Transactions on Automatic Control*, Vol. AC-30, No. 9, pp. 861-867, 1985.

[68] C. A. Desoer and J. D. Schulman, "Zeros and Poles of Matrix Transfer Functions and Their Dynamical Interpretation", *IEEE Transactions on Circuits and Systems*, Vol. CAS-21, No. 1, pp. 3 - 8, January 1974.

[69] C. A. Desoer and M. Vidyasagar, *Feedback Systems: Input-Output Properties*, New York: Academic Press, 1975.

[70] C. A. Desoer and Y. T. Wang, "The robust non-linear servomechanism problem", *International Journal of Control*, Vol. 29, No. 5, pp. 803-828, 1979.

[71] S. Di Gennaro, "Output attitude tracking for flexible spacecraft", *Automatica*, Vol. 38, pp. 1719-1726, 2002.

[72] W. E. Dixon, D. M. Dawson, E. Zergeroglu and F. Zhang, "Robust tracking and regulation control for mobile robots," *Int. J. Robust and Nonlinear Control*, Vol. 10, pp. 199-216, 2000.

[73] J.L. Domínguez-García, M.I. García-Planas, "Output Controllability and Steady-Output Controllability Analysis of Fixed Speed Wind Turbine", PHYSCON 2011, León, Spain, September, 5-September, 8, pp. 1-5, 2011.

[74] V. Dragan, T. Morozan, A.-M. Stoica, *Mathematical Methods in Robust Control of Linear Stochastic Systems*, New York: Springer Science+Business Media, 2010.

[75] C. Edwards and S. K. Spurgeon, "Robust output tracking using a sliding mode controller/observer scheme", *International Journal of Control*, Vol. 64, No. 5, pp. 967-983, 1996.

[76] C. Edwards and S. K. Spurgeon, "Sliding mode output tracking with application to a multivariable high temperature furnace problem," *Int. J. Robust and Nonlinear Control*, Vol. 7, pp. 337-351, 1997.

[77] B. Etkin, *Dynamics of Flight-Stability and Control*, New York: John Wiley, 1982.

[78] O. I. Elgerd, *Control Systems Theory*, New York: McGraw-Hill Book Company, 1967.

[79] A. Emami-Naeini and G. F. Franklin, "Deadbeat control and tracking of discrete-time systems", *IEEE Transactions on Automatic Control*, Vol. AC-27, No. 1, pp. 176-180, 1982.

[80] F. W. Fairman, *Linear Control Theory: The State Space Approach*, Chichester, England: John Wiley $ Sons, 1998.

[81] P. Falb, *Methods of Algebraic Geometry in Control Theory: Multivariable Linear Systems and Projective Algebraic Geometry Part II*, Boston: Birkhauser, 1999.

[82] L. Fang, P. J. Antsaklis, L. Montestruque, B. McMickell, M. Lemmon, Y. Sun, H. Fang, I. Koutroulis, M. Haenggi, M. Xie, and X. Xie, "A wireless dead reckoning pedestrian tracking system," *ACM Workshop on Applications of Mobile Embedded Systems*, Boston, MA, pp. 1 -3, 2004.

[83] L. Fang and P. J. Antsaklis, "Decentralized formation tracking of multi-vehicle systems with consensus-based controllers," Chapter 15 in *Advances in Unmanned Aerial Vehicles; State of the Art and the Road to Autonomy,* Ed. K. P. Valavanis, Berlin: Springer, pp. 455-471, 2007.

[84] L. Fang, P. Antsaklis, "Decentralized formation tracking of multi-vehicle systems with nonlinear dynamics," *14th Mediterranean Conference on Control and Automation, (MED '06)*, Universitá Politecnica delle Marche, Ancona, pp. 1-6, June 28-30, 2006.

[85] D. R. Fannin, W. H. Tranter, and R. E. Ziemer, *Signals & Systems Continuous and Discrete*, fourth edition, Englewood Cliffs, NJ: Prentice Hall, 1998.

[86] Y. Feng and M. Yagoubi, *Robust control of Linear Descriptor Systems*, Berlin: Springer, 2017.

[87] P. M. G. Ferreira, "Tracking with sensor failures," *Automatica*, Vol. 38, pp. 1621-1623, 2002.

[88] I. Flügge-Lotz and C. F. Taylor, "Synthesis of a nonlinear control system", *IRE Transactions on Automatic Control*, Vol. 1, No. 1, pp. 3-9, May 1956

[89] T. E. Fortmann and K. L. Hitz, *An Introduction to Linear Control Systems*, New York: Marcel Dekker, Inc., 1977.

[90] T. I. Fossen, *Guidance and Control of Ocean Vehicles*, Chichester: John Wiley & Sons, 1994.

[91] R. A. Freeman and P. V. Kokotović, "Tracking controllers for systems linear in the unmeasured states," *Automatica*, Vol. 32, No. 5, pp. 735-746, 1996.

[92] L.-C. Fu and T.-L. Liao, "Globally stable robust tracking of nonlinear systems using variable structure control and with an application to a robotic manipulator", *IEEE Transactions on Automatic Control*, Vol. 35, No. 12, pp. 1345-1350, December 1990.

[93] Z. Gajic and M.T. Lim, *Optimal Control of Singularly Perturbed Linear system and Applications*, Boca Raton, FL: CRC Press, 2001.

[94] F. R. Gantmacher, *The Theory of Matrices*, Vol. 1, New York: Chelsea Publishing Co., 1960, 1974.

[95] F. R. Gantmacher, *The Theory of Matrices*, Vol. 2, New York: Chelsea Publishing Co., 1960, 1974.

[96] R. A. García and C. E. D'Attellis, "Trajectory tracking in nonlinear systems via nonlinear reduced-order observers," *Int. J. Control*, Vol. 62, No. 3, pp. 685-715, 1995.

[97] E. Gershon, U. Shaked and I. Yaesh, "H_∞ tracking of linear continuous-time systems with stochastic uncertainties and prev," *Int. J. Robust and Nonlinear Control*, Vol. 14, No. 7, pp. 607-626, 2004.

[98] A. Gharbi, M. Benrejeb and P. Borne, "Tracking error estimation of uncertain Lur'e Postnikov systems", *International Conference on Control, Decision and Information Technologies (CoDIT)*, Metz, France, pp. 537 - 540, December 2014.

[99] E. G. Gilbert, "Controllability and Observability in Multivariable Control Systems", *SIAM Journal of Control*, Ser.A, Vol. 1, 1963, pp. 128 - 151.

[100] I. Gohberg, P. Lancaster, L. Rodman, *Matrix Polynomials*, New York: Academic Press, 1982

[101] O. R. González and P. J. Antsaklis, "Internal models in regulation, stabilization and tracking," *Proceedings of 28th Conference on Decision and Control*, Tampa, Florida, pp. 1343-1348, 1989.

[102] G. C. Goodwin, S. F. Graebe and M. E. Salgado, *Control System Design*, New Jersey USA, London UK: Prentice Hall - Pearson, 2001.

[103] O. M. Grasselli, S. Longhi, A. Tornambè, "Robust output regulation and tracking for linear periodic systems under structured uncertainties", *Automatica*, Vol. 32, No. 7, pp. 1015-1019, 1996.

[104] T. J. Greattinger and B. H. Krogh, "On the computation of reference signal constraints for guaranteed tracking performance," *Automatica*, Vol. 18, No. 6, pp. 1125-1141, 1992.

[105] J. W. Grizzle, M. D. Di Benedetto and F. Lamnabhi-Lagarrigue, "Necessary conditions for asymptotic tracking in nonlinear systems", *IEEE*

Transactions on Automatic Control, Vol. AC-39, No. 9, pp. 1782-1794, September 1994.

[106] Lj. T. Grujić, "Adaptive tracking control for a class of plants with uncertain parameters and non-linearities", *International Journal of Adaptive Control and Signal Processing*, Vol. 2, pp. 49-71, 1988.

[107] Lj. T. Grujić, "Algebraic conditions for absolute tracking control of continuous-time Lurie systems", *Proceedings of the Conference on Linear Algebra in Signals, Systems, and Control*, (Boston, MA, August 12-14, 1986), published as *Linear Algebra in Signals, Systems, and Control*, Eds. B. N. Datta, C. R. Johnson, M. A. Kaashoek, R. J. Plemmons, and E. D. Sontag, Philadelphia, PA: SIAM, pp. 535-555, 1988.

[108] Lj. T. Grujić, "Algebraic conditions for absolute tracking control of Lurie systems", *Int. J. Control*, Vol. 48, No. 2, pp. 729-754, 1988.

[109] Lj. T. Grujić, "Algorithms for CAD of Continuous-Time Non-Stationary Non-Linear Tracking Systems via the Output-Space", *ACTA Press*, Anaheim, pp. 58-61, 1985.

[110] Lj. T. Grujić, "Algorithms for CAD of Discrete-Time Non-Stationary Non-Linear Tracking Systems via the State-Space", *ACTA Press*, Anaheim, pp. 135-138, 1985.97.

[111] Ly. T. Grouyitch, *Automatique : dynamique linéaire*, Notes de cours, Belfort : Ecole Nationale d'Ingénieurs de Belfort, 1997.

[112] Lj. T. Grujić, *Automatique–Dynamique Linéaire*, Lecture Notes, Belfort: Ecole Nationale d'Ingénieurs de Belfort, 1994–1996.

[113] Ly. T. Grouyitch, *Automatique: dynamique linéaire*, Lecture Notes, Belfort: University of Technology Belfort–Montbeliard, 1999–2000.

[114] Lj. T. Grujić, *Automatique - Dynamique Linéaire*, Lecture Notes, Belfort: Université de Technologie de Belfort–Montbéliard, 2000–2003.

[115] Lj. T. Grujić (Ly. T. Gruyitch), *Continuous Time Control Systems*, Lecture notes for the course "DNEL4CN2: Control Systems", Durban: Department of Electrical Engineering, University of Natal, South Africa, 1993.

[116] Lj. T. Grujić, "Exponential quality of time-varying dynamical systems: stability and tracking", Ch. 5, in *Advances in Nonlinear Dynamics*, Vol. 5, Eds. S. Sivasundaram and A. A. Martynyuk, Amsterdam: Gordon and Breach Science Publishers Ltd., pp. 51-61, 1997.

[117] Lj. T. Grujić, "Natural Trackability and Control: Multiple Time Scale Systems", *Preprints of the IFAC-IFIP-IMACS Conference: Control of Industrial Systems*, **2**, Pergamon, Elsevier, London, pp. 111-116, 1997.

[118] Lj. T. Grujić, "Natural trackability and control: Multiple time scale systems", *Proceedings of the IFAC Conference: Control of Industrial Systems*, (Ed's. Lj. T. Grujić, P. Borne, A. El Moudni and M. Ferney), Vol. 2, Pergamon, Elsevier, London, pp. 669-674, 1997.

[119] Lj. T. Grujić, "Natural Trackability and Control: Perturbed Robots", *Preprints of the IFAC-IFIP-IMACS Conference: Control of Industrial Systems*, **3**, Belfort, France, pp. 691-696, 1997.

[120] Lj. T. Grujić, "Natural trackability and control: perturbed robots", *Proceedings of the IFAC Conference: Control of Industrial Systems*, (Ed's. Lj. T. Grujić, P. Borne, A. El Moudni and M. Ferney), Vol. 3, Pergamon, Elsevier, London, pp. 1641-1646, 1997.

[121] Lj. T. Grujić, "Natural trackability and tracking control of robots", *IMACS-IEEE-SMC Multiconference CESA'96: Symposium on Control, Optimization and Supervision*, Vol. 1, Lille, France, pp. 38-43, 1996.

[122] Lj. T. Grujić, "Non-linear singularly perturbed tracking systems", *Proc. AMSE Conference on Modelling and Simulation*, Paris, pp. 116-123, 1982.

[123] Lj. T. Grujić, "On general solutions of non-linear tracking for stationary systems", *AI 83 IASTED Symposium*, Lille, pp. 49-53, 1983.

[124] Lj. T. Grujić, "On large-scale systems stability", in *Computing and computers for control systems*, Eds. P. Borne et al., J. C. Baltzer AG, Scientific Publishing Co., IMACS, pp. 201-206, 1989.

[125] Lj. T. Grujić, "On non-linear tracking domain estimates: Continuous-time systems", *AI 83 IASTED Symposium*, Lille, pp. 65-66, 1983.

[126] Lj. T. Grujić, "On non-linear tracking domain estimates: Discrete-time systems", *AI 83 IASTED Symposium*, Lille, pp. 59-63, 1983.

[127] Lj. T. Grujić, "On non-linear tracking phenomena and problems", *AI 83 IASTED Symposium*, Lille, pp. 45-48, 1983.

[128] Lj. T. Grujić, "On the non-linear tracking systems theory: I-Phenomena, concepts and problems via ouptut space", *Automatika*, Zagreb, Vol. 27, No. 1-2, pp. 3-8, 1986.

[129] Lj. T. Grujić, "On the non-linear tracking systems theory: II-Phenomena, concepts and problems via state space", *Automatika*, Zagreb, Vol. 27, No. 1-2, pp. 9-16, 1986.

[130] Lj. T. Grujić, "On the non-linear tracking systems theory: III-Liapunov-like approach via the output-space: Continuous-time", *Automatika*, Zagreb, Vol. 27, No. 3-4, pp. 99-104, 1986.

[131] Lj. T. Grujić, "On the non-linear tracking systems theory: IV-Liapunov-like approach via the state-space: Continuous-time", *Automatika*, Zagreb, Vol. 27, No. 3-4, pp. 105-116, 1986.

[132] Lj. T. Grujić, "On the non-linear tracking systems theory: V-Liapunov-like approach via the output-space: Discrete-time", *Automatika*, Zagreb, Vol. 27, No. 5-6, pp. 197-202, 1986.

[133] Lj. T. Grujić, "On the non-linear tracking systems theory: VI-Liapunov-like approach via the state-space: Discrete-time", *Automatika*, Zagreb, Vol. 27, No. 5-6, pp. 203-211, 1986.v

[134] Lj. T. Grujić, "On the theory and synthesis of non-linear non-stationary tracking singularly perturbed systems", *Control Theory and Advanced Technology*, MITA Press, Tokyo, Japan, Vol. 4, No. 4, pp. 395-409, 1988.

[135] Lj. T. Grujić, "On the theory of nonlinear systems tracking with guaranteed performance index bounds: Application to robot control", *Proceedings of the 1989 IEEE International Conference on Robotics and Automation*, Scottsdale, AZ: IEEE-Computer Society Press, Vol. 3, pp. 1486-1490, May 14-19, 1989.

[136] Lj. T. Grujić, "On the tracking problem for nonlinear systems", in *Applied Control*, Ed. S. G. Tzafests, New York, NY: Marcel Dekker, pp. 325-343, 1993.

[137] Lj. T. Grujić, "Phenomena, concepts and problems of automatic tracking: Continuous-time stationary non-linear systems with variable inputs" (in Serbo-Croatian), *Proc. of the First International Seminar "AUTOMATON and ROBOT"*, USAUM Srbije i "OMO", Belgrade, pp. 307-330, 1985.

[138] Lj. T. Grujić, "Phenomena, concepts and problems of automatic tracking: Discrete-time stationary non-linear systems with variable inputs" (in Serbo-Croatian), *Proc. of the First International Seminar "AUTOMATON and ROBOT"*, USAUM Srbije i "OMO", Belgrade, pp. 401-422, 1985.

[139] Lj. T. Grujić, "Possibilities of Linear System Design on the Basis of Conditional Optimization in Parameter Plane", Part I, *Automatika: Theoretical supplement*, Zagreb (Yugoslavia-Croatia), Vol. 2, No. 1-2, pp. 49-60, 1966.

[140] Lj. T. Grujić, "Stability versus tracking in automatic control systems" (in Serbo-Croatian), *Proc. JUREMA 29*, Part 1, Zagreb, pp. 1-4, 1984.

[141] Lj. T. Grujić, "Synthesis of automatic tracking systems and the output-space: Continuous-time stationary non-linear systems with variable inputs" (in Serbo-Croatian), *Proc. of the First International Seminar "AUTOMATON and ROBOT"*, USAUM Srbije i "OMO", Belgrade, pp. 331-370, 1985.

[142] Lj. T. Grujić, "Synthesis of automatic tracking systems and the state-space: Continuous-time stationary non-linear systems with variable inputs" (in Serbo-Croatian), *Proc. of the First International Seminar "AUTOMATON and ROBOT"*, USAUM Srbije i "OMO", Belgrade, pp. 371-400, 1985.

[143] Lj. T. Grujić, "Tracking control obeying prespecified performance index", *Proc. 12th World Congress on Scientific Computation*, IMACS, Paris, France, pp. 332-336, July 18-22, 1988.

[144] Lj. T. Grujić, "Tracking versus stability: Theory", *Proc. 12th World Congress on Scientific Computation*, IMACS, Paris, pp. 319-327, July 18-22, 1988.

[145] Lj. T. Grujić, "Tracking with prespecified index limits: Control synthesis for non-linear objects", *Proc. II International Seminar and Sympo-*

sium: "AUTOMATON and ROBOT", SAUM and IEE, Belgrade, pp. S-20-S-52, 1987.

[146] Lj. T. Grujić, A. A. Martynyuk and M. Ribens-Pavella, *Large-Scale Systems under Structural and Singular Perturbations* (in Russian, Kiev: Naukova Dumka, 1984), Berlin, Germany: Springer Verlag, 1987.

[147] Lj. T. Grujić and W. P. Mounfield, *Natural Tracking Controller*, US Patent No 5,379,210, Jan. 3, 1995.

[148] Lj. T. Grujić and W. P. Mounfield, Jr., "Natural Tracking Control of Linear Systems", *Proceedings of the 13th IMACS World Congress on Computation and Applied Mathematics*, Eds. R. Vichnevetsky and J. J. H. Miller, Trinity College, Dublin, Ireland, Vol. 3, pp. 1269-1270, July 22 - 26, 1991.

[149] Lj. T. Grujić and W. P. Mounfield, "Natural Tracking Control of Linear Systems", in *Mathematics of the Analysis and Design of Process Control*, Ed. P. Borne, S.G. Tzafestas and N.E. Radhy, Elsevier Science Publishers B. V., IMACS, pp. 53-64, 1992.

[150] Lj. T. Grujić and W. P. Mounfield, "Natural Tracking PID Process Control for Exponential Tracking", *American Institute of Chemical Engineers Journal*, **38**, No. 4, pp. 555-562, 1992.

[151] Lj. T. Grujić and W. P. Mounfield, "PD-Control for Stablewise Tracking with Finite Reachability Time: Linear Continuous Time MIMO Systems with State-Space Description", *International Journal of Robust and Nonlinear Control*, England, Vol. 3, pp. 341-360, 1993.

[152] Lj. T. Grujić and W. P. Mounfield, "PD Natural Tracking Control of an Unstable Chemical Reaction", *Proc. 1993 IEEE International Conference on Systems, Man and Cybernetics*, Le Touquet, Vol. 2, pp. 730-735, 1993.

[153] Lj. T. Grujić and Mounfield W. P., "PID Natural Tracking Control of a Robot: Theory", *Proc. 1993 IEEE International Conference on Systems, Man and Cybernetics*, Le Touquet, Vol. 4, pp. 323-327, 1993.

[154] Lj. T. Grujić and W. P. Mounfield, "Ship Roll Stabilization by Natural Tracking Control: Stablewise Tracking with Finite Reachability Time", *Proc. 3rd IFAC Workshop on Control Applications in Marine Systems*, Trondheim, Norway, pp. 202-207, 10-12 May, 1995.

[155] Lj. T. Grujić and W. P. Mounfield, "Stablewise Tracking with Finite Reachability Time: Linear Time-Invariant Continuous-Time MIMO Systems", *Proc. of the 31st IEEE Conference on Decision and Control*, Tucson, Arizona, pp. 834-839, 1992.

[156] Lj. T. Grujić and W. P. Mounfield, Jr., "Tracking Control of Time-Invariant Linear Systems Described by IO Differential Equations", *Proceedings of the 30th IMACS Conference on Decision and Control*, Brighton, England, Vol. 3, pp. 2441-2446, December 11-13, 1991.

[157] Lj. T. Grujić and B. Porter, "Continuous-time tracking systems incorporating Lur'e plants with single non-linearities", *Int. J. Systems Science*, Vol. 11, No. 2, pp. 177-189, 1980.

[158] Lj. T. Grujić and B. Porter, "Discrete-time tracking systems incorporating Lur'e plants with mutliple non-linearities", *Int. J. Systems Science*, Vol. 11, No. 12, pp. 1505-1520, 1980.

[159] Ly. T. Gruyitch, *Advances in the Linear Dynamic Systems Theory*, Tamarac, FL: Llumina Press, 2013.

[160] Ly. T. Gruyitch, "Aircraft natural control synthesis: Vector Lyapunov function approach," *Actual Problems of Airplane and Aerospace Systems: Processes, Models, Experiments*, Vol. 2, No. 6, Kazan, Russia, and Daytona Beach, FL, pp. 1-9, 1998.

[161] Ly. T. Gruyitch, "A physical principle and consistent Lyapunov methodology: time-invariant nonlinear systems", *Proc. International Conference on Advances in Systems, Signals, Control and Computers*, 1, Durban, South Africa, pp. 42 - 50, 1998.

[162] Ly. T. Gruyitch, *Conduite des systèmes*, Lecture Notes: Notes de cours SY 98, Belfort: University of Technology Belfort-Montbéliard, 2000, 2001.

[163] Ly. T. Gruyitch, "Consistent Lyapunov Methodology for Exponential Stability: PCUP Approach", in *Advances in Stability Theory at the End of the 20th Century*, Ed. A. A. Martynyuk, London: Taylor and Francis, pp. 107-120, 2003.

[164] Ly. T. Gruyitch, *Contrôle commande des processus industriels*, Lecture Notes: Notes de cours SY 51, Belfort: University of Technology Belfort-Montbeliard, 2002, 2003.

[165] Ly. T. Gruyitch, *Einstein's Relativity Theory. Correct, Paradoxical, and Wrong,* Trafford, Victoria, Canada, http://www.trafford.com/06-2239, 2006.

[166] Ly. T. Gruyitch, "Exponential stabilizing natural tracking control of robots: theory", *Proceedings of the Third ASCE Specialty Conference on Robotics for Challenging Environments,* held in Albuquerque, New Mexico, USA, (Ed's. Laura A. Demsetz, Raymond H. Bryne and John P. Wetzel), Reston, Virginia, USA: American Society of Civil Engineers (ASCE), pp. 286-292, April 26-30, 1998.

[167] Ly. T. Gruyitch, *Galilean-Newtonean Rebuttal to Einsteins Relativity Theory,* Cambridge International Science Publishing, Cambridge UK, 2015.

[168] Ly. T. Gruyitch, "Gaussian generalisations of the relativity theory fundaments with applications", *Proceedings of the VII International Conference: Physical Interpretations of Relativity Theory,* Ed. M. C. Duffy, British Society for the Philosophy of Science, London, pp. 125-136, September 15-18, 2000.

[169] Ly. T. Gruyitch, "Global natural θ-tracking control of Lagrangian systems", *Proceedings of the American Control Conference,* San Diego, CA, pp. 2996-3000, June 1999.

[170] Ly. T. Gruyitch, *Linear Continuous-time Systems,* Boca Raton, FL: CRC Press/Taylor & Francis Group, 2017.

[171] Ly. T. Gruyitch, *Linear Control Systems. I: Observability and Controllability of General Linear Systems,* in print, Boca Raton, FL: CRC Press, 2018.

[172] Ly. T. Gruyitch, "Natural control of robots for fine tracking", *Proceedings of the 38th Conference on Decision and Control,* Phoenix, Arizona, USA, pp. 5102-5107, December 1999.

[173] Ly. T. Gruyitch, "Natural tracking control synthesis for Lagrangian systems", *V International Seminar on Stability and Oscillations of Nonlinear Control Systems,* Russian Academy of Sciences, Moscow, pp. 115-120, June 3-5, 1998.

[174] Ly. T. Gruyitch, "New Development of vector Lyapunov functions and airplane control synthesis", Chapter 7 in *Advances in Dynamics and Control*, Ed. S. Sivasundaram, Boca Raton, FL: Chapman & Hall/CRC, pp. 89-102, 2004.

[175] Ly. T. Gruyitch, *Nonlinear Systems Tracking*, Boca Raton, FL: CRC Press/Taylor & Francis Group, 2016.

[176] Ly. T. Gruyitch, "On tracking theory with embedded stability: control duality resolution", *Proceedings of the 40th IEEE Conference on Decision and Control*, Orlando, FL, pp. 4003-4008, December 2001.

[177] Ly. T. Gruyitch, "Physical Continuity and Uniqueness Principle. Exponential Natural Tracking Control", *Neural, Parallel & Scientific Computations*, 6, pp. 143-170, 1998.

[178] Ly. T. Gruyitch, "Robot global tracking with finite vector reachability time", *Proceedings of the European Control Conference*, Karlsruhe, Germany, Paper # 132, pp. 1-6, 31 August - 3 September 1999.

[179] Ly. T. Gruyitch, "Robust prespecified quality tracking control synthesis for 2D systems", *Proc. International Conference on Advances in Systems, Signals, Control and Computers*, **3**, Durban, South Africa, pp. 171-175, 1998.

[180] Ly. T. Gruyitch, *Systèmes d'asservissement industriels*, Lecture Notes: Notes de cours SY 40, Belfort: Universite de Technologie de Belfort-Montbeliard, 2001.

[181] Ly. T. Gruyitch, *Time and Consistent Relativity. Physical and Mathematical Fundamentals*, Waretown N.J and Oakville ON: Apple Academic Press, Inc., 2015.

[182] Ly. T. Gruyitch, *Time. Fields, Relativity, and Systems*, Coral Springs, FL: Llumina, 2006.

[183] Ly. T. Gruyitch, *Time and Time Fields. Modeling, Relativity, and Systems Control*, Trafford, Victoria, Canada, 2006.

[184] Ly. T. Gruyitch, "Time, Relativity and Physical Principle: Generalizations and Applications", *Proc. V International Conference: Physical Interpretations of Relativity Theory*, (Ed. M. C. Duffy), pp. 134-170, London, 11-14 September, 1998; (also in: *Nelinijni Koluvannya*, Vol. 2, No. 4, pp. 465-489, Kiev, Ukraine, 1999).

[185] Ly. T. Gruyitch, "Time, Systems and Control", invited, submitted and accepted paper has had 50 pages, its abstract was published in *Abstracts of the papers of the VIII International seminar "Stability and oscillations of nonlinear control systems"*, Ed. V. N. Thai, Moscow: IPU RAN, June 2-4, 2004.

[186] Ly. T. Gruyitch, "Time, Systems, and Control: Qualitative Properties and Methods", Chapter 2 in *Stability and Control of Dynamical Systems with Applications*, Editors D. Liu and P. J. Antsaklis, pp. 23-46, Boston: Birkhäuser, pp. 23-46, 2003.

[187] Ly. T. Gruyitch, "Time and Uniform Relativity Theory Fundaments", *Problems of Nonlinear Analysis in Engineering Systems*, 7, No. 2(14), Kazan, Russia, pp. 1-29, 2001.

[188] Ly. T. Gruyitch, *Tracking Control of Linear Systems*, Boca Raton, FL: CRC Press/Taylor & Francis Group, 2013.

[189] Ly. T. Gruyitch, "Vector Lyapunov function synthesis of aircraft control", *Proceedings of INPAA-98: Second International Conference on Nonlinear Problems in Aviation & Aerospace*, Ed. Seenith Sivasundaram, ISBN: 0 9526643 1 3, Cambridge UK: European Conference Publications, Vol. 1, pp. 253-260, 1999.

[190] Ly. T. Gruyitch and W. Pratt Mounfield, Jr., "Absolute output natural tracking control: MIMO Lurie systems", *Proceedings of the 14th Triennial World Congress*, Beijing, P. R. China, Pergamon - Elsevier Science, Vol. C, pp. 389-394, July 5-9, 1999.

[191] Ly. T. Gruyitch and W. Pratt Mounfield, Jr., "Constrained natural tracking control algorithms for bilinear DC shunt wound motors", *Proceedings of the 40th IEEE Conference on Decision and Control*, Orlando, FL, pp. 4433-4438, December 2001.

[192] Ly. T. Gruyitch and W. Pratt Mounfield, Jr., "Elementwise stablewise tracking with finite reachability time: linear time-invariant continuous-time MIMO systems", *International Journal of Systems Science*, Vol. 33, No.4, pp. 277-299, 2002.

[193] Ly. T. Gruyitch and W. P. Mounfield, Jr., "Robust elementwise exponential tracking control: IO linear systems", *Proceedings of th 36th IEEE Conference on Decision and Control*, San Diego, CA, pp. 3836-3841, December 1997.

[194] Ly. T. Gruyitch and W. Pratt Mounfield, Jr., "Stablewise absolute output natural tracking control with finite reachability time: MIMO Lurie systems", *CD Rom Proceedings of the 17th IMACS World Congress, Invited session IS-2 : Tracking theory and control of nonlinear systems*, Paris, France, pp. 1-17, July 11-15, 2005; *Mathematics and Computers in Simulation*, Vol. 76, pp. 330 - 344, 2008.

[195] Ly. T. Gruyitch, J.-P. Richard, P. Borne, J.-C. Gentina, *Stability Domains*, Boca Raton, FL: Chapman & Hall/CRC, CRC Press company, 2004.

[196] M. Haidekker, *Linear Feedback Controls*, London: Elsevier, 2013.

[197] M. L. J. Hautus, "Controllability and observability conditions of linear autonomous systems", *Nederlandse Akademie Vanwettenschappen*, Series A, V72, 1969, pp. 443-448.

[198] E. Hendricks, O. Sørensen and H. Paul, *Linear Systems Control*, Berlin: Springer, 2008.

[199] J. P. Hespanh, *Linear Systems Theory*, Princeton, NJ: Princeton University Press, 2009.

[200] C. H. Houpis and S. N. Sheldon, *Linear Control System Analysis and Design with MATLAB®*, Boca Raton, FL: CRC Press, Taylor & Francis Group, 2014.

[201] R. O. Hughes, "Optimal control of sun tracking solar concentrators", *Journal of Dynamic Systems, Measurement, and Control*, Vol. 101, No. 2, pp. 157-161, 1979.

[202] D. G. Hull, *Optimal Control Theory for Applications*, New York: Springer Verlag, 2003.

[203] E. Jarzębowska, *Model-Based Tracking Control of Nonlinear Systems*, Boca Raton, FL: Chapman and Hall/CRC Press, 2012.

[204] T. Kaczorek, *Polynomial and Rational Matrices: Applications in Dynamical Systems Theory*, Berlin: Springer, 2007.

[205] T. Kailath, *Linear Systems*, Englewood Cliffs, NJ: Prentice-Hall, Inc., 1980.

[206] R. E. Kalman, "Algebraic structure of linear dynamical systems, I. The module of Σ," *Proc. National Academy of Science: Mathematics*, USA NAS, Vol. 54, pp. 1503-1508, 1965.

[207] R. E. Kalman, "Canonical structure of linear dynamical systems", *Proceedings of the National Academy of Science: Mathematics*, USA NAS, Vol. 48, pp. 596-600, 1962.

[208] R. E. Kalman, "Mathematical description of linear dynamical systems", *J.S.I.A.M. Control*, Ser. A, Vol. 1, No. 2, pp. 152-192, 1963.

[209] R. E. Kalman, "On the General Theory of Control Systems", *Proceedings of the First International Congress on Automatic Control*, London: Butterworth, 1960, pp. 481-491.

[210] R. E. Kalman, P. L. Falb and M. A. Arbib, *Topics on Mathenatical System Theory*, NY: Mc Graw Hill, 1969.

[211] R. E. Kalman, Y. C. Ho and K. S. Narendra, "Controllability of Linear Dynamical Systems", *Contributions to Differential Equations*, Vol.1, No. 2, pp. 189-213, 1963.

[212] D. E. Kirk, *Optimal Control Theory: An Introduction.* Englewood Cliffs, NJ.: Prentice-Hall 1970, Dover republication 2004.

[213] B. Kisačanin and G. C. Agarwal, *Linear Control Systems: With Solved Problems and MATLAB Examples*, NY: Kluwer Academic Press/Plenum Publishers, 2001.

[214] A. Kökösy, *Poursuite Pratique de Systemes de commande Automatique des Robots Industriels*, Ph. D. dissertation, Belfort, France: University of Belfort-Montbeliard, 1999.

[215] A. Kőkősy, "Practical tracking with settling time: Bounded control for robot motion", *Proceedings of the 14th IFAC Triennial World Congress*, Beijing, P. R. China, C-2a-11-4, pp. 377–382, 1999.

[216] A. Kőkősy, "Practical tracking with settling time: Criteria and algorithms", *Proceedings of the VI International SAUM Conference on Systems, Automatic Control and Measurement*, Nish, Serbia, Yugoslavia, pp. 296–301, September 28–30, 1998.

[217] A. Kőkősy, "Practical tracking with vector settling and vector reachability time", *Proceedings of the IFAC Workshop on Motion Control*, Grenoble, France, pp, 297–302, Sept. 21–23, 1998.

[218] A. Kőkősy, "Robot control: Practical tracking with reachability time", *Proceedings of the International Conference on System, Signals, Control, Computers*, Vol. 3, Durban, RSA, pp. 186-190, 1998.

[219] B. C. Kuo, *Automatic Control Systems*, Englewood Cliffs, NJ: Prentice-Hall, Inc., 1967.

[220] B. C. Kuo, *Automatic Control Systems*, Englewood Cliffs, NJ: Prentice-Hall, Inc., 1987.

[221] H. Kwakernaak and R. Sivan, *Linear Optimal Control Systems*, New York: Wiley-Interscience, 1972.

[222] P. Lancaster and M. Tismenetsky, *The Theory of Matrices*, San Diego: Academic Press, 1985.

[223] H. Lauer, R. Lesnick and L. E. Matson, *Servomechanism Fundamentals*, New York, NY: McGraw-Hill Book Company, Inc., 1947.

[224] D. V. Lazitch, *Analysis and Synthesis of Practical Tracking Automatic Control* (in Serb), D. Sci. Dissertation, Faculty of Mechanical Engineering, University of Belgrade, Belgrade, Serbia, 1995.

[225] D. V. Lazitch, "Uniform exponential practical automatic control tracking" (in Serb), *Proceedings of the V Conference on Systems, Automatic Control and Measurement (SAUM)*, Novi Sad, pp. 68-70, October 2-3, 1995.

[226] D. V. Lazitch, "Uniform practical automatic control tracking" (in Serb), *Proceedings of the V Conference on Systems, Automatic Control and Measurement (SAUM)*, Novi Sad, pp. 53-57, October 2-3, 1995.

[227] D. V. Lazitch, "Uniform practical automatic control tracking with the vector reachability time" (in Serb), *Proceedings of the V Conference on Systems, Automatic Control and Measurement (SAUM)*, Novi Sad, pp. 63-67, October 2-3, 1995.

[228] D. V. Lazitch, "Uniform practical automatic control tracking with the vector settling time", (in Serb), *Proceedings of the V Conference on*

Systems, Automatic Control and Measurement (SAUM), Novi Sad, pp. 58-62, October 2-3, 1995.

[229] J. R. Leigh, *Functional Analysis and Linear Control Theory*, Mineola, NY: Dover Publications, Inc., 1980, 2007.

[230] A. M. Lyapunov, *The General Problem of Stability of Motion*, (in Russian), Kharkov Mathematical Society, Kharkov, 1892; in Academician A. M. Lyapunov: "Collected Papers", U.S.S.R. Academy of Science, Moscow, II, pp. 5-263, 1956. French translation: "Problème général de la stabilité du mouvement", *Ann. Fac. Toulouse*, 9, pp. 203 - 474; also in: *Annals of Mathematics Study*, No. 17, Princeton University Press, 1949. English translation: *Intern. J. Control*, 55, pp. 531-773, 1992; also the book, Taylor and Francis, London, 1992.

[231] L. A. MacColl, *Fundamental Theory of Servomechanisms*, New York: D. Van Nostrand Company, Inc., 1945.

[232] J. M. Maciejowski, *Multivariable Feedback Systems*, Wokingham: Addison-Wesley Publishing Company, 1989.

[233] G. Marsaglla, *Bounds for the rank of the sum of two matrices*, Seattle: Mathematics Research Laboratory, Boeing Scientific Research Laboratories: Mathematical Note No, 3bA, D1-82-0343, pp. 1-13, April 1964. http://www.dtic.mil/dtic/tr/fulltext/u2/600471.pdf

[234] G. Marsaglla, *Bounds for the Rank of the Sum of Matrices*, Seattle: Mathematics Research Laboratory, Boeing Scientific Research Laboratories, pp. 455-462, April 1964. http://www.ic.unicamp.br/~meidanis/PUB/Doutorado/2012-Biller/Marsaglia1964.pdf

[235] V. M. Matrosov, "To the theory of stability of motion" (in Russian), *Prikl. Math. Mekh.* Vol. 26, No. 5, pp. 885-895, 1962.

[236] V. M. Matrosov, *Vector Lyapunov Function Method: Analysis of Dynamical Properties of Nonlinear Systems* (in Russian), Moscow, Russia: FIZMATLIT, 2001.

[237] V. M. Matrosov, "Vector Lyapunov functions in the analysis of nonlinear interconnected systems", *Proc. Symp. Mathematica*, Bologna, Italy, Vol. 6, pp. 209-242, 1971.

[238] E. J. McShane, *Integration*, Princeton, NJ: Princeton University Press, 1944.

[239] J. L. Melsa and D. G, Schultz, *Linear Control Systems*, New York: McGraw-Hill Book Company, 1969.

[240] A. N. Michel and C. J. Herget, *Algebra and Analysis for Engineers and Scientists*, Boston, MA: Birkhäuser, 2007.

[241] A. N. Michel and R. K. Miller, *Qualitative Analysis of Large-Scale Dynamical Systems*, New York, NY, USA: Academic Press, 1977.

[242] D. E. Miller and E. J. Davison, "The self-tuning robust servomechanism problem", *IEEE Transactions on Automatic Control*, Vol. AC-34, No. 5, pp. 511-523, May 1989.

[243] R. K. Miller and A. N. Michel, *Ordinary Differential Equations*, New York, NY: Academic Press, 1982.

[244] B. R. Milojković and Lj. T. Grujić, *Automatic Control*, the textbook in Serbo-Croatian, Belgrade: Faculty of Mechanical Engineering, University of Belgrade, 1977.

[245] B. R. Milojković and L. T. Grujić, *Automatic control*, in Serbo-Croatian, Belgrade: Faculty of Mechanical Engineering, 1981.

[246] N. Minamide, "Design of a deadbeat adaptive tracking system", *International Journal of Control*, Vol. 39, No. 1, pp. 63-81, 1984.

[247] W. P. Mounfield, Jr. and Lj. T. Grujić, "High-Gain Natural Tracking Control of Linear Systems", *Proceedings of the 13th IMACS World Congress on Computation and Applied Mathematics*, Eds. R. Vichnevetsky and J. J. H. Miller, Trinity College, Dublin, Ireland, Vol. 3, pp. 1271-1272, 1991.

[248] W. P. Mounfield, Jr. and Lj. T. Grujić, "High-Gain Natural Tracking Control of Time-Invariant Systems Described by IO Differential Equations", *Proceedings of the 30th Conference on Decision and Control*, Brighton, England, pp. 2447-2452, 1991.

[249] W. P. Mounfield, Jr. and Lj. T. Grujić, "High-Gain Natural Tracking Control of Linear Systems", *Proceedings of the 13th IMACS World Congress on Computation and Applied Mathematics*, Eds. R. Vichnevetsky and J. J. H. Miller, Trinity College, Dublin, Ireland, Vol. 3, pp. 1271-1272, July 22-26, 1991.

[250] W. P. Mounfield, Jr. and Lj. T. Grujić, "High-gain PI natural tracking control for exponential tracking of linear single-output systems with state-space description", *RAIRO- Automatique, Productique, Informatique Industrielle (APII)*, Vo. 26, pp. 125-146, 1992.

[251] W. P. Mounfield and Lj. T. Grujić, "High-Gain PI Control of an Aircraft Lateral Control System", *Proc. 1993 IEEE International Conference on Systems, Man and Cybernetics*, Le Touquet, Vol. 2, pp. 736-741, 1993.

[252] W. P. Mounfield, Jr. and Lj. T. Grujić, "High-gain PI natural tracking control for exponential tracking of linear MIMO systems with state-space description", *International Journal of Control*, Vol. 25, No. 11, pp. 1793-1817, 1994.

[253] W. P. Mounfield and Lj. T. Grujić, "Natural tracking control for exponential tracking : lateral high-gain PI control of an aircraft system with state-space description", *Neural, Parallel & Scientific Computations*, Vol. 1, No. 3, pp. 357-370, 1993.

[254] W. P. Mounfield and Lj. T. Grujić, "PID - Natural Tracking Control of a Robot: Application", *Proc. 1993 IEEE International Conference on Systems, Man and Cybernetics*, Le Touquet, Vol. 4, pp. 328-333, 1993.

[255] W. P. Mounfield and Lj. T. Grujić, " Robust Natural Tracking Control for Multi-Zone Space Heating Systems", *Proc. 14th IMACS World Congress*, Vol. 2, 841-843, 1994.

[256] W. P. Mounfield, Jr. and Ly. T. Gruyitch, "Control of aircrafts with redundant control surfaces: stablewise tracking control with finite reachability time", *Proceedings of the Second International Conference on Nonlinear Problems in Aviation and Aerospace*, European Conference Publishers, Cambridge, Vol. 2, 1999, pp. 547-554, 1999.

[257] W. P. Mounfield, Jr. and Ly. T. Gruyitch, "Elementwise stablewise finite reachability time natural tracking control of robots", *Proceedings of the 14th Triennial World Congress*, Beijing, P. R. China, Pergamon - Elsevier Science, Vol. B, pp. 31-36, July 5-9, 1999.

[258] T. Nambu, *Theory of Stabilization for Linear Boundary Control Systems*, Boca Raton, FL: CRC Press, 2016.

[259] D. B. Nauparac, *Analysis of the Tracking Theory on the Real Electro-hydraulic Servosystem* (in Serb), M. Sci. Thesis, Faculty of Mechanical Engineering, University of Belgrade, Belgrade. Serbia, 1993.

[260] N. N. Nedić and D. H. Pršić, "Pneumatic and hydraulic time varying desired motion control using natural tracking control", *SAI-Avtomatika i Informatika '2000*, Sofia, Vol. 3, pp. 37-40, 24-26 ocktomvri, 2000.

[261] N. N. Nedić and D. H. Pršić, "Pneumatic position control using natural tracking law", *IFAC Workshop on Trends in Hydraulic and Pneumatic Components and Systems*, Chicago, IL, pp. 1-13, Nov. 8-9, 1994.

[262] N. N. Nedić and D. H. Pršić, "Time variable speed control of pump controlled hydraulic motor using natural tracking control", *IFAC-IFIP-IMACS International Conference on Control of Industrial Systems*, Belfort, France, pp. 197-202, May 20-22, 1997.

[263] Sir Isaac Newton, *Mathematical Principles of Natural Philosophy - BOOK I. The Motion of Bodies*, Chicago: William Benton, Publisher, Encyclopedia Britannica, Inc., *(The first publication: 1687)* 1952.

[264] K. Ogata, *State Space Analysis of Control Systems*, Englewood Cliffs, NJ: Prentice Hall, 1967.

[265] K. Ogata, *Modern Control Engineering*, Englewood Cliffs, NJ: Prentice Hall, 1970.

[266] D. H. Owens, *Feedback and Multivariable Systems*, Stevenage, Herts: Peter Peregrinus Ltd., 1978.

[267] *Polynomial and Polynomial Matrix Glossary*, http://www.polyx.cz/glossary.htm, 2017.

[268] B. Porter, "Fast-sampling tracking systems incorporating Lur'e plants with multiple nonlinearities", *Int. J. Control*, Vol. 34, No. 2, pp. 333-344, 1981.

[269] B. Porter, "High-gain tracking systems incorporating Lur'e plants with multiple nonlinearities", *Int. J. Control*, Vol. 34, No. 2, pp. 345-358, 1981.

[270] B. Porter, "High-gain tracking systems incorporating Lur'e plants with multiple switching nonlinearities", *VIII IFAC World Congress*, Kyoto, Japan, Session 3, pp. I-72-I-77, 1981.

[271] B. Porter and Lj. T. Grujić, "Continuous-time tracking systems incorporating Lur'e plants with multiple non-linearities", *Int. J. Systems Science*, Vol. 11, No. 7, pp. 827-840, 1980.

[272] B. Porter and Lj. T. Grujić, "Discrete-time tracking systems incorporating Lur'e plants with single non-linearities", *Third IMA Conference on Control Theory*, Academic Press, London, pp. 115-133, 1981.

[273] D. Prshitch and N, Neditch, "Pneumatic cylinder control by using natural control" (in Serb), *HIPNEF'93*, Belgrade, Serbia, pp. 127-132, 1993.

[274] H. M. Power and R. J. Simpson, *Introduction to Dynamics and Control*, London: McGraw-Hill Book Company (UK) Limited, 1978.

[275] Z. Qu and D. M. Dawson, *Robust Tracking Control of Robot Manipulators*, Piscataway, NJ: IEEE Press, 1995.

[276] K. V. Ramachandra, *Kalman Filtering Techniques for Radar Tracking*, New York: Marcel Dekker, 2000

[277] Z. Retchkiman, "The problem of output tracking for nonlinear systems in the presence of dingular points," *Proc. American Control Conference*, Vol. 3, pp. 2975-2979, San Francisco, CA, June 2-4, 1983.

[278] Z. Retchkiman, J. Alvarez, and R. Castro, "Asymptotic output tracking through singular points for nonlinear systems: Stability, disturbance rejection and robustness," *Int. J. Robust and Nonlinear Control*, Vol. 5, pp. 553 -572, 1995.

[279] Z. B. Ribar, D. V. Lazic and M. R. Jovanovic, "Application of practical exponential tracking in fluid transportation industry", *Proceedings of the 14th International Conference on Material Handling and Warehousing*, Belgrade, Serbia, pp. 5.65-5.70, December 11-12, 1996.

[280] Z. Ribar, R. Jovanović, and D. Sekulić, "Fuzzy control of a hydraulic servosystem based on practical tracking algorithms," *Proc. Bath Workshop on Power Transmission and Motion Control*, Eds. C. R. Burrows and K. A. Edge, London: Professional Engineering Publishing, 2000, pp. 73-87.

[281] Z. B. Ribar, M. R. Yovanovitch and R. Z. Yovanovitch, "Application of practical linear tracking in process industry" (in Serb), *Proceedings*

of the XLI Conference ETRAN, Zlatibor, Serbia, Notebook 1, pp. 444-447, June 3-6, 1997.

[282] H. H. Rosenbrock, "Some properties of relatively prime polynomial matrices", *Electronic Letters*, Vol. 4, 1968, pp. 374-375.

[283] H. H. Rosenbrock, *State-Space and Multivariable Theory*, London: Thomas Nelson and Sons Ltd., 1970.

[284] H. Seraji, "Transfer function matrix", *Electronic Letters*, Vol. 23, No.4, pp. 256-257, 1987.

[285] Y. B. Shtessel, "Nonlinear nonminimum phase output tracking via dynamic sliding manifolds", *J. Franklin Inst.*, 335B, No. 5, pp. 841-850, 1998.

[286] Y. B. Shtessel, "Nonlinear output tracking in conventional and dynamic sliding manifolds", *IEEE Transactions on Automatic Control*, Vol. 42, No. 9, pp. 1282-1286, September 1997.

[287] D. D. Šiljak, *Large-Scale Dynamic Systems: Stability and Structure*, New York: North Holland, 1978.

[288] C. Silvestre, A. Pascoal and I. Kaminer, "On the design of gain-scheduled trajectory tracking controllers," *Int. J. Robust and Nonlinear Control*, Vol. 12, pp. 797-839, 2002.

[289] A. Sinha, *Linear Systems: Optimal and Robust Control*, Boca Raton, FL: CRC Press, Taylor & Francis Group, 2007.

[290] R. E. Skelton, *Dynamic Systems Control: Linear Systems Analysis and Synthesis*, New York: John Wiley & Sons, 1988.

[291] J. J. Slotine and S. S. Sastry, "Tracking control of non-linear systems using sliding surfaces, with application to robot manipulators," *Int. J. Control*, Vol. 38, No. 2, pp. 465-492, 1983.

[292] Y. D. Song, "Adaptive motion tracking control of robot manipulators-Non-regresssor based approach," *Int. J. Control*, Vol. 63, No. 1, pp. 41-54, 1996.

[293] E. D. Sontag, "Further facts about input to state stabilization", *IEEE Transactions on Automatic Control*, Vol. 35, No. 3, pp. 473-476, 1990.

[294] S. K. Spurgeon and X. Y. Lu, "Output tracking using dynamic sliding mode techniques", *International Journal of Robust and Nonlinear Control*, Vol. 7, pp. 407-427, 1997.

[295] L. D. Stone, C. A. Barlow, and T. L. Corwin, *Bayesian Multiple Target Tracking*, Norwood, MA: Artech House, 1999.

[296] T. Sugie and M. Vidyasagar, "Further results on the robust tracking problem in two-degree-of-freedom control systems", *Systems & Control Letters*, Vol. 13, pp. 101-108, 1989.

[297] D. D. Šiljak, *Feedback Systems Synthesis via Squared Error Conditional Optimization*, D. Sc. Dissertation (in Serbo-Croatian), Departement of Electrical Engineering, University of Belgrade, Belgrade (Yugoslavia-Serbia), 1963.

[298] A. I. Talkin, "Adaptive servo tracking", *IRE Transactions on Automatic Control*, Vol. 6, No. 2, pp. 167-172, May 1961.

[299] L. Tan, *A Generalized Framework of Linear Multivariable Control*, Kidlington-Oxford: Butterworth-Heinemann, 2017.

[300] S. Tarbouriech, C. Pittet and C. Burgat, "Output tracking problem for systems with input saturations via nonlinear integrating actions", *Int. J. Robust and Nonlinear Control*, Vol. 10, pp. 489-512, 2000.

[301] B. O. S. Teixeira, M. A. Santillo, R. S. Erwin and D. S. Bernstein, "Spacecraft tracking using sampled-data Kalman filters", *IEEE Control Systems Magazine*, Vol. 28, No. 4, pp. 78-94, August 2008.

[302] H. T. Toivonen and J. Pensar, "A worst-case approach to optimal tracking control with robust performance", *Int. J. Control*, Vol. 65, No. 1, pp. 17-32, 1996.

[303] D. E. Torfs, R. Vuerinckx, J. Swevers and J. Schoukens, "Comparison of two feedforward design methods aiming at accurate trajectory tracking of the end point of a flexible robot arm," *IEEE Trans. Control Systems Technology*, Vol. 6, No. 1, pp. 2-14, January 1998.

[304] H. L. Trentelman, A. A. Stoorvogel, M. Hautus, *Control Theory for Linear Systems*, London, Berlin: Springer, 2001, 2012.

[305] J. Tsiniaas and J. Karafyllis, "ISS property for time-varying systems and application to partial-static feedback stabilization and asymptotic tracking," *IEEE Trans. Automatic Control*, Vol. 44, No. 11, pp. 2179-2184, November 1999.

[306] A. F. Vaz and E, J. Davison, "The structured robust decentralized servomechanism problem for interconnected systems", *Automatica*, Vol. 25, No. 2, pp. 267-272, 1989.

[307] S.-S. Wang and B-S Chen, "Simultaneous deadbeat tracking controller synthesis", *International Journal of Control*, Vol. 44, No. 6, pp. 1579-1586, 1986.

[308] P. E. Wellstead and P. Zanker, "Servo self-tuners", *International Journal of Control*, Vol. 30, No. 1, pp. 27-36, 1979.

[309] J. C. West, *Textbook of Servomechanisms*, London: English Universities Press, 1953.

[310] D. M. Wiberg, *State Space and Linear Systems*, New York: McGraw-Hill Book Company, 1971.

[311] R. L. Williams II and D. A. Lawrence, *Linear State-Space Control Systems*, Hoboken, NJ: John Wiley & Sons, Inc., 2007.

[312] W. A. Wolovich, *Linear Multivariable Systems*, New York: Springer-Verlag, 1974.

[313] W. M. Wonham, *Linear Multivariable Control. A Geometric Approach*, Berlin: Springer-Verlag, 1974.

[314] A. R. Woodyatt, M. M. Seron, J. S. Freudenberg, and R. H. Middleton, "Cheap control tracking performance," *Int. J. Robust and Nonlinear Control*, Vol. 12, pp. 1253-1273, 2002.

[315] H. Yamane and B. Porter, "Synthesis of limit-tracking incorporating linear multivariable plants," *Int. J. Systems Science*, Vol. 25, No. 12, pp. 2095-2111, 1994.

[316] G.-H. Yang, X.-G. Guo, W.-W. Che and W. Guan, *Linear Systems: Non-Fragile Control and Filtering*, Boca Raton, FL: CRC Press, 2017.

[317] J.-M. Yang and J.-H. Kim, "Sliding mode control for trajectory tracking nonholonomic wheeled mobile robots", *IEEE Transactions on Robotics and Automation*, Vol. 15, No. 3, pp. 578-587, June 1999.

[318] B.-T. Yazdan, *Introduction to Linear Control Systems*, New York: Academic Press, 2017.

[319] T. Yoshikawa and T. Sugie, "Analysis and synthesis of tracking systems considering sensor dynamics," *Int. J. Control*, Vol. 41, No. 4, pp. 961-971, 1985.

[320] T. Yoshikawa, T. Sugie and H. Hanafusa, "Synthesis of robust tracking systems with specified transfer matrices", *Int. J. Control*, Vol. 43, No. 4, pp. 1201-1214, 1986.

[321] T. Yoshizawa, *Stability Theory by Lyapunov's Second Method*, Tokyo: Mathematical Society of Japan, 1966.

[322] K. Youcef-Toumi and O. Ito, "A time delay controller for systems with unknown dynamics," *J. Dynamic Systems, Measurement, and Control*, Vol. 112, pp. 133-142, 1990.

[323] K. Youcef-Toumi and O. Ito, "Controller design for systems with unknown nonlinear dynamics," *Proc. 1987 American Control Conference*, Minneapolis, MN, Vol. 2, pp. 836-844, June 10-12, 1990.

[324] M. R. Yovanovitch, *Practical Tracking Automatic Control of the Axial Piston Hydraulic Motors* (in Serb), M. Sci. Thesis, Faculty of Mechanical Engineering, University of Belgrade, Belgrade, Serbia, 1998.

[325] R. Zh. Yovanovitch, *Fuzzy Tracking Control Algorithms of Electrohydraulic Servosystems* (in Serb), D. Sci. Dissertation, Faculty of Mechanical Engineering, University of Belgrade, Belgrade, Serbia, 2011.

[326] R. Zh. Yovanovitch and Z. B. Ribar, "Fuzzy practical exponential tracking of an electrohydraulic servosystems," Faculty of Mechanical Engineering (FME), Belgrade, Serbia, *FME Trans.*, Vol. 39, pp. 9-15, 2011.

[327] W.-S. Yu and Y.-H. Chen, "Decoupled variable structure control design for trajectory tracking on mechatronic arms", *IEEE Transactions on Control Systems Technology*, Vol. 13, No. 5, pp. 798-806, September 2005.

[328] S. Y. Zhang and C. T. Chen, "Design of compensators for robust tracking and disturbance rejection", *IEEE Transactions on Automatic Control*, Vol. AC-30, No. 7, pp. 684-687, July 1985.

[329] J. Zhao and I. Kanellakopoulos, "Flexible backstepping design for tracking and disturbance attenuation", *International Journal of Robust and Nonlinear Control*, Vol. 8, pp. 331-348, 1998.

[330] Y.-S. Zhong, "Robust output tracking control of SISO plants with multiple operating points and with parametric and unstructured uncertainties," *Int. J. Control*, Vol. 75, No. 4, pp. 219-241, 2002.

[331] Y. Zhu, D. Dawson, T. Burg, and J. Hu, "A cheap output feedback tracking controller with robustness: The RLFJ problem," *Proc. 1996 IEEE International Conference on Robotics and Automation*, Minneapolis, pp. 939-944, April 1996.

Part VI

INDEX

Author Index

Subject Index